1810 P

There was light in a vacuum-enclosed filamen of carbon, when fed an electrical current. It took a genius to find just how to do it. Thomas Edison's discovery (1879) led to the first power stations. Charles Parsons' steam turbine (1884) and the great Niagara Falls Hydroelectric station (1893) were pioneering events in the evolution of electrical engineering.

Automotive engineering really began to roll about 1885, when Gottlieb Daimler's first internal-combustion car appeared Henry Ford's genius was to create mass-productior —Model T's for the millions. Rudolph Diesel developed his famed work-horse engine in 1892. Civil engineers set out to build roads fast enough to keep up with the drivers. That race isn't over yet.

STUDY ASSIGNMENTS	DATE Assigned	DATE DUE
4: 1-13	8-14-80	8-19-80
4: 13-16	8-21	
4: 17-19		

DR. OBENDORF —
P. 96 is all wrong.
Pg 115, STEP 3

MARION NANCY LEE

DESCRIPTIVE GEOMETRY

SECOND EDITION

JAMES H. EARLE
Texas A&M University

DESCRIPTIVE GEOMETRY

SECOND EDITION

ADDISON-WESLEY PUBLISHING COMPANY

Reading, Massachusetts · Menlo Park, California
London · Amsterdam · Don Mills, Ontario · Sydney

Chapter opening illustrations are reproduced through the courtesy of Mattel, Inc. ("Decision" photo) and the National Aeronautics and Space Administration (all other photos).

Endpaper drawings are reproduced through the courtesy of Keuffel & Esser Company.

 Printed in the United States of America. Published simultaneously in Canada. Library of Congress Catalog Card No. 76-55640.

ISBN 0-201-01776-8
BCDEFGHIJ-HA-79

Dedicated to my mother,
Edna Webb Earle

Preface

Design is a major function of the engineer and technologist, and descriptive geometry and engineering graphics are the fundamental tools of the design process. Descriptive geometry is presented in this textbook as a problem-solving tool and as a means of developing solutions to technical problems. A generous number of photographs of products and equipment are included to show some of the many applications of descriptive geometry to various projects.

Our treatment of descriptive geometry motivates the student by exposing him or her to engineering examples taken from real-life situations. Instead of boring the student with synthetic projects, the approach taken in this text is intended to stimulate interest in engineering and technology as creative professions.

Many of the illustrations and the accompanying text have been designed to enable the student to grasp key principles on his or her own; thus, less of the instructor's time is required. Throughout the book, a second color is used in the illustrations to highlight significant steps and notes. Also, the more complex problems are presented by the *step method*, whereby the steps leading to the solution of a problem are presented in sequence, with the instructional text closely related to each step. This method of presentation was tested throughout a semester's work; 2800 student samples were taken during the study.* The step method was found to be 20 percentage points superior to the conventional textbook approach. These results prompted the author to introduce the step method in this volume.

Sufficient material is included in this text for a full course in descriptive geometry for the engineering and technology student. The design process is introduced early in the book and is referred to throughout.

The problems provided at the end of each chapter permit the student to test his or her mastery of the principles covered in that chapter. However, it is highly recommended that a laboratory manual be used in conjunction with this book. Too much lay-out time is required when the chapter problems are used. Printed problem sheets in most laboratory manuals are much more efficient and can increase the content covered in a course by as much as 100 percent. Problem books that can be used with this textbook are *Geometry for Engineers 1* and *Geometry for Engineers 2.* Both can be obtained from Creative Publishing Company, Box 9292,

---

* James H. Earle, *An Experimental Comparison of Three Self-Instruction Formats for Descriptive Geometry.* Unpublished dissertation, Texas A&M University, College Station, Texas, 1964.

College Station, Texas 77840, Phone 713-846-7907.

This edition of *Descriptive Geometry* has incorporated many new illustrations and a considerable revision of the text material. Much of the content has been rearranged to make the book more functional, with a better continuity. Some of the changes are immediately apparent; others, while more subtle in nature, also contribute substantially to the book's usefulness. Metric dimensions are included to assist in making the transition from the English to the metric system. Additional tables have been provided in the Appendix.

Thanks are due to the hundreds of industries who provided photographs, drawings, and examples included in this book. Appreciation is also due to Professor Michael P. Guerard, formerly of Texas A&M University, for his assistance in preparing the section on nomography. Credits would not be complete without mentioning the encouragement, confidence, and assistance given to the author by the staff of Addison-Wesley Publishing Company.

College Station, Texas J. H. E.
June 1977

Contents

1

Introduction to Descriptive Geometry and Design

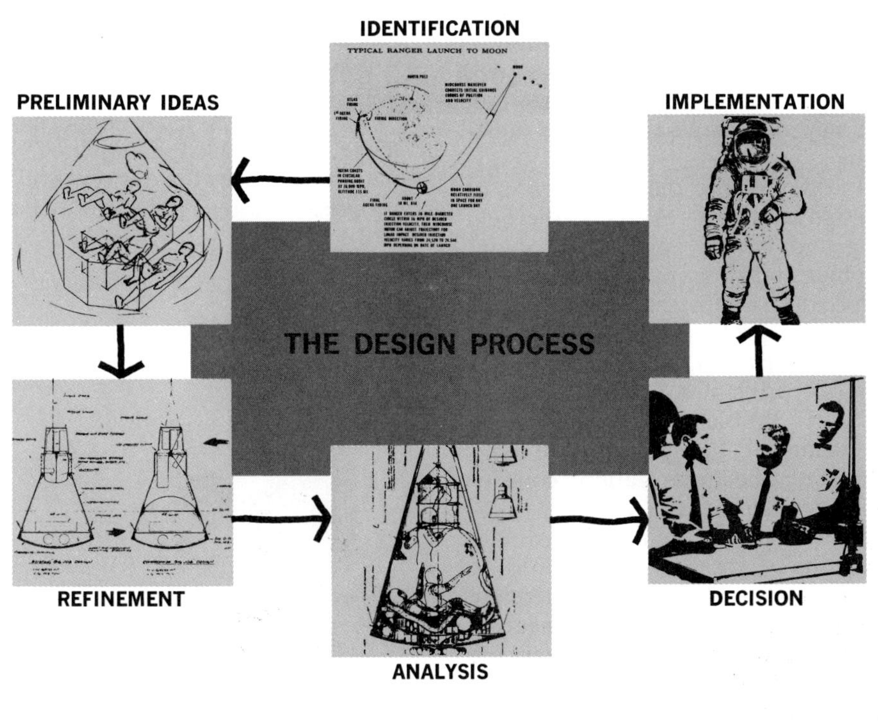

1–1 INTRODUCTION

Engineering and technology have made significant contributions to the advancement of our standard of living—probably more than any other profession. Essentially all our daily activities are assisted by products, systems, and services made possible by the engineer. Our utilities, heating and cooling equipment, automobiles, machinery and consumer products have been provided at an economical rate to the bulk of our national population by the engineering profession.

The engineer and technician must function as members of a team composed of other related, and sometimes unrelated, disciplines. Many engineers have been responsible for innovations of life-saving mechanisms used in medicine which were designed in cooperation with members of the medical profession. Other engineers are technical representatives or salesmen who explain and demonstrate applications of technical products to a specialized segment of the market. Even though there is a wide range of activities within the broad definition of engineering, the engineer is basically a *designer.* This is the activity that most distinguishes him from other associated members of the technological team.

This book is devoted to the introduction of elementary design concepts related to the field of engineering and to the application of engineering graphics and descriptive geometry to the design process. Examples are given which have an engineering problem at the core, and which require organization, analysis, problem solving graphical principles, communication, and skill Fig. 1–1. Problems which require a minimum of technical background are presented, emphasizing the organization, conceptualization, and development of a design solution with graphical methods used as the primary method of solution. Extensive illustrations of engineering applications are included to relate the theoretical principles to actual engineering situations. These illustrations will introduce the student to various fields of engineering and technology and will familiarize him with the wide variety of applications of graphical principles in the design and solution of problems.

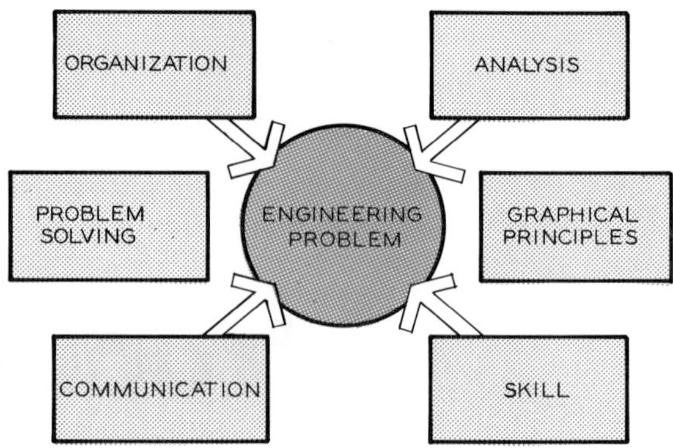

Fig. 1–1. Problems in this text require a total engineering approach with the engineering problem as the central theme.

Creativity and imagination are encouraged by this textbook as essential ingredients in the engineer's professional activities. All principles presented are structured to emphasize the importance of innovation and experimentation in solving elementary engineering problems. Albert Einstein, the famous physicist, said that "Imagination is more important than knowledge, for knowledge is limited, whereas imagination embraces the entire world . . . stimulating progress, or, giving birth to evolution . . . " (Fig. 1–2).

Fig. 1–2. Albert Einstein, the famous physicist, said. "Imagination is more important than knowledge . . . .

1–2 ENGINEERING GRAPHICS

Engineering graphics is usually considered to be the total field of graphical problem solving and includes two major areas of specialization, descriptive geometry and working drawings. Other areas that can be utilized for a wide variety of scientific and engineering applications are also included within the field. These are nomography, graphical mathematics, empirical equations, technical illustration, vector analysis, data analysis, and other graphical applications associated with each of the different engineering industries. Engineering graphics should not be confused with drafting, since it is considerably more extensive than the communication of an idea in the form of a working drawing. Graphical methods are the primary means of creating a solution to a problem requiring innovations not already available to the designer. Graphics is the designer's method of thinking, solving, and communicating his ideas throughout the design process. Man's progress can be attributed to a great extent to the area of engineering graphics. Even the simplest of structures could not have been designed or built without drawings, diagrams, and details that explained their construction (Fig. 1–3). For many years technical drawings, such as they were, were confined to two dimensions, usually a plan view. Supplemental sketches and pictorials were used to expain other dimensions of the project being depicted. Gradually, graphical methods were developed to show three related views of an object to simulate its three-dimensional representation. A most significant development in the engineering graphics area was descriptive geometry as introduced by Gaspard Monge (Fig. 1–4).

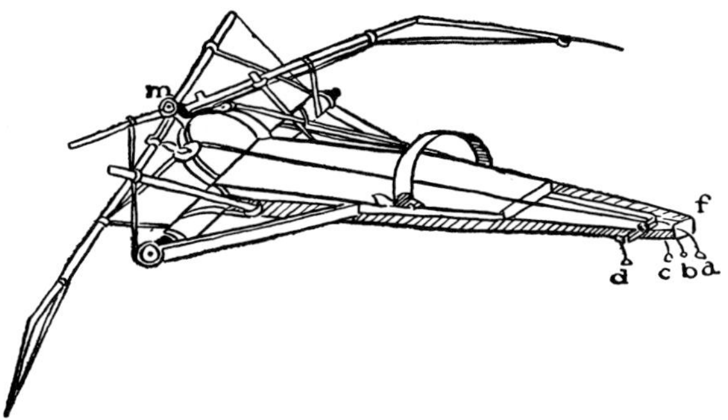

Fig. 1–3. Leonardo da Vinci developed many creative designs through the use of graphical methods.

Fig. 1–4. Gaspard Monge, the "father of descriptive geometry."

Descriptive Geometry. Gaspard Monge (1746–1818) is considered the "father of descriptive geometry." Young Monge used this graphical method of solving design problems related to fortifications and battlements while a military student in France. He was scolded by his headmaster for not solving a problem by the usual, long, tedious mathematical process traditionally used for problems of this type. It was only after long explanations and comparisons of the solutions of both methods that he was able to convince the faculty that his graphical methods could be used to solve the problem in considerably less time. This was such an improvement over the mathematical solution that it was kept a military secret for fifteen years before it was allowed to be taught as part of the technical curriculum. Monge became a scientific and mathematical aide to Napoleon during his reign as general and emperor of France.

Descriptive geometry has been simplified from the "indirect" method introduced by Monge to the "direct" method used today. In the "indirect" method, the first angle of projection is used primarily, with the front view projected above the top view, and the projections are revolved onto the projection planes to obtain the desired relationships (Fig. 1–5). The "direct" method utilizes the third angle, with the top view projected over the front view, and auxiliary views are projected directly to auxiliary planes in succession until the required geometric relationships are found.

Descriptive geometry can be defined as the projection of three-dimensional figures onto a two-dimensional plane of paper in such a manner as to allow geometric manipulations to determine lengths, angles, shapes, and other descriptive information concerning the figures. The type of problems lending themselves to solution by descriptive geometry, although very common, are usually considerably more difficult to solve by mathematics. The simple determination of the angle between planes is a basic descriptive geometry problem, but is difficult to determine mathematically when the plane of the angle does not appear true size in the given views.

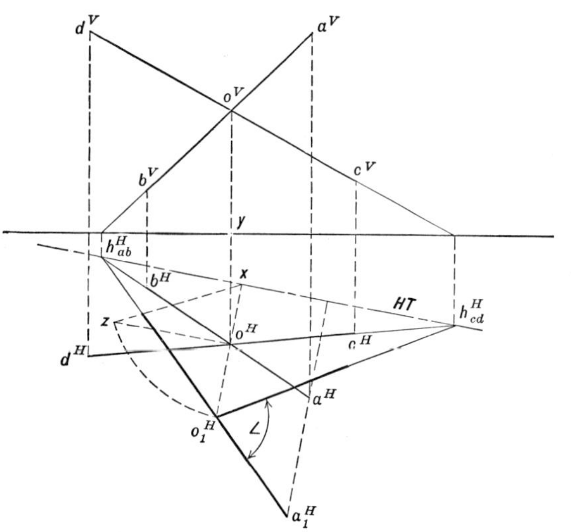

Fig. 1–5. An indirect solution to a descriptive geometry problem utilizing the Mongean method. (Courtesy of C. H. Schumann, Jr., *Descriptive Geometry*, 3rd edition, Van Nostrand, 1938.)

1–3 INTRODUCTION TO DESIGN

Engineering graphics in general and descriptive geometry in particular are important in the design of any product or system. These are the methods used by the designer in communicating with himself and with others. He uses these methods to develop and solve many problems that would be difficult to solve by other methods.

The terms "design" and "design process" will be used often in this book to call attention to the application of the principles discussed here to the solution of engineering and technological problems. This is to encourage you to develop your imagination and to become aware of the process used in developing original solutions to problems. Consideration of the role of descriptive geometry in the design process will make the principles more meaningful to you.

The following article will briefly discuss the *design process* to give you an overview of the procedures that are followed. The design process is not a rigid formula but a flexible approach to the solution of a problem. Better results are usually obtained when the designer follows a somewhat disciplined procedure rather than using an unstructured approach.

1–4 THE DESIGN PROCESS

The act of devising an original solution to a problem by a combination of principles, resources, and products is design. Design is the most distinguishing responsibility that separates the engineer from the scientist and the technician. His solutions may involve a combination of existing components in a different arrangement to provide a more efficient result or they may involve the development of an entirely new product; but in either case his work is referred to as the act of designing. This process is not an inspirational phenomenon that is experienced by only a few, but is the result of a systematic, disciplined approach to the needs of the problem.

The design process is the usual pattern of activities that are followed by the designer in arriving at the solution of a technological problem. Many combinations of steps have been suggested to enable one to achieve design objectives. This book emphasizes a six-step design process that is a composite of the sequences most commonly employed in solving problems. The six steps are (1) problem identification, (2) preliminary ideas, (3) problem refinement, (4) analysis, (5) decision, and (6) implementation (Fig. 1–6). Although the designer will sequentially work from step to step, he may recycle to previous steps as he progresses.

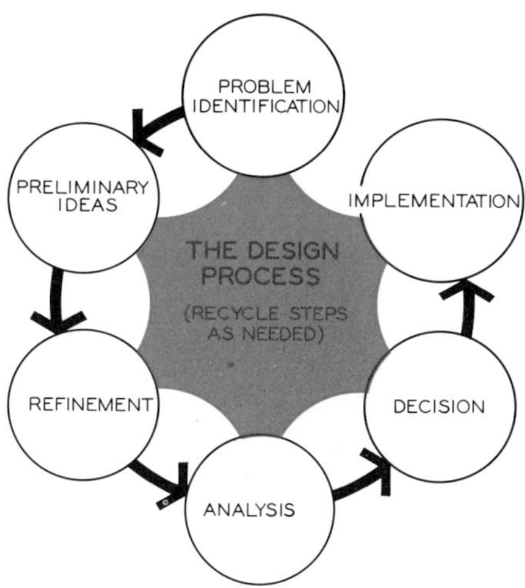

Fig. 1–6. The steps of the design process.

Problem Identification. Most technological problems are not clearly defined at the outset; consequently they must be identified before an attempt is made to solve the problem (Fig. 1–7). For example, a prominent concern today is air pollution. Before this problem can be solved, you must identify what air pollution is and what causes it. Is pollution caused by automobiles, factories, atmospheric conditions that harbor impurities, or geographic features that contain impure atmospheres?

When you enter a bad street intersection where traffic is unusually congested, do you identify the reasons for it being congested? Are

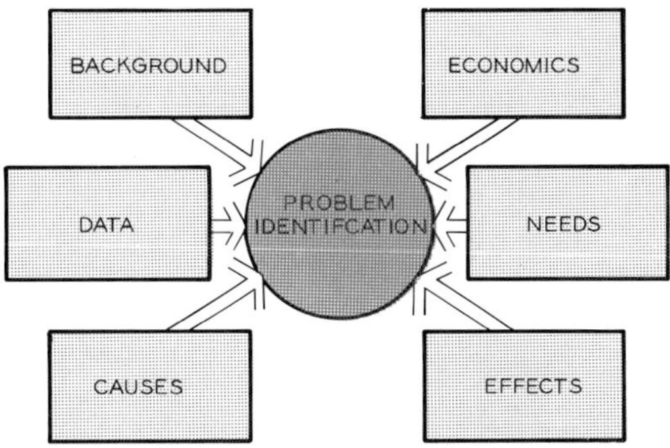

Fig. 1–7. Problem identification requires the accumulation of as much information concerning the problem as possible before a solution is attempted by the designer.

there too many cars, are the signals poorly synchronized, or are there visual obstructions resulting in congested traffic? The answers to these questions are helpful in identifying the problem that must be solved.

Problem identification requires considerable study beyond a simple problem statement like "solve air pollution." You will need to gather data of several types: field data, opinion surveys, historical records, personal observations, experimental data, and physical measurements and characteristics (Fig. 1–7). This will result in an accumulation of facts and numbers that should be recorded and interpreted both graphically and in written form.

Preliminary Ideas. Once the problem has been identified, the next step is to accumulate as many ideas for solution as possible (Fig. 1–8). Preliminary ideas can be gathered by individual or group approaches. Preliminary ideas should be sufficiently broad to allow for unique solutions that could revolutionize present methods. All ideas should be recorded in written form. Many rough sketches of preliminary ideas should be made and retained as a means of generating original ideas and stimulating the design process. Ideas and comments should be noted on the sketches as a basis for further preliminary designs.

Preliminary ideas can be gathered from several commonly used methods, including

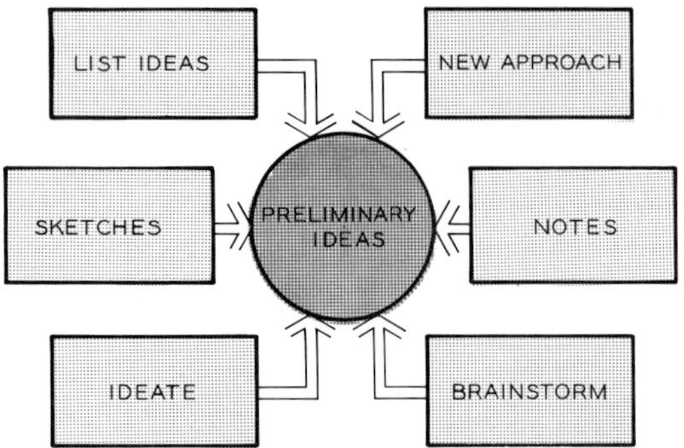

Fig. 1–8. Preliminary ideas are developed after the identification step has been completed. All possibilities should be listed and sketched to give the designer a broad selection of ideas from which to work.

brainstorming, market analysis, or research of present designs. These methods are explained in detail in later chapters. All work is most useful if completed in graphical form for easy analysis.

Problem Refinement. Several of the better preliminary ideas are selected for further refinement to determine their true merits. Rough sketches are converted to scale drawings that will permit space analysis, critical measurements, and the calculation of areas and volumes affecting the design (Fig. 1–9). Consideration is given to spatial relationships, angles between planes, lengths of structural members, intersections of surfaces and planes. Information of this type is necessary to determine the feasibility of manufacture and the physical characteristics of a design. Descriptive geometry is a very valuable tool for determining information of this type, and it precludes the necessity for tedious mathematical and analytical methods. Engineering graphics is employed to construct the necessary views of the design so that it can be analyzed for its spatial characteristics with descriptive geometry.

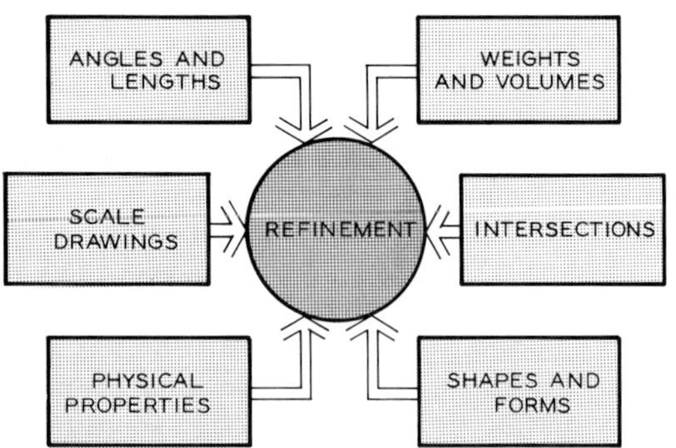

Fig. 1–9. Refinement begins with the construction of scale drawings of the better preliminary ideas. Descriptive geometry and graphical methods are used to find necessary geometric characteristics.

An example of a problem of this type is illustrated in the landing gear of the lunar vehicle shown in Fig. 1–10. It was necessary for the designer to make many freehand sketches of the design and finally a scale drawing to establish clearances with the landing surface. The configuration of the landing gear was drawn to scale in the descriptive views of the landing craft. It was necessary, at this point, to determine certain fundamental lengths, angles, and specifications that are related to the fabrication of the gear. The length of each leg of the landing apparatus and the angles between the members at the point of junction had to be found to design a connector, and the angles the legs made with the body of the spacecraft had

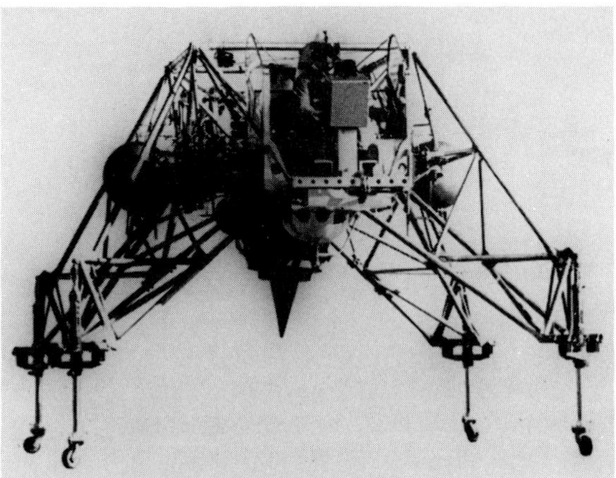

Fig. 1–10. The refinement of the lunar vehicle required the use of descriptive geometry and other graphical methods. (Courtesy of Ryan Aeronautics, Inc.)

to be known in order to design these joints. All of this information was easily and quickly determined with the use of descriptive geometry.

Analysis. Analysis is the step of the design process where engineering and scientific principles are used most (Fig. 1–11). Analysis involves the study of the best designs to determine the comparative merits of each with respect to cost, strength, function, and market appeal. Graphical principles can also be applied to analysis to a considerable extent. The determination of forces is somewhat simpler with graphical vectors than with the analytical method. Functional relationships between moving parts will also provide data that can be obtained graphically more easily than by analytical methods. Graphical solutions to analytical problems offer a readily available means of checking the solution, therefore reducing checking time. Graphical methods can also be applied to the conversion of functions of mechanisms to a graphical format that will permit the designer to convert this action into an equation form that will be easy to utilize. Data can be gathered and graphically analyzed that would otherwise be difficult to analyze by mathematical means. For instance, empirical curves that do not fit a normal equation are often integrated graphically when the mathematical process would involve unwieldy equations.

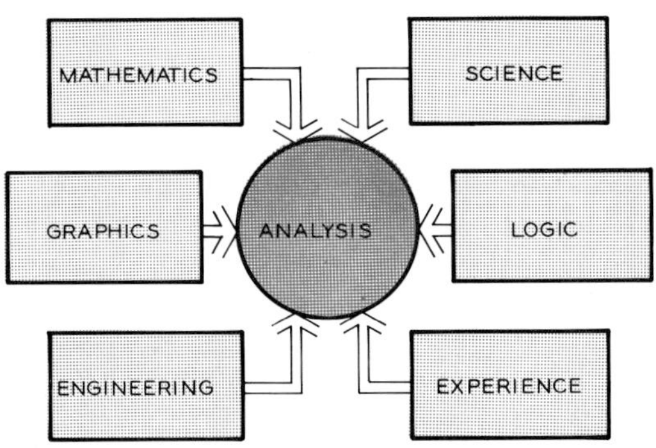

Fig. 1–11. The analysis phase of the design process is the application of all available technological methods from science to graphics in evaluating the refined designs.

Graphical methods are vital supplements to the engineering sciences when applied to the analysis procedure. These methods should be well understood by the engineer, technician, or designer, to afford him every available aid to effectively solve a problem in the minimum time. Human factors are the basis for a design and are given as an introduction to this important area of consideration.

Decision. A decision must be made at this stage to select a single design that will be accepted as the solution of the design problem (Fig. 1–12). Each of the several designs that have been refined and analyzed will offer unique features, and it will probably not be possible to include all of these in a single final solution. In many cases, the final design is a compromise that offers as many of the best features as possible.

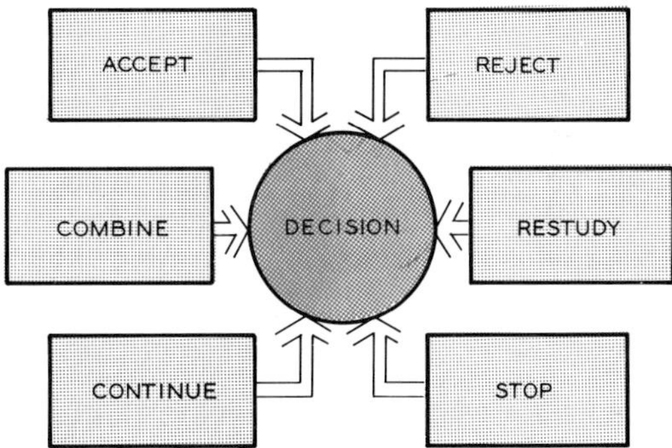

Fig. 1–12. Decision is the selection of the best design or design features to be implemented.

The decision may be made by the designer on an independent, unassisted basis, or it may be made by a group of associates. Regardless of the size of the group making the decision as to which design will be accepted, graphics is a primary means of presenting the proposed designs for a decision. The outstanding aspects of each design usually lend themselves to presentation in the form of graphs which compare costs of manufacturing, weights, operational characteristics, and other data that would be considered in arriving at the final decision. Pictorial sketches or formal pictorials are excellent methods of graphically studying different designs before arriving at a decision.

Implementation. The final design concept must be presented in a workable form after the best design has been selected and decided upon. This type of presentation refers primarily to the working drawings and specifications that are used as the actual instruments for the fabrication of the product, whether it is a small piece of hardware or a bridge (Fig. 1–13). Engineering graphics fundamentals must be used to convert all preliminary designs and data into the language of the manufacturer who will be responsible for the conversion of the ideas into a reality. Workmen must have complete detailed instructions for the manufacture of each single part, measured to a thousandth of an inch to facilitate its proper manufacture. Working drawings must be sufficiently detailed and explicit to provide a legal basis for a contract which will be the document for the contractor's bid on the job.

Plans are usually executed by draftsmen and technicians who are specialists in this area. The designer or engineer must be sufficiently knowledgeable in graphical presentation to be able to supervise the preparation of working drawings even though he may not be involved in the mechanics of producing them. He must approve all plans and specifications prior to their release for production. This responsibility necessitates that he be well-rounded in all aspects of graphical techniques to enable him to approve the plans with assurance. This step of the design process is less creative than the subsequent steps, but it is no less important than any other step.

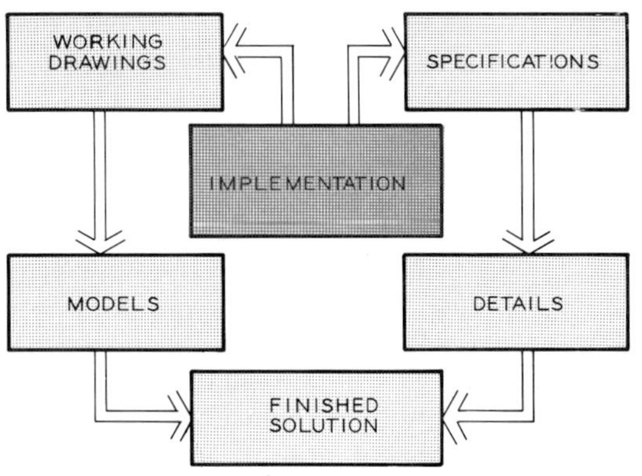

Fig. 1–13. Implementation is the final step of the design process, where drawings and specifications are prepared from which the final product can be constructed.

1–5 SUMMARY

The design process introduced in this chapter is composed of six steps. These steps are (1) problem identification, (2) preliminary ideas, (3) problem refinement, (4) analysis, (5) decision, and (6) implementation. These six steps must be applied to all design problems to some degree, whether in this sequence or others. The process of designing can take many forms and be approached by a number of methods. The procedure that works best for an individual is the one he should use.

Graphical methods are necessary tools of the designer striving for a problem solution. These methods have definite applications to every aspect of the design process; therefore graphics and descriptive geometry principles form an indispensable part of the design process. The designer must actually think with a pencil in developing creative designs. Graphics is used more extensively throughout the design process than any other single tool at the designer's disposal.

PROBLEMS

Problems should be presented on $8\frac{1}{2}'' \times 11''$ paper, grid or plain, using the format presented in Article 3–6. All notes, sketches, drawings, and graphical work should be neatly prepared in keeping with good practices, as introduced in this volume. Written matter should be legibly lettered using $\frac{1}{8}''$ guidelines.

1. List engineering achievements that have demonstrated a high degree of creativity in the following areas: (a) the household, (b) transportation, (c) recreation, (d) educational facilities, (e) construction, (f) agriculture, (g) power, (h) manufacture.

2. Make an outline of your plan of activities for the weekend. Indicate areas in your plans that you feel display a degree of creativity or imagination. Explain why.

3. Write a short report on the engineering achievement or the man who you feel has exhibited the highest degree of creativity. Justify your selection by outlining the creative aspects of your choice. Your report should not exceed three typewritten pages.

4. Test your creativity in recognizing needs for new designs. List as many improvements for the typical automobile as possible. Make suggestions for implementing these improvements. Follow this same procedure in another area of your choice.

5. Make a list of new products that have been introduced within the last five years with which you are familiar.

6. Make a list of products and systems that you would anticipate for life on the moon.

7. Assume that you have been assigned the responsibility for organizing and designing a go-kart installation on your campus. This must be a self-supporting enterprise. Write a paragraph on each of the six steps of the design process to explain how the steps would be applied to the problem. For example, what action would you take to identify the problem?

8. You are responsible for designing a motorized wheelbarrow to be marketed for home use. Write a paragraph on each of the six steps of the design process to explain how the steps would be applied to the problem. For example, what action would you take to identify the problem?

9. List and explain a sequence of steps that you feel would be adequate for the design process, yet different from the six given in this chapter. Your version of the design process may contain as many of the steps discussed here as you desire.

10. Can you design a device for holding a fishing pole in a fishing position while you are fishing in a rowboat? This could be a simple device that will allow you freedom while performing other chores in the boat. Make notes and sketches to describe your design.

11. Assume that you are responsible for designing a car jack that would be more serviceable than present models. Review the six steps of the design process and make a brief outline of what you would do to apply these steps to your attempt to design a jack. Write the sequential steps and the methods that would be used to carry out each step. List the subject areas that would be used for each step and indicate the more difficult problems that you would anticipate at each step. Keep your outline brief, but thorough Freehand letter your paper.

12. As an introductory problem to the steps of the design process, design a door stop that could be used to prevent a door from slamming into a wall. This stop could be attached to the floor or the door and should be as simple as possible. Make sketches and notes as necessary to give tangible evidence that you have proceeded through the six steps and label each step. Your work should be entirely freehand and rapid. Do not spend longer than 30 minutes on this problem. Indicate any information you would need in a final design approach that may not be accessible to you now.

13. List areas that you must consider during the problem identification phase of a design project for the following products: a new skillet design for the housewife, a lock for a bicycle, a handle for a piece of luggage, an escape from prison, a child's toy, a stadium seat, a desk lamp, an improved umbrella, a hotdog stand.

14. Make a series of rough, freehand sketches to indicate your preliminary ideas for the solution of the following problems: a functional powdered soap dispenser for washing hands, a protector for a football player with an injured elbow, a method of positioning the cross-bar at a pole vault pit, a portable seat for waiting in long lines, a method of protecting windshields of parked cars during freezing weather, a pet-proof garbage can, a bicycle rack, a door knob, a seat to support a small child in a bathtub.

15. Evaluate the sketches made in Problem 4 above and briefly outline in narrative form the information that would be needed to refine your design into a workable form. Use freehand lettering, striving for a neat, readable paper.

16. Many automobiles are available on the market. Explain your decision for selecting the one that would be most appropritae for the activities listed below: a trip on a sightseeing tour in the mountains, a hunting trip in a wooded area for several days, a trip from coast to coast, the delivering of groceries, a business trip downtown. List the type of vehicle, model, its features and why you made your decision to select it.

2

Application of Descriptive Geometry

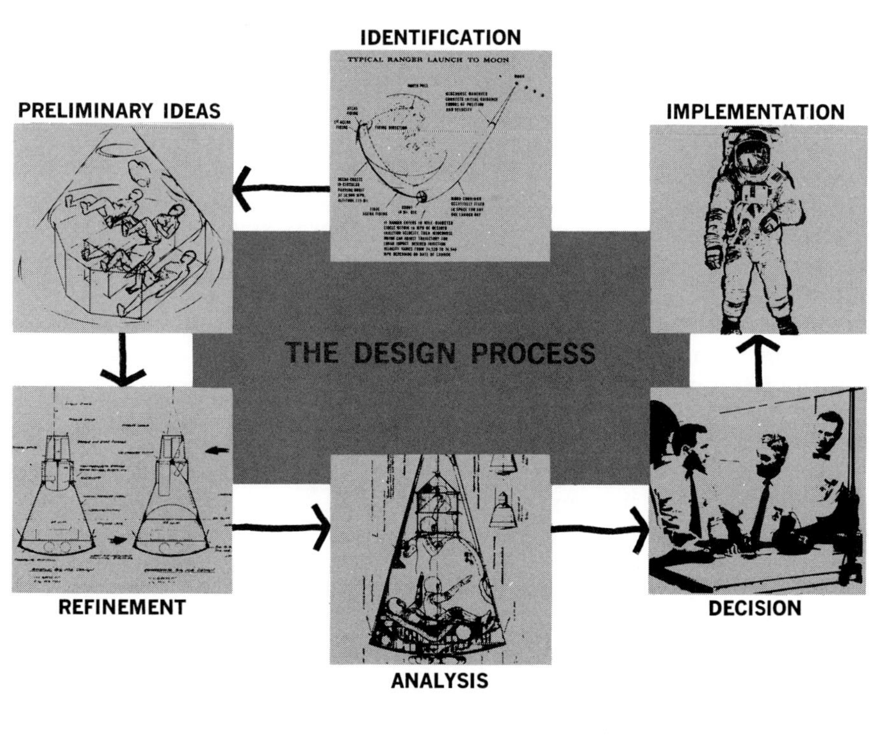

Fig. 2–1. The designer's first step in the refinement phase of his preliminary ideas is the preparation of scale drawings. (Courtesy of Chrysler Corporation.)

2–1 GENERAL

Descriptive geometry is the graphical discipline that has the greatest application to the refinement step of the design process. The designer begins refinement only after he has accumulated a sufficient number of preliminary ideas in the form of sketches and notes. In design refinement it is necessary to make instrument drawings that are rendered to scale, and to provide an accurate check on critical dimensions and measurements that were sketched during the early stages of the design process. When clearances or other measurements are critical, free hand sketches can be misleading. A scale drawing will give a true picture of the dimensions in question (Fig. 2–1).

Design refinement is the initial departure from the unrestricted freedom of creativity and imagination. Any design is subject to the limitations imposed by practicality of function and operation. Therefore, several better ideas must be selected and refined so that a comparison can be made during analysis and decision with regard to the final design solution to be implemented.

The designer must begin the analysis and decision functions to some extent during design refinement. Unless he makes a general analysis of the functional capabilities of the preliminary ideas, he will have to refine all of his designs, which will require considerable time if he has drawn a number of preliminary ideas. Consequently, the designer must develop an ability to form opinions of preliminary ideas as they are conceived—but without becoming negative and restricting his freedom of imagination.

2–2 DETERMINATION OF PHYSICAL PROPERTIES

The refinement stage of the design process is concerned primarily with the physical properties and general limitations that are evident prior to a formal analysis of a design. For example, scale drawings were made of three proposed configurations for the refinement of the "Big Joe" spacecraft (Fig. 2–2). These scale drawings evolved from many preliminary sketches and design features of experimental vehicles previously tested to determine the most desirable characteristics. Scale drawings of this type are helpful in developing the final shape and dimensions of a design. The functions and activities of the astronauts to be housed in the craft will have considerable influence on the size, volume, and general configuration of the capsule. To determine the weight of the craft, it is necessary to know the surfaces areas of the vehicle parts and the types of material used. Interior components and other equipment must be known also, as well as the approximate weight of the passengers. The volume of the craft must be determined to

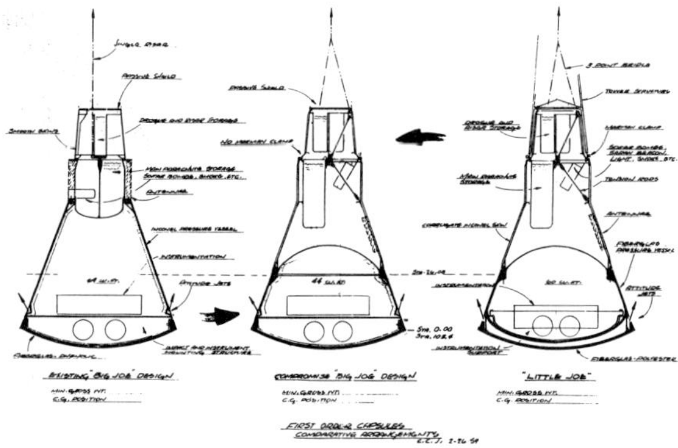

Fig. 2–2. These scale drawings were used to refine the final design of the "Big Joe" spacecraft and to incorporate desirable features from other systems. (Courtesy of NASA.)

Fig. 2–3. The method of construction of a stereo portable phonograph was determined through the refinement of preliminary concepts. (Courtesy of ALCOA.)

Fig. 2–4. The assembled phonograph is shown here in its completed form with the dual-purpose speakers detached for stereo listening. (Courtesy of ALCOA.)

ensure that sufficient space is available for the accessory equipment required during flight.

The calculation of practically any given physical properties begins with basic geometric elements—points, lines, areas, volumes, and angles. The measurements of these elements are determined as a design is refined prior to the preparation of working drawings from which the object can be constructed. The refined design is not necessarily a working drawing, but it is a scale drawing from which an accurate appraisal can be made.

Design refinement may involve a three-dimensional problem requiring spatial analysis. It may also involve planning for the use of stock components, as in the portable stereo phonograph shown in Fig. 2–3 in its final form. After preliminary sketches were made of this product, scale drawings were prepared to assist in refining the details of construction to be as economical and efficient as possible. Stock materials were used to reduce the expense of special fabrication. As the problem was refined, a modular system of construction was developed that would be both attractive and economical (Fig. 2–4). After refinement of the structural system, a complete final working drawing was prepared to implement construction.

A problem of the spatial analysis variety is the gas processing system shown in Fig. 2–5.

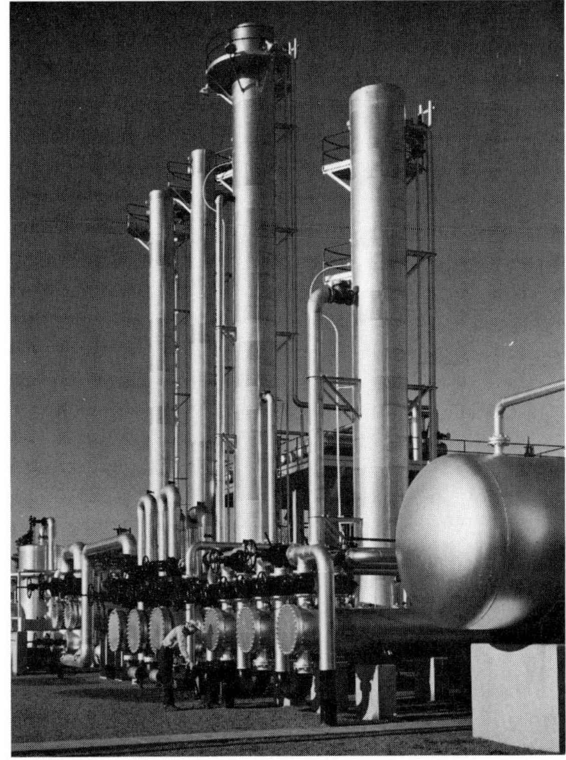

Fig. 2–5. This gas processing plant is an example of a system requiring considerable spatial analysis and the application of descriptive geometry principles. (Courtesy of Exxon Corporation.)

The general layout of the various units of the system can be drawn as a schematic diagram to indicate the sequence of the process. Refinement requires the determination of the sizes of the vessels, their volume, the land area required, and the fabrication details of the vessels and pipes. Since the vessels have a variety of shapes, and since the pipes and vessels intersect at different angles, the problem will require a three-dimensional analysis. The lengths of pipes must be found, vessels developed as patterns on flat metal stock, and intersections between pipes and surfaces determined.

Access to this information will enable the designer to formulate a final design with the necessary details for construction. Since stock components will be used as often as possible to reduce costs, the designer should revise his design during refinement to take advantage of available components. Pipe fittings are designed to join pipes at several standard angles, and a refinery system should be designed so that connections match these standard angles. This is possible with the application of descriptive geometry.

2–3 APPLICATION OF DESCRIPTIVE GEOMETRY

Descriptive geometry is the study of points, lines, and surfaces in three-dimensional space. This area of study has many applications to the refinement of a preliminary design and its analysis. Descriptive geometry can be applied to engineering problems that would be difficult to solve by other engineering methods.

An example of a problem requiring the use of descriptive geometry to determine geometric characteristics is a surgical light. The development of the light is a twofold problem, involving both the illumination and the geometry of the unit.

A major concern of the designer is the positioning of the light source so that maximum light falls on the work area with the minimum obstruction. A sketch of a well-adapted lamp is shown in Fig. 2–6. Note that the light emitted from specially designed reflectors converges to a small beam at shoulder level to minimize shadows cast by interference of the surgeon's shoulders, arms, and hands. The particular design that was developed, with a focal range between 30″ and 60″, is shown in Fig. 2–7.

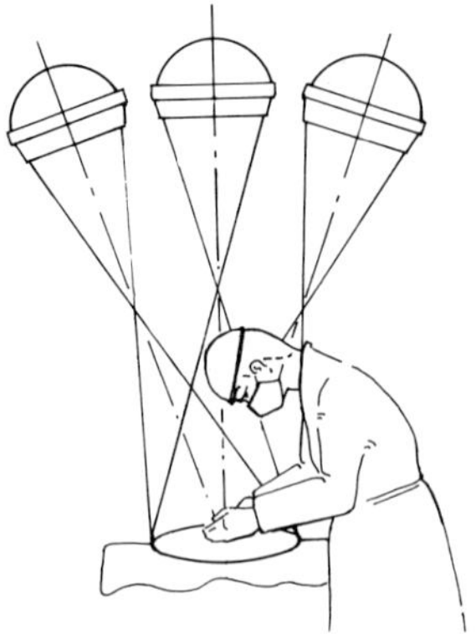

Fig. 2–6. A well-adapted surgical lamp emits light that passes around the surgeon's shoulders with a minimum of shadow. (Courtesy of Sybron Corporation.)

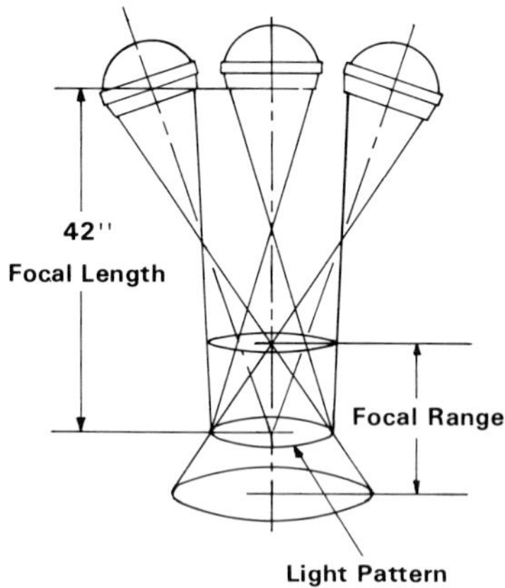

Fig. 2–7. The focal range of this surgical lamp is between 30″ and 60″. (Courtesy of Sybron Corporation.)

Scale drawings of the design are drawn so that the geometry of the fixture can be studied in detail. Measurements, angles, areas, and other geometry can be determined from these drawings (Fig. 2–8).

Figure 2–9 is an example of a refinement drawing that shows critical dimensions. Eventually this geometry will be tested by construction of a working model to confirm the information found in the refinement drawings.

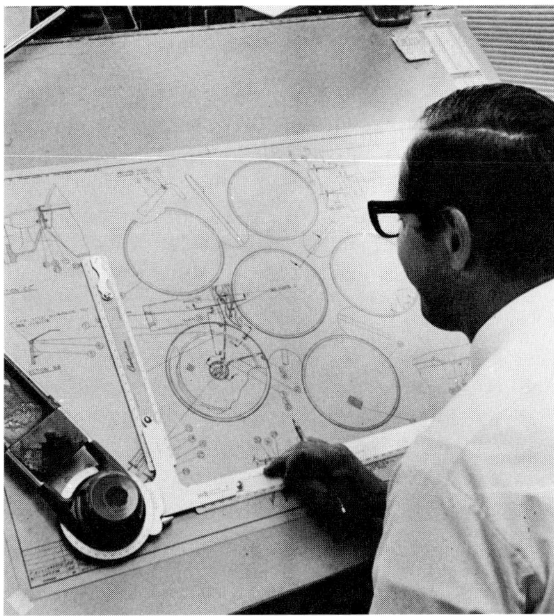

Fig. 2–8. The geometry of the surgical lamp can be studied through the use of scale drawings as a step toward refinement of preliminary ideas. (Courtesy of Sybron Corporation. Photograph by Brad Bliss.)

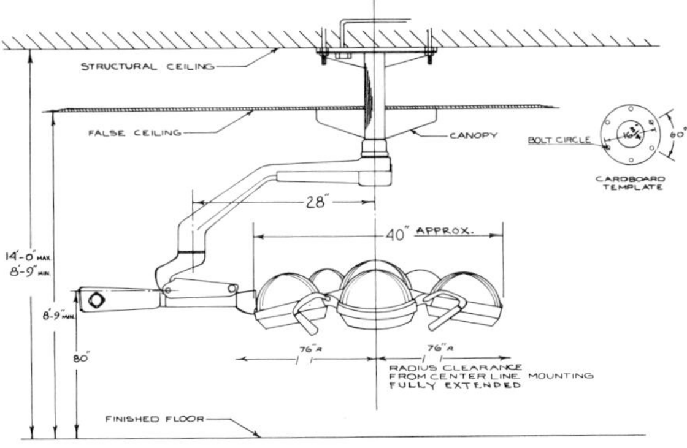

Fig. 2–9. The overall dimensions of the final design are shown in this refinement drawing. (Courtesy of Sybron Corporation.)

Fig. 2–10. This 10-foot-diameter underwater sphere could not have been designed without descriptive geometry methods. (Official photograph, U.S. Navy.)

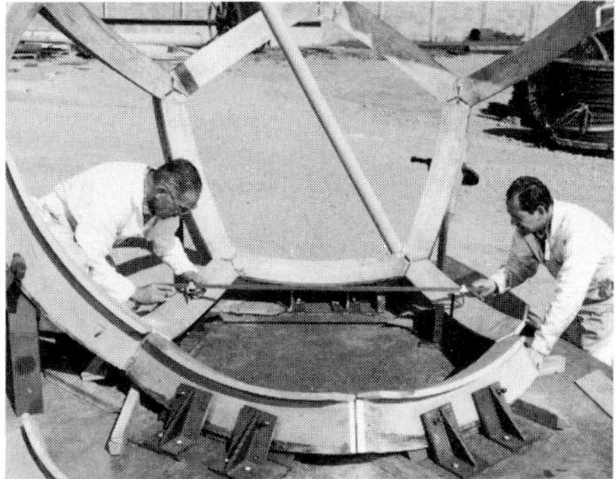

Fig. 2–11. The complex joints of the structural members were designed with the use of descriptive geometry. (Official photograph, U.S. Navy.)

Fig. 2–12. Jigs used to assemble the frame were designed through the use of descriptive geometry. (Official photograph, U.S. Navy.)

Another example of a problem refined by descriptive geometry is a structural frame for the 10-foot-diameter underwater sphere (Fig. 2–10). Before working drawings could be made, the physical properties and dimensions of the spherical pentagons had to be determined through a series of auxiliary views. The determination of the angles between the members was necessary before the joints could be detailed to give the snug fit illustrated in Fig. 2–11. A further refinement was the development of the necessary jigs for holding the components during assembly (Fig. 2–12). The solution of this problem would be essentially impossible without the principles of descriptive geometry.

A layout drawing for refining an automotive design is shown in Fig. 2–13. Here the designer's task was to determine the limitations of the size of the fender opening so as to allow minimum clearance between the wheel and the fender. The wheel is turned to its maximum

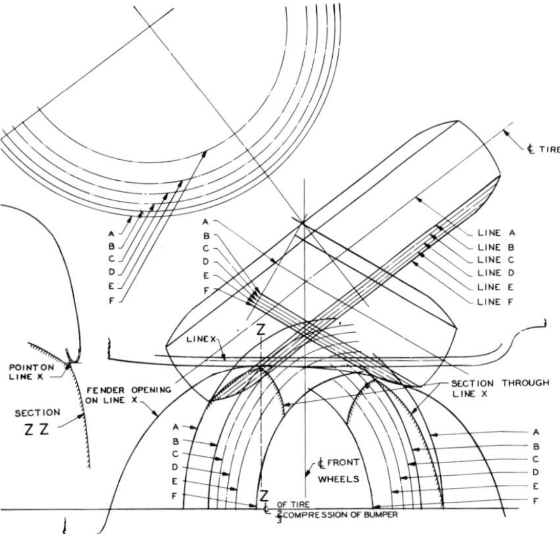

Fig. 2–13. The clearance between a fender and a tire is used to determine the fender opening of an automobile by the application of descriptive geometry and graphical methods. (Courtesy of Chrysler Corporation.)

steering angles to locate lines of interference, which will determine the minimum opening of the front fender. The left side of the layout shows a section of the fender opening turn-under. Projections are made to determine interferences at critical locations. The shaded tire sections indicate the conditions which determine the fender opening.

2–4 PRESENTATION OF DESCRIPTIVE GEOMETRY

The following six chapters are devoted to the introduction of those fundamentals of descriptive geometry that are most commonly applied to the refinement of design problems.

Problems have been reduced to the fundamental elements of points, lines, and surfaces for simplification of the basic principles. All actual engineering problems can be reduced to these elements and solved in the same manner as the example problems. Identification of the type of problems encountered is the initial step of problem solution. If the solution desired is the determination of the angle between two planes of a design, the problem can be resolved to the two planes in question and solved as the application of this principle. For example, the angle between two planes of an automobile windshield is found in Fig. 2–14 when two orthographic views have been obtained.

Since the refinement phase of the design process provides the transition from a preliminary idea to the necessary specifications and information required for the preparation of working drawings, it is, in essence, the problem-solving portion of a design project.

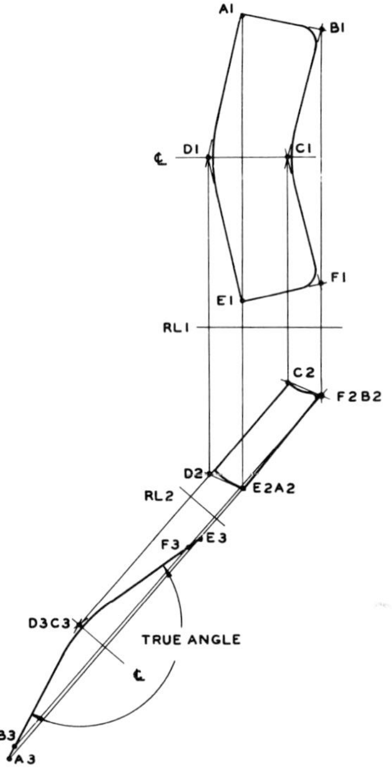

Fig. 2–14. The angle between the planes of a windshield can be determined by simplifying the problem to the fundamental planes involved. (Courtesy of Chrysler Corporation.)

2–5 REFINEMENT OF AN ENGINEERING DESIGN

An example design problem encountered in a research and development program was to develop a design for a Mobile Laboratory that would travel on the moon's surface. The vehicle was to provide a 14-day, 250-mile lunar operational range capability for two men. The Bendix Corporation was responsible for the four-wheel traction drive mechanism (TDM) and front-wheel steering drive mechanism (SDM) designs.

Many preliminary sketches were prepared to provide a selection of various possible solutions to this problem. These were evaluated to determine the most appropriate design for the project needs. Figure 2–15 shows the conceptual design layout of the left front wheel TDM and SDM incorporating DC series motors, two-ratio electric clutch systems and final stage hermetically sealed nutator transmission. The TDM units were designed to be mounted within each wheel axle. Although this drawing was drawn to scale, it is not a working drawing, since dimensions and sufficient information to construct the assembly are not given. This is merely a drawing used to refine the preliminary sketches. Note that several oblique angles are incorporated in the design, which will require solution by descriptive geometry.

A pictorial of the refined design introduced in Fig. 2–15 is shown in Fig. 2–16 to better describe the concept to the customer. It is easier to understand the relationships of the parts of the assembly in this partially sectioned pictorial. This final preliminary concept was accepted and the contract was awarded.

A follow-up contract was awarded for the design and fabrication of a mobility test article (MTA) vehicle intended for earth-testing the MOLAB mobility system. The mobility test article TDM and SDM hardware was to provide the same mobility characteristics as the pro-

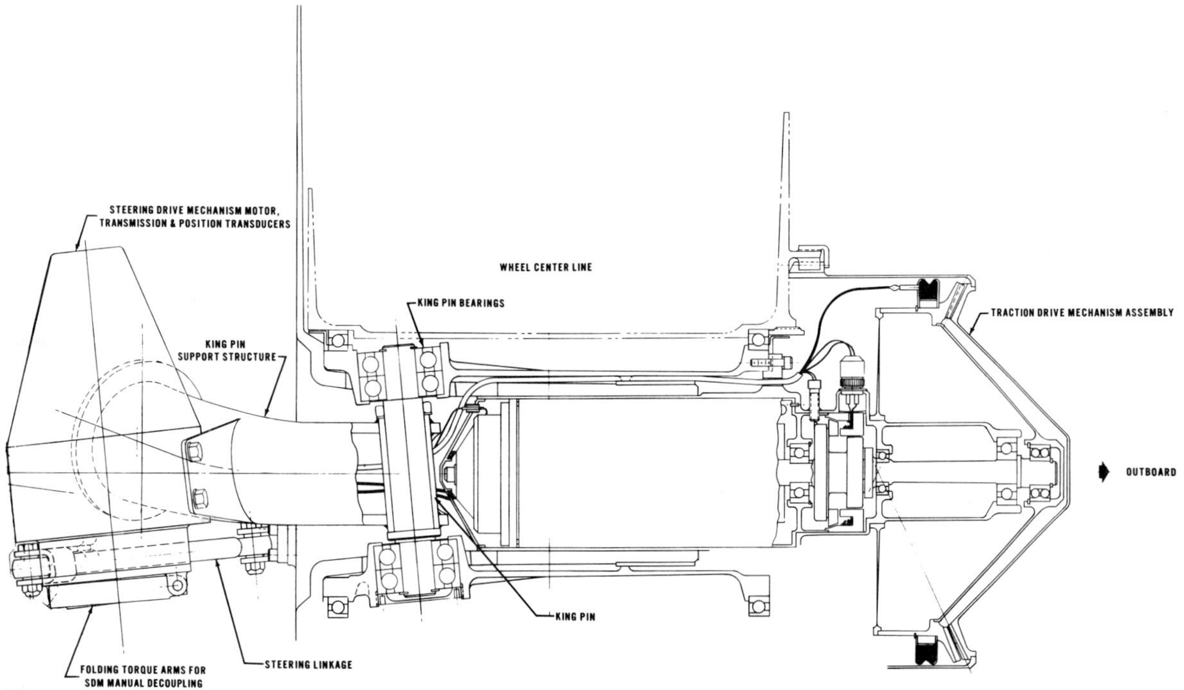

Fig. 2–15. A preliminary refinement drawing of a left front wheel for a vehicle designed to travel on the moon's surface. Dimensions and details of construction can be omitted on refinement drawings. (Courtesy of Bendix Corporation.)

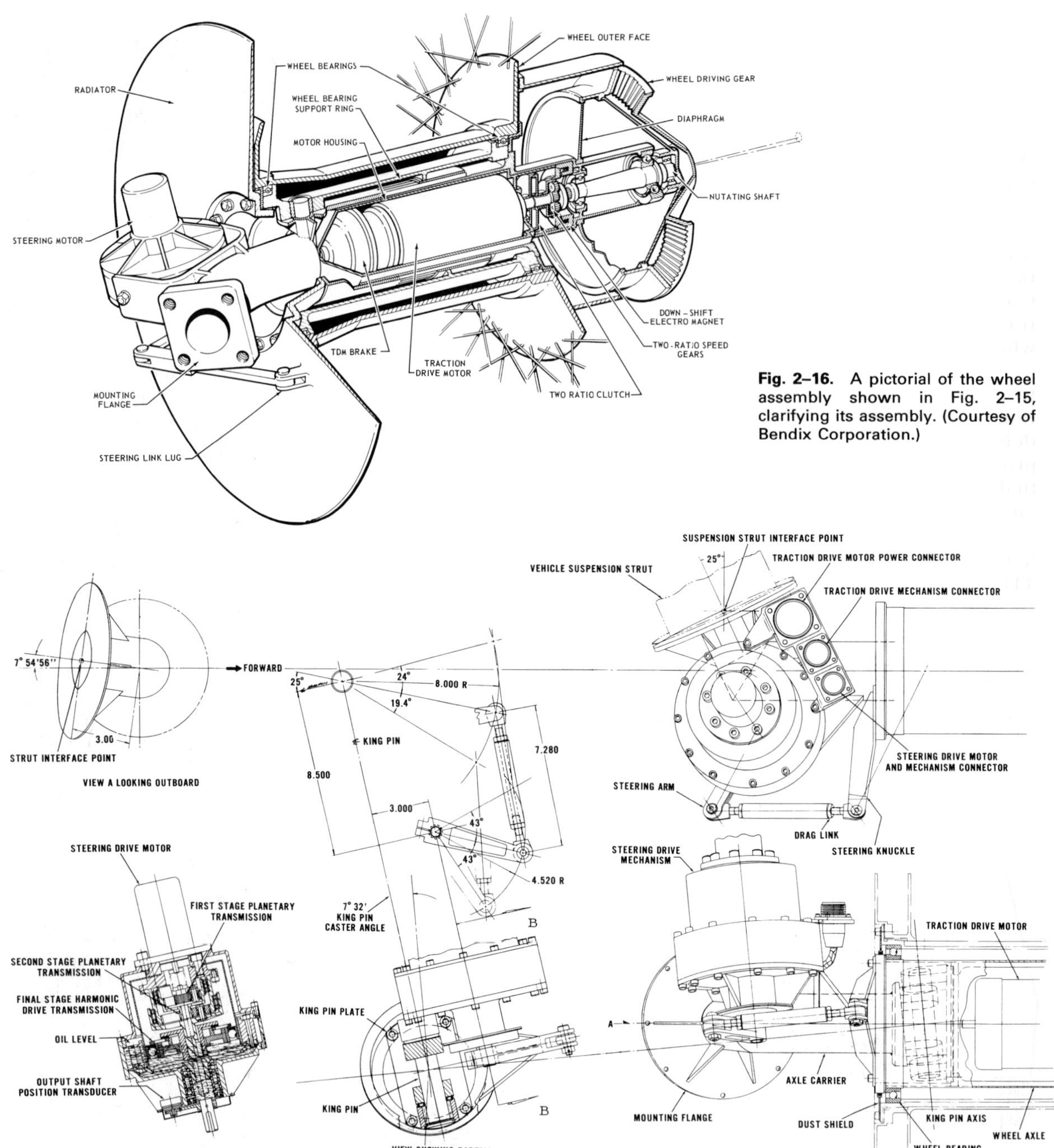

Fig. 2–16. A pictorial of the wheel assembly shown in Fig. 2–15, clarifying its assembly. (Courtesy of Bendix Corporation.)

Fig. 2–17. The final refinement of the wheel mechanism is achieved through the application of descriptive geometry to determine physical properties of its linkage system. Additional auxiliary views were also used to finalize the design. Note that this assembly is more complicated than the initial refinement in Fig. 2–15. (Courtesy of Bendix Corporation.)

posed MOLAB designs within a limited cost and delivery schedule. The final refinement of the design is shown in Fig. 2–17. This drawing shows the MTA hardware design for the left front wheel TDM and SDM systems. Two adjacent orthographic views were drawn to enable the designer to project auxiliary views using descriptive geometry so that the critical dimensions of the steering linkage could be determined. Descriptive geometry was also used to establish the position of the mounting flange so that it could be attached properly to the vehicle suspension strut.

A photograph of the final deliverable hardware is shown in Fig. 2–18. This design problem is typical of those encountered in industry. This design of an assembly required the application of spatial relations and descriptive geometry principles to determine critical relationships.

Fig. 2–18. A photograph of the completed, deliverable wheel assembly. (Courtesy of Bendix Corporation.)

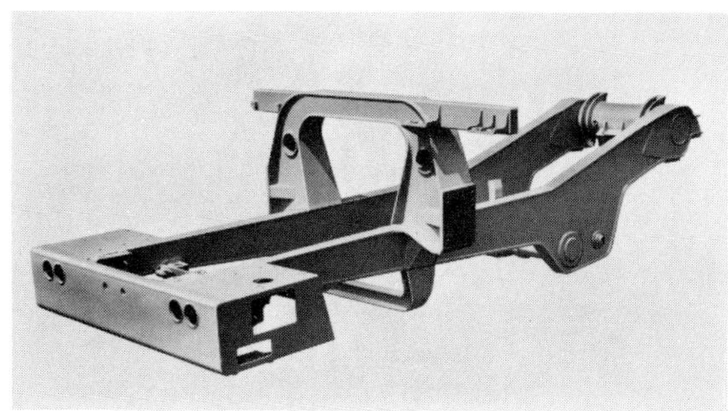

Fig. 2–19. Descriptive geometry principles were utilized in arriving at the final configuration of this heavy-duty truck frame. (Courtesy of LeTourneau-Westinghouse Company.)

Another problem of this type is the design of a frame for a heavy-duty truck (Fig. 2–19). This frame is composed of intersecting planes and surfaces that must be refined so that they may be analyzed for strength by the application of engineering principles. The true size of oblique surfaces must be found along with the angles between the intersecting planes and flanges. This information must be known before complete working drawings can be made. Again, descriptive geometry is the primary method used to obtain this information.

2–6 SUMMARY

Refinement is the process of developing several preliminary ideas into scale drawings to determine more of their physical characteristics, dimensions, angles, and other relationships that will affect their acceptance. Refinement drawings must be drawn accurately to scale so that graphical methods may be used to best advantage. These drawings may be a combination of schematics, diagrams, orthographic views, and pictorials to help the designer develop his preliminary concepts.

Descriptive geometry principles can be used to determine the physical properties important to the design refinement. Angles, lengths, shapes, and sizes can be found graphically by this method with less difficulty than is experienced when attempting to find this same information by mathematical methods. The solutions of problems of this type should be noted and dimensioned to provide an easy reference at a later date when the design is analyzed. Refinement drawings are added to the accumulation of worksheets and preliminary design work previously filed in the design binder or envelope.

Several ideas are refined to give the designer a broader selection of possibilities for the design solution. He should not develop and refine preliminary ideas that are obviously inferior, but he should refine as many ideas as he feels have desirable features. The refinement process will give him a better opportunity to study these designs in greater detail and, therefore, to be better prepared to support his final decision. The final decision should be postponed until after the analysis phase of the design process, which follows the refinement step.

PROBLEMS

Problems should be presented on $8\frac{1}{2}'' \times 11''$ paper, grid or plain, using the format introduced in Article 3–6. Each grid square represents $\frac{1}{4}''$. All notes, sketches, drawings, and graphical work should be neatly prepared in keeping with good practices as covered in this volume. Written matter should be legibly lettered using $\frac{1}{8}''$ guidelines.

1. When refining a design for a folding lawn chair, what physical properties would a designer need to determine? What physical properties would be needed for the following items: a TV-set base, a golf cart, a child's swing set, a portable typewriter, an earthen dam, a shortwave radio, a portable camping tent, a warehouse dolly used for moving heavy boxes?

2. Why should scale drawings be used in the refinement of a design rather than freehand sketches? Explain.

3. List five examples of problems that involve spatial relationships that could be solved by the application of descriptive geometry. Explain your answers.

4. Make a freehand sketch of two oblique planes that intersect. Indicate by notes and algebraic equations how you would determine the angle between these planes mathematically.

5. What is the difference between a working drawing and a refinement drawing? Explain your answer and give examples.

6. How many preliminary designs should be refined when this step of the design process is reached? Explain.

7. Make a list of refinement drawings that would be needed to develop the installation and design of a 100-foot radio antenna. Make rough sketches indicating the type of drawings needed with notes to explain their purposes.

8. Make a list of refinement drawings that would be needed to refine a preliminary design for a rear-view mirror that will attach to the out-side of an automobile.

9. After a refinement drawing has been made and the design is found to be lacking in some respects, so that it is eliminated as a possible solution, what should be the designer's next step? Explain.

10. Would a pictorial be helpful as a refinement drawing? Explain your answer.

11. List several design projects that an engineer or technician in your particular field of engineering would probably be responsible for. Outline the type of refinement drawings that would be necessary in projects of this type.

3

Drawing Standards

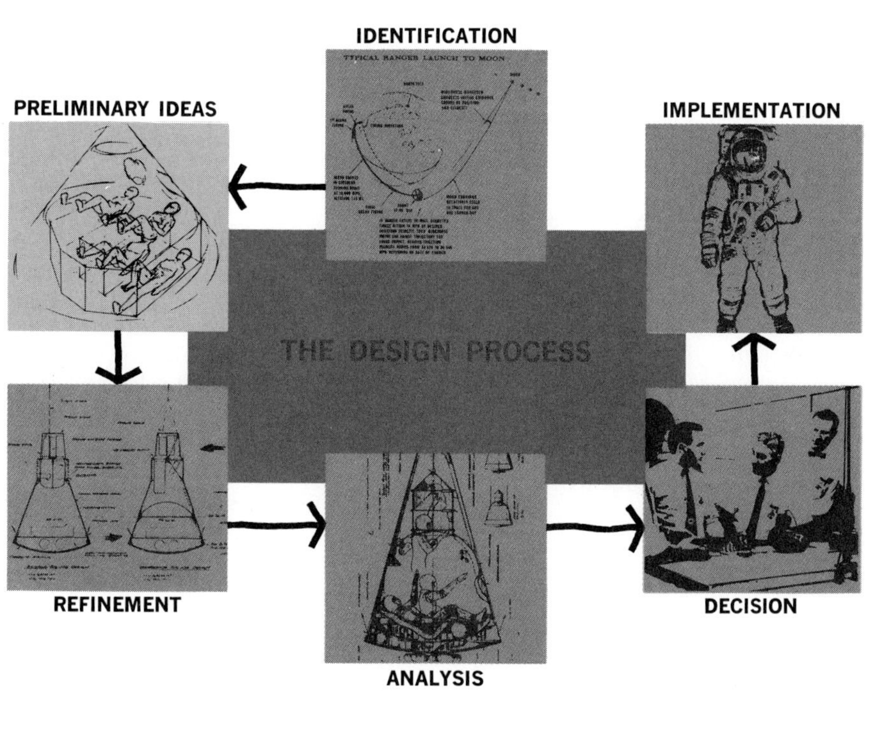

3–1 GENERAL

Descriptive geometry can be mastered only through the solution of problems that utilize principles of orthographic projection and good drafting techniques. This chapter reviews the basic drawing practices that should be followed in solving descriptive geometry problems.

The selection of the proper pencil and how it is sharpened and used is very basic to all descriptive geometry and graphics problems. Similarly, good lettering is necessary to give a drawing a professional appearance.

English and metric scales are reviewed in this chapter, since all problems must be laid out using given dimensions and specifications. The student should become familiar with both systems of measurement.

Suggestions for noting and laying out problems conclude the chapter. The general principles mentioned here should be applied when solving problems presented elsewhere in this book.

3–2 PENCIL SELECTION

Since all drawing begins with a pencil, the selection of the proper pencil is an important step. Each pencil is identified by numbers and letters which specify the hardness of the lead. The following is a list of pencils and their suggested uses:

Designation	*Weight*	*Use*
7B	Soft	Sketching and artistic applications
6B	Soft	
5B	Soft	
4B	Soft	
3B	Soft	
2B	Soft	
B	Medium	Sketching
HB	Medium	Sketching
F	Medium	Object lines
H	Medium	Object lines
2H	Medium	Center lines
3H	Medium	Center lines
4H	Hard	For highly technical construction and accurate measurements
5H	Hard	
6H	Hard	
7H	Hard	
8H	Hard	
9H	Hard	

Cross-sectional views of the three extreme pencil grades listed are shown in Fig. 3–1. Note that the soft pencils have larger lead diameters than the hard pencils.

The pencil should be sharpened to a conical point using a sandpaper board or a mechanical sharpener, as shown in Fig. 3–2. The point should be sharper for thin lines than for bold, heavy lines.

3–3 LETTERING

Lettering is a primary means of communicating engineering information and specifying methods of construction. In many cases, lettering may be more critical to a project than the drawings, since the drawings may consist of schematics with notes relating details and specifications. If lettering is poor, it may cause a misrepresentation of important information and failure of the design. Lettering must be both easy to execute and legible.

Practically all engineering lettering will be single-stroke Gothic, vertical or inclined. The usual height of the letters is $\frac{1}{8}''$ on working drawings and graphs. An F or HB pencil lead is used for lettering on the average drawing surface.

Guidelines should be used for all lettering, whether a single word or a long note.

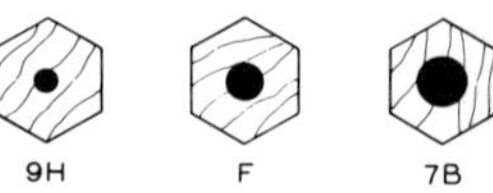

Fig. 3–1. As the pencil grades get progressively softer from 9H to 7B, the diameter of the lead becomes larger.

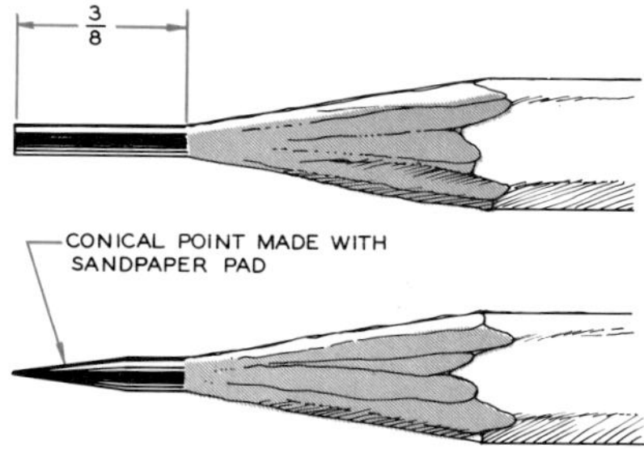

Fig. 3–2. The point of the drafting pencil should be sharpened to expose about 3/8″ of lead. Sharpen the lead to a slim conical point with a sandpaper pad or a mechanical sharpener.

Guidelines should be drawn very lightly with a hard pencil, 3H or 4H, just dark enough to serve the purpose without distracting from the finished lettering.

The *Braddock-Rowe lettering triangle* is an instrument that is widely used in constructing guidelines for lettering (Fig. 3–3). Sets of holes are designated with numerals which represent thirty-seconds of an inch. For example, 8 represents $\frac{8}{32}''$ or $\frac{1}{4}''$, which is the height of capital letters made with this set of holes. The triangle is used in conjunction with a straightedge held firmly in position, and the triangle placed snugly against the edge. A sharp 4H pencil is placed in the holes selected until the point is in contact with the paper. The pencil is guided across the drawing while the triangle is kept in contact with the straightedge as the light guidelines are drawn. This operation is repeated for the other holes until a sufficient number of guidelines are drawn.

The inclined slot in the triangle can be used for drawing diagonal lines across the horizontal guidelines to aid inclined lettering.

The *Ames lettering instrument* has a moveable disc that can be turned to an index that indicates the height of the capital letters desired. In Fig. 3–4, the disc is set on 8, which means $\frac{8}{32}''$ or $\frac{1}{4}''$. (The numbers on the index represent thirty-seconds of a inch.) A pencil is inserted in the set of holes selected and the guidelines are drawn in the same manner as with the Braddock-Rowe triangle. Notice that fractions are twice as tall as single numerals.

Capital and lower-case letters are illustrated in Fig. 3–5, with strokes suggested for their construction. Capital letters are drawn inside squares with a horizontal guideline at mid-height to give a visual guide in the construction of the letters. Capital letters are used almost entirely on most engineering drawings. The height of the letters is usually $\frac{1}{8}''$ on most drawings and guidelines are always used for uniformity. Strokes may vary with the individual, but those presented here will offer assistance to the beginner who is developing his technique.

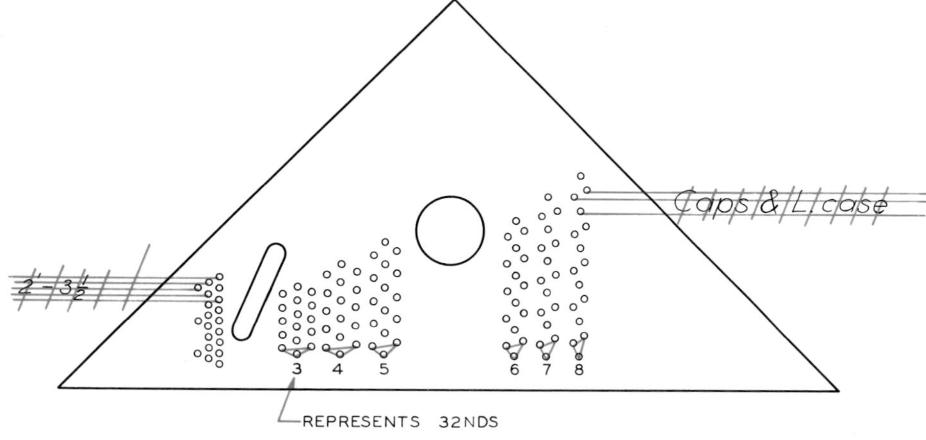

Fig. 3–3. The Braddock-Rowe triangle serves as a lettering guide and as a triangle. The numbers under the series of holes represent thirty-seconds of an inch. For example, 8 represents 8/32″ or 1/4″. The holes at the left are used for guidelines for fractions. The slot is used for inclined lettering. (Courtesy of Braddock Instrument Company.

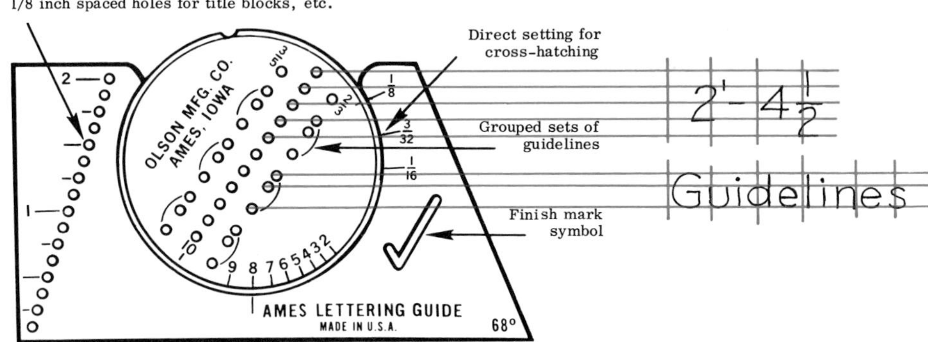

Fig. 3–4. The Ames lettering instrument can be used for the construction of guidelines in much the same manner as the Braddock-Rowe triangle. An adjustable disc can be set on the letter height desired as seen at the bottom of the disc, where the numbers represent thirty-seconds of an inch. The disc is shown set at 8/32″ or 1/4″. (Courtesy of Olson Manufacturing Company.)

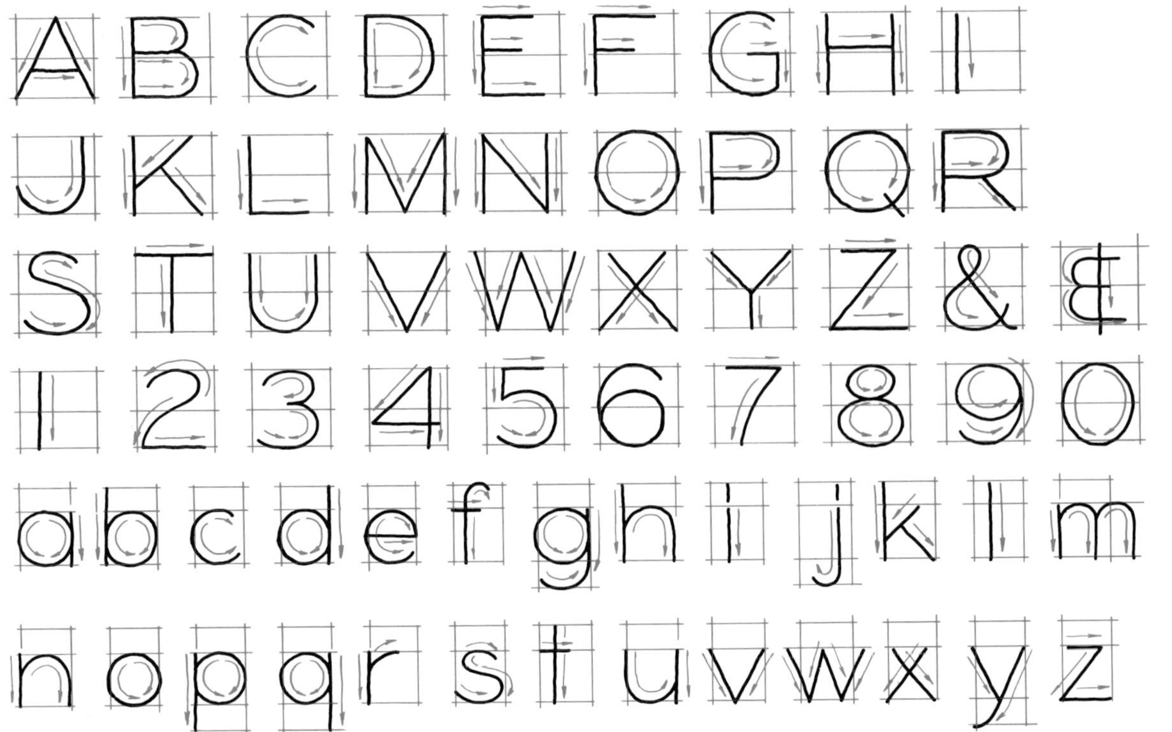

Fig. 3–5. Engineering single-stroke Gothing lettering: vertical upper- and lower-case letters with suggested strokes.

The horizontal guidelines for the vertical, lower-case letters shown in Fig. 3–5 are based on four-unit vertical spacing. The body of the letters is two units high through the central portion of the guidelines, and one unit is left above and below for letters that extend above and/or below the body height. Lower-case letters are more often used in engineering notes and computations than on working drawings.

Both vertical letters and inclined letters are acceptable as standard forms of lettering. However, the same type of lettering should be used

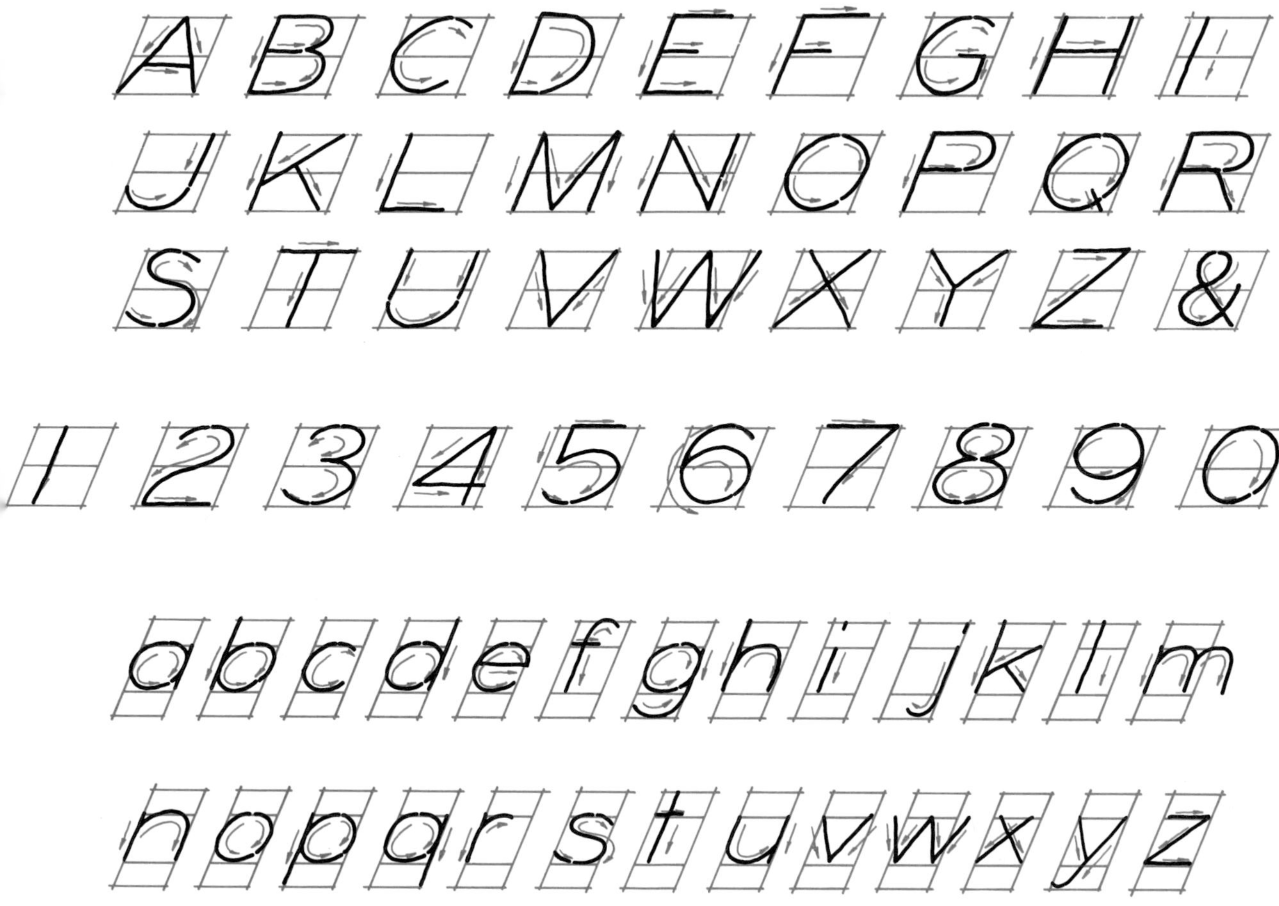

Fig. 3–6. Engineering single-stroke Gothic lettering: inclined upper- and lower-case letters with suggested strokes.

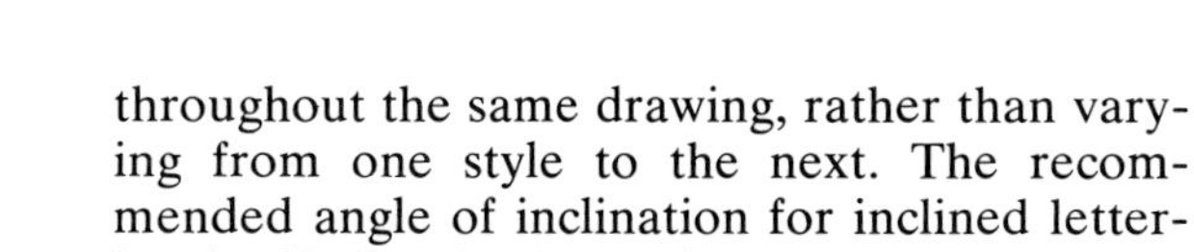

Fig. 3–7. Fractions are drawn twice as tall as whole numbers. The holes at the left are found on Braddock-Rowe lettering triangles. The numerals do not touch the crossbar of the fraction.

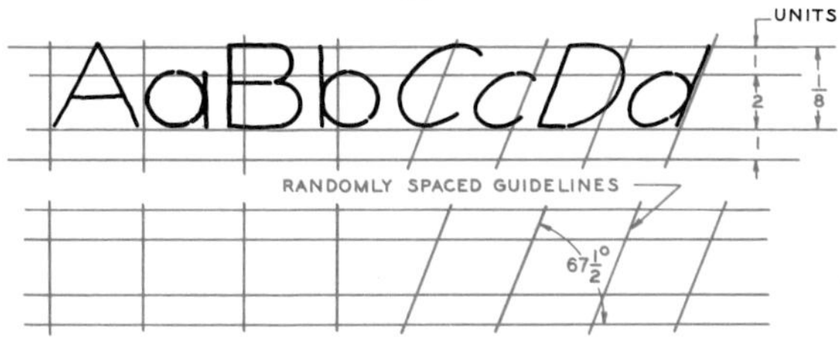

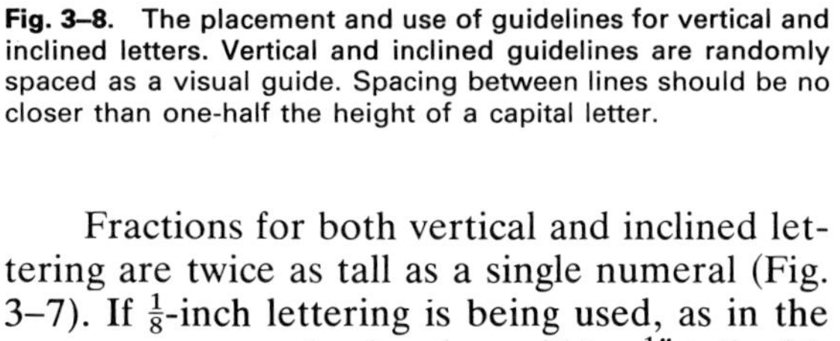

Fig. 3–8. The placement and use of guidelines for vertical and inclined letters. Vertical and inclined guidelines are randomly spaced as a visual guide. Spacing between lines should be no closer than one-half the height of a capital letter.

throughout the same drawing, rather than varying from one style to the next. The recommended angle of inclination for inclined lettering is 67.5°. The letter forms for both capital and lower-case inclined letters are shown in Fig. 3–6.

Fractions for both vertical and inclined lettering are twice as tall as a single numeral (Fig. 3–7). If $\frac{1}{8}$-inch lettering is being used, as in the customary case, the fraction will be $\frac{1}{4}''$ tall with the crossbar placed at the center. The numerals used in the fraction are drawn so that they do not touch the crossbar; they are slightly less than $\frac{1}{8}''$ in height. It is very helpful to the improvement of lettering if randomly spaced vertical guidelines are drawn as visual guides for all lettering.

A comparison of upper- and lower-case letters is given in Fig. 3–8, where both inclined and vertical styles are shown. The body of lower-case letters is two-thirds the height of capital letters. The space between lines of lettering should be approximately the same as the height of the letter or a little less.

3–4 TRADITIONAL SCALES

Essentially all engineering drawings will require the measurement of lengths, sizes, and other linear dimensions as an integral part of all construction. This requires a thorough understanding of the scales used to make the drawings and their various applications. A number of scales are available in a variety of scale designations from metric scales to fractional divisions. The most common scales used in engineering drawing are the architects' scale and the engineers' scale. This article will cover the use of traditional scales based on the inch as the unit of measurement.

Architects' Scale. The architects' scale is used to dimension and scale items normally encountered by the architect, which include building designs, cabinet work, interior plumbing, and electrical layouts. In general, most indoor measurements are measured in feet and inches with an architects' scale. It is obvious that few drawings can be made full size; consequently, scales are used to draw the layouts at proportional, but reduced sizes. For example, the architects' scale has the following graduations:

Full size	$\frac{3}{8}'' = 1'$-$0''$
$3'' = 1'$-$0''$	$\frac{1}{4}'' = 1'$-$0''$
$1\frac{1}{2}'' = 1'$-$0''$	$\frac{3}{32}'' = 1'$-$0''$
$1'' = 1'$-$0''$	$\frac{1}{8}'' = 1'$-$0''$
$\frac{3}{4}'' = 1'$-$0''$	$\frac{3}{16}'' = 1'$-$0''$
$\frac{1}{2}'' = 1'$-$0''$	

Since the architect makes measurements in feet and inches, the architects' scale is graduated into inches on one end and feet on the other. The 16 scale is the scale used for making full-size dimensions with each inch divided into sixteenths of an inch. This is the scale found on most yardsticks and tape measurements used by the workmen who will be constructing the finished components from architectural drawings.

Scale: Full Size. The use of the 16 scale for measuring a full-size line is shown in Fig. 3–9A. One end of the line is placed at the zero end of the scale, and the reading is made at the other end to the nearest $\frac{1}{16}''$. In this example, it can be seen that the length of the line is $3\frac{1}{8}''$.

Scale: $1'' = 1'$-$0''$. When an architects' scale is used to express feet and inches, it is necessary to measure the line from the nearest whole-foot graduation and let the excess length extend into the inch graduations at the end of the scale. This may be at the left or the right end, depending on which scale is being used, since each edge of the architects' scale is used for two scales. The example in Fig. 3–9B gives the measurement of $2'$-$3\frac{1}{2}''$. Note that the right end of the line being measured is aligned with the whole foot unit, $2'$. The other end gives a reading of $3\frac{1}{2}''$.

Scale: $\frac{3}{8}'' = 1'$-$0''$. This scale is used when $\frac{3}{8}''$ is used to represent $12''$ on a drawing. In Fig. 3–9C a line is measured to be $7'$-$5''$ long. Note that the 7 does not appear on the scale, but if it did, it would be in the same position as the number 11, which is a unit on the scale running in the oposite direction. Care must be taken not to confuse the numbering on an architects' scale.

Scale: $\frac{1}{2}'' = 1'$-$0''$. The $\frac{1}{2}''$ scale is used to represent one foot or $12''$ by a $\frac{1}{2}''$ distance. A measurement of $5'$-$8\frac{1}{2}''$ is scaled in Fig. 3–9D. Care must be taken to avoid confusing $5'$-$8\frac{1}{2}''$ with $8'$-$8\frac{1}{2}''$. The inch divisions at the end of the scale are not numbered because the divisions are too small to be labeled.

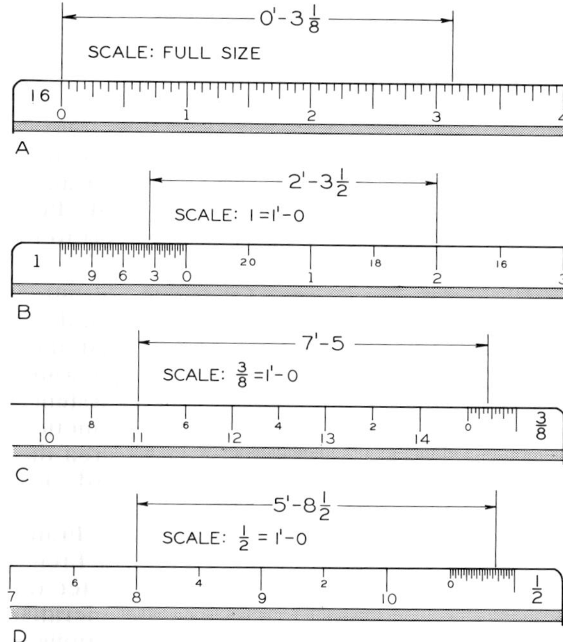

Fig. 3–9. Measurement of lines using the architects' scale.

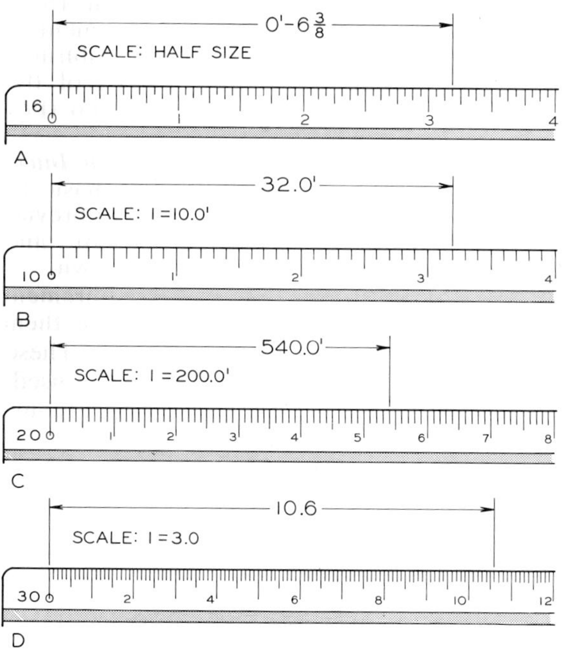

Fig. 3–10. Measurement of lines using the 16-scale and the engineers' scale.

Scale: Half Size. A drawing may be drawn half size to advantage in some cases. The 16 scale is used for this purpose. Since the drawing is to be made half size, a $6\frac{3}{8}''$ measurement will be drawn $3\frac{3}{16}''$ long (Fig. 3–10A). In other words, the measurement to be drawn is divided by 2 and drawn with a 16 scale. When a measurement is being taken from a half-size drawing, the measurement is scaled and doubled to give the actual length.

The Engineers' Scale. The engineers' scale is a decimal scale on which each division is divided into multiples of 10 units. This scale is commonly used for engineering drawings of structures or projects that are erected outdoors, such as street systems, drainage systems, and other systems having measurements of rather long dimensions associated with topography and map drawing. Since the divisions are in decimals, it is easy to perform multiplication and division without the complicated conversion of feet and inches as would be the case if the architects' scale were used. Areas and volumes can be found with the minimum of difficulty by use of the engineers' scale. The engineers' scale is divided into scales denoted by 10, 20, 30, 40, 50, and 60, which means that an inch is divided into this number of parts on each scale, respectively. The graduations permit the following scales:

10 scale:	1″ = 0.1″,	1″ = 1′,	1″ = 10′
20 scale:	1″ = 20′,	1″ = 200′,	etc.
30 scale:	1″ = 300′,	1″ = 3000′,	etc.
40 scale:	1″ = 40′,	1″ = 400′,	etc.
50 scale:	1″ = 50′,	1″ = 500′,	etc.
60 scale:	1″ = 600′,	1″ = 6000′,	etc.

Many other combinations may be obtained by increasing or reducing the scales by multiples of 10. For example, the 10 scale can be used for 1″ = 0.0001″ or 1″ = 10,000′ by simply moving the decimal point.

10 Scale. In Fig. 3–10B, the 10 scale is used to measure a line at the scale 1″ = 10′. The zero index is placed at one end of the line and the measurement is read directly at the other end. A line is measured to have a length of

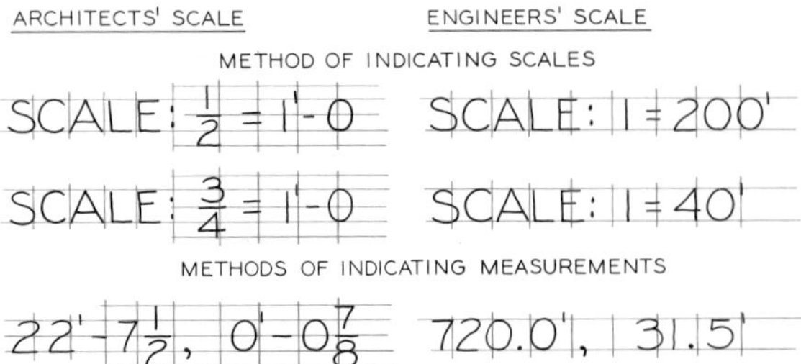

Fig. 3–11. The methods of indicating scales and measurements with the architects' and engineers' scales.

32.0′. Note that the dimensions are written in inches with a decimal. If a line were measured with an engineers' scale to give an even measurement of 2 inches, it would be written as 2.0″ to indicate that it was an accurate reading rather than a rounded-off measurement.

20 Scale. Each inch is divided into 20 divisions on the 20 scale. Using a scale of 1″ = 200′ a line is measured to have a length of 540.0′ in Fig. 3–10C.

30 Scale. A line in Fig. 3–10D is measured to be 10.6″ long. Note that the 30 scale was used for this measurement.

The method of indicating scales on drawings is shown in Fig. 3–11. Notice that common fractions are twice as tall as capital letters. Also shown are examples of measurements that have been made with both types of scales. Zeros are used for feet and inch measurements when these values are less than one foot or one inch. Common fractions are used for measurements made with the architects' scale, and decimal fractions are used with the engineers' scale.

3–5 THE METRIC SYSTEM—SI

The English system of measurements has been used in the United States, Britain, and Canada since these countries were established. Presently a movement is underway to convert to the more universal metric system.

The English system was based on arbitrary units of the inch, foot, cubit, yard, and mile, as shown in Fig. 3–12. It is apparent that there was little relationship between one measurement and the next; consequently this system is cumbersome to use when simple arithmetic is performed. For example, finding the area of a rectangle that is 25 inches by $6\frac{3}{4}$ yards is a complex problem.

The metric system was proposed by France in the fifteenth century. In 1793, the French National Assembly agreed that the meter (m) would be one ten-millionth of the meridian quadrant of the earth (Fig. 3–13). Fractions of the meter were to be decimal fractions. Debate continued until an international commission officially adopted the metric system in 1875. Since a slight error in the first measurement of the meter was found, the meter was established as equal to 1,650,763.73 wavelengths of the orange-red light given off by krypton-86 (Fig. 3–13).

The metric system is called *System International* and is abbreviated SI. The basic SI units are shown in Fig. 3–14 with their abbreviations. It is important that lower-case and upper-case abbreviations be used as shown.

Several practical units of measurement have been derived (Fig. 3–15) to make them easier to use in many practical situations. These are unofficial units of SI that are widely used. Note that degrees Celsius (centigrade) is recom-

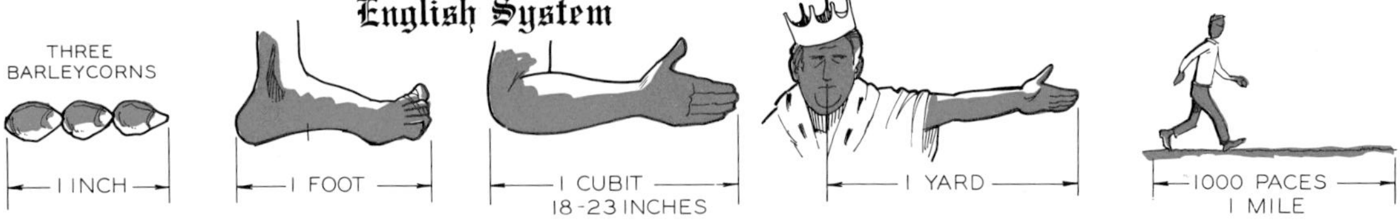

Fig. 3–12. The units of measurement used in the English system were based on unrelated, inaccurate dimensions.

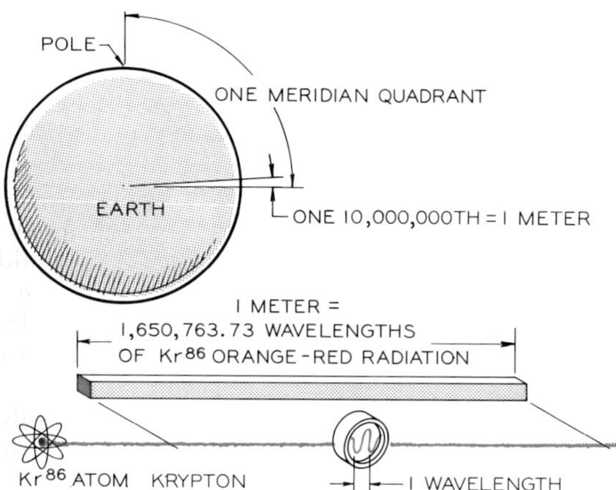

Fig. 3–13. The origin of the meter was based on the dimensions of the earth, but it has since been based on the wavelength of krypton-86. A meter is 39.37 inches.

SI UNITS			DERIVED UNITS		
LENGTH	METER	m	AREA	SQ METER	m^2
MASS	KILOGRAM	kg	VOLUME	CU METER	m^3
TIME	SECOND	s	DENSITY	KILOGRAM/CU MET	kg/m^3
ELECTRICAL CURRENT	AMPERE	A	PRESSURE	NEWTON/SQ MET	N/m^2
TEMPERATURE	KELVIN	K			
LUMINOUS INTENSITY	CANDELA	cd			

Fig. 3–14. The basic SI units and their abbreviations. The derived units are units that have come into common usage.

PARAMETER	PRACTICAL UNITS		SI EQUIVALENT
TEMPERATURE	DEGREES CELSIUS	°C	0°C = 273.15 K
LIQUID VOLUME	LITER	l	$l = dm^3$
PRESSURE	BAR	BAR	BAR = 0.1 MPa
MASS WEIGHT	METRIC TON	t	$t = 10^3$ kg
LAND MEASURE	HECTARE	ha	$ha = 10^4 m^2$
PLANE ANGLE	DEGREE	°	$1° = \pi/180$ RAD

Fig. 3–15. These practical metric units are a few of those that are widely used because they are easier to deal with than the official units of SI.

VALUE		PREFIX	SYMBOL
1,000,000	$= 10^6 =$	MEGA	M
1,000	$= 10^3 =$	KILO	k
100	$= 10^2 =$	HECTO	h
10	$= 10^1 =$	DEKA	da
1	$= 10^0 =$		
.1	$= 10^{-1} =$	DECI	d
.01	$= 10^{-2} =$	CENTI	c
.001	$= 10^{-3} =$	MILLI	m
.000 001	$= 10^{-6} =$	MICRO	μ

Fig. 3–16. The prefixes and abbreviations used to indicate the decimal placement for SI measurements.

mended over the official temperature measurement, Kelvin. When using Kelvin, the freezing and boiling temperatures are 273.15°K and +373.15°K, respectively. Pressure is measured in bars, where one bar is equal to 0.1 megapascal or 100,000 pascals.

Many SI units have prefixes to indicate placement of the decimal. The more common prefixes and their abbreviations are shown in Fig. 3–16. Some of the abbreviations are lowercase letters and others are upper-case. You can see that a centimeter (cm) is one one-hundredth of a meter, and a kilometer (km) is one thousand meters.

Several comparisons of English and SI units are given in Fig. 5–17.

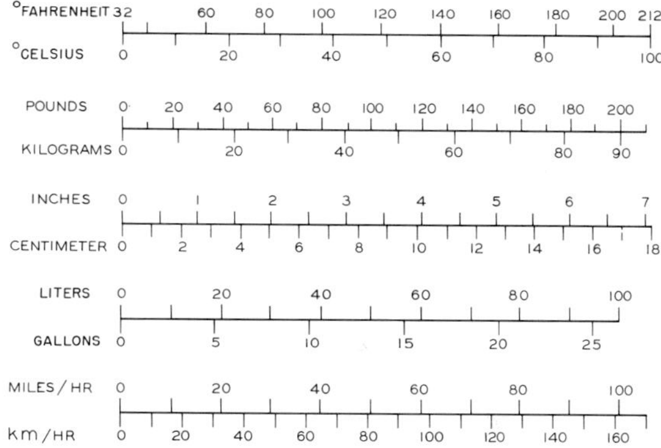

Fig. 3–17. A comparison of metric units with those used in the English system of measurement.

Metric Scales. The basic unit of measurement on an engineering working drawing is the millimeter (mm), which is one one-thousandth of a meter. These units are understood unless other SI units are indicated on a drawing. Scales are indicated with two numbers separated by a colon such as 1:2, 1:400, 1:5, and so on. The first numeral is 1 and the second number indicates the ratio of enlargement or reduction.

1:1 Scale. This scale is full size. The numbered calibrations represent centimeters, which are equivalent to ten millimeters (Fig. 3–18A). A measurement of 59 mm is shown in this example. This same scale can be used to repre-

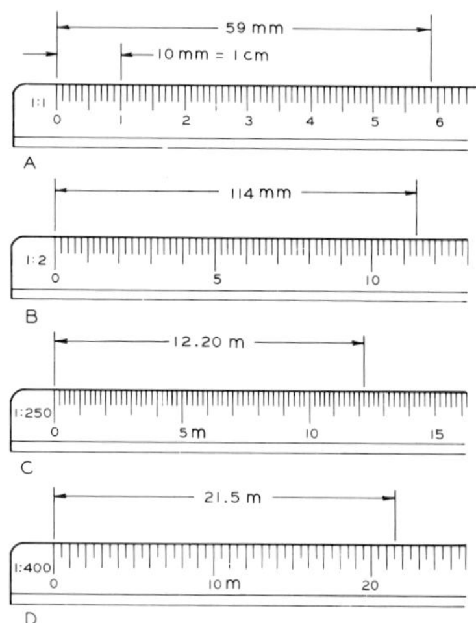

Fig. 3–18. Measurements made with metric scales. The millimeter is the basic measurement of SI, but other units may be used if specified.

sent 1 : 10, 1 : 100, 1 : 0.10, and so forth by moving the decimal point.

1:2 Scale. This scale is used where two millimeters is represented by a measurement of one millimeter, which results in a half-size drawing. A measurement of 114 mm is given in Fig. 3–18B.

1:250 Scale. This scale is used where 250 millimeters (0.25 meter) is represented by one millimeter. Again the decimal point can be moved to represent 1:2.5. A measurement of 12.2 m is given in Fig. 3–18C; this measurement would be 0.122 m if the scale had been 1:2.5.

1:400 Scale. In Fig. 3–18D, an example measurement is made using the 1:400 scale. This measurement is 21.5 m or 21,500 mm.

Other Scales. Many other SI scales are used: 1:3, 1:5, 1:60, and so on. The scale ratios mean that one unit represents the number of units shown to the right of the colon. For example, 1:20 means that one millimeter equals 20 mm, or one centimeter equals 20 cm, or one meter represents 20 m.

Scale Conversion. Tables for converting inches to millimeters are given in Appendix 4; however this conversion can be performed by multiplying decimal inches by 25.4 to obtain millimeters. For example, 1.5 inches would be $1.5 \times 25.4 = 38.1$ mm. To convert an engineers' scale to be near that of a metric scale, you must multiply by 12. For example, Scale: 1″ = 100′ means that 1″ = 1200″, which would be expressed metrically as 1:1200. The common fractions of the architects' scale must be eliminated to convert to a metric ratio. For example, Scale: $\frac{1}{8}''$ = 1′-0″ is the same as $\frac{1}{8}''$ = 12″ or 1″ = 96″. This is near to the metric scale of 1:100. Many of the scales used in the metric system cannot be converted to exact equivalents of the English system.

Several general rules of expressing SI units and scales are given in Fig. 3–19.

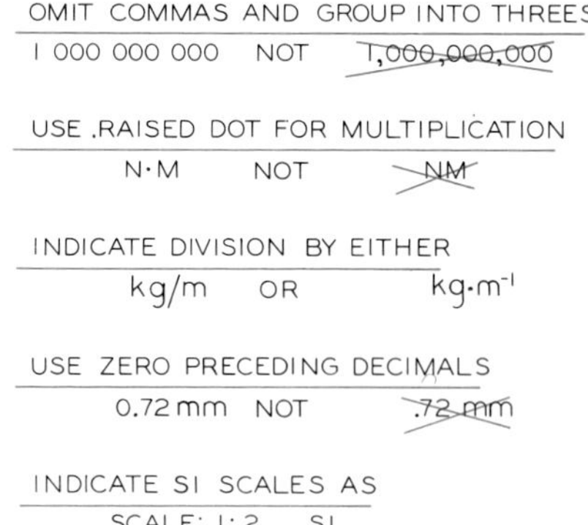

Fig. 3–19. General rules to be used in indicating SI measurements and scales.

3–6 SOLUTION OF PROBLEMS

You may be assigned problems at the end of each chapter or you may be assigned problems from a laboratory manual, which greatly reduces lay-out time. In either case, you should become familiar with several techniques of solving and noting these problems in order to produce professional quality drawings with the greatest clarity.

The following rules of solving descriptive geometry problems are illustrated in Fig. 3–20.

Adherence to these rules will assist you in producing high quality drawings.

Lettering. All points, lines, and planes should be labeled using $\frac{1}{8}''$ letters. Always use guidelines for lettering, both horizontal and vertical or inclined guidelines. Lines should be labeled at each end. Planes should be labeled at each corner. Either letters or numerals can be used.

Points. A point in space should be indicated by two short dashes to form a cross. The approximate length of one of the dashes should be $\frac{1}{8}''$.

Points on a Line. If a point on a line is to be indicated, a short perpendicular dash should be used to mark the point, *not* a dot. Label the point with a letter or numeral.

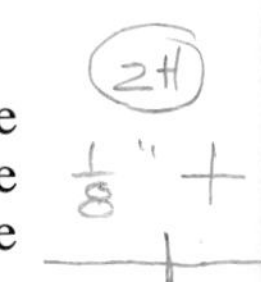

MARK POINTS WITH WITH A CROSS

LABEL ALL POINTS USING GUIDELINES & $\frac{1}{8}$ LETTERS OR NUMERALS

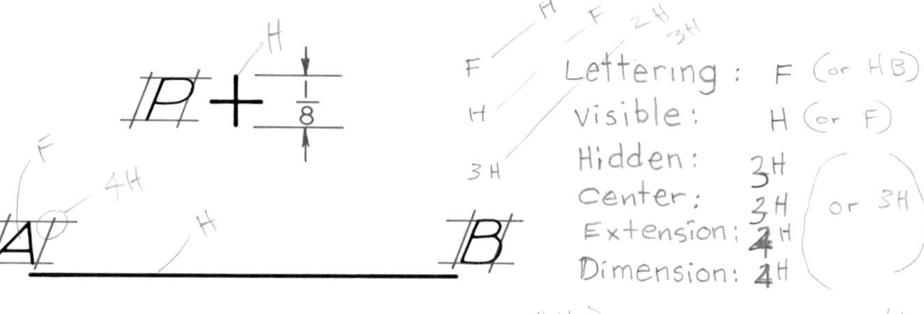

USE A ⊥ SLASH TO MARK A POINT ON A LINE

1 C 2 $\frac{1}{8}$

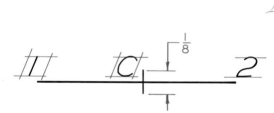

LABEL ALL REFERENCE LINES (Fold Lines)

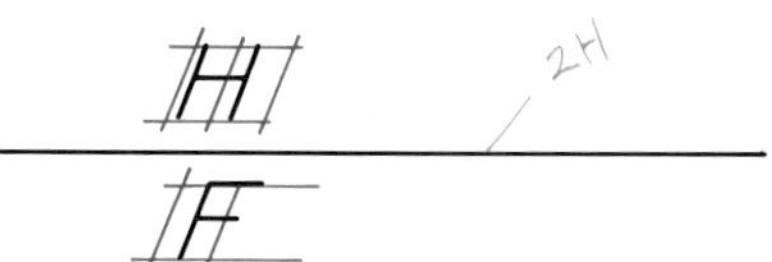

LABEL TRUE LENGTH LINES

MARK PIERCING POINTS & SHOW VISIBILITY

5 6 D P E 7

GAP

THINNER HIDDEN LINES

Fig. 3–20. Rules that should be followed in solving descriptive geometry problems.

Reference Lines. Reference lines are thin black lines that should be labeled in accordance with the text in Chapter 4.

Object Lines. Lines used to represent points, lines, and planes should be drawn considerably heavier than reference lines, with an H or F pencil. When the lines are hidden, they should be thinner than the visible-object lines.

True Length Lines. Lines that appear true length should be labeled as such by a full note, TRUE LENGTH, or the abbreviation, TL.

True Size Planes. Planes that appear true size and true shape should be labeled by a note, TRUE SIZE, or by the abbreviation, TS.

Projection Lines. Projectors that are used in constructing the solution to a problem should be drawn very precisely with a 4H pencil. These should be thin *gray* lines, just dark enough to be visible. These need not be erased after the problem is solved.

The application of a number of these rules is illustrated in Fig. 3–21. You will notice that some of the lettering and numbering is aligned with inclined lines and reference lines to which the labeling applies, while other lettering is not aligned but parallel to the edge of the paper. You may use either technique or a combination of the two.

Most problems at the end of each chapter are to be solved on $8\frac{1}{2}'' \times 11''$ paper, using instruments or drawing freehand as specified. A grid paper printed with a $\frac{1}{4}''$ grid can be used in laying out the problems. Or, if plain paper is to be used, the layout can be made with a 16 scale (architects' scale). The grid of the problems in each chapter represents $\frac{1}{4}''$ intervals that can be

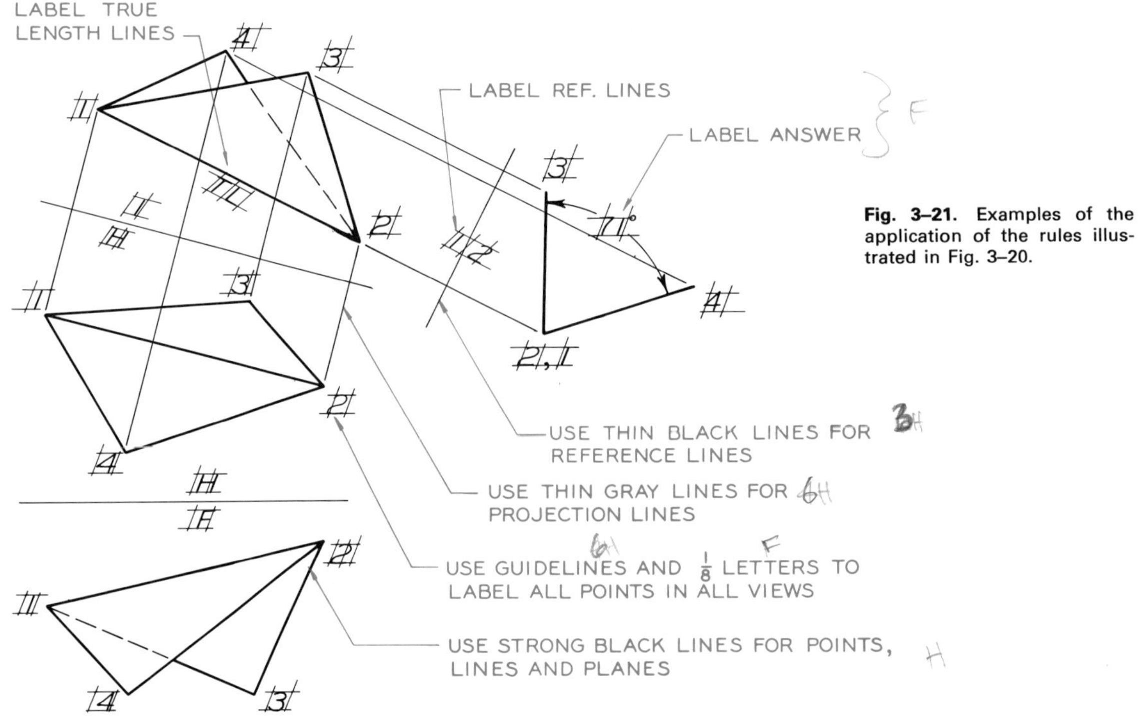
LABEL TRUE LENGTH LINES
LABEL REF. LINES
LABEL ANSWER
USE THIN BLACK LINES FOR REFERENCE LINES
USE THIN GRAY LINES FOR PROJECTION LINES
USE GUIDELINES AND 1/8 LETTERS TO LABEL ALL POINTS IN ALL VIEWS
USE STRONG BLACK LINES FOR POINTS, LINES AND PLANES

Fig. 3–21. Examples of the application of the rules illustrated in Fig. 3–20.

counted and transferred to a similar grid or scaled on plain paper.

Each problem sheet should be endorsed as shown in Fig. 3–22. Guidelines should be drawn with a straightedge to aid in lettering, using $\frac{1}{8}''$ letters.

Answers to essay-type questions should be lettered, using approved, single-stroke, Gothic lettering. Each page should be numbered and stapled in the upper left corner if turned in for review by the instructor. All solved problems should be maintained in a notebook for future reference.

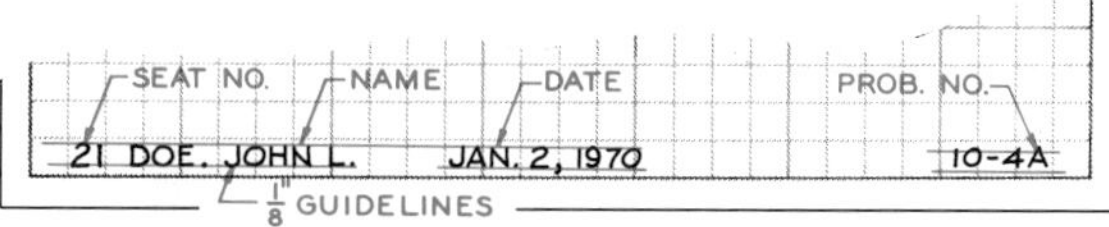

FIGURE 3–22

4

Spatial Relationships

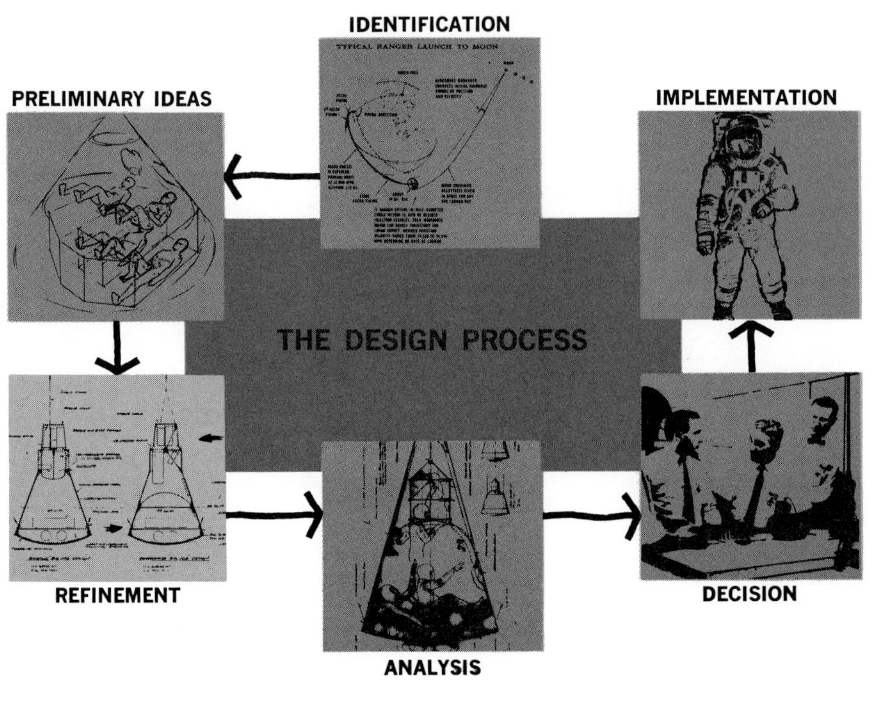

4–1 ORTHOGRAPHIC PROJECTION

The preparation of engineering drawings that must be used by large numbers of people working on a common project in a variety of geographical locations requires a universal system of presentation. This universal system is *orthographic projection.* We shall now take a closer look at the underlying theory supporting this standardized form of presentation.

Orthographic projection may be defined as a method of representing three-dimensional objects through the use of views which are projected perpendicularly onto planes of projection with parallel projectors (Fig. 4–1). The three mutually perpendicular projection planes, called principal planes, are shown as they would be positioned in space in Fig. 4–2A. Part B of the figure shows the transformation of the three principal planes into one common plane (part C). This common plane is the sheet of paper on which the engineer, designer, or draftsman must represent a three-dimensional object that may vary in size from a small bolt to a large

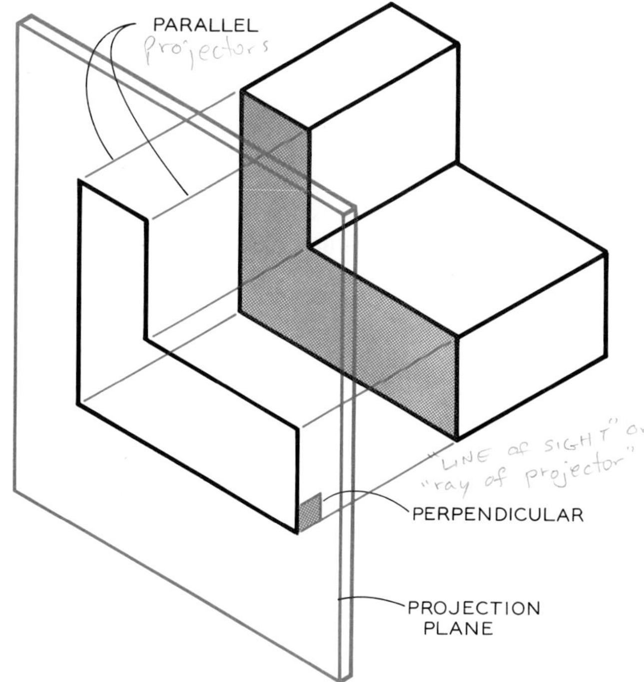

Fig. 4–1. Orthographic projection is defined as the projection of a view onto a projection plane with parallel projectors. The projectors are perpendicular to the projection plane.

FIGURE 4–2. THE PRINCIPAL PROJECTION PLANES

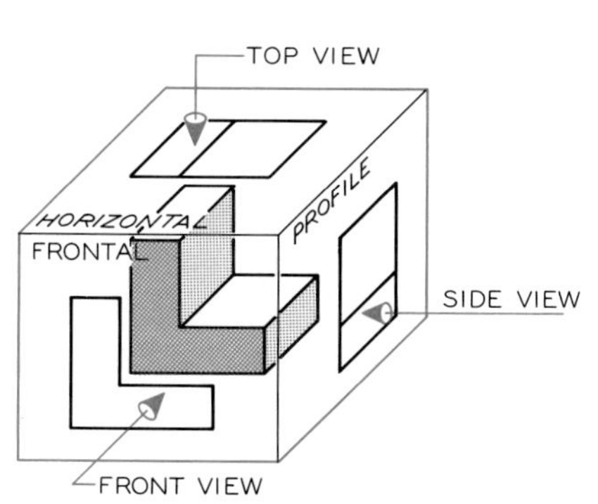

A. The three principal projection planes used in orthographic projection can be thought of as planes of a glass box.

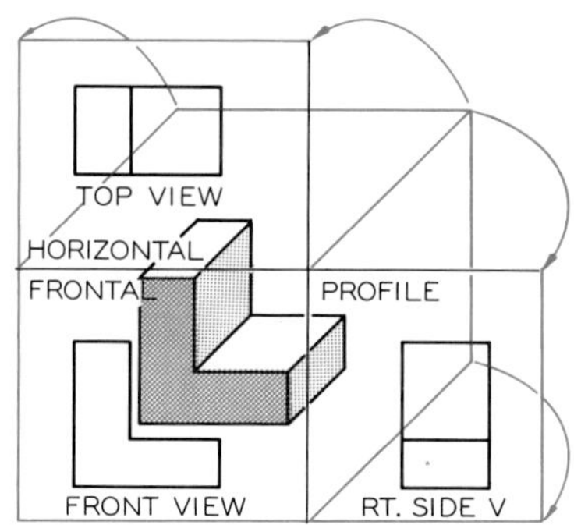

B. Views of an object are projected onto the projection planes, which are opened into the plane of the drawing surface.

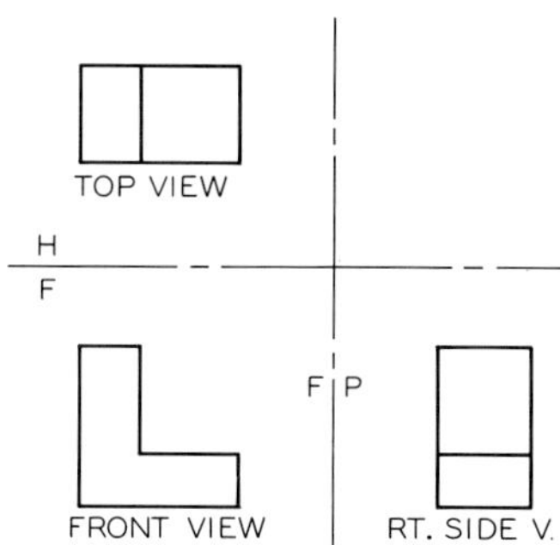

C. The outlines of the planes are omitted. The fold lines are drawn and labeled in this manner when descriptive geometry problems are solved.

bridge girder. Since projection planes are infinite in size and therefore have no perimeters, there is no need to indicate the outlines of the projection box. However, the intersections of the planes, or fold lines, are usually drawn as shown in part C to aid in the solution of descriptive geometry problems. Note that the three principal planes—*horizontal*, *frontal*, and *profile*—are represented by means of the single letters H, F, and P placed on their respective sides of the fold lines. This system of notation will be used throughout this book.

Figure 4–2 illustrates the relationship between the three principal planes pictorially and orthographically. It should be noted that the *front* view is projected onto the *frontal* projection plane; the *side* view onto the *profile* projection plane; and the *top* view onto the *horizontal* projection plane. This system allows three-dimensional objects to be represented by means of related, two-dimensional views in a manner which will be developed in this chapter.

4–2 SIX PRINCIPAL VIEWS

Some objects cannot be fully represented through the three views mentioned in the first article, but require separate views projected from each side. The system of orthographic projection permits six principal views of a given object to be drawn. Figure 4–3A suggests how an imaginary box formed by the six principal planes is opened to form one common plane, as shown in part B of the figure. It should be observed that the top and bottom planes are both horizontal planes and are labeled with the letter H; the front and rear views are both frontal planes and are noted with the letter F, and the left and right side views are both profile planes and are labeled with the letter P.

4–3 DIRECTIONAL RELATIONSHIPS

Certain verbal terms are used to describe spatial relationships both in orthographic projec-

FIGURE 4–3. SIX VIEWS OF AN OBJECT

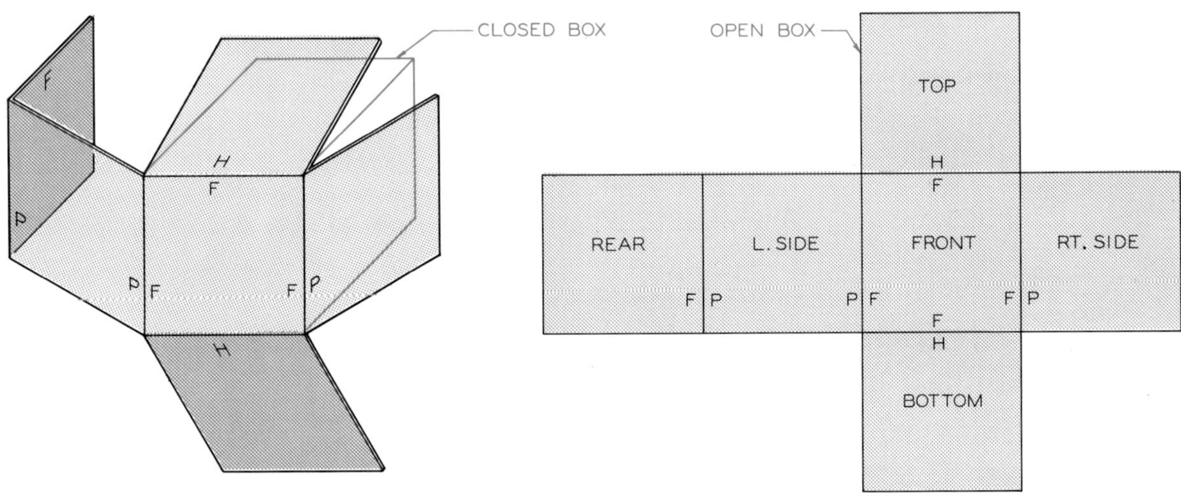

A. The six projection planes of the imaginary glass box can be opened into a single plane to give six principal views of an object.

B. This is the standard arrangement for a six-view drawing. Note how the fold lines are labeled using letters to represent the three principal planes, F, H, and P.

tion and in general discussion. The more commonly used terms are forward, back, left, right, up, and down. Combinations of these directions will allow an object to be generally located in space.

The parallel directions of up and down are illustrated in Fig. 4–4A. These directions are perpendicular to the horizontal plane on which the top view is projected. Since the arrows which indicate up and down directions are vertical, they would appear as points on the horizontal plane if shown. Part B of the figure shows how the directional arrows would project in orthographic projection on a single drawing surface. Note that the directions are perpendicular to the horizontal plane and parallel to the frontal and profile planes in the orthographic layout, just as they are in the pictorial. The directional arrows projected onto the frontal and profile planes are the same length since they are vertical.

Left and right directions are shown pictorially in Fig. 4–5A. Both of these parallel directions are perpendicular to the profile plane. If the projection planes were revolved into the conventional position, the directions would project as shown in part B. The directional arrows are parallel to the horizontal plane and perpendicular to the profile plane. To locate a point that is to the right or left of a given point, we must use the front or top view, since these arrows appear as points in the profile view.

Forward and backward directions are shown pictorially in Fig. 4–6A. The parallel directional arrows are parallel to the horizontal

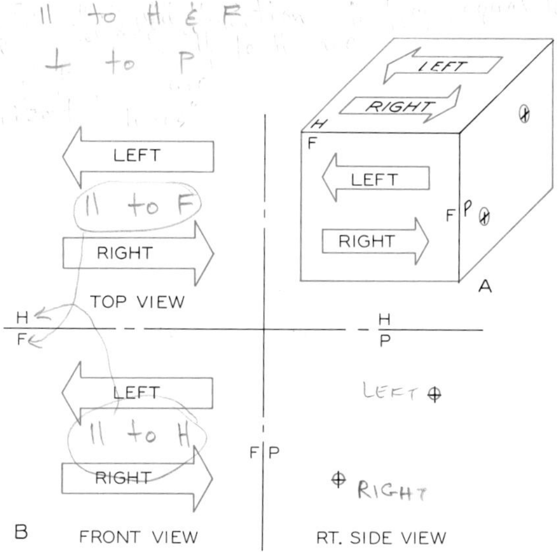

Fig. 4–5. The directions of left and right can be seen in the top and front views of orthographic drawings.

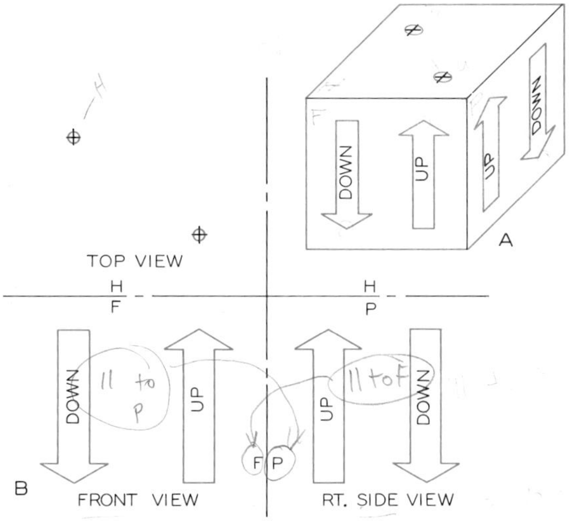

Fig. 4–4. The directions of up and down can be seen in the front and side views of orthographic drawings.

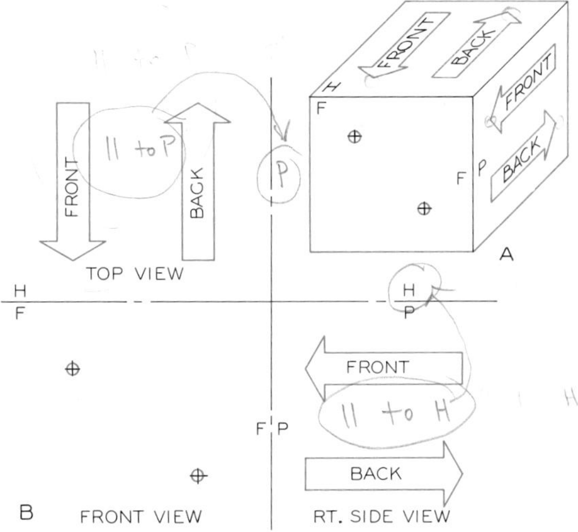

Fig. 4–6. The directions of forward and backward can be seen in the top and side views of orthographic drawings.

and profile planes. These relationships are also illustrated in the usual three-view arrangement for orthographic projection in part B. Location of a point in space with respect to forward or backward directions must be established in either the horizontal view or the profile view, since they cannot be established in the front view.

Any two of the three basic groups of directions will establish the location of a point on a principal plane.

4-4 ORTHOGRAPHIC PROJECTION OF A POINT

The point is the basic geometric element that is used to establish all other elements regardless of their degree of complication. A point is a theoretical location in space and has no dimensions. However, a series of points can establish areas, volumes, and lengths, which are the basis of our physical world.

A point in space must be projected perpendicularly onto at least two principal planes to establish its true position. Figure 4–7A is a pictorial representation of a point projected onto each of the principal planes.

The three orthographic projections or views of point 2 are shown in Fig. 4–7B. The point is at the same distance below the horizontal projection plane in the front view as it is in the right side view, and it is located directly below the top view of point 2. Similarly, the top and side views of the point are at the same distance from the edge view of the frontal plane. These relationships are shown in pictorial form in Fig. 4–7A. Note that the projector lines between the three views of point 2 are perpendicular to the principal planes (fold lines) in both of these illustrations.

A point can be located easily from a verbal description which uses units of measurement taken from the fold lines. Assume that the following coordinates were given: (1) 11 units left of the profile plane, (2) 8 units below the horizontal plane, and (3) 10 units back of the frontal plane. The measurement of 11 units to the left can be established in the top and front views as shown in Fig. 4–8. The measurement of 8 units below the horizontal plane isolates the exact position of the frontal projection of point *A*. The right side view of point *A* will also lie on the projector 8 units below the horizontal plane. The third coordinate, 10 units

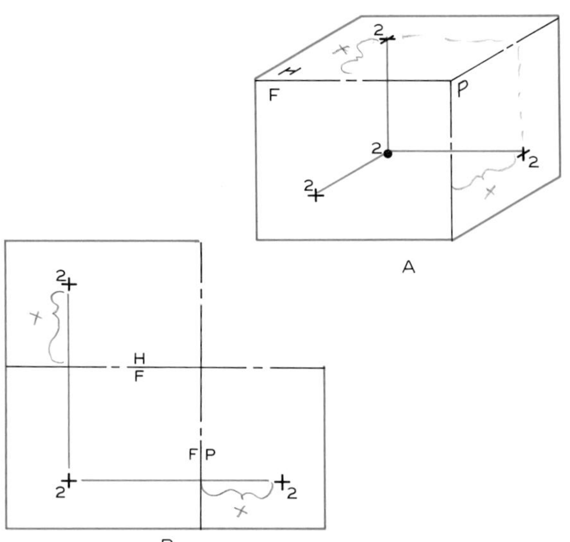

Fig. 4–7. Three views of a point projected onto the three principal planes: horizontal, frontal, and profile.

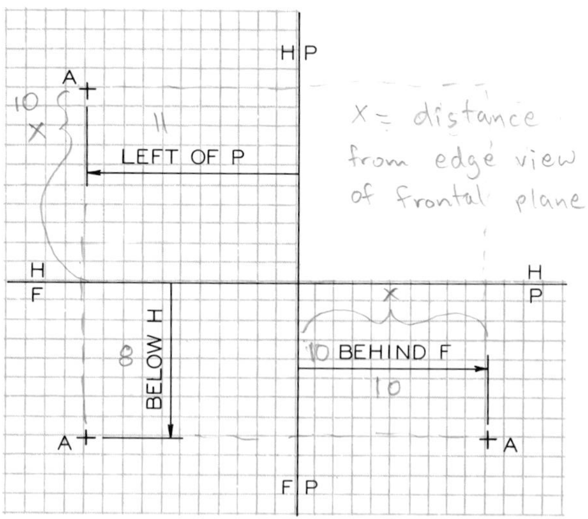

Fig. 4–8. Location of a point 8 units below the horizontal, 10 units back of the frontal, and 11 units to the left of the profile.

back of the frontal plane, will complete the location of the top and side views of the point.

In the top or horizontal view, the frontal and profile planes appear as edges. In the frontal view the horizontal and profile planes appear as edges, and in the profile view the frontal and horizontal planes appear as edges. This relationship permits two coordinates to be plotted in each view, since they are measured perpendicularly from planes that appear as edges.

4–5 LINES

A line is a straight path between two points in space. It can appear in three forms: (1) as a point, (2) as a true-length line, or (3) as a foreshortened line (Fig. 4–9). Line 1–2 appears foreshortened in each view in Fig. 4–10. The line of sight is always perpendicular to a true-length line or, in other words, a true-length line is parallel to the plane on which it is projected. These relationships will be discussed further in the following examples.

Oblique Lines. An *oblique line* is a line that is neither perpendicular nor parallel to a principal projection plane, as shown in Fig. 4–10. When line 1–2 is projected onto the horizontal, frontal, and profile planes, it appears foreshortened as represented in Fig. 4–10B. The two endpoints must be established and then connected to represent the line. An oblique or foreshortened line is the general case of a line.

Principal Lines. If a line is parallel to a principal plane, it is referred to as a *principal line.* There are three principal lines, (1) horizontal, (2) frontal, and (3) profile, since there are three principal planes to which they can be parallel. The principal lines are true length in the view where the principal plane with which they are parallel appears true size. For example, horizontal lines are true length in the top view, which is the view that shows the true size of the horizontal plane.

4–6 HORIZONTAL LINES

A *horizontal line* is parallel to the horizontal plane, as illustrated in Fig. 4–11A. A horizontal line may be shown in an infinite number of positions in the top view, provided that it is parallel to the horizontal plane in the frontal and profile projections, as shown in Fig. 4–11A. The observer's line of sight is perpendicular to the horizontal plane in the top view, and is also perpendicular to all horizontal lines. Therefore, horizontal lines project true length in the top view (horizontal view).

Note that observation of the top view of any given line cannot reveal whether the line is horizontal. This can be established from the front and side views. A line is horizontal if the projection of the line in these views is parallel to the H–F fold line. A line projecting as a point in either of these views is also horizontal. A line appearing as a point in the front view is a combination horizontal and profile

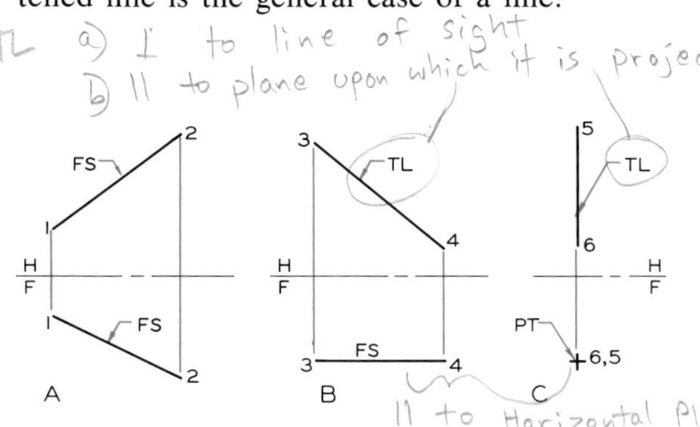

Fig. 4–9. A line in orthographic projection can appear as a point (PT), foreshortened (FS), or true length (TL).

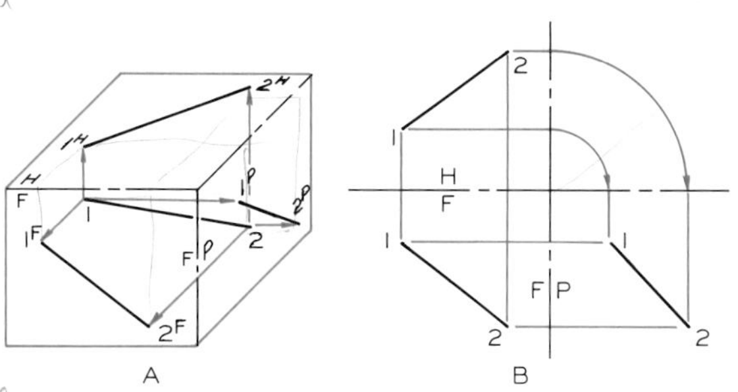

Fig. 4–10. Three orthographic views of an oblique line in space.

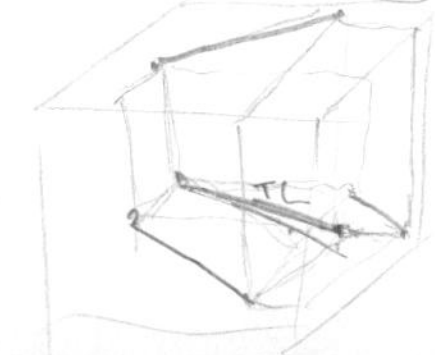

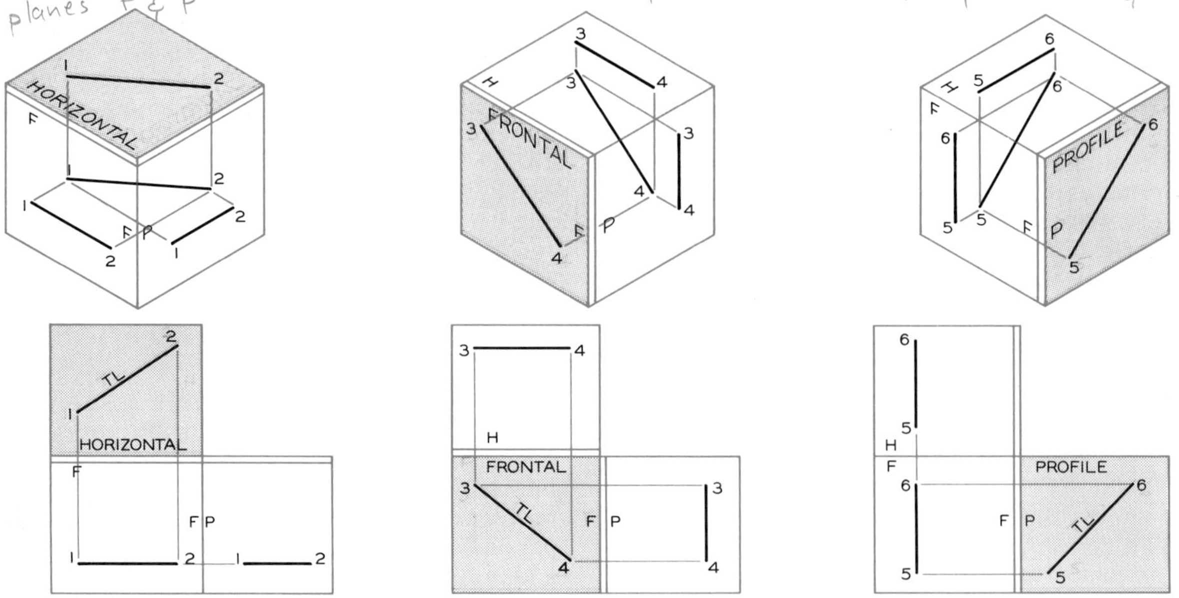

A. HORIZONTAL LINE B. FRONTAL LINE C. PROFILE LINE

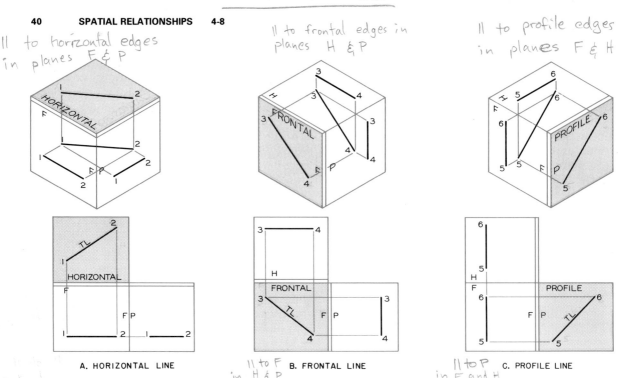

Fig. 4–11. Projections of the three principal lines: horizontal, frontal, and profile. Each of these is parallel to one of the principal planes and is true length when projected onto that plane.

line, while a point view of a line in the profile view indicates a combination horizontal and frontal line.

4–7 FRONTAL LINES

Recall that the front view of an object is projected onto a principal plane called the frontal plane. A line parallel to this plane is a principal line and is called a *frontal line.* The frontal plane appears true shape in the front view, since the observer's line of sight is perpendicular to it (Fig. 4–11B). His line of sight will also be perpendicular to any line parallel to the frontal plane. This line is projected as true length in the front view.

A frontal line is projected parallel to the frontal plane in both the top and side views, where the frontal plane appears as an edge. It is possible to see in the top and side views that line 3–4 is a frontal line, but in the front view it is not possible. A line that appeared as a point in the top view would be a combination frontal and profile line. A line projected as a point in the side view would be a combination horizontal and frontal line.

4–8 PROFILE LINES

The side view is projected onto the principal plane called the profile plane, which appears true shape in the side view. Line 5–6, which is parallel to the profile plane, is a principal line called a *profile line* (Fig. 4–11C). Since the profile plane is shown true shape in the side view, a profile line will be projected true length in the side view.

The relationship of three views of a profile line is shown in Fig. 4–11C. Line 5–6 is parallel to the profile plane in both of the views in which the profile plane appears as an edge, i.e., the top and front views. A line projecting as a point in the top view will be a combination profile and frontal line, while a line projecting as a point in the front view will be a combination profile and horizontal line.

4-9 LOCATION OF A POINT ON A LINE

A line is composed of an infinite number of points. The solution of descriptive geometry problems requires that the locations of specific points on lines and surfaces in space be determined. Figure 4–12 gives the top and front views of line 1–2. Since the endpoints of line 1–2 are located on projectors that are perpendicular to the fold lines, any point on the line can be found by applying the projection principles illustrated in the figure. For example, if point *O* is located on the line in the top view, the front view of the point may be found by projecting it perpendicular to the H–F fold line until point *O* is found on the line in the front view.

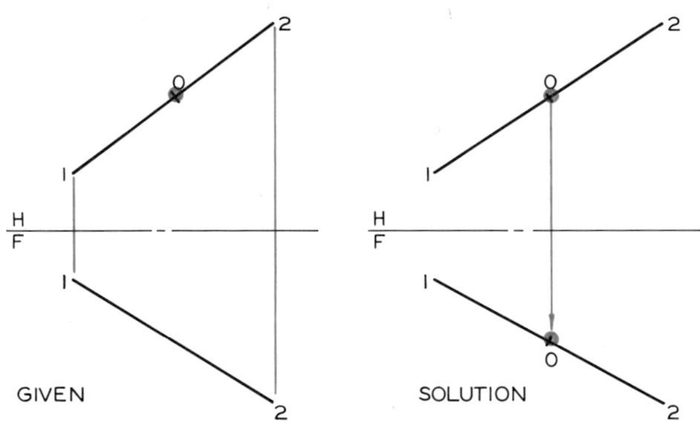

Fig. 4–12. A point on a line in the top view can be projected to the front view of the line.

Any point on the line can be found in the same manner. If the point is located at the midpoint of the line in any view, it will appear at the mid-point of the line in any other projection, although the line can vary in its projected length. Any other ratio of divisions of a line will remain constant in other views as well.

4-10 NONINTERSECTING LINES

Lines may cross in many views, but crossing lines are not necessarily intersecting lines. Lines 3–4 and 5–6 cross in the top and front views in

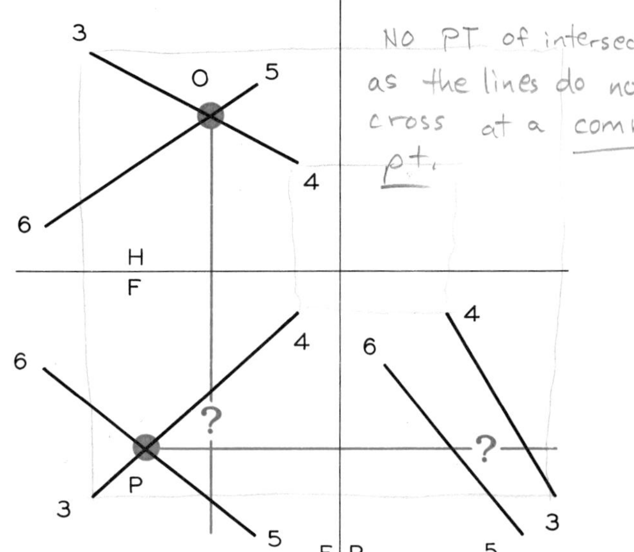

Fig. 4–13. Crossing lines do not intersect unless a projector from the point of intersection in one orthographic view passes through the point of intersection in the adjacent views. These lines do not intersect.

Fig. 4–13, although when the side view is inspected, it is obvious that these lines do not intersect.

A point of intersection is a common point that lies on both lines, and must therefore project to both lines in all views as did point *O* in Fig. 4–12. Even if the side view were not given in Fig. 4–13, it would be possible to determine whether the lines intersected by projecting the crossing point, *O*, to the front view. Since the lines do not cross at a common point along this projection line in the front view, it is apparent that point *O* is not common to both lines at a single point, and thus there is no point of intersection.

On the other hand, Fig. 4–14 illustrates two lines, 5–6 and 7–8, that do intersect at a common point, point *O*, which is the intersection of the projections from all views, as was the case in Fig. 4–12. It is necessary to have at least two views before it is possible to determine whether two crossing lines intersect.

Fig. 4–14. These crossing lines actually intersect because point *O* is an orthographic projection of the point of intersection in the adjacent views.

4–11 VISIBILITY OF CROSSING LINES

Two lines, *AB* and *CD*, are shown crossing in the top and front views of Fig. 4–15. It is obvious that these lines do not intersect by application of principles outlined in Article 4–10. However, we want to determine which of the lines is visible in each view at the points of crossing.

The crossing point in the front view is projected to the top view in step 1. This projector intersects line *AB* before line *CD*, indicating that line *CD* is farther back. This establishes line *AB* as being visible in the front view, since the horizontal view depicts true distances from the frontal plane and since the line that is closest to the front view would therefore be the one that is visible.

The visibility of the top view is determined by projecting the crossing point from the top view to the front view shown in step 2 of Fig. 4–15. This projector intersects line *CD* first, establishing line *CD* as being higher than or above line *AB* and therefore visible in the top view.

FIGURE 4–15. VISIBILITY OF LINES

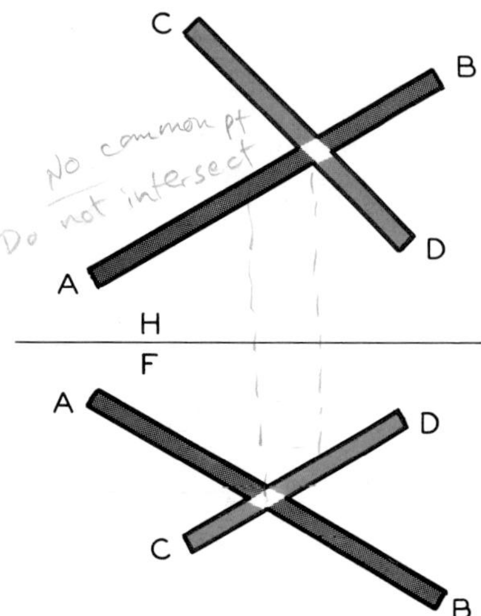

Given: The top and front views of lines *AB* and *CD*.

Required: Find the visibility of the lines in both views.

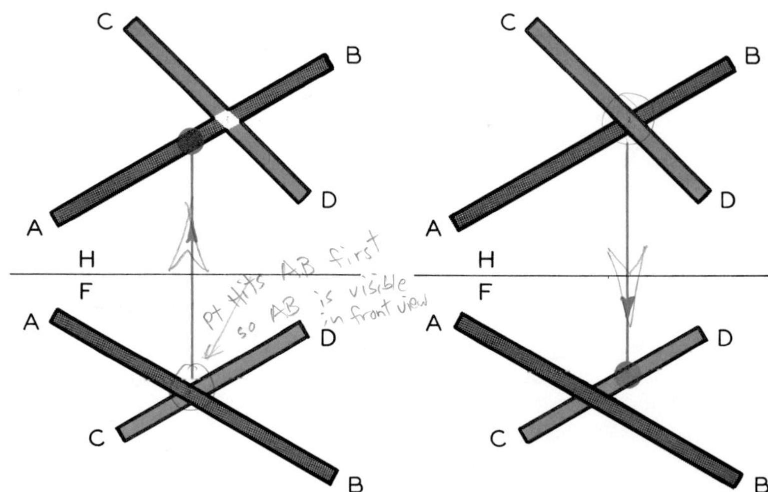

Step 1: Project the point of crossing from the front view to the top view. This projector encounters *AB* before *CD*; therefore, line *AB* is in front and is visible in the front view.

Step 2: Project the point of crossing from the top view to the front view. This projector encounters *CD* before *AB*; therefore, line *CD* is above *AB* and is visible in the top view.

Visibility in a given view cannot be established by that view only. It is necessary to determine visibility by inspecting the preceding view, as outlined in this example.

4–12 VISIBILITY OF A LINE AND A PLANE

The principle of visibility of intersecting lines applies to an intersecting line and plane in much the same manner as outlined in Article 4–11. The given part of Fig. 4–16 shows that plane 1–2–3 and line *AB* cross in the top and front views.

Line *AB* crosses two lines of the plane, lines 1–3 and 2–3, in the front view of step 1. To determine the visibility in the front view, we project these crossing points to the top view. In both cases, the projectors intersect the lines of the plane before they intersect line *AB*, which means that the plane is closer than the line in the front view. Therefore the portion of the line that crosses the plane in the front view is invisible and shown as a hidden line.

The visibility of the top view is found similarly by projecting the crossing points in the top view to the front view (step 2). These projectors intersect line *AB* before lines 1–3 and 2–3. Consequently, line *AB* is higher than the plane in the top view and is drawn as being visible.

4–13 PLANES

Whereas lines have only one dimension, length, planes have two dimensions that establish an area. Planes may be considered as infinite in certain problems. However, in most solutions segments of planes are used for convenience.

Four methods of representing planes are shown in Fig. 4–17. These are (A) three points

FIGURE 4–16. VISIBILITY OF A LINE AND A PLANE

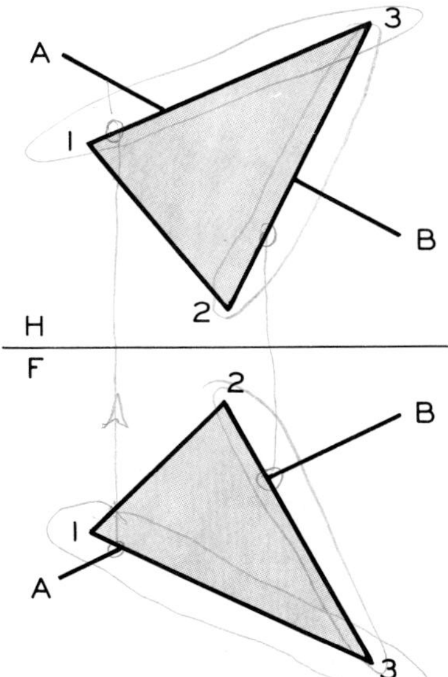

Given: The front and top views of plane 1–2–3 and line *AB*.

Required: Find the visibility of the plane and the line in both views.

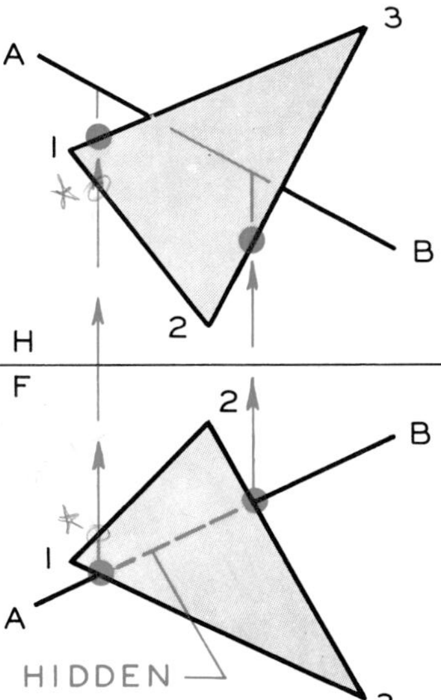

Step 1: Project the two points where *AB* crosses the outer sides of the plane in the front view to the top view. These projectors encounter lines of the plane (1–3 and 2–3) first; therefore, the plane is in front of the line, making the line invisible in the front view.

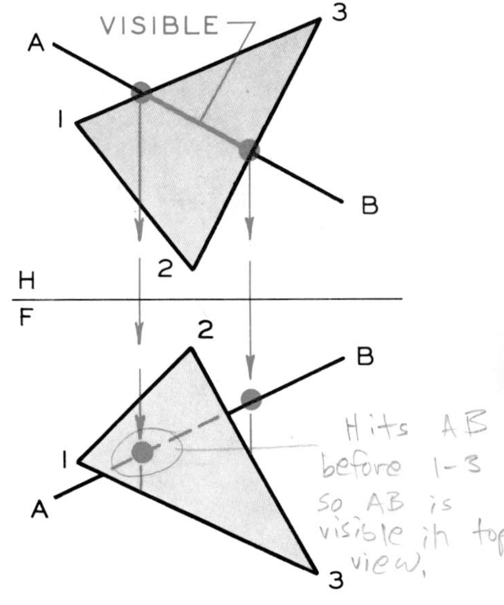

Step 2: Project the two points where *AB* crosses the outer sides of the plane in the top view to the front view. These projectors encounter line *AB* first; therefore, the line is higher than the plane, and the line is visible in the top view.

FIGURE 4–17. REPRESENTATIONS OF A PLANE

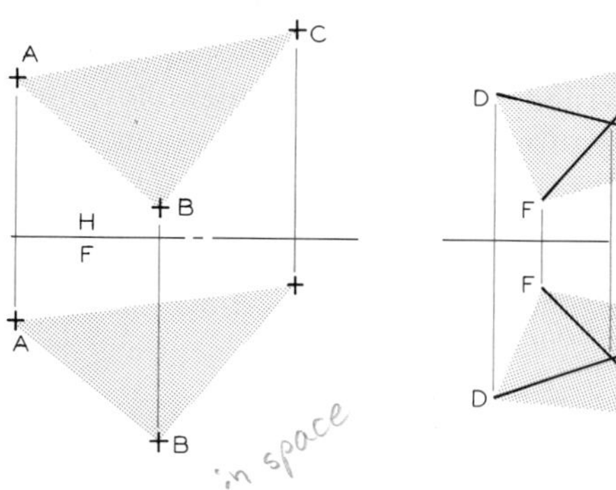

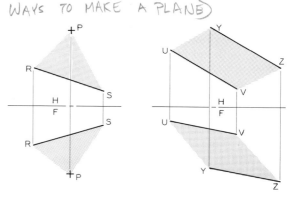

A. Three points not in a straight line. **B.** Two intersecting lines. **C.** A line and a point not on the line. **D.** Two parallel lines.

not in a straight line, (B) two intersecting lines, (C) a point and a line, and (D) two parallel lines. The areas of the planes established by these methods need not be limited by the bounds of the points or lines used. The elements merely establish the necessary locations to orient the plane in space so it can be used in solving problems involving planes.

Planes may be projected in one of the following forms (Fig. 4–18): (A) as an edge, (B) as true size, or (C) foreshortened. Planes parallel to one of the three principal projection planes—horizontal, frontal, and profile—are principal planes that will appear true size in a principal view. Each of these and the exception, the oblique plane, are discussed below.

Oblique Plane. An oblique plane is a plane that is not parallel to a principal projection plane in any view, as shown in Fig. 4–19. Its projections may appear as lines or as foreshorteded areas which are smaller than its true size. Three orthographic views of plane 1–2–3 are shown in Fig. 4–19. This is the general case of a plane. Each of the vertex points is found in the same manner in each view as though it were an individual point.

Frontal Plane. A frontal plane is parallel to the frontal projection plane, as shown pictorially in Fig. 4–20A. This principal plane appears true size in the front view and as an edge in the

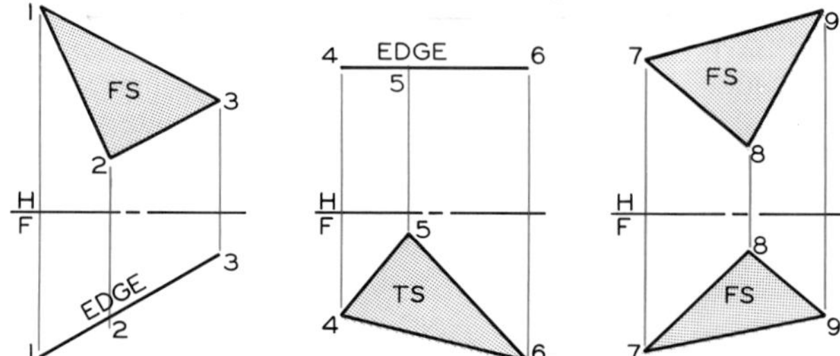

Fig. 4–18. A plane in orthographic projection can appear as an edge, true size (TS), or foreshortened (FS).

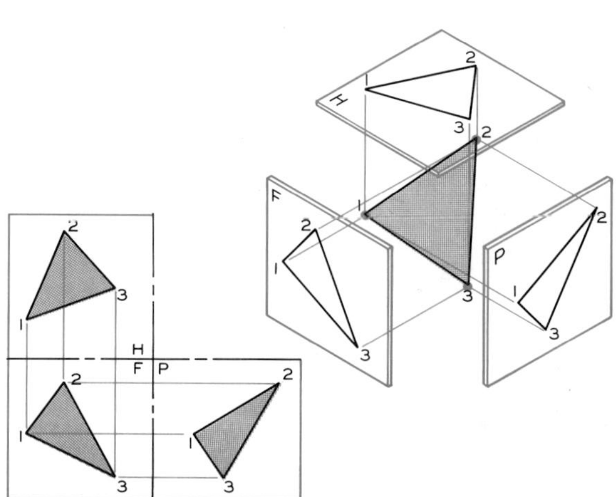

Fig. 4–19. The three projections of an oblique plane.

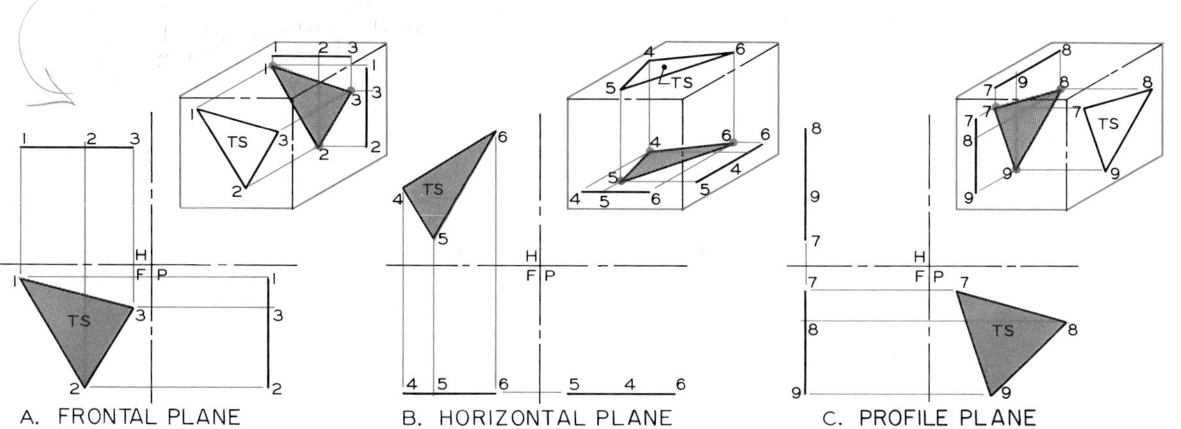

Fig. 4–20. Projections of the three principal planes: frontal, horizontal, and profile. Each of these is parallel to one of the principal projection planes and appears true size on that plane.

top and side views. The edge views of plane 1–2–3 are shown parallel to the frontal plane in Fig. 4–20A. There are an infinite number of shapes a frontal plane may have in the front view, but the top view and side views must be edges that are parallel to the frontal plane.

Horizontal Plane. A horizontal plane is parallel to the horizontal projection plane, as shown in Fig. 4–20B. A horizontal plane is a principal plane and it appears true size in the top view. Three orthographic views of horizontal plane 4–5–6 are shown in Fig. 4–20B. If the plane appears as an edge in both the front and side views and is parallel to the H–F fold line in the same views, it is a horizontal plane. Observation of the top view of a plane is not sufficient to determine whether it is a horizontal plane.

Profile Plane. The third principal plane is the profile plane, which is parallel to the profile projection plane (Fig. 4–20C). Plane 7–8–9 is true size in the profile view, or side view. Note that the plane appears as an edge in the top and front views and that these edges are parallel to the edge view of the profile plane.

4–14 PROJECTION OF A LINE ON A PLANE

Plane 1–2–3–4 is given in Fig. 4–21 with line *AB* drawn on the plane in the front view. We are required to locate line *AB* on the plane in the top view.

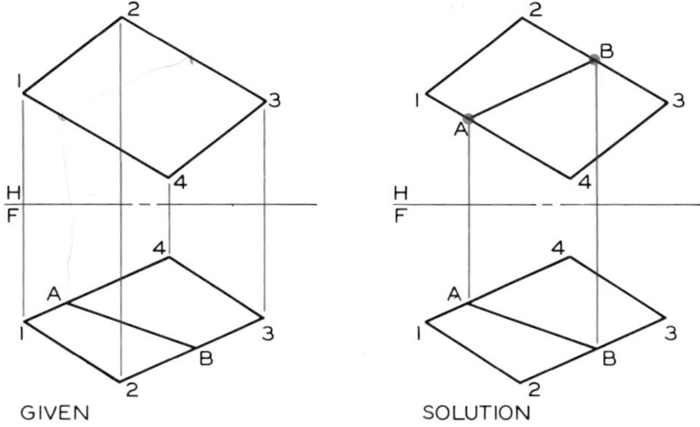

Fig. 4–21. To find the top view of line *AB* on the plane, points *A* and *B* are projected to lines 1–4 and 2–3, respectively.

Point *A* lies on line 1–4, while point *B* lies on line 2–3; thus it is possible to find these points in the top view by projecting them as shown in the solution. The connection of points *A* and *B* will establish the line on plane 1–2–3–4 in the top view. This is an application of the principle covered in Article 4–9.

4–15 LOCATION OF A POINT ON A PLANE

Many problems require that points be located on a plane in several views. Plane *ABC* (Fig. 4–22) has a point indicated on its surface in the front view. We desire to locate the top view of this point as well. A line is drawn in any

FIGURE 4–22. PROJECTION OF A POINT ON A PLANE

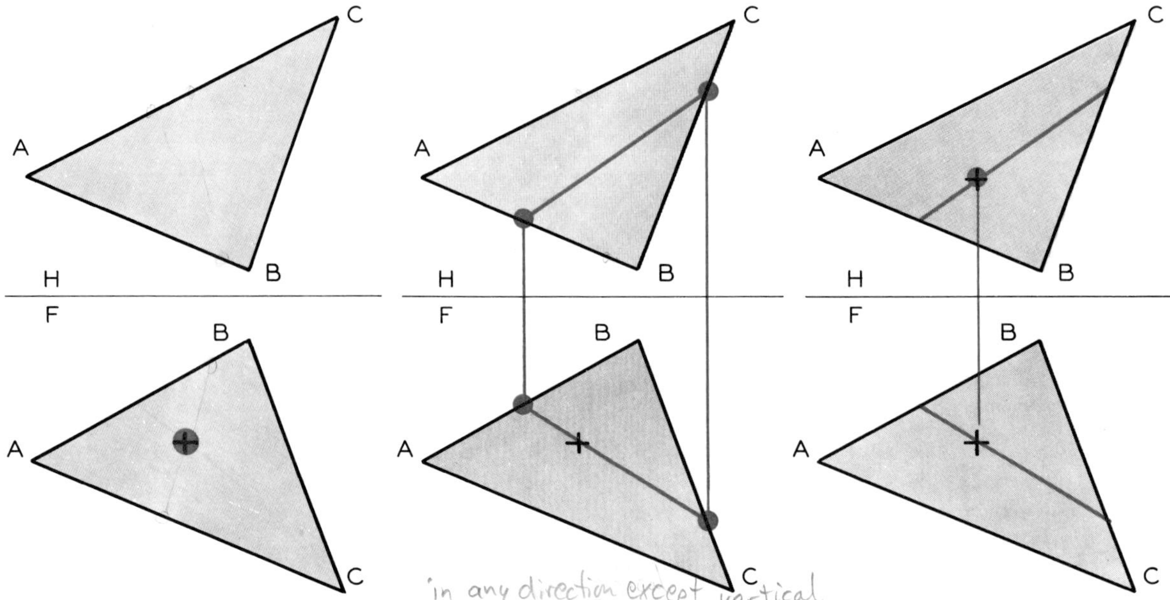

Given: The top and front views of a plane and a point on the front view.

Required: Find the top view of the point on the plane.

Step 1: Draw a line through the point in the front view and project the line to the top view.

Step 2: To locate the point, project the point from the front view to the line in the top view.

direction other than vertical through the point in the given view (step 1). The points where this line intersects the edges of the plane in the front view are projected to these same respective lines in the top view. The point can be located by projecting from the front view to this line in the top view (step 2). Review Article 4–9 for a stronger understanding of this principle.

4–16 PRINCIPAL LINES ON A PLANE

Principal lines—horizontal, frontal, and profile—may be found in any view of a plane by application of the previously discussed principles. Principal lines are essential to the system of successive auxiliary views, which will be studied in later chapters.

Two horizontal lines are shown on plane 1–2–3 in Fig. 4–23A. These were found by constructing lines on plane 1–2–3 that are parallel to the horizontal projection plane. The top views are found by projecting to the plane in the top view. These lines lie on the plane and are true length in the top view since they are horizontal.

Frontal lines which are parallel to the frontal projection plane are located in the top view of Fig. 4–23B. The front view of the lines is found by projection. Frontal lines are true length in the front view.

Profile lines must be located in the front or top view and projected to the side view, as shown in Fig. 4–23C. Profile lines are true length in the side views.

It should be observed that principal lines on any view of an oblique plane are parallel. For example, many profile lines could have been drawn in Fig. 4–23C, but all would have been parallel and would have appeared true length in the side views.

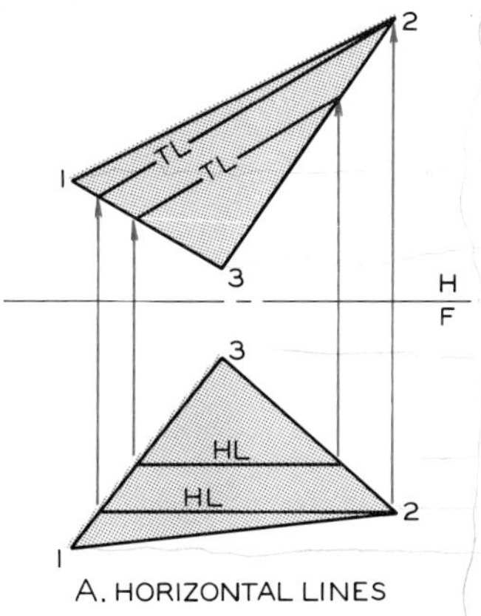

A. HORIZONTAL LINES

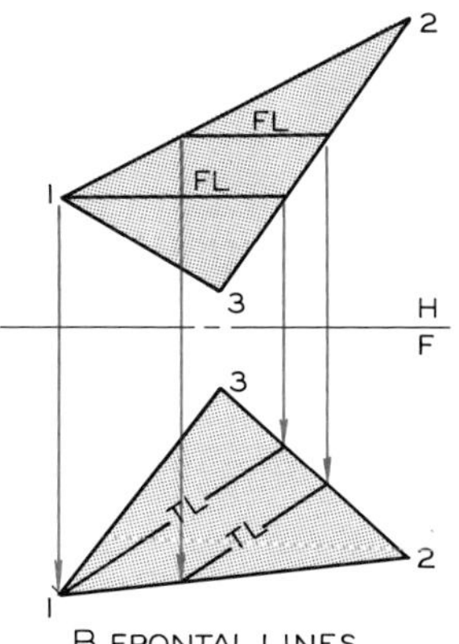

B. FRONTAL LINES

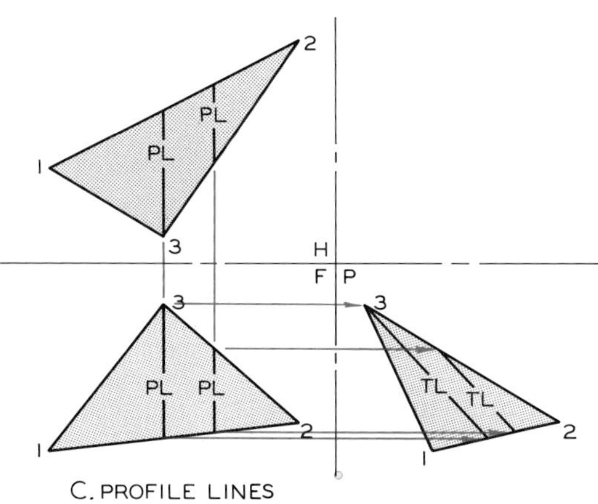

C. PROFILE LINES

Fig. 4–23. Construction of the principal lines—horizontal, frontal, and profile—on a given plane.

4–17 PARALLELISM

In the solution of spatial problems it is often necessary to know whether lines or planes are parallel. This information can be determined by orthographic projection.

Two lines that are parallel will be projected parallel in all views, except in the view where both lines appear as points. Lines *AB* and *CD* in Fig. 4–24 appear oblique in three views, but they are also parallel in each; therefore the lines are parallel in space.

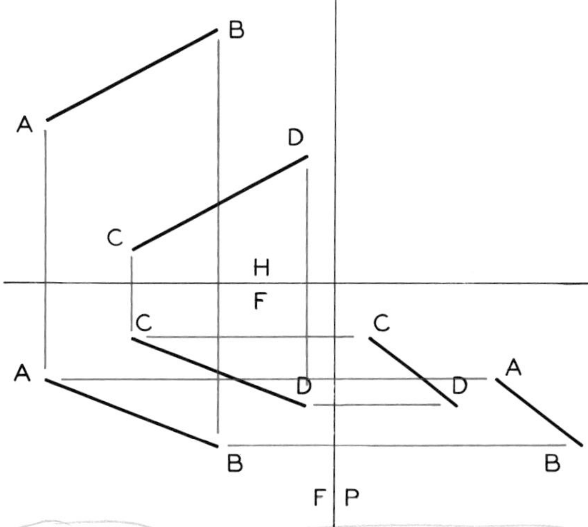

Fig. 4–24. When two lines are parallel, they will project as parallel in all orthographic views.

When only one view of two lines is available, it cannot be assumed that the lines are parallel even though they are projected as parallel in this view. More than one view is necessary to determine whether two lines are parallel.

Figure 4–25 illustrates how this principle may be applied to a problem. Given is line 3–4 and point *O*. We are required to construct a line equal in length and parallel to 3–4, with its midpoint at *O*. Since the midpoint of a line will be the midpoint of any projection of that line, the top view of the line is drawn through point *O* with its midpoint as shown in step 1. The projection of this line is the same length as the projection of line 3–4. The front view of the line is drawn parallel to the front projection of

FIGURE 4–25. CONSTRUCTION OF A LINE PARALLEL TO A LINE

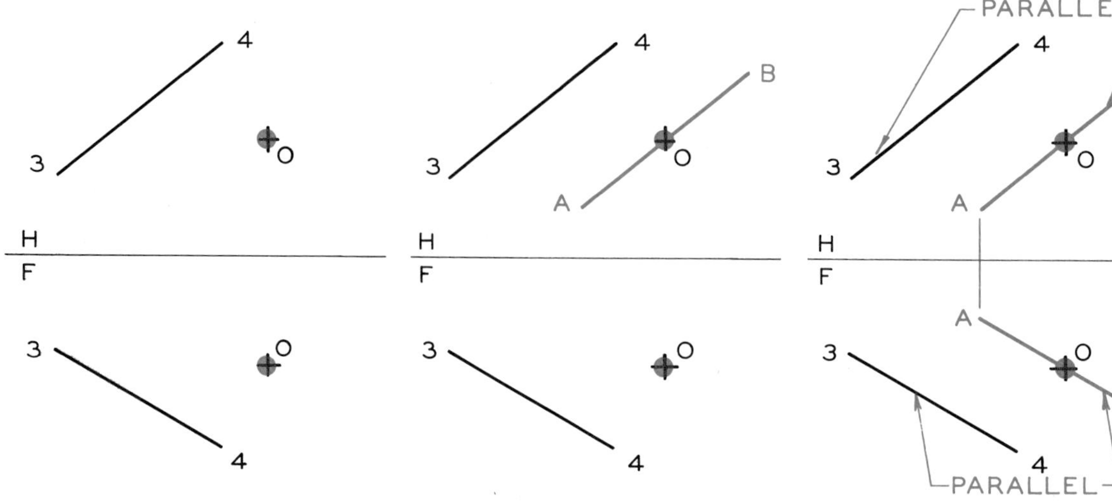

Given: The top and front views of line 3–4 and point *O*.

Required: Construct a line with its midpoint at *O* that is parallel to line 3–4.

Step 1: Draw line *AB* parallel to the top view of line 3–4 with its midpoint at *O*.

Step 2: Draw the front view of line *AB* parallel to the front view of line 3–4 through point *O*, which is an orthographic projection of the top view.

line 3–4 in step 2. The ends of the line are established by projecting from the top view. The frontal projections of the two lines are also equal in length, and the resulting line is parallel to the given line 3–4.

4–18 PARALLELISM OF PLANES

Two planes are parallel when intersecting lines in one plane are parallel to intersecting lines in the other, as shown pictorially in Fig. 4–26. Orthographic projections of parallel planes are shown in Fig. 4–27. Note that the same two sets of lines are parallel in the top view of part A as in the front view. These lines happen to be exterior lines of the planes concerned, but this does not have to be the case. The planes could be dissimilar in shape as shown in the example in part B. Plane 7–8–9 and plane *ABCD* were found to be parallel by drawing parallel lines on each plane in one view and projecting them to the other. If these lines are parallel in this view also, then the planes are parallel. When two planes appear as parallel edges in one view, such as in part C, they are parallel in space.

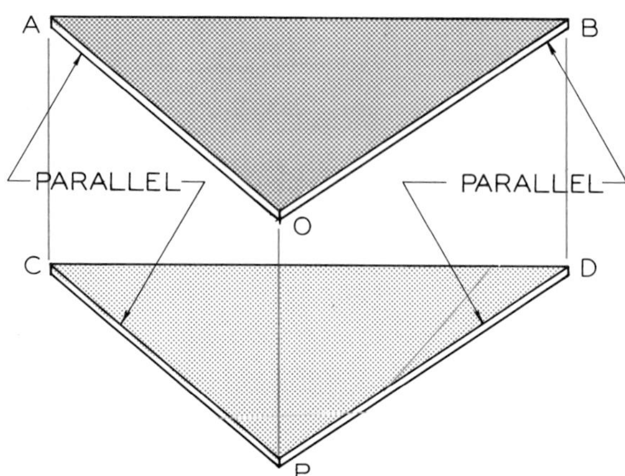

Fig. 4–26. Two planes are parallel when two intersecting lines in one plane are parallel to two lines in the other plane.

The problem in Fig. 4–28 requires that a plane be constructed through point *O* parallel to plane 1–2–3. Line *AB* is drawn parallel to line 1–2 of the plane in the top and front views in step 1. Line *CD* is then drawn through point *O* parallel to line 2–3 in both views in step 2.

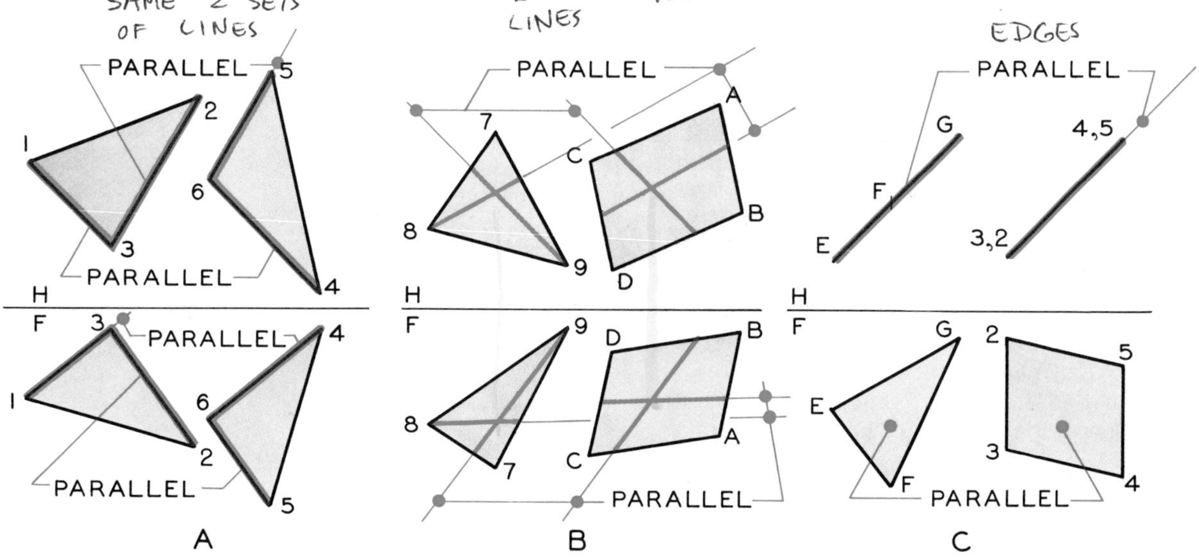

SAME 2 SETS OF LINES

INTERSECTING LINES

EDGES

PARALLEL 5 2 1 6 3 PARALLEL 4 H F 3 PARALLEL 4 1 6 2 PARALLEL 5 A

PARALLEL A 7 C B 8 9 D H F 9 D B 8 A C 7 PARALLEL B

PARALLEL G 4,5 F E 3,2 H F G 2 5 E 3 F PARALLEL 4 C

Fig. 4–27. Examples of planes that are parallel. 2 Planes are parallel when 2 intersecting lines in one plane are parallel to 2 intersecting lines in the other.

FIGURE 4–28. CONSTRUCTION OF A PLANE THROUGH A POINT PARALLEL TO A PLANE

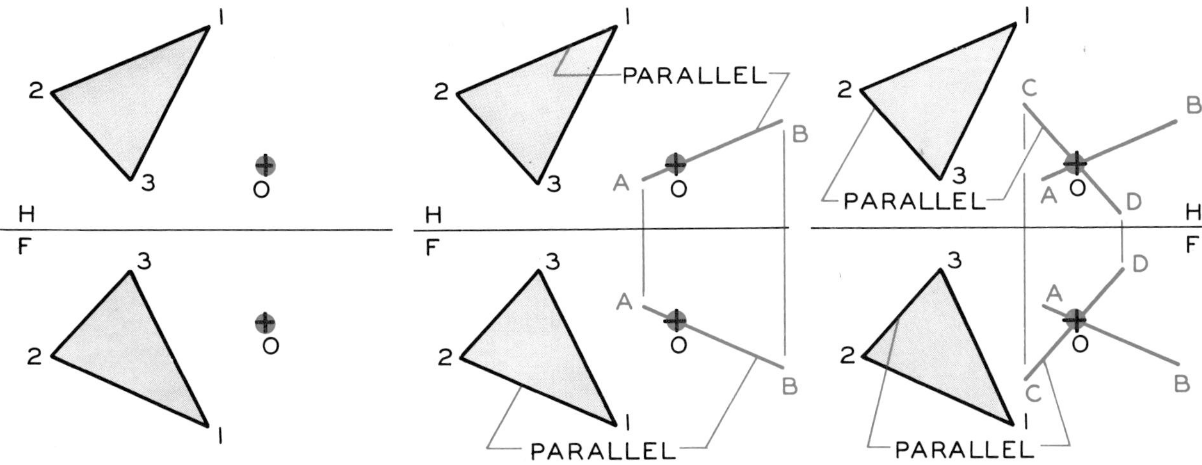

Given: The top and front views of plane 1–2–3 and point *O*.

Required: Construct a plane through point *O* parallel to plane 1–2–3.

Step 1: Draw line *AB* parallel to line 1–2 in both views.

Step 2: Draw line *CD* parallel to line 2–3 in both views. Plane *ADBC* is parallel to plane 1–2–3.

need a set of intersecting lines

Since these lines intersect at point *O*, they form a plane. This plane is parallel to plane 1–2–3, since two intersecting lines of one plane are parallel to two intersecting lines of the other.

4–19 PARALLELISM OF A LINE AND A PLANE

A line is parallel to a plane if it is parallel to any line in that plane. Two orthographic views of a line and a plane are shown in Fig. 4–29. Line *AB* is parallel to line 1–3 in the top view and in the front view, thus establishing these lines as parallel; therefore line *AB* is parallel to plane 1–2–3.

The problem given in Fig. 4–30 requires that a line drawn through point *O* be parallel to the plane formed by the two intersecting lines 1–2 and 3–4. The line is constructed in step 1 through point *O* and parallel to line 1–2 in both views. This line is parallel to the plane, since it is parallel to a line in the plane. This line could have also been drawn parallel to line 3–4.

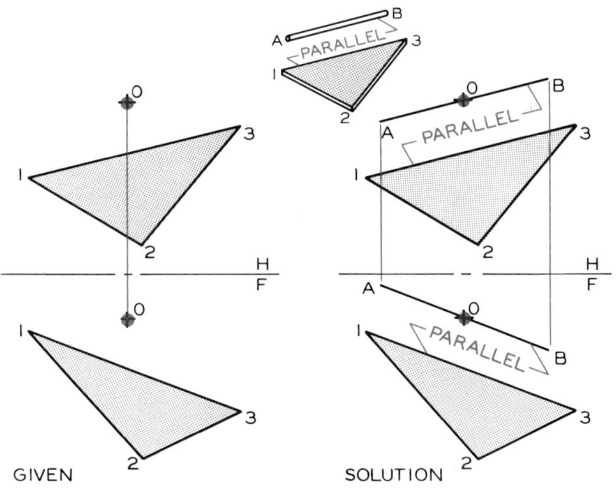

Fig. 4–29. Construction of a line through point *O* parallel to plane 1–2–3. Line *AB* is drawn parallel to line 1–3 in the front and top views.

FIGURE 4–30. CONSTRUCTION OF A LINE PARALLEL TO A PLANE

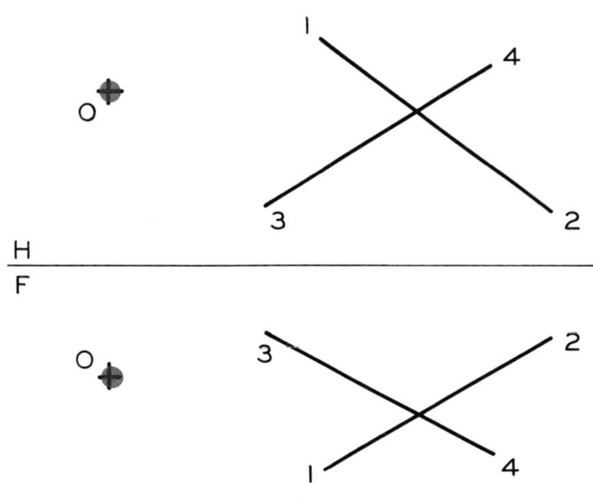

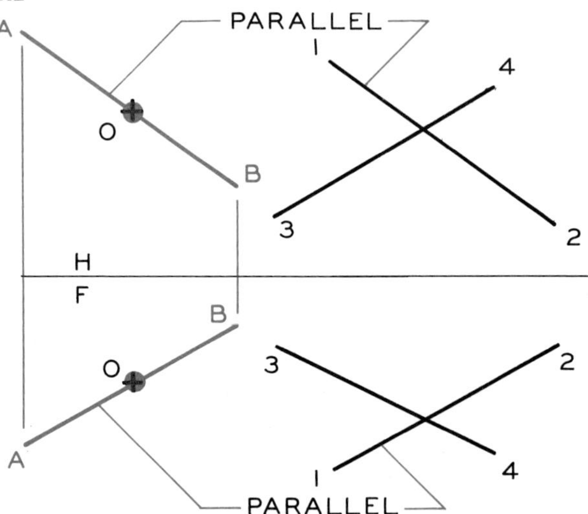

Given: The top and front views of point *O* and plane 1–2–3–4.

Required: Draw the two views of a line through point *O* that is parallel to the plane.

Solution: Line *AB* is drawn parallel to one of the lines in the plane, line 1–2 in this case, in both views. The line could have been drawn parallel to any line in the plane.

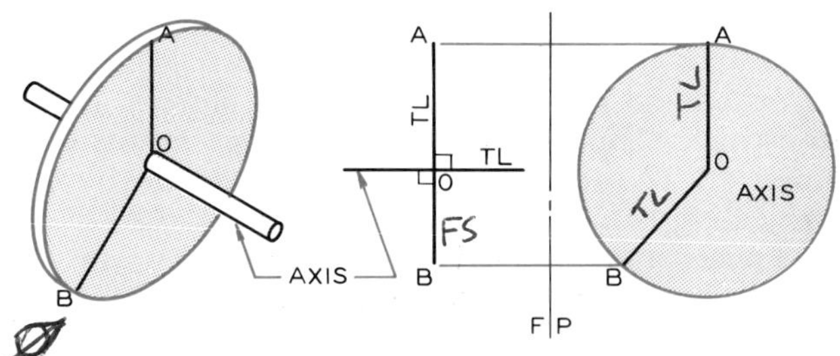

Fig. 4–31. Perpendicular lines will be projected as perpendicular in a view where one or both of the lines appear true length.

4–20 PERPENDICULARITY OF LINES

The engineer will design many mechanisms that are composed of perpendiculars, whether they are lines or planes. It is therefore necessary that perpendicularity be understood sufficiently in order that it may be recognized or drawn when it occurs.

Figure 4–31 illustrates pictorially and orthographically the basic rules of perpendicularity. *Two perpendicular lines will be projected as perpendicular in any view where one or both are true length.* A line may be revolved around another line in an infinite number of positions and still be perpendicular to the other line, the axis.

LAW OF PERPENDICULARITY

In the orthographic view, the lines *OA* and *OB* are projected such that they are perpendicular to the true-length axis. The axis and line *OA* are both true length, but line *OB* is not true length. However, line *OB* is projected as perpendicular to the axis because the axis is true length, thereby satisfying the previously stated rule.

When two lines are perpendicular, but neither is true length, they will not project with a true 90° angle. The plane of the 90° angle will appear foreshortened and distorted.

4–21 A LINE PERPENDICULAR TO A PRINCIPAL LINE

Given in Fig. 4–32 are frontal line 1–2 and point *O*. Construct the top and front views of a line through point *O* that intersects line 1–2 and is perpendicular to it.

FIGURE 4–32. CONSTRUCTION OF A LINE PERPENDICULAR TO A PRINCIPAL LINE

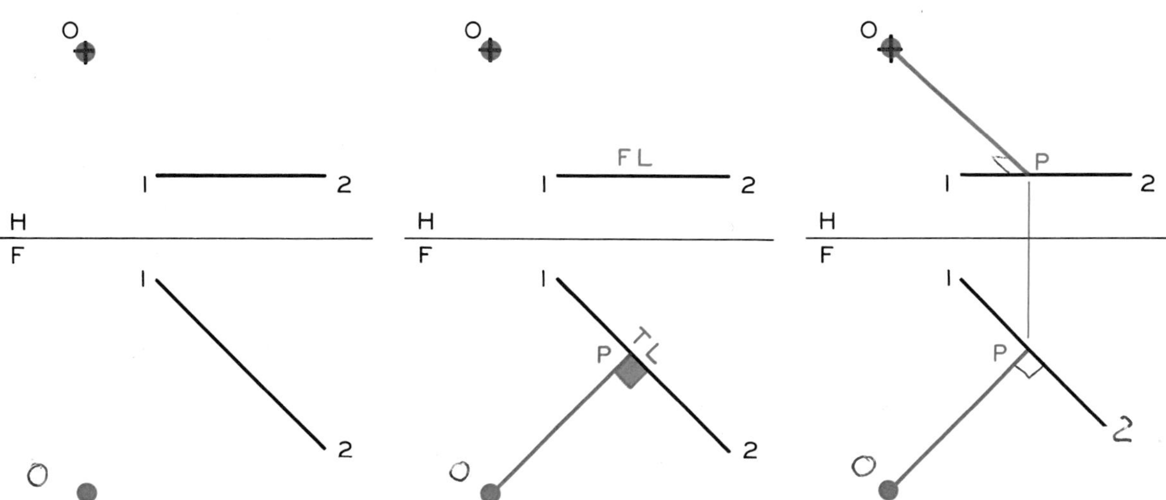

Given: The top and front views of principal line 1–2 and point *O*.

Required: Construct a line from point *O* perpendicular to line 1–2.

Step 1: Draw a line perpendicular to the true-length line 1–2 in the front view.

Step 2: Locate point *P* on line 1–2 in the top view and connect this point with point *O*.

Perpendicular when they form a 90° ∡ and one line is T.L.

Line 1–2 is a principal line, a frontal line, and is consequently true length in the front view. By applying the rule of perpendicularity from Article 4–20, it is possible to construct line *OP* in the front view perpendicular to the true-length line. Since point *P* lies on the line, it may be found in the top view by projecting above its front view to line 1–2, as shown in step 2. These lines do not appear as perpendicular in the top view since neither are true length in this view.

4–22 A LINE PERPENDICULAR TO AN OBLIQUE LINE

The top and front views of an oblique line, 3–4 are given in Fig. 4–33. We are required to construct a line through the midpoint of line 3–4 that would be perpendicular to it.

It is necessary that a true-length line be constructed before a perpendicular can be found. Thus in step 1, a horizontal line *OP* is drawn through the midpoint of the front view of line 3–4 to some convenient length. Line *OP* will be projected true length in the top view since it is horizontal. It may be drawn in any direction and still be true length; therefore, it is constructed perpendicular to line 3–4 through point *O*. The top view of point *P* is found by projecting from the front view (step 2). These two lines are perpendicular because they are perpendicular in the view where one of them is true length.

4–23 PERPENDICULARITY INVOLVING PLANES

A plane or a line can be constructed perpendicular to another plane by applying the principles of the previous articles. A line is perpendicular to a plane if it is perpendicular to two intersecting lines on that plane, as illustrated in Fig. 4–34A.

Also, a plane is perpendicular to another plane if a line in one plane is perpendicular to the other plane. This is illustrated in Fig. 4–34B.

FIGURE 4–33. CONSTRUCTION OF A LINE PERPENDICULAR TO AN OBLIQUE LINE

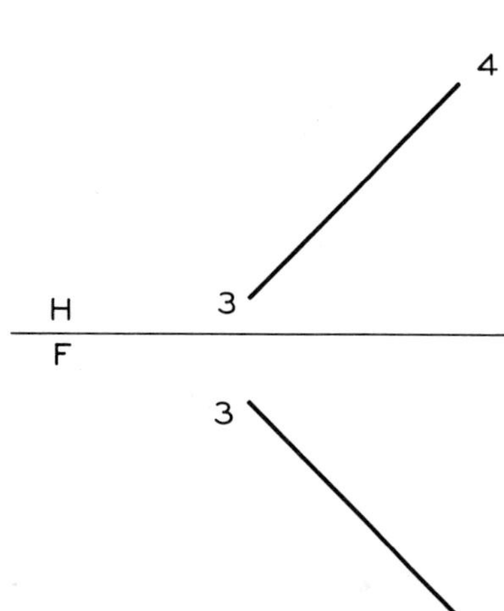

Given: The top and front views of an oblique line 3–4.

Required: Construct a line from the midpoint of the oblique line 3–4 that is perpendicular to it.

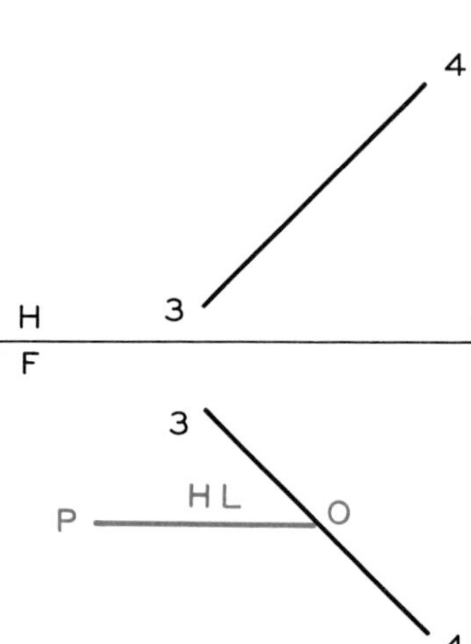

Step 1: Construct a horizontal line from the midpoint of the front view of line 3–4.

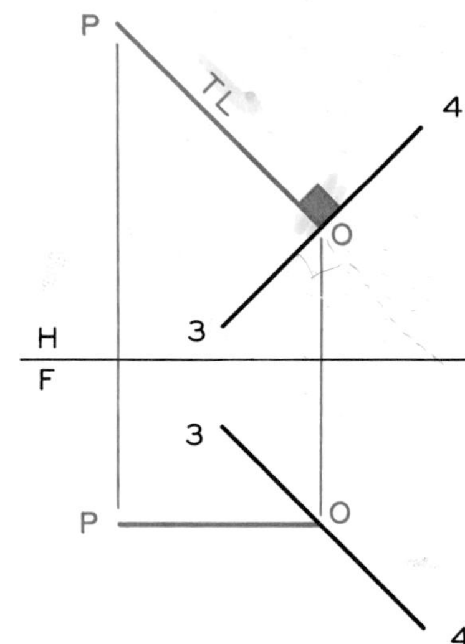

Step 2: Project point *O* to the top view of the line and draw line *OP* (which is true length in the top view) perpendicular to line 3–4.

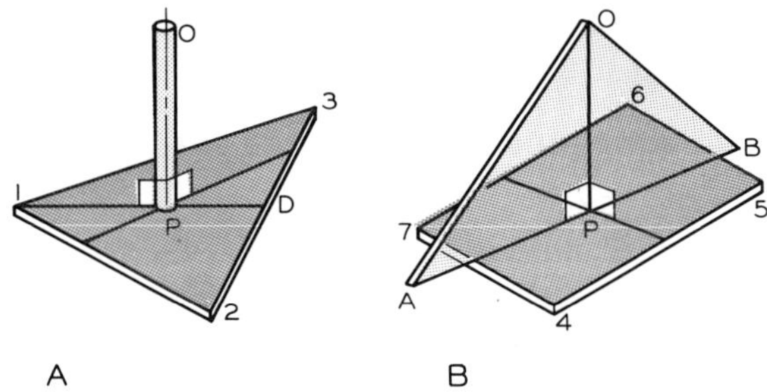

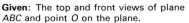

Fig.4–34. (A) A line is perpendicular to a plane if it is perpendicular to two intersecting lines on the plane. (B) A plane is perpendicular to another plane if the plane contains a line that is perpendicular to the other plane.

4–24 A LINE PERPENDICULAR TO A PLANE

Plane *ABC* is given in Fig. 4–35 with point *O* located on the plane. We are required to construct a perpendicular to the plane through point *O*. It is possible to find a true-length line on any view of a plane by constructing a principal line, as covered previously. A true-length line is found in the front view by constructing a frontal line through point *O* in the top view and projecting it to the front view (step 1). This line is true length and is a line on the plane, consequently, line *OP* can be drawn through point *O* perpendicular to the frontal line. If line *OP* is to be perpendicular to another line as well, we can draw a horizontal line on the plane in the front view (step 2) and project it to the top view, where it appears true length. The top view of *OP* is constructed perpendicular to this line to establish the top view of line *OP*.

This line is perpendicular to the plane because it is perpendicular to two intersecting lines on the plane. This relationship is apparent where lines on the plane are shown true length.

FIGURE 4–35. CONSTRUCTION OF A LINE PERPENDICULAR TO A PLANE

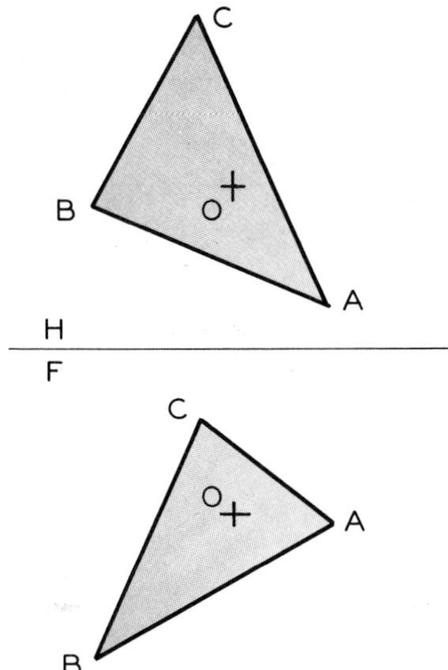

Given: The top and front views of plane *ABC* and point *O* on the plane.

Required: Construct a line from point *O* that is perpendicular to plane *ABC*.

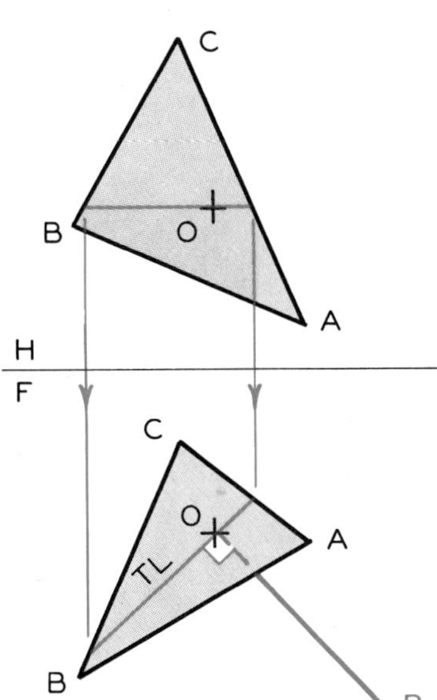

Step 1: Construct a frontal line through point *O* in the top view to find a true-length view of the line in the front view. Construct a perpendicular to this line.

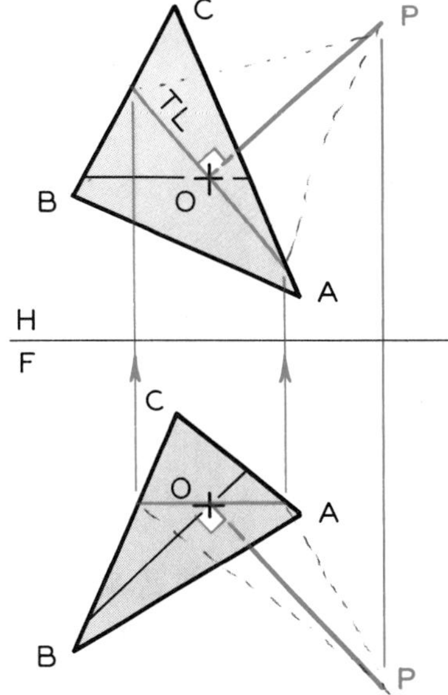

Step 2: Construct a horizontal line through point *O* in the front view to find the true-length line in the top view. Draw a perpendicular to this line.

4–25 A PLANE PERPENDICULAR TO AN OBLIQUE LINE

Line 1–2 and point *O* are given in Fig. 4–36. We are required to construct a plane through point *O* that will be perpendicular to line 1–2.

A plane may be established through point *O* by drawing two intersecting lines that intersect at point *O*. These lines will be true length if they are principal lines, which will permit perpendicularity to be established. Frontal line *AB* is drawn through point *O* in the top view and projected to the front view, where it is true length (step 1). It is drawn perpendicular to line 1–2, which gives one line in the plane perpendicular to line 1–2.

A second line, *CD*, is drawn as a horizontal line (step 2) in the front view and it is projected to the top view, where it will be true length and perpendicular to line 1–2. Line 1–2 is perpendicular to two intersecting lines in the plane, and we have now constructed a plane perpendicular to the given line.

4–26 PERPENDICULARITY OF PLANES

Planes may be perpendicular to other planes in many technological problems. The rule for determining perpendicularity of planes is a combination of the previously covered principles of perpendicularity.

A plane is perpendicular to another plane if a line in one plane is perpendicular to the other plane. This is illustrated in Fig. 4–37B, where plane *OAB* is perpendicular to plane 4–5–6–7, because if line *OP* is perpendicular to two intersecting lines on a plane, then it is perpendicular to the plane. It can be seen from these principles that plane *OAB* is perpendicular to plane 4–5–6–7.

A plane and a line are given in Fig. 4–38. We are required to construct a plane passing

FIGURE 4–36. CONSTRUCTION OF A PLANE THROUGH A POINT PERPENDICULAR TO AN OBLIQUE LINE

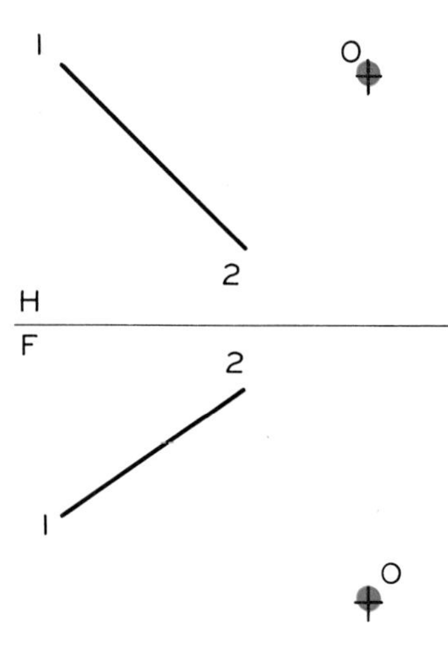

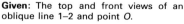

Given: The top and front views of an oblique line 1–2 and point *O*.

Required: Construct a plane through point *O* that is perpendicular to oblique line 1–2.

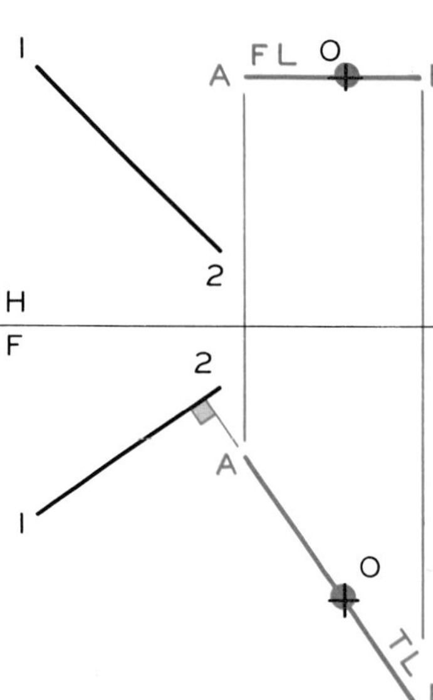

Step 1: Draw frontal line *AB* in the top view and perpendicular to the front view of line 1–2. Line *AB* is true length in the front view.

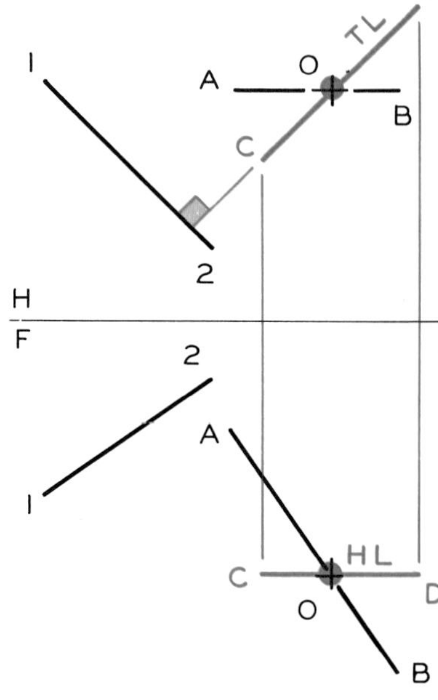

Step 2: Draw horizontal line *CD* in the front view and perpendicular to the top view of line 1–2. Line *CD* is true length in the top view. Plane *ABCD* is perpendicular to line 1–2.

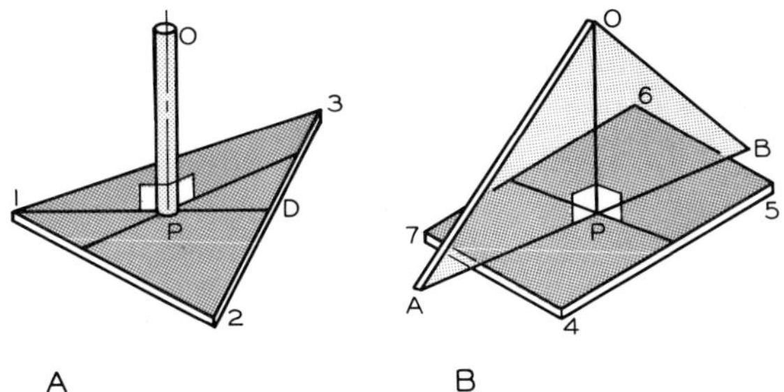

Fig.4–37. (A) A line is perpendicular to a plane if it is perpendicular to two intersecting lines on the plane. (B). A plane is perpendicular to another plane if the plane contains a line that is perpendicular to the other plane.

through line *AB* that is perpendicular to plane 1–2–3.

An infinite number of planes can be established through line *AB* by intersecting it with another line, since two intersecting lines form a plane. If the line drawn to intersect line *AB* were drawn perpendicular to plane 1–2–3, the plane formed would be perpendicular to plane 1–2–3.

A true-length line is found in the front view of plane 1–2–3 in step 1 by constructing a frontal line in the top view and projecting it to the front view. Line *CD* is drawn through a convenient point on line *AB* perpendicular to the extension of the true-length line in the front view. A horizontal line is constructed in the front view of step 2, and projected to the top view, where it is true length. The top-view projection of line *CD* is drawn perpendicular to this true-length line which goes through the top view of the point on line *CD*.

Line *CD* has been constructed perpendicular to two intersecting lines on the plane and is known to be perpendicular to the plane. Since the line *CD* intersects line *AB*, it forms a plane containing a line perpendicular to plane 1–2–3, which results in two perpendicular planes.

FIGURE 4–38. CONSTRUCTION OF A PLANE THROUGH A LINE PERPENDICULAR TO A PLANE

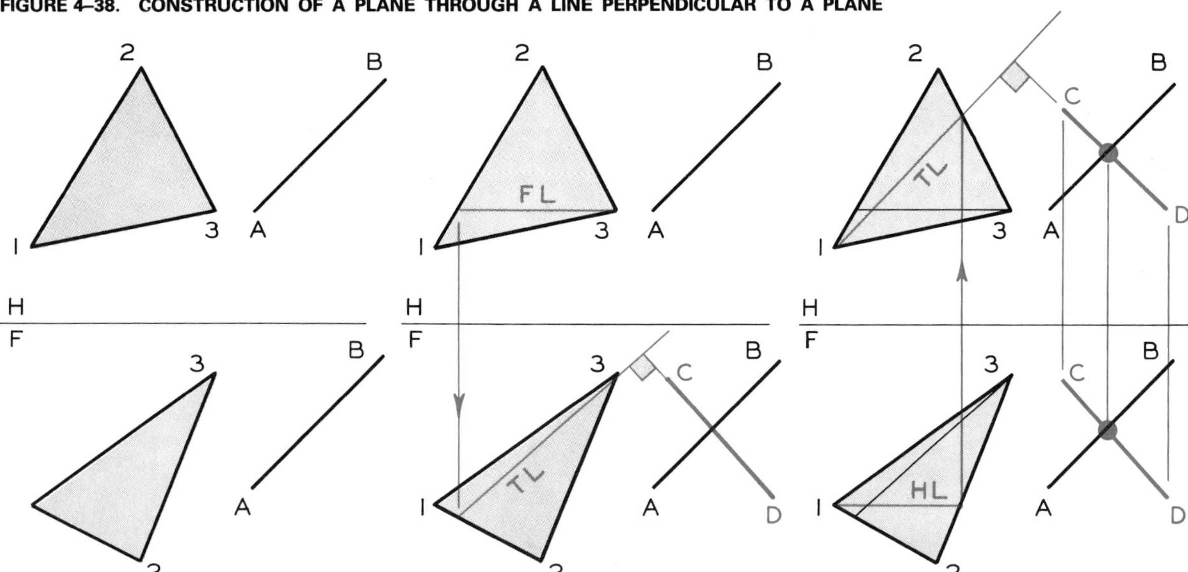

Given: The top and front views of plane 1–2–3 and line *AB*.

Required: Construct two views of a plane that passes through line *AB* and is perpendicular to the given plane.

Step 1: A frontal line is drawn in the top view and found true length in the front view. Line *CD* is drawn through any point on *AB* to be perpendicular to the true-length line.

Step 2: A horizontal line is drawn on the plane and is found true length in the top view. Line *CD* is drawn through the point of intersection on line *AB* to be perpendicular to the true-length line in the top view.

These two planes do not actually intersect as they are represented in this illustration; however, they are perpendicular and would intersect at a 90° angle if both were extended to their line of intersection.

4–27 SUMMARY

The principles of orthographic projection covered in this chapter are fundamentals that will be applied in the development of spatial problems involving descriptive geometry principles in succeeding chapters. A review of the basic elements—the point, line, and plane—and their relationship to each other in space is the basis of the solution of all graphical and descriptive geometry problems.

It is important that a solid understanding be gained of the principal projection planes in this chapter before continuing further. Principal lines and principal planes are related to the projection planes to which they are parallel. The principal lines and planes are (1) horizontal, (2) frontal, and (3) profile. Principal lines are true length in the view where they are parallel to the projection plane being viewed; principal planes are true size in this view also.

Relationships such as parallelism and perpendicularity are common in essentially all engineering problems. The designer can prepare more accurate and efficient plans if he understands these projections.

The succeeding chapters will discuss the principles of projection as they relate to the auxiliary-view method of problem solution. The projection of auxiliary views is possible through a thorough understanding of the basic projection principles covered here. Frequent reference should be made to this chapter when necessary, to review projection principles that are used in other solutions.

PROBLEMS

Problems for this chapter can be constructed and solved on $8\frac{1}{2}'' \times 11''$ sheets as illustrated by the accompanying figures. Given that each grid represents $\frac{1}{4}''$, lay out and solve the problems on grid or plain paper. All reference planes and points should be labeled in all cases, using $\frac{1}{8}''$ letters and guidelines.

1. Use Fig. 4–39 for all parts of this problem. (A) Draw the missing view of point *A*. Locate point *B* from point *A* 3 units forward, 2 units to the right, and 2 units below. Show this in three views. (B) Find the missing view of point *C*. With respect to point *C*, locate in three views point *D* that is 4 units forward, 2 units to the right, and 3 units above. (C) Draw three views of line *EF*. Point *F* is 4 units in front, 5 units to the right, and 3 units below point *E*. (D) Draw three views of line *GH*. Point *H* is 4 units behind, 3 units below, and 5 units to the right of point *G*. (E) Line *IJ* is a horizontal line 3 units below the horizontal plane. Draw the line in all views and label its true-length view. (F) Line *KL* is a frontal line 4 units behind the frontal plane. Draw three views of the line and label its true-length view.

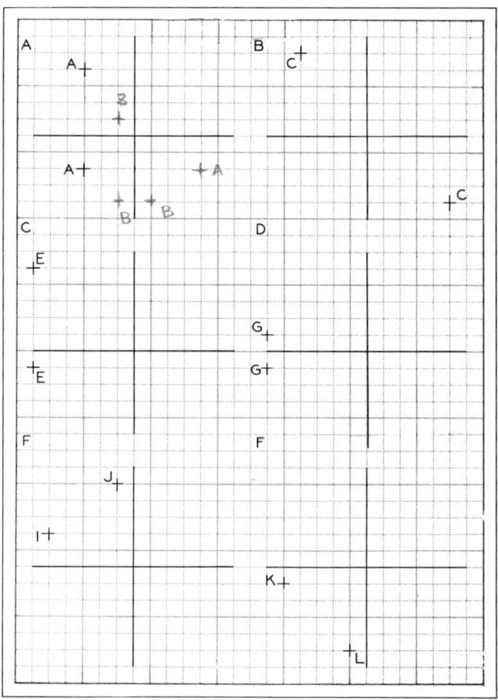

Fig. 4–39. Projections of a point.

2. Use Fig. 4–40 for all parts of the problem. (A) Draw three views of line 1–2 with point 1 located 4 units to the left, 2 units below, and 3 units in front of point 2. (B) Draw three views of line 1–2 given that point 2 is located 4 units to the right, 2 units above, and 4 units in front of point 1. (C) Draw three views of frontal line 3–4 with point 4 located 3 units below point 3. Label its true-length view. (D) Draw three views of horizontal line 5–6 with point 6 located 4 units to the right and 3 units behind point 5. Label its true-length view. (E) Draw three views of horizontal line 7–8. Label its true-length view. (F) Draw three views of frontal line 1–2 and label its true-length view.

3. Use Fig. 4–41 for all parts of this problem. (A) Draw three views of profile line *AB* and label its true-length view. (B) Locate the midpoint of line *CD* in three views. (C) Divide line *EF* into three equal parts and show the divisions in three views. (D) Draw the side view of the two lines in part D. Determine whether they intersect. (E) Construct three views of the lines *KL* and *IJ* so that they will be intersecting lines. (F) Construct line *PO* in three views such that it will intersect line *MN*.

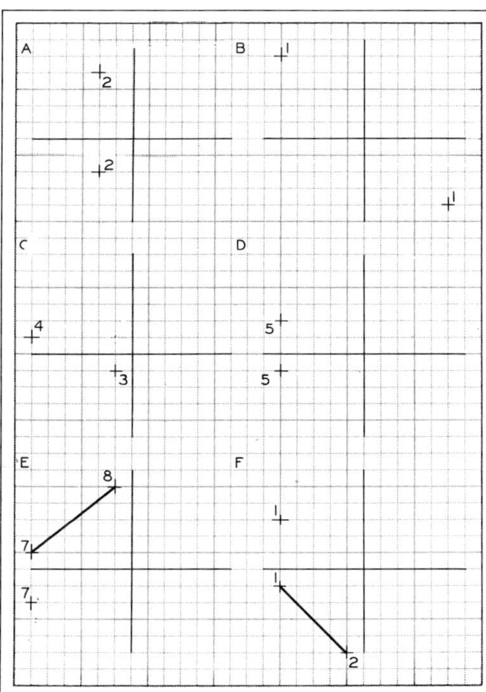

Fig. 4–40. Projections of principal lines.

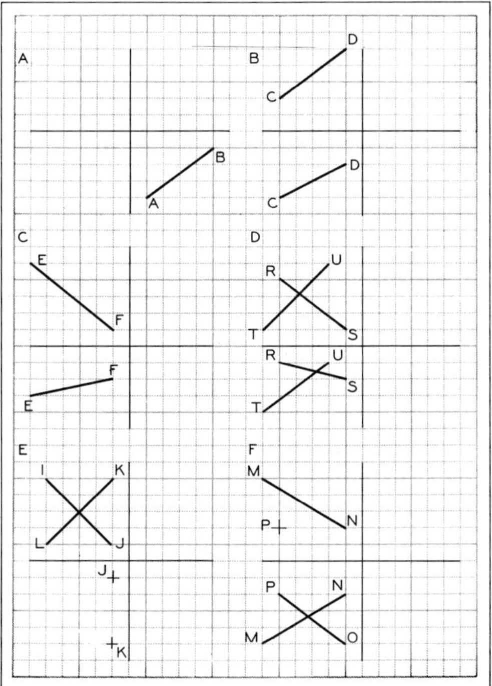

Fig. 4–41. Spatial relationships of lines.

4. Use Fig. 4–42 for all parts of the problem. (A) Draw three views of profile line 1–2. Draw three views of a line 1″ long that intersects 1–2 at its midpoint and appears as a point in the profile view. (B) Draw three views of a line $\frac{1}{2}$″ long that intersects the given line at its midpoint and appears as a point in the front view. (C) Draw the side view of the given lines and determine whether they are intersecting lines. (D) Construct line *OP* that passes under the given line and does not intersect it. (E) Draw three views of the line and the plane and indicate visibility. (F) Draw three views of the line and the plane and indicate visibility.

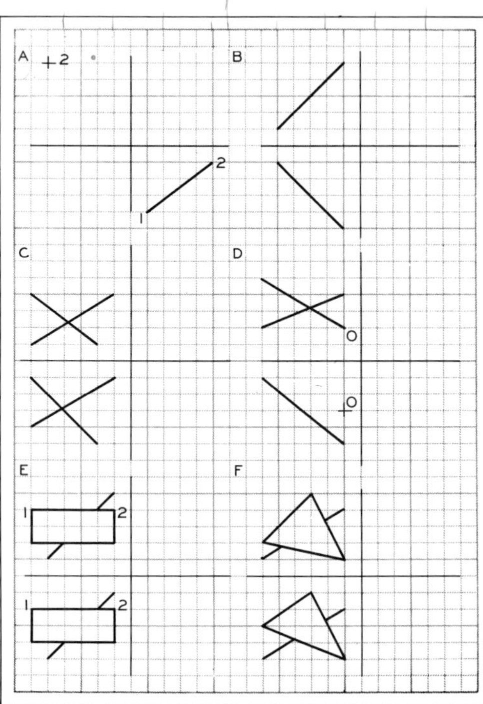

Fig. 4–42. Intersecting lines and visibilty.

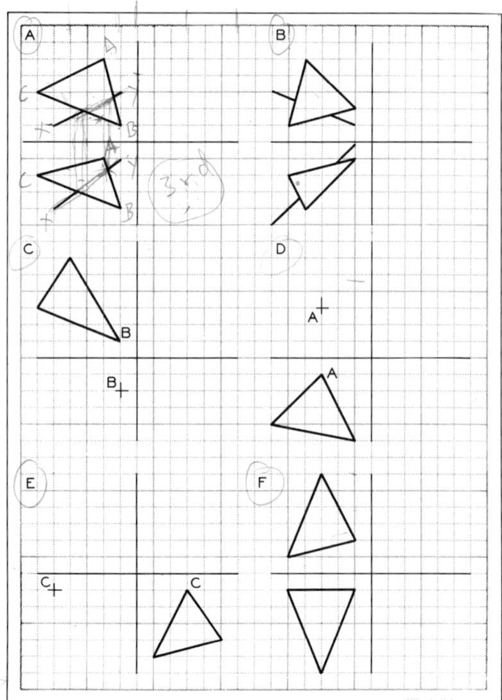

Fig. 4–43. Relationships of lines and planes.

5. Use Fig. 4–43 for all parts of this problem. (A) Draw three views of the line and the plane and indicate visibility. (B) Draw three views of the line and the plane and indicate visibility. (C) Construct three views of the horizontal plane and label its true-size view. (D) Construct three views of the frontal plane and label its true-size view. (E) Construct three views of the profile plane and label its true-size view. (F) Draw three equally spaced frontal lines on the plane in three views. Draw the side view.

6. Use Fig. 4–44 for all parts of this problem. (A) Draw two equally spaced horizontal lines on the plane and show them in three views. (B) Draw three equally spaced profile lines on the plane and show them in three views. (C) Construct line *JP* which lies on a plane that slopes upward to the left. Show the line and the plane in the side view also. (D) Construct line *KR* in a plane that slopes downward and backward. Show the line and the plane in the side view also. (E) Construct a line through point *A* that is parallel to the plane. (F) Construct a line through point *A* that is parallel to the plane.

7. Use Fig. 4–45 for all parts of this problem. (A) Construct a line through point *A* that is parallel to the line and equal in length. (B) Construct a line through point *B* that is parallel to the plane formed by the line and point. (C) Construct a line from point 2 that is perpendicular to the line at its midpoint. (D) Construct a line from point 3 in the front view that will be perpendicular to the front view of the line. Show this line in three views. Is it perpendicular to the line? (E) Construct a line from point 4 that will be perpendicular to the plane. (F) Construct a plane through point *O* that will be perpendicular to the line.

8. Use Fig. 4–46 for all parts of this problem. (A) Draw three views of the horizontal plane. Construct a line $\frac{1}{2}''$ long perpendicular to it and show the plane and the line in three views. (B)

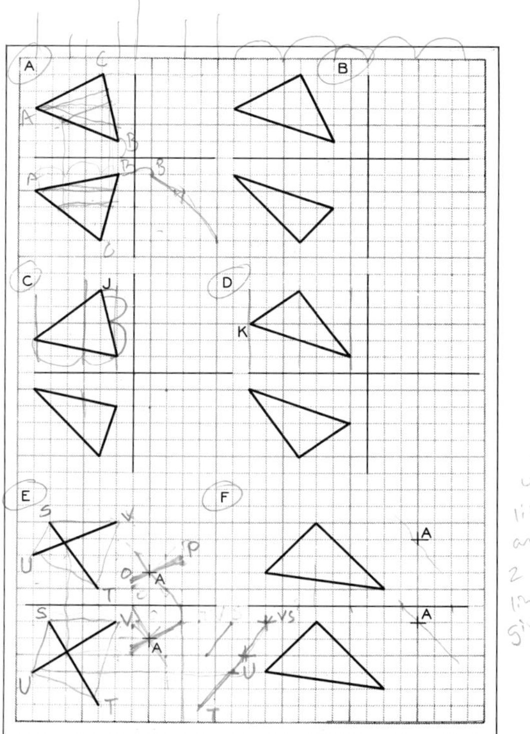

Fig. 4–44. Principal planes and spatial relationships.

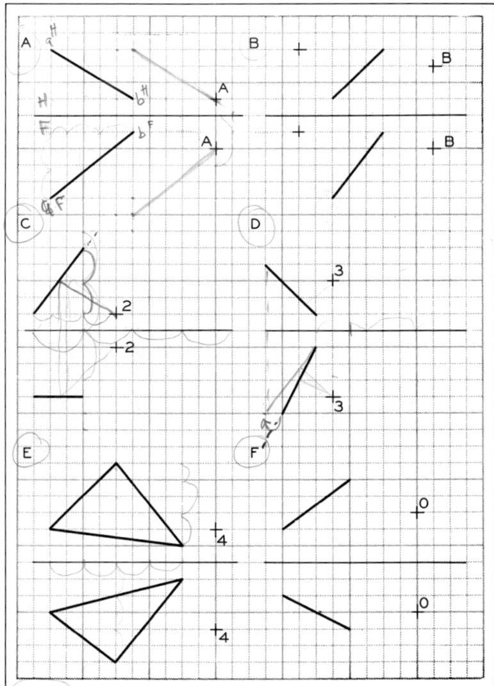

Fig. 4–45. Perpendicularity problems.

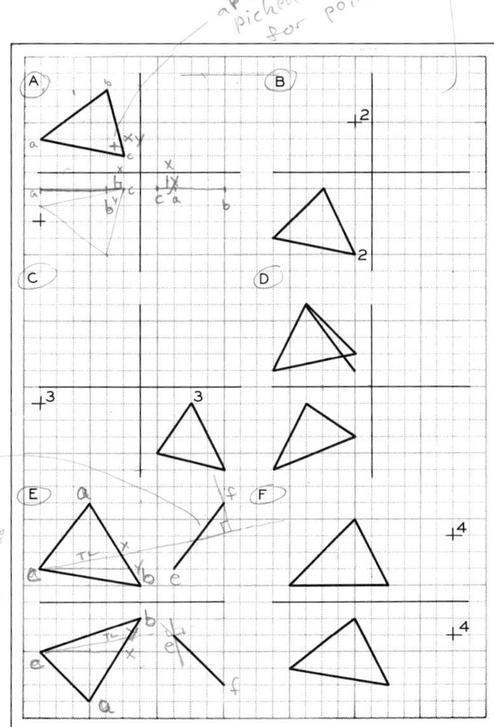

Fig. 4–46. Perpendicularity problems.

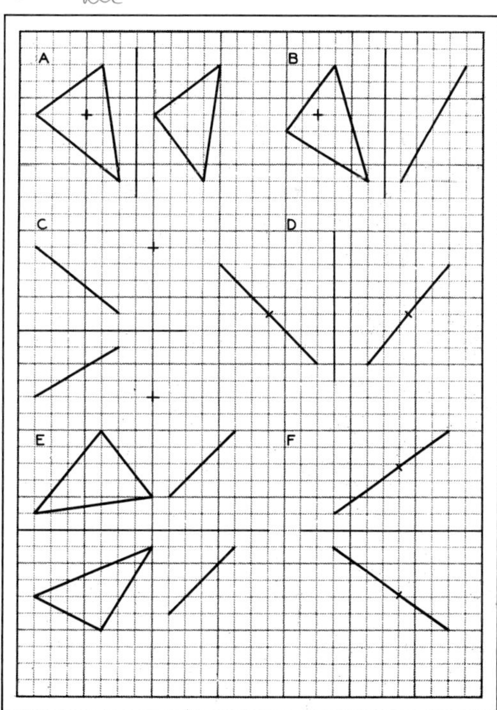

Fig. 4–47. Combination problems.

Draw three views of a line $\frac{1}{2}''$ long that is perpendicular to and intersecting the frontal plane on its back side. Show the line and the plane in three views. (C) Construct three views of a line $\frac{1}{2}''$ long that is perpendicular to and intersecting the profile plane on its right side. Show the line and the plane in three views. (D) Construct three views of the plane and line with the line lying on the plane as shown in part D of the figure. (E) Construct a plane through the line that is perpendicular to the plane given in part E of the figure. (F) Construct a plane through point 4 that is perpendicular to the plane given in part F of the figure.

9. Use Fig. 4–47 for all parts of this problem. (A) Construct a plane that passes through the point on the plane and is perpendicular to the plane. (B) Construct a plane that passes through the point on the plane and is perpendicular to the plane. (C) Construct a plane through the point that is perpendicular to the line. (D) Construct a line from the point on a line that is perpendicular to the line on the upward side. (E) Construct a plane through the line that is perpendicular to the plane. (F) Construct a plane that passes through the point on the line and is perpendicular to the line.

5

Primary Auxiliary Views

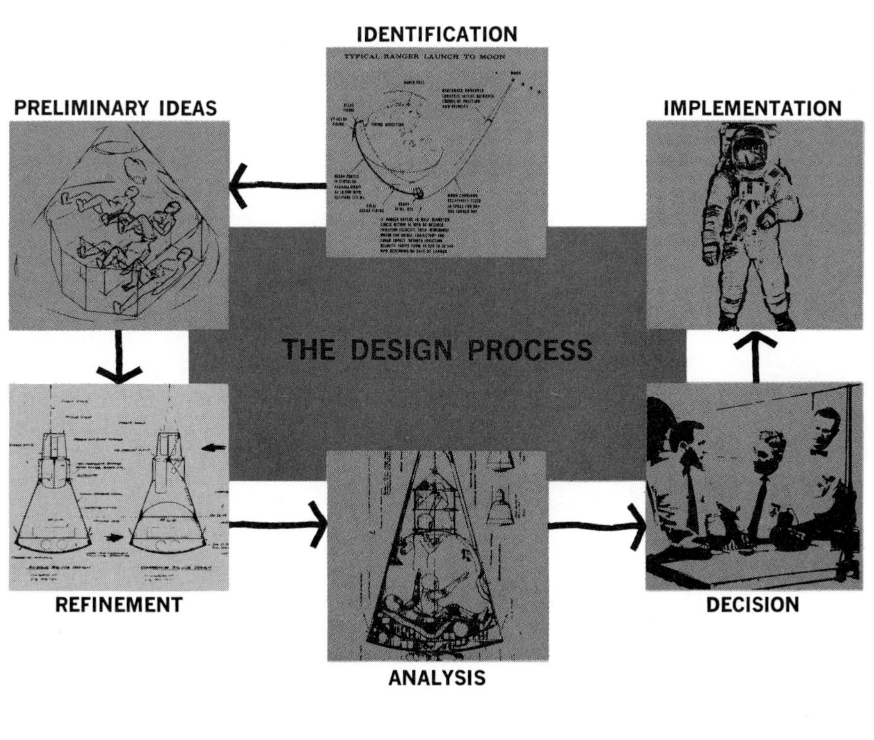

5–1 INTRODUCTION

Chapter 4 reviewed principles of orthographic projection as applied to the principal views of points, lines, and planes in space. Although orthographic projection in principal views offers solutions to many spatial problems, auxiliary projections are necessary to analyze many designs for critical information that would be difficult to obtain by other means. Distances, lengths, angles, sizes, and areas must be determined during the refinement of preliminary designs to permit analysis in the next phase of the design process.

An example of a simple design problem appears in Fig. 5–1, which shows an exhaust pipe designed for installation in an automobile. It was necessary to determine the bend angles of the pipe and its length while providing the necessary clearance with other interior components. Design of the support brackets required that angular measurements, distances, and similar information be found by descriptive geometry methods prior to the preparation of the finished specifications.

The Mariner spacecraft shown in Fig. 5–2 illustrates a number of applications of primary auxiliary views and secondary auxiliary views, which will be covered in the chapter that follows. The dimensions of each structural member must be found prior to the analysis of the frame for strength. Similarly, the dimensions of the members must be known to compute the weight of the vehicle, which is a critical aspect of space travel. The angles between the members must be obtained in order that the connecting joints may be designed. Knowledge of the angles between planes is necessary for fabrication of this system and for the determination of the true size of the planes formed by structural members. Graphical methods and descriptive geometry form the practical approach to solving realistic technological problems of the types shown in Figs. 5–1 and 5–2.

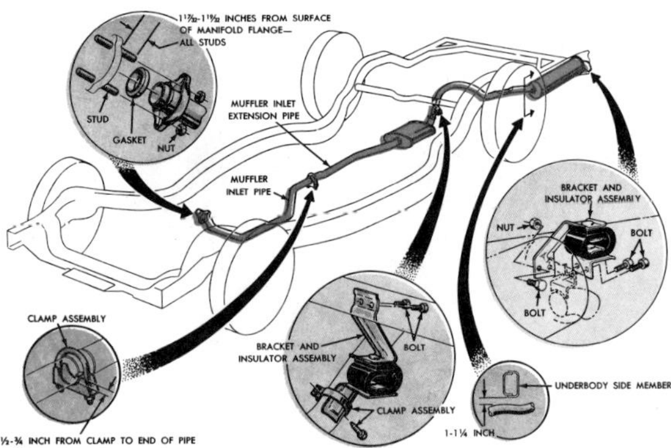

Fig. 5–1. The exhaust system of this automobile is an example of a spatial problem requiring the application of primary auxiliary views for solution. (Courtesy of Ford Motor Company.)

Fig. 5–2. The structural frame of the Mariner spacecraft illustrates many spatial relationships that must be determined during its design. (Courtesy of NASA.)

5–2 TRUE SIZE OF INCLINED SURFACES

A plane that appears as an oblique edge in a principal view may be found true size in a primary auxiliary view. Such a plane will appear foreshortened in adjacent principal views from which the oblique edge view is projected.

Primary auxiliary views are projected from any of the three principal views—the horizontal, frontal, or profile. A view projected from a horizontal view will require a horizontal reference plane; one projected from a front view will require a frontal reference plane; one projected from a side view will require a profile reference plane.

True Size of an Inclined Surface—Frontal Reference Plane. A plane that appears as an edge in the front view (Fig. 5–3) can be found true size in a primary auxiliary view projected from the front view. Reference line F–1 is drawn parallel to the edge view of the inclined plane in the front view at any convenient location. It is drawn to conform to the pictorial of the figure and to relate with the orthographic views in step 1. Note that the primary auxiliary plane is perpendicular to the frontal plane and that the line of sight is perpendicular to the auxiliary plane and parallel to the frontal plane. When an observer views the object in the direction indicated by the line of sight, he will see the frontal projection plane as an edge, and consequently, he will see measurements perpendicular to the frontal plane true length. These dimensions are those of the depth, represented here by *D*, which is perpendicular to the edge view of the frontal plane in the top and side views (step 1).

The inclined plane is projected from the front view perpendicular to the F–1 line. Depth *D* is measured in the top or side views and transferred to the auxiliary view with dividers. Each of the four corners of the surface is found in this manner and then all four corners are connected with one another to give the true size of the inclined plane.

True Size of an Inclined Surface—Horizontal Reference Plane. The inclined plane in Fig. 5–4 is inclined to the frontal and profile planes and is perpendicular to the horizontal plane. It will appear true size when projected onto an auxiliary plane which is parallel to the inclined plane and perpendicular to the horizontal plane.

Reference line H–1 is drawn parallel to the edge view of the inclined plane in step 1 of the figure. When the line of sight is perpendicular to an auxiliary plane projected from the horizontal (top) view, the horizontal plane will appear as an edge, and the height dimension *H* will appear true length. Each corner of the inclined plane is projected perpendicularly to the auxiliary plane and located by transferring the *H* distance from the front view to the auxiliary view with dividers. The four corner points are connected to give the true-size view of the inclined plane.

FIGURE 5–3. TRUE SIZE OF PLANE—FRONTAL REFERENCE LINE

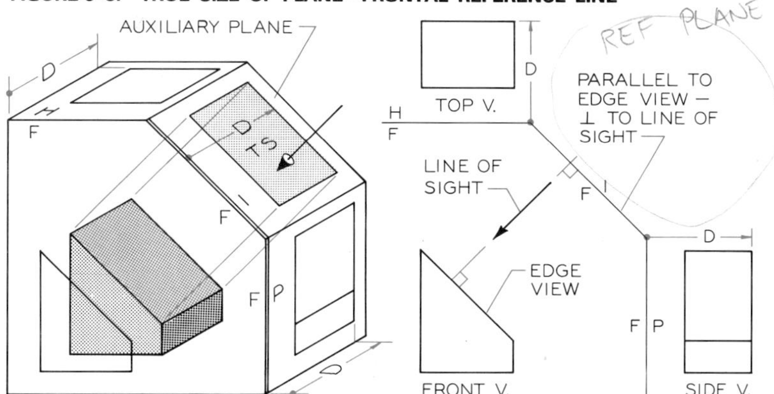

Given: An object with an inclined surface shown pictorially.

Required: Find the inclined surface true size by auxiliary views.

Step 1: Construct a line of sight perpendicular to the edge view of the inclined surface. Draw reference line F–1 parallel to the edge.

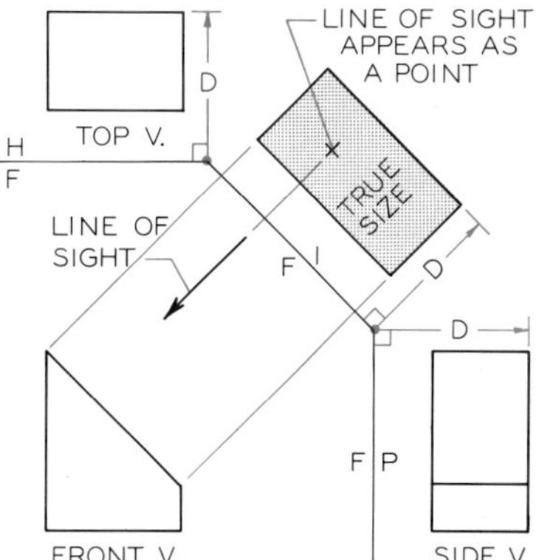

Step 2: Project the four corners of the inclined edge parallel to the line of sight. Locate corners by measuring perpendicularly from the frontal plane with depth dimensions *D*. The surface appears true size.

FIGURE 5–4. TRUE SIZE OF PLANE—HORIZONTAL REFERENCE LINE

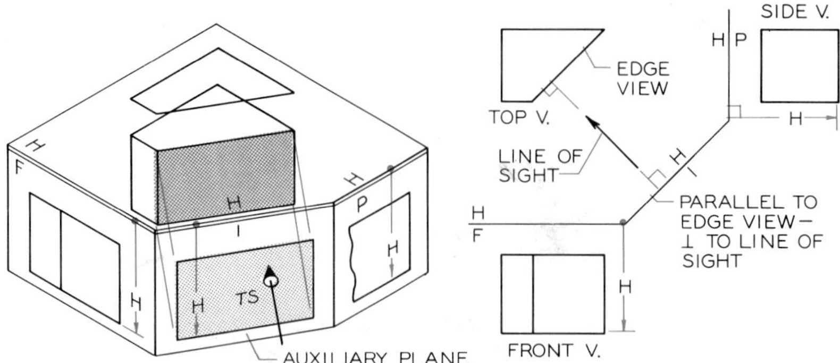

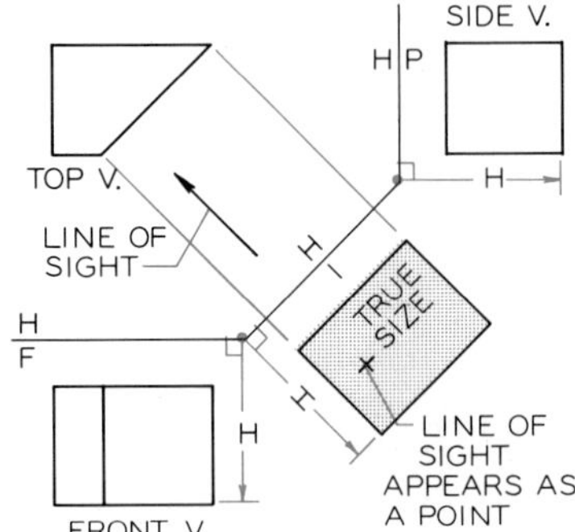

Given: An object with an inclined surface shown pictorially.

Required: Find the inclined surface true size by auxiliary views.

Step 1: Construct a line of sight perpendicular to the edge view of the inclined surface. Draw reference line H–1 parallel to the edge.

Step 2: Project the four corners of the inclined edge parallel to the line of sight. Locate corners by measuring perpendicularly from the horizontal plane with height dimensions *H*.

True Size of an Inclined Surface—Profile Reference Plane. The inclined plane in Fig. 5–5 appears as an edge in the side view, which means it is inclined to the horizontal and frontal planes and perpendicular to the profile plane. Auxiliary line P–1 is drawn parallel to the edge view of the inclined surface in the side view at some convenient location. An observer whose line of sight is perpendicular to the auxiliary plane will see the profile plane as an edge. Consequently, dimensions of width *W* will appear true length in the auxiliary view. The observer also sees the profile plane as an edge when he views the front view, so the width *W* dimensions are true length in this view. Therefore each measurement of *W* can be transferred from the front view to the auxiliary view to establish the corners of the inclined surface.

FIGURE 5–5. TRUE SIZE OF PLANE—PROFILE REFERENCE LINE

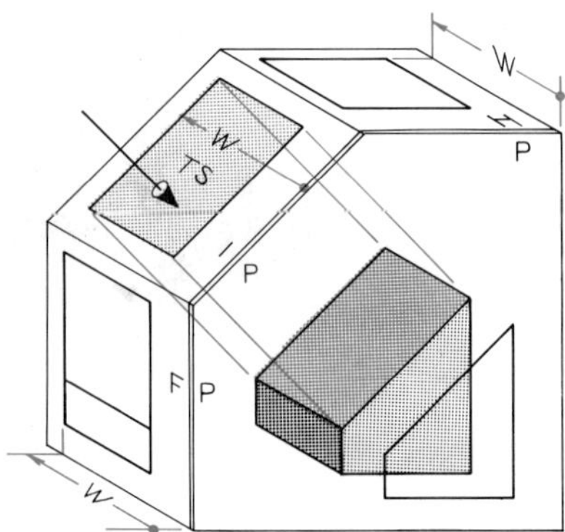

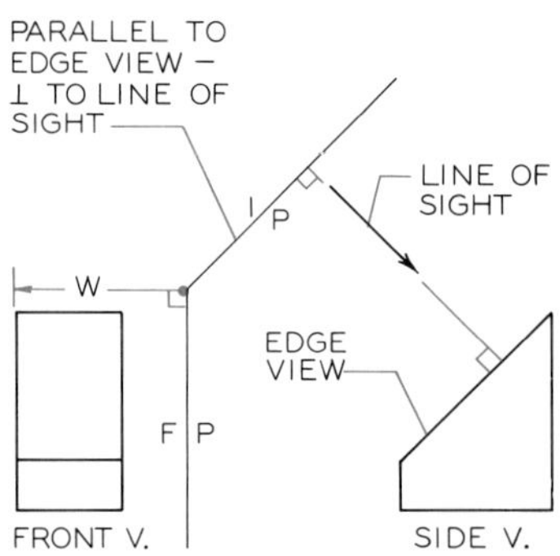

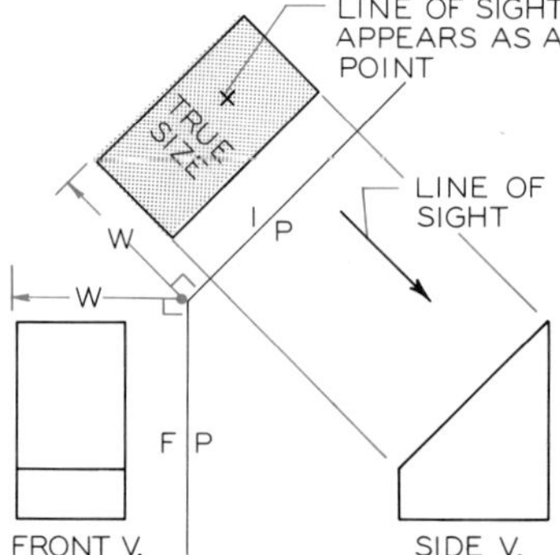

Given: An object with an inclined surface shown pictorially.

Required: Find the inclined surface true size by auxiliary views.

Step 1: Construct a line of sight perpendicular to the edge view of the inclined surface. Draw reference line P–1 parallel to the edge.

Step 2: Project the four corners of the inclined edge parallel to the line of sight. Locate corners by measuring perpendicularly from the profile plane with width dimensions *W*.

Fig. 5–6. The bed of this Model 45 Haulpak truck was designed through the use of auxiliary views to determine sizes of oblique planes. (Courtesy of LeTourneau-Westinghouse Company.)

The corners are connected to give the true-size view of the inclined surface.

The truck bed shown in Fig. 5–6 is composed of oblique planes that can be found true size by auxiliary views. Any two adjacent orthographic views of these planes can be used to find each plane true size by following the previously covered principles. It is necessary only for the planes to appear as edges in a principal view.

5–3 PRIMARY AUXILIARY VIEW OF A POINT

Point 3 is shown pictorially in Fig. 5–7A, where it is projected onto the horizontal, frontal, and auxiliary planes. Note that the auxiliary plane is perpendicular to the horizontal plane; consequently the observer will see the horizontal plane as an edge when his line of sight is perpendicular to the auxiliary plane. Distances that are perpendicular to the horizontal plane will appear true length when the horizontal plane appears as an edge. Therefore point 3 can be located in the auxiliary view by measuring its distance *H* from the horizontal plane in the front view and transferring this distance to the auxiliary plane, where the horizontal plane also appears as an edge.

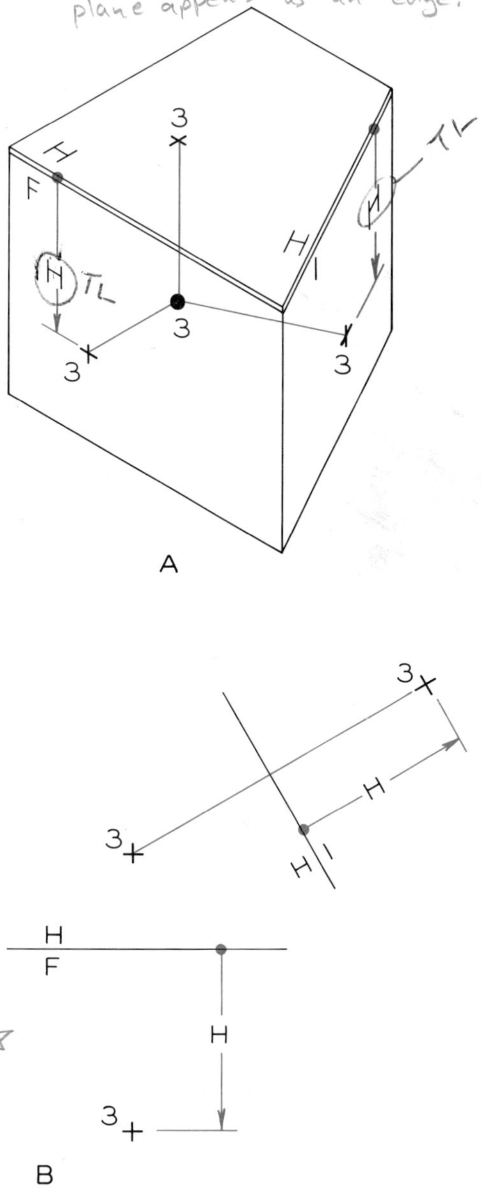

Fig. 5–7. Primary auxiliary view of a point shown pictorially and orthographically.

The orthographic construction of the primary auxiliary view is illustrated in part B of the figure. It should be observed that there are an infinite number of positions for the auxiliary plane projected from the top view, but in every case point 3 would lie at the distance *H* from the horizontal plane. Similarly, an auxiliary view could have been projected from the front view through the use of an auxiliary plane that was perpendicular to the frontal plane, as introduced in Article 5–2.

5–4 PRIMARY AUXILIARY VIEW OF A LINE

The projections of line 1–2 are shown pictorially in Fig. 5–8A. The line is projected onto the horizontal, frontal, and auxiliary planes with the primary auxiliary plane constructed parallel to the frontal projection of line 1–2. When the observer's line of sight is perpendicular to the auxiliary plane, the frontal plane will appear as an edge, and all dimensions perpendicular to the frontal plane will be projected true length. Depth dimension *D* is perpendicular to the frontal plane and is therefore used as the measurement to construct the auxiliary view of line 1–2. Point 2 is the same distance from the frontal plane in the auxiliary view as it is in the horizontal projection (top view).

The orthographic construction of the primary auxiliary view is shown in part B of the figure. The auxiliary plane is drawn parallel to the front view of line 1–2 so that the observer's line of sight will be perpendicular to the line when it is perpendicular to the auxiliary plane.

Figure 5–9 separates the sequential steps required to find the true length of an oblique line. It is beneficial to letter all reference planes using the notation suggested in the example illustrations. The reference line drawn between

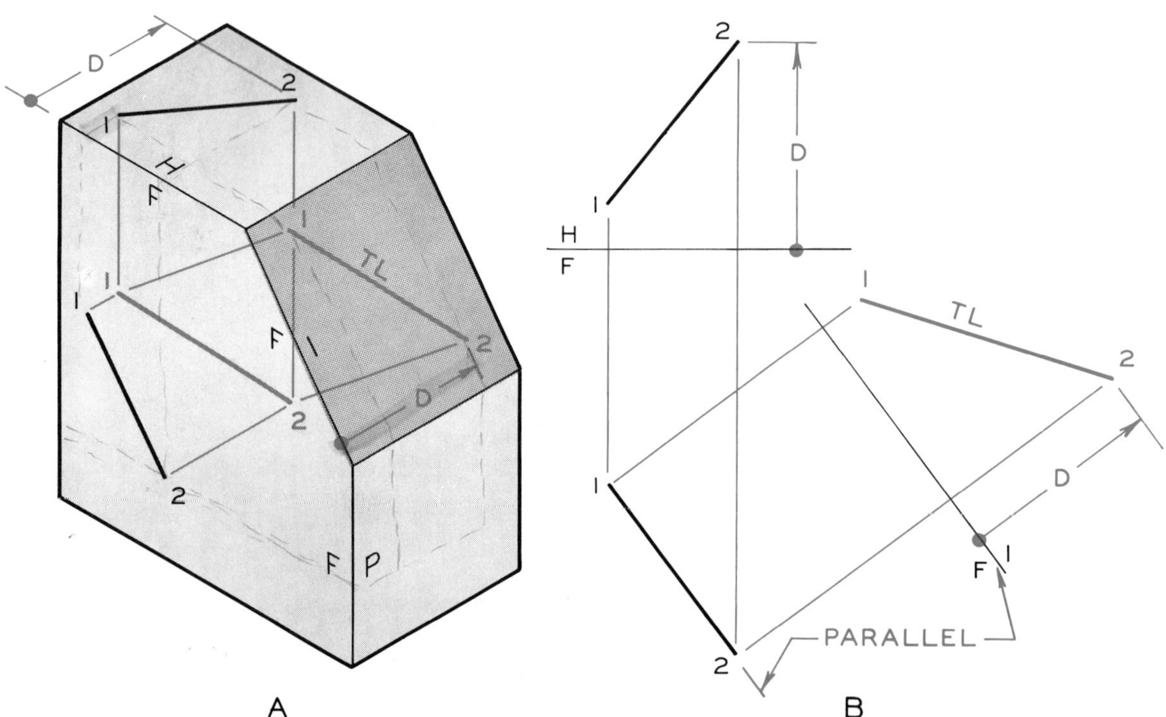

Fig. 5–8. Primary auxiliary view of a line.

FIGURE 5–9. TRUE LENGTH OF A LINE BY PRIMARY AUXILIARY VIEW

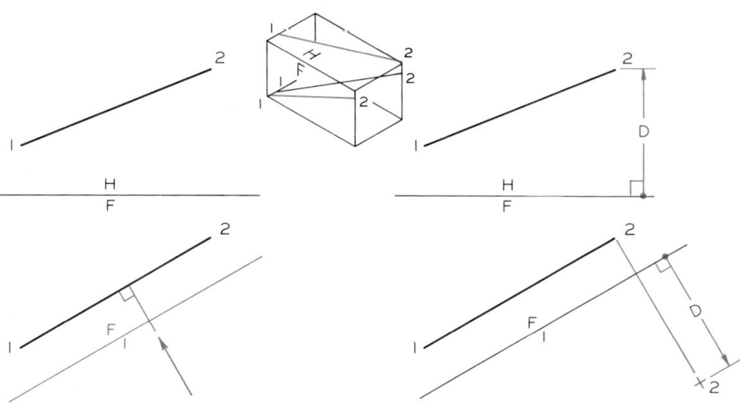

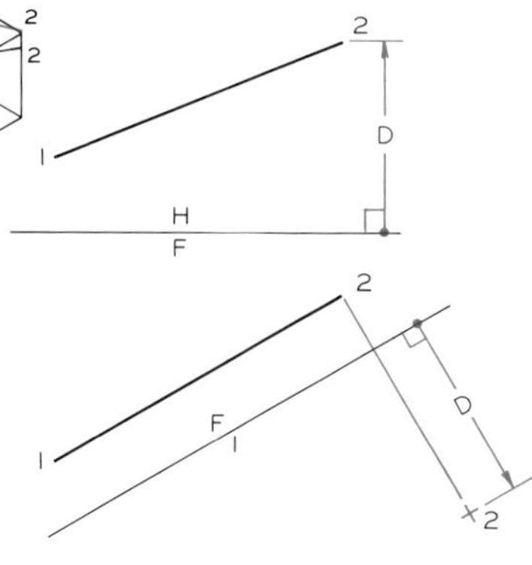

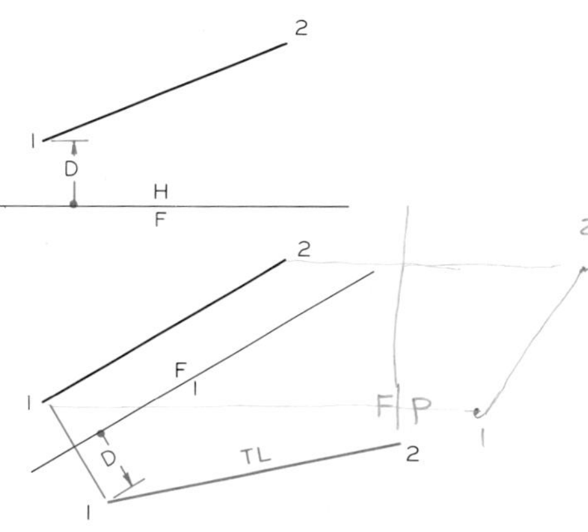

Step 1: To find line 1–2 true length, construct the line of sight perpendicular to the line. Draw reference line F–1 parallel to the line.

Step 2: Project point 2 perpendicularly from the front view. Dimension *D* from the top view locates point 2 in the auxiliary view.

Step 3: Point 1 is located by transferring dimension *D* from the top view. Line 1–2 appears true length in this view.

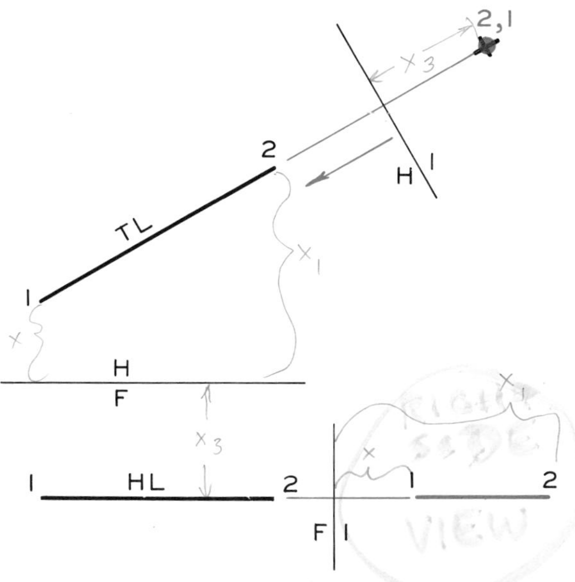

Fig. 5–10. A primary auxiliary view which is projected parallel to a true-length line gives a point view of the line.

Fig. 5–11. The diagonal structural members of the Saturn 1-B launch vehicle could have been designed by using primary auxiliary views to determine their true lengths. (Courtesy of NASA.)

the principal plane and the primary auxiliary plane is a line representing the line of intersection between the primary and auxiliary planes, as shown in Fig. 5–8. A primary view projected from a front view has a reference line labeled F–1, from the horizontal view, H–1, and from the profile view, P–1.

The point view of a line can be found in a primary auxiliary view when the line is true length in a principal view. Line 1–2 in Fig. 5–10 is horizontal in the front view, which makes the top view true length. When auxiliary line H–1 is drawn perpendicular to the direction of the top view of line 1–2, the resulting auxiliary view will project as a point view of line 1–2. Note that the auxiliary view, which is projected from the frontal projection of line 1–2, does not result in a point view of the line, but instead gives a foreshortened view. This particular projection is actually a right side view.

The true length of the diagonal structural members of the Saturn 1–B launch vehicle (Fig. 5–11) can be found by primary auxiliary views with a high degree of accuracy. These structural members would not appear true length in principal views. Numerous other examples of oblique lines requiring auxiliary-view solution can be seen in the structural framework in the background of this illustration.

5–5 TRUE LENGTH BY ANALYTICAL GEOMETRY

The analytical approach to determining the true length of a line is illustrated in Fig. 5–12, where line 3–4 appears true length in the front view. The line can be measured graphically for its true length, or else its true length can be found by application of the Pythagorean theorm. The Pythagorean theorem states that the hypotenuse of a right triangle is equal to the square root of the sum of the squares of the other two sides.

An oblique line that does not project true length in the principal views requires the manipulation of three coordinates—*X*, *Y*, and *Z*. Such a line is shown pictorially in Fig. 5–13. Note that the *X*- and *Y*-coordinates are pro-

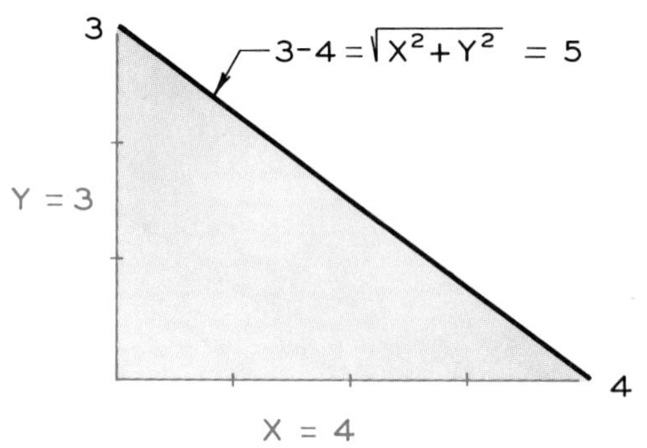

Fig. 5–12. The true length of a frontal line can be found analytically by the Pythagorean theorem.

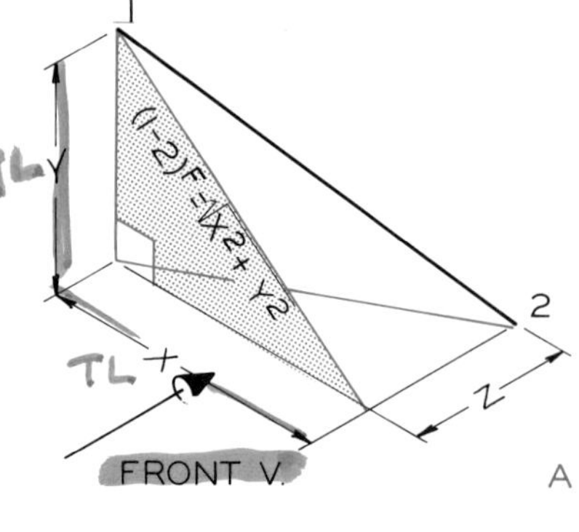

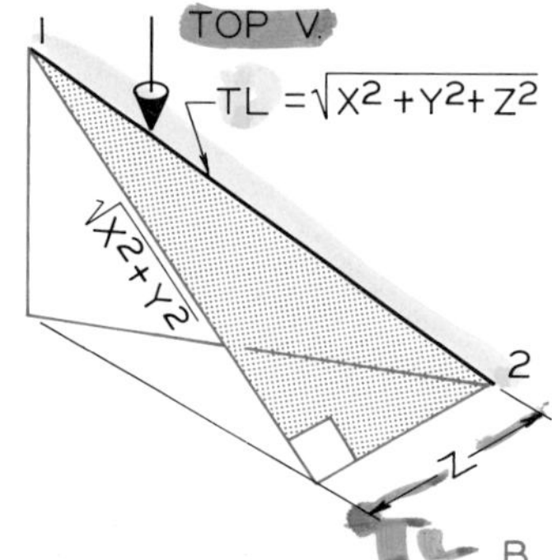

Fig. 5–13. The *X*-, *Y*-, and *Z*-coordinates used for finding the true length of an oblique line.

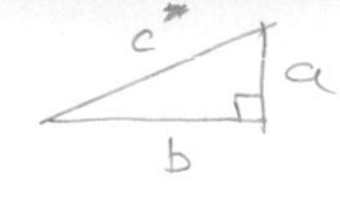

FIGURE 5–14. FINDING THE TRUE LENGTH OF AN OBLIQUE LINE—ANALYTICAL METHOD

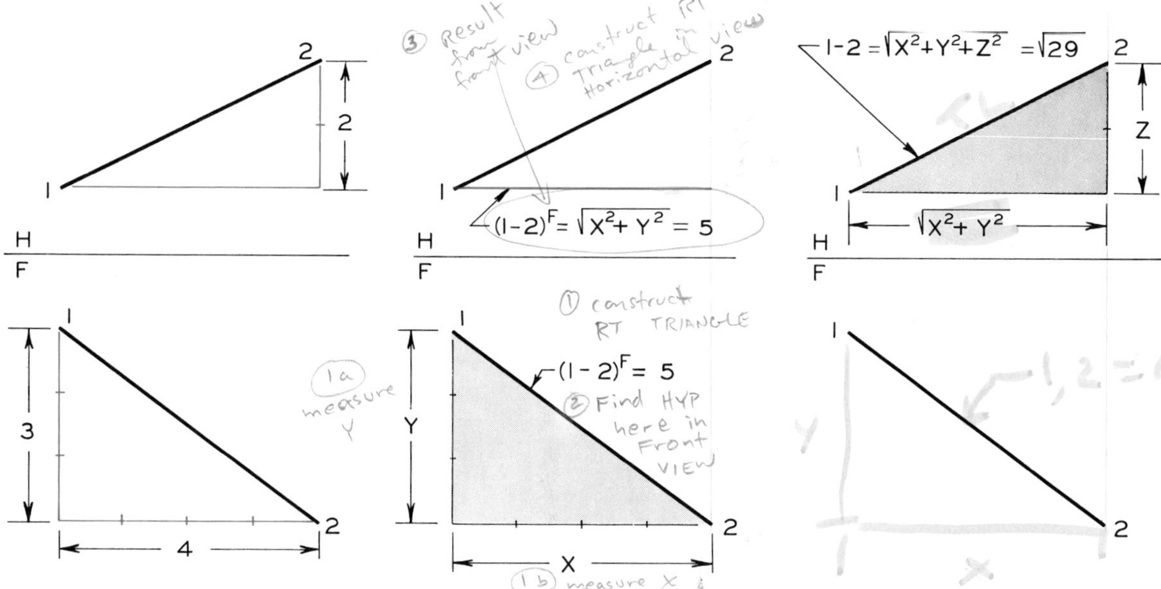

Step 1: Right triangles are constructed with line 1–2 as the hypotenuse in the top and front views. The coordinates, or legs of the right triangles, are drawn parallel and perpendicular to the H–F reference line.

Step 2: The true length of the frontal projection of line 1–2 is found by application of the Pythagorean theorem as though the line were true length in the front view. The frontal projection of line 1–2 is found to be 5 units in length by substituting the units of 3 and 4 as the *X*- and *Y*-coordinates. The resulting length can be visualized by referring to the pictorial of the line in Fig. 5–13.

Step 3: The true length of the line can be found by combining the true length of the frontal projection with the true length of the *Z*-coordinate in the top view. The total equation of line 1–2 becomes $\sqrt{29}$.

jected true length in the front view while the *Z*-coordinate is true length in the top view. Two views are needed to determine the true length of an oblique line by analytical methods, just as two views were needed to find a graphical solution.

The steps for determining the true length of line 1–2 using the analytical method are illustrated in Fig. 5–14.

It can be seen by comparison that the analytical methods for determination of the true length of a line are very similar to the primary auxiliary method, where the projection of a line is used as the basis for an additional view in which the missing coordinate appears true length. Both systems of spatial analysis should be used in combination to promote accuracy and to provide a means for a more thorough analysis.

5–6 TRUE LENGTH BY A TRUE-LENGTH DIAGRAM

The true length of line 1–2, or any oblique line, can be found by a true-length diagram such as that illustrated in Fig. 5–15. This is not an auxiliary view method, but a knowledge of primary auxiliary views is necessary to understand this method.

FIGURE 5–15. TRUE-LENGTH DIAGRAM

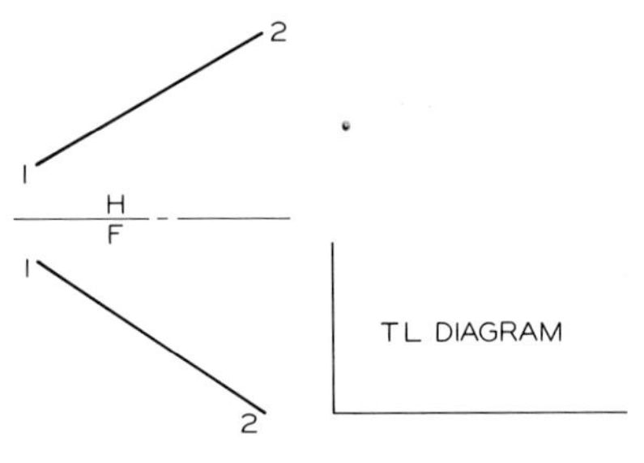

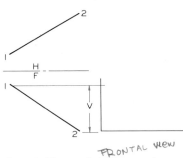

Given: Two views of line 1–2.

Required: Find 1–2 true length in the TL diagram.

Step 1: The vertical distance between the ends of line 1–2 is transferred to the vertical leg of the TL diagram.

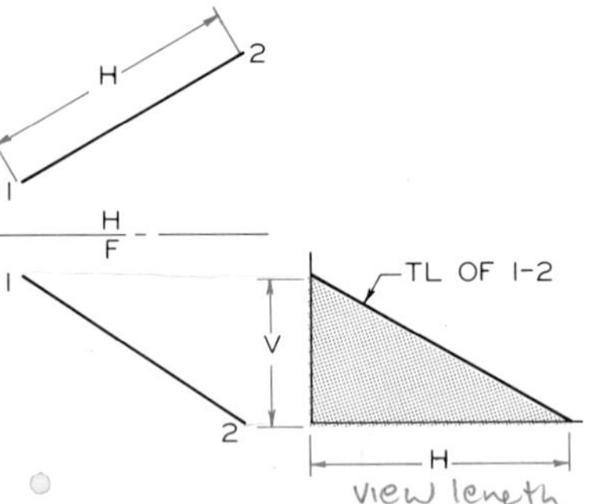

Step 2: The horizontal distance between the ends of line 1–2 is transferred from the top view to the other leg of the TL diagram.

The true-length diagram method does not give a direction for the line, but merely its true length. In general, a true-length diagram can be projected from any two adjacent views, with one measurement being the distance measured between the two endpoints in a direction perpendicular to the reference plane between the two views. The other measurement is the projected length of the line in the other adjacent view.

5–7 ANGLE BETWEEN A LINE AND A PRINCIPAL PLANE

When a primary auxiliary view is drawn, an edge view of a principal plane will appear in the resulting auxiliary view. When the auxiliary view is projected from the front view, the frontal plane will appear as an edge (Fig. 5–16A); when it is projected from the horizontal view, the horizontal plane will appear as an edge (Fig.

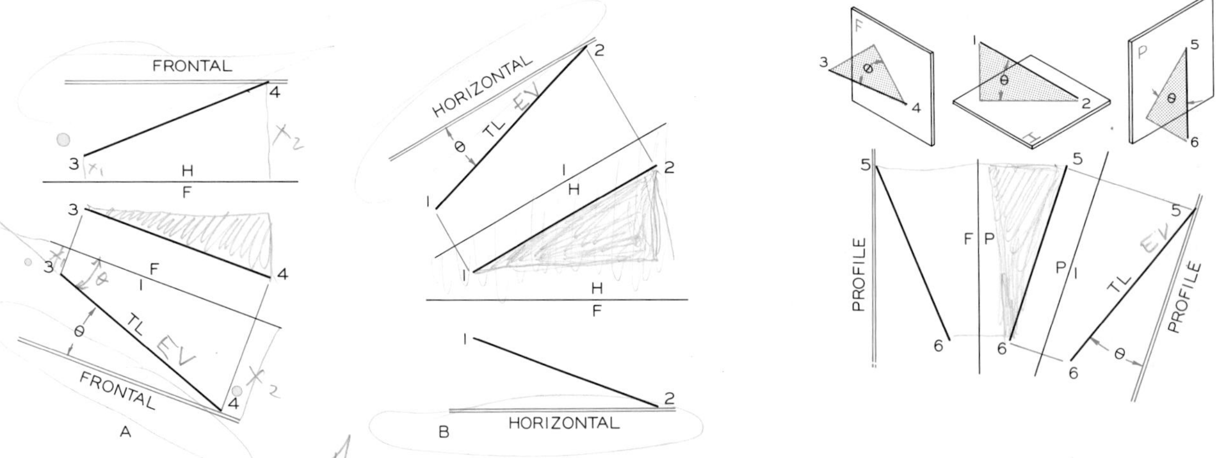

Fig. 5–16. The true angle that a line makes with a principal plane can be measured in a primary auxiliary view where a line is true length. The angle with the frontal is shown at A, with the horizontal at B, and with the profile at C.

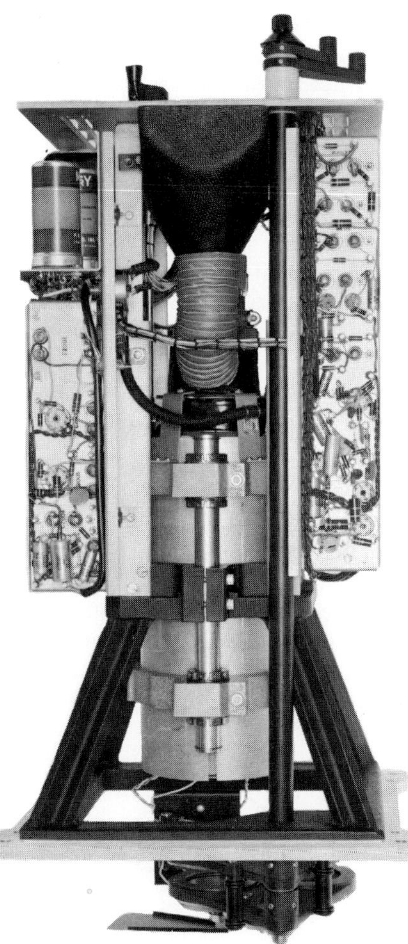

Fig. 5–17. The angle of the diagonal base supports with the horizontal base of the image orthicon camera could have been determined by an auxiliary view. (Courtesy of ITT Industrial Laboratories.)

5–16B); when it is projected from the profile view, the profile plane will appear as an edge (Fig. 5–16C). The true angle between a principal plane and a line can be found in the view where the principal plane in question appears as an edge and the line is true length.

The structural mount used to attach the image orthicon camera (Fig. 5–17) to a telescope is an example of the need to determine the angle between a line and a principal plane. In this case, the line represents the diagonal base supports, while the horizontal plane is the plane of the base. This angle must be known to fabricate the design.

5–8 SLOPE OF A LINE

Slope is defined as the angle a line makes with the horizontal plane. It may be specified by either of the three methods shown in Fig. 5–18; it can be indicated as *slope angle, percent grade,* or *slope ratio.*

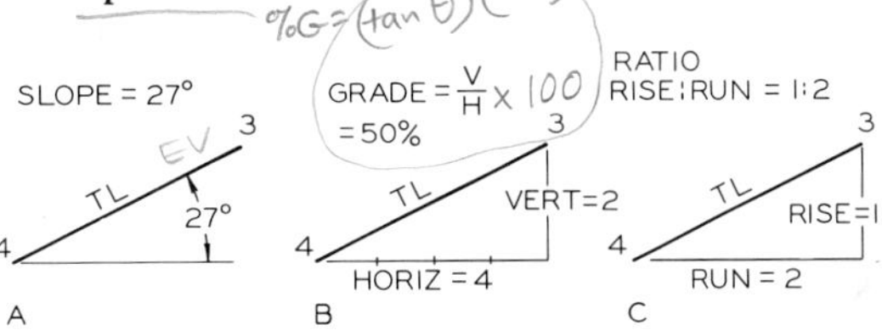

Fig. 5–18. The slope of a line can be indicated by (A) slope angle, (B) percent grade, and (C) slope ratio.

Slope Angle. Line 1–2 in Fig. 5–19 is found true length in a primary auxiliary view projected from the top view. This gives the line true length and the horizontal plane as an edge where the slope can be measured as 30°. The true-length auxiliary must be projected from the top view to find the horizontal plane as an edge.

Percent Grade. The percent grade of a line is the ratio of its horizontal (*run*) to its vertical (*rise*) from one end to the other. Two methods of finding the percent grade of a line are shown in Fig. 5–20. These are special cases, since the lines are true length in the front view where the horizontal plane will automatically appear as an edge.

In part A, the vertical rise is divided by the horizontal run to give a percent grade of −58%. It is given a minus sign since it slopes downhill from *A* to *B*. If the direction had been specified from *B* to *A*, the percent grade would have been positive (+). Note that the percent grade is actually the *tangent* of θ, but it is converted to a percent by multiplying by 100.

In part B, the percent grade of the line from *C* to *D* is found by using a graphical technique that eliminates the mathematics used

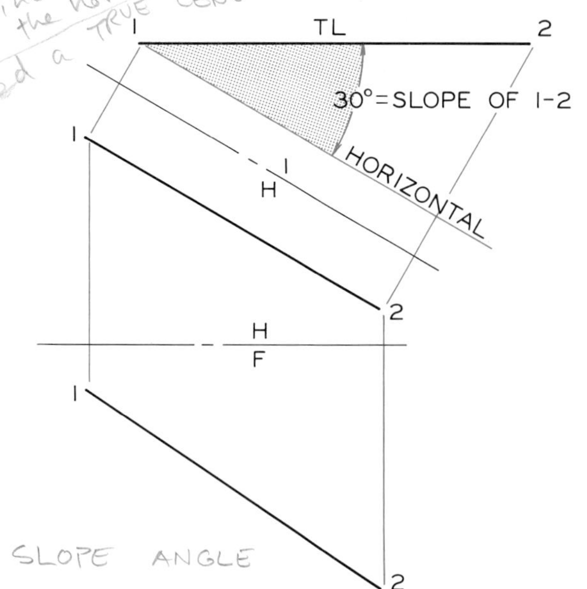

Fig. 5–19. The slope of a line can be found in a view where the line appears true length and the horizontal plane appears as an edge. The angle between them can be measured in an auxiliary view. projected from the top view.

in part A. An engineers' or metric scale is used to measure 10 units along the horizontal and a vertical distance perpendicular to the 10 units. The vertical distance can be measured and the percent grade found by moving the decimal point one place to the right. Since the line is specified from *C* to *D*, it is given a positive grade of +58% because it slopes uphill.

Slope Ratio. Slope ratio is the same as percent grade except for the method of expressing inclination. The first number of the ratio is always one, such as 1:10, 1:200, and so on. The first number of the ratio (one) is always equal to the rise, and the second number is equal to the run (Fig. 5–18C).

Percent Grade of an Oblique Line. Oblique line 3–4 is shown pictorially in Fig. 5–21A. Since it is not true length in the front view, it must be found true length in an auxiliary view projected from the top view so that the angle between the horizontal and the true length can be found (part B). Using your engineers' scale,

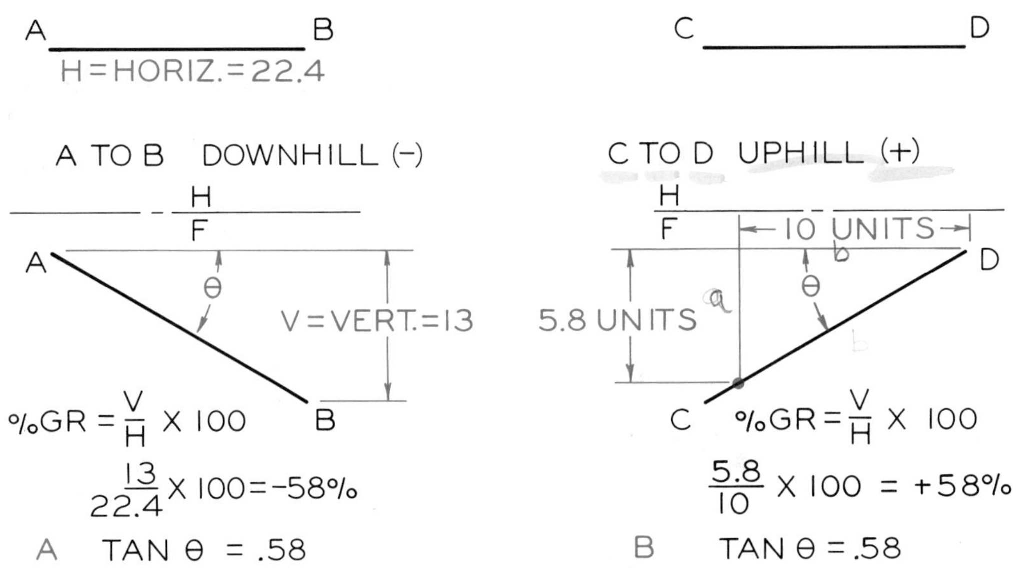

Fig. 5–20. The percent grade of a line is the vertical rise divided by the horizontal run, converted to a percent. Percent grade can be found also by converting the tangent of the slope angle to a percent.

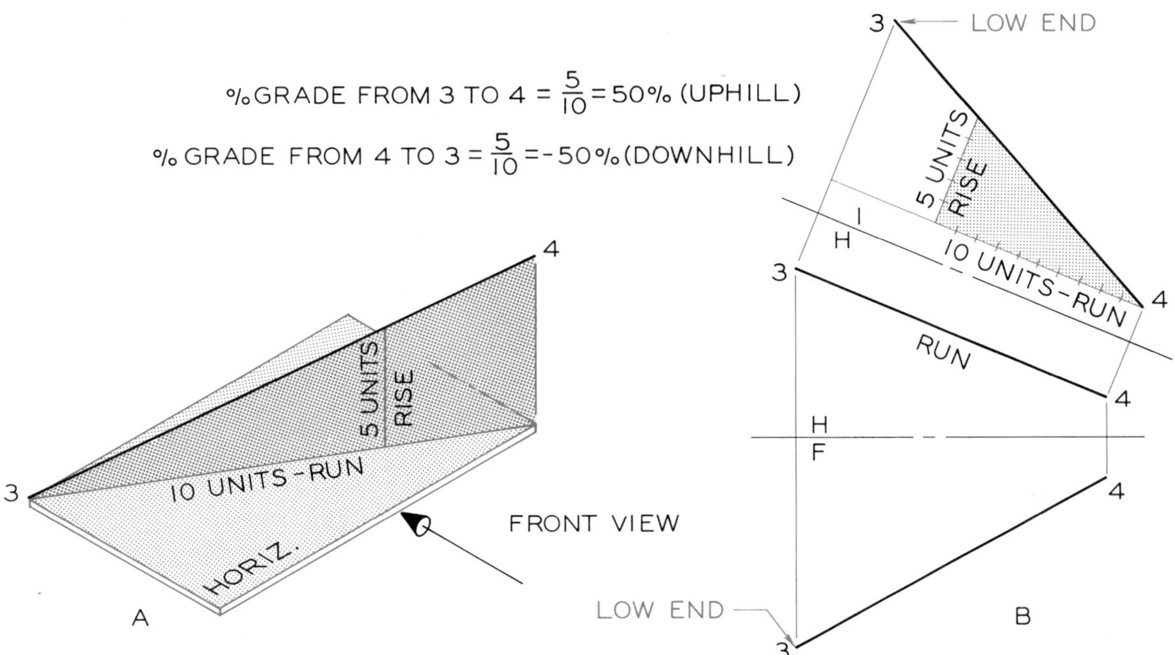

Fig. 5–21. The percent grade of an oblique line.

find the grade by laying off 10 units along the horizontal and then measure the vertical distance perpendicularly from the end of this measurement. Scale this line to find a 50% grade. From 3 to 4 the grade is positive (+); from 4 to 3 it is negative (−).

All gravity-flow drainage systems must be analyzed to determine the slopes and percent grades within the system. In the pipeline in Fig. 5–22, for example, the slope was calculated from field data to determine the length and operational effectiveness of the pipeline.

Fig. 5–22. The slope that a pipe line is to follow at various intervals must be established during its design. (Courtesy of Trunkline Gas Transmission Company.)

5–9 COMPASS BEARINGS OF A LINE

In civil engineering and geological applications, lines are often drawn from verbal information and field notes. Lines may also represent paths of motion in navigation, where verbal instructions are given by voice from a remote source. A commonly accepted method of locating a line verbally is by using the points of a compass.

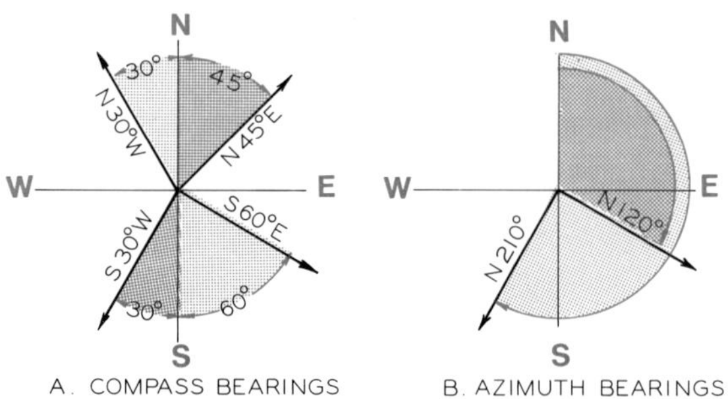

Fig. 5–23. Compass bearings of four lines (A) and the azimuth bearings of two lines (B).

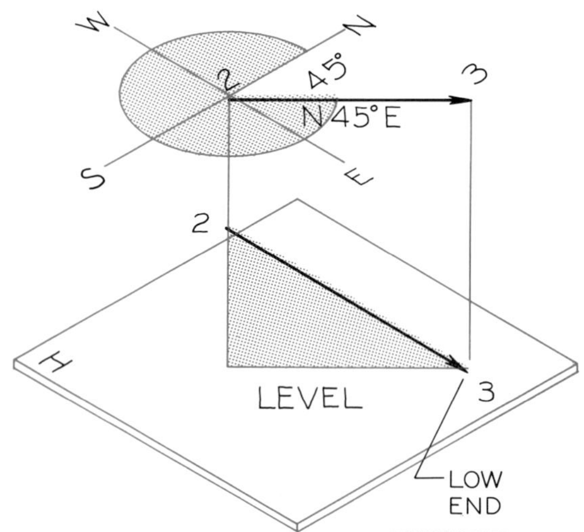

Fig. 5–25. The compass bearing of a line is measured in the top view toward the low end.

Figure 5–23A gives the compass bearings of four lines. Note that the bearings begin with the north or south direction in all cases. The bearing of a line 30° to the west of north is given as North 30° West, or N 30° W. A line making 60° with the south point of a compass is given as South 60° East, or S 60° E. Since a compass can be read only when held horizontally, the bearings of a line can be determined in the horizontal (top) view only. A bearing is a horizontal direction.

Figure 5–23B is an example of an azimuth, which is measured from the north point of a compass in a clockwise direction. Azimuth readings are used to avoid the confusion that might be caused by reference to the four points of a compass. An azimuth of 120° is the direction making 120° with north. This is the same bearing as S 60° E shown in Fig. 5–23. The azimuth of 210° is equivalent to S 30° W. A topography crew (Fig. 5–24) must take many bearings to survey an area.

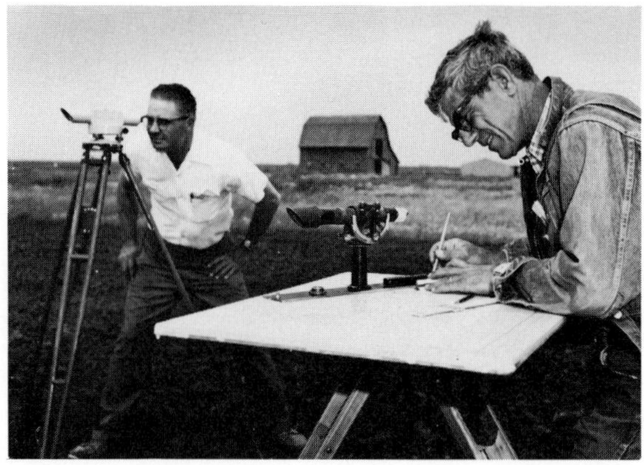

Fig. 5–24. A topography crew must use many bearings to record an irrigation area. (Courtesy of the Bureau of Reclamation, U.S. Department of the Interior.)

Although bearings and azimuths are horizontal directions, these terms are often used to specify the construction of a drainage system or the design of a piping system whose pipes are inclined with the horizontal plane. In these cases, the bearings usually refer to the direction of the line with respect to the lower end, if not specified. Figure 5–25 illustrates that line 2–3 has a bearing of N 45° E, since the line lies in a vertical plane with point 3 being the low end of the line.

The slope of a line bearing S 60° W in Fig. 5–26 is found by projecting an auxiliary view from the top view in order that the horizontal plane may be seen as an edge. Slope (the angle a line makes with the horizontal) in this case is found to be 34°. Had the low end been point *B*, the bearing of line *AB* would have been N 60° E instead.

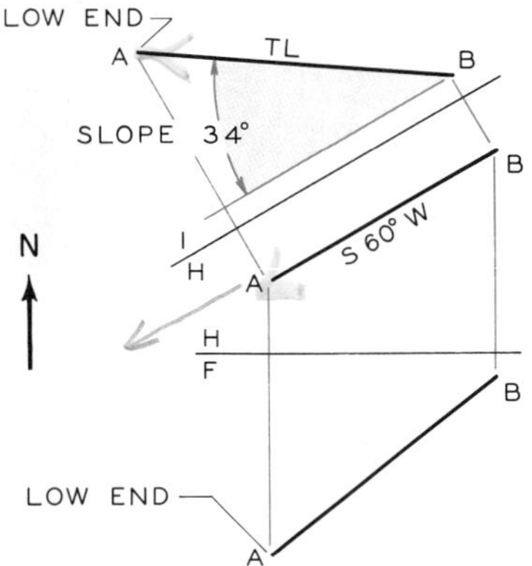

Fig. 5–26. Determining the slope and bearing of a line by an auxiliary view.

5–10 CONTOUR MAPS AND PROFILES

Since the surface of the earth is rarely uniform or level, some system is required for graphically representing irregularities in a drawing. A contour map is a widely accepted method employed by engineers to represent irregular shapes and surfaces of the earth. A pictorial view and a conventional map view of a contour map are shown in Fig. 5–27. The following definitions should be understood prior to further discussion.

Contour Lines. Contour lines are lines that represent constant elevations from a common horizontal datum plane, such as sea level. Contour lines can be thought of as the intersection of horizontal planes with the surface of the earth. The interval of spacing in Fig. 5–27 is 10′.

Contour Maps. A contour map represents the irregularities on the surface of the earth with a network of contour lines, as shown in Fig. 5–27. Contour lines do not cross each other on a contour map. The closer the contour lines are to each other in the contour map, the steeper the terrain.

Profile. A profile is a vertical section through the surface of the earth which describes the contour of the earth's surface at any desired location. Two profiles are shown in Fig. 5–27. When applied to topography, a vertical section is called a profile regardless of the direction in which the view is projected. Contour lines appear as the edge views of equally spaced horizontal planes in the profiles. The true representation of a profile is drawn such that the vertical scale is equal to the scale of the contour map; however, this scale may be increased to emphasize changes in elevation that would not otherwise be apparent.

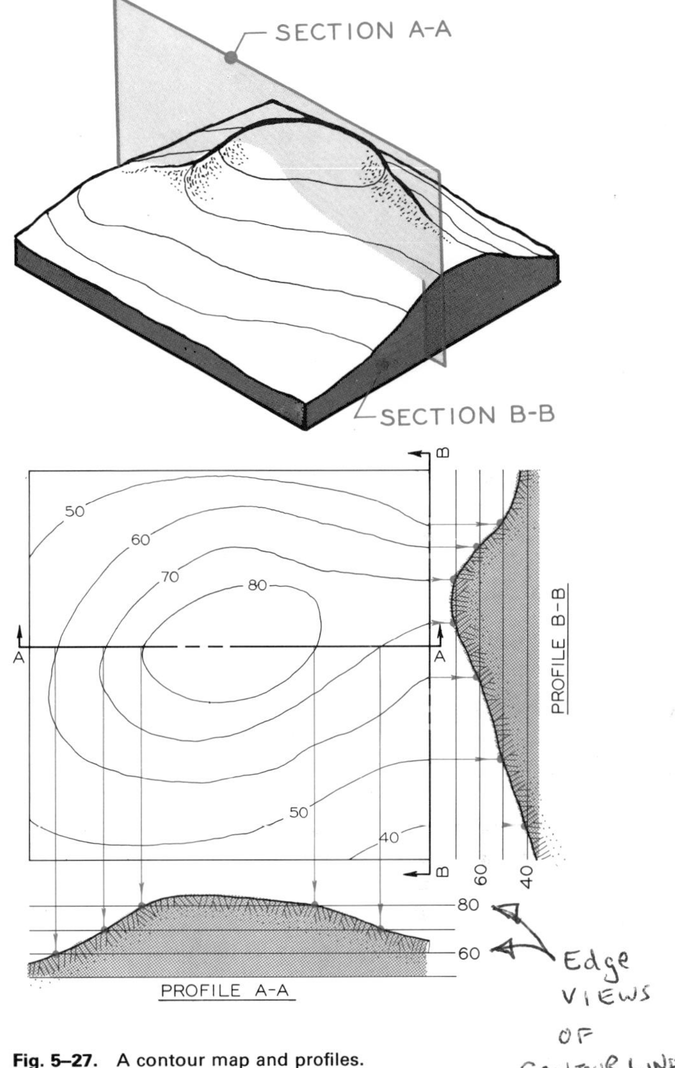

Fig. 5–27. A contour map and profiles.

Contoured Surfaces. Contoured lines are also used to describe irregular shapes other than the surface of the earth. Examples are airfoils,

Fig. 5–28. Contours apply to irregular-shaped designs as well as the earth's surface. The method of developing templates for an irregular surface such as the body of this experimental vehicle is called *lofting.* (Courtesy of Goodyear Corporation.)

Fig. 5–29. Geologists study the surface of the earth by viewing separate photographs which give a three-dimensional view of the terrain. (Courtesy of Exxon Corporation.)

irregular-shaped castings, automobile bodies, and household appliances. The vehicle shown in Fig. 5–28 was depicted by a contoured layout to represent its shape prior to manufacture. When applied to manufactured objects, this technique is called *lofting.* Contours are shown in three principal views to fully describe an irregular shape such as a ship's hull.

Geologists study irregularities on the surface of the earth through a three-dimensional viewer or *stereoscope* (Fig. 5–29). Each of the two photographs viewed through the stereoscope must be made with a separate camera lens which is calibrated to match the lens of the viewer. This three-dimensional analysis, which is called *photogrammetry,* can be used to study the contour of the surface and to determine contour lines.

5–11 VERTICAL SECTIONS

A vertical section may be used to show the variation of the earth's surface; this section may be referred to as a profile. In Fig. 5–30, vertical sections are passed through the top views of an underground pipe system that begins at point 1 and ends at point 2.

The top view of the cutting planes coincides with the top view of the pipes. Auxiliary views are projected from the top view to find the surface of the earth in steps 1 and 2. As specified in the problem, the ends of the pipes are located 15 feet below the surface at points 1, 2, and 3. The percent grade of the pipes is found in step 3 to complete the problem.

The same scale was used for constructing the profiles that were used in drawing the contour map in this case. This makes it possible to measure true lengths and angles. The percent grade and the compass bearing of each line is labeled on the contour map.

5–12 PLAN-PROFILES

A plan-profile is a combination drawing that includes a plan with contours and a vertical section called a profile. A plan-profile is used to show a drainage system underground from manhole 1 to manhole 3 in Figs. 5–31 and 5–32.

The profile section is drawn with an exaggerated vertical scale to emphasize the variations in the surface of the earth and the grade of the pipe. Although the vertical scale is usually exaggerated, it need not be; it could be drawn using the same scale as is used in the plan.

Manhole 1 is projected to the profile section, but the other points are not orthographic projections (Fig. 5–31). Distances H_1 and H_2 are transferred to the profile with dividers so that the horizontal distances will be true length

FIGURE 5–30. VERTICAL SECTIONS

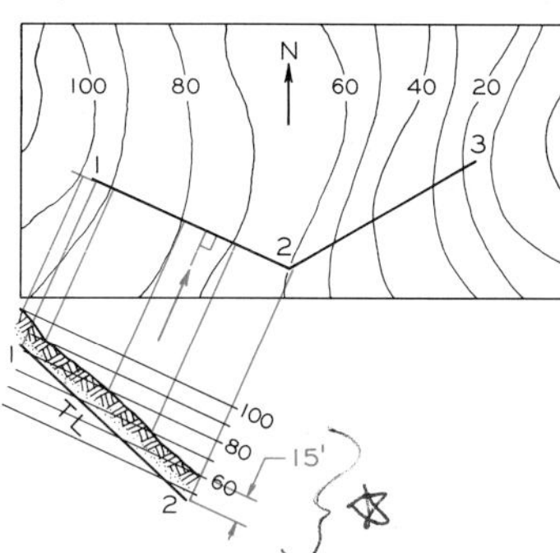

Step 1: To locate pipes 15 feet under the surface from point 1 to 2 to 3, vertical sections are passed through the pipes in the plan view. A profile is taken perpendicularly from pipe 1–2 to find the surface of the ground, the ends of the pipe 15 feet under the surface, and pipe 1–2 true length.

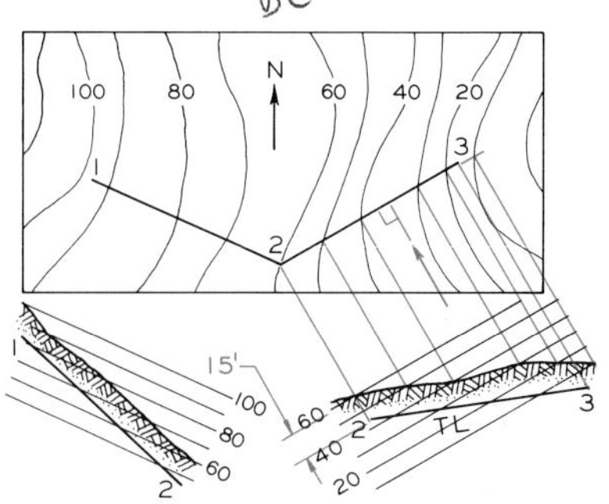

Step 2: A second profile is projected perpendicularly from the plan view of pipe 2–3 to find the surface and pipe 2–3 true length. Notice that the cutting plane is extended beyond point 3 in the top view to provide a more descriptive profile section.

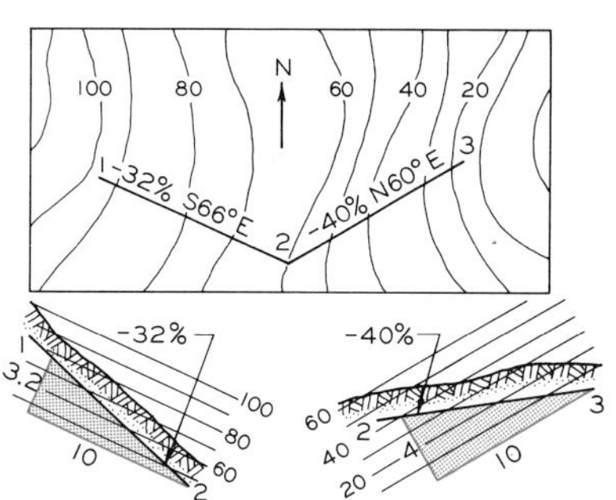

Step 3: The percent grades of pipes 1–2 and 2–3 are found by construction in the profiles. These are negative since the pipes run downhill from 1 to 3. The percent grades and bearings are used to label the top view of the pipes. The elevations of each point can be measured in the profile sections.

FIGURE 5–31. PLAN-PROFILE

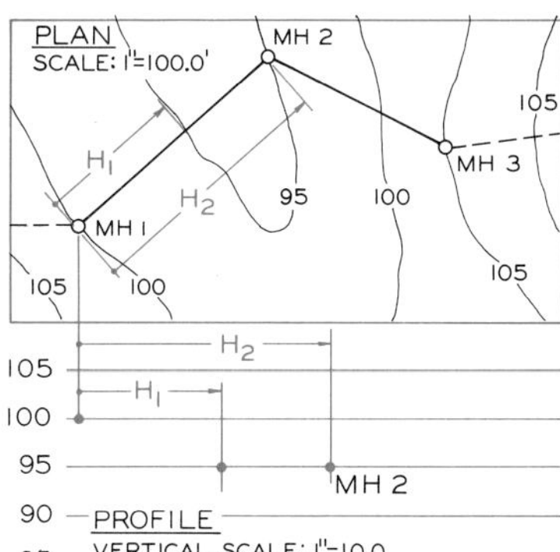

Given: The plan-profile shown.

Required: Find the profile of the earth over the underground drainage system.

Step 1: Distances H_1 and H_2 from manhole 1 are transferred to their respective elevations in the profile. This is not an orthographic projection.

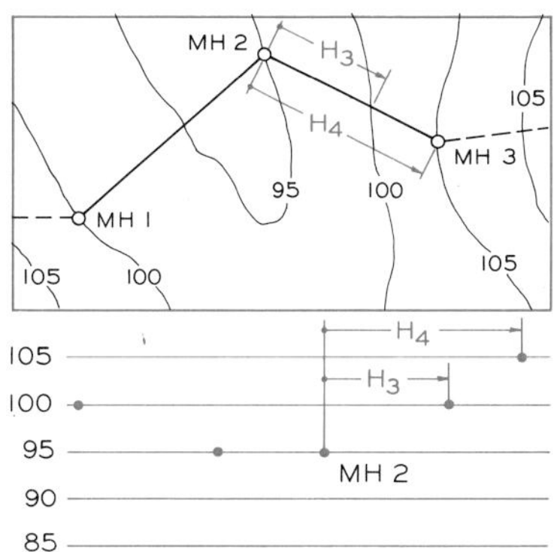

Step 2: Distances H_3 and H_4 are measured from manhole 2 in the plan and are transferred to their respective elevations in the profile. These points represent elevations of points on the earth above the pipe.

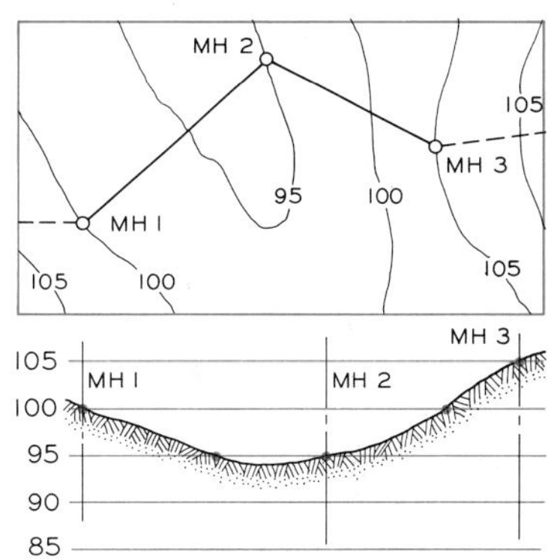

Step 3: The five points are connected with a freehand line and the drawing is crosshatched using the symbol given in Chapter 3 to represent the earth's surface. Center lines are drawn to show the locations of the three manholes that will be located in Fig. 5–32.

in this view. The points where the contour lines cross the pipe in the top view are transferred to their respective elevations in the profile. All these points are connected to give the profile of the earth over the underground pipe and the location of the manhole center lines.

It is easy to calculate the vertical drop from one end to the other of a pipe, since the slope is given in percent grade (Fig. 5–32). The vertical drop from manhole 1 to manhole 2 is found to be 5.20′ by multiplying the horizontal distance of 260.00′ by −2.00% grade. This drop is subtracted from the depth of manhole 1 to find the elevation of manhole 2 to be 89.80′.

Since the pipes intersect at manhole 2 at an angle, the flow of the drainage is disrupted at the turn; consequently a drop of 0.20′ is given from the inlet to the outlet across the floor of

FIGURE 5–32. PLAN-PROFILE, MANHOLE LOCATION

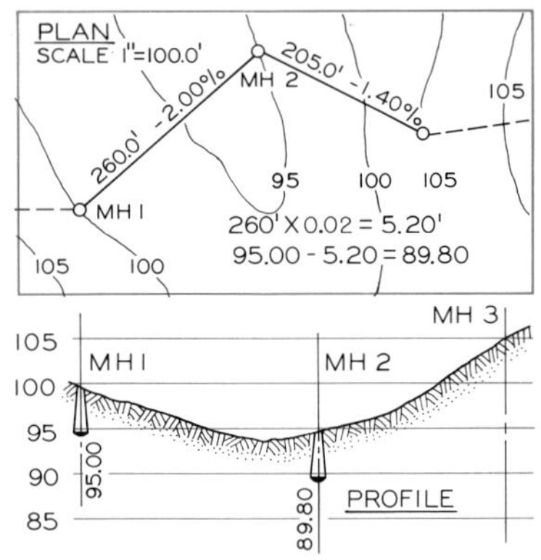

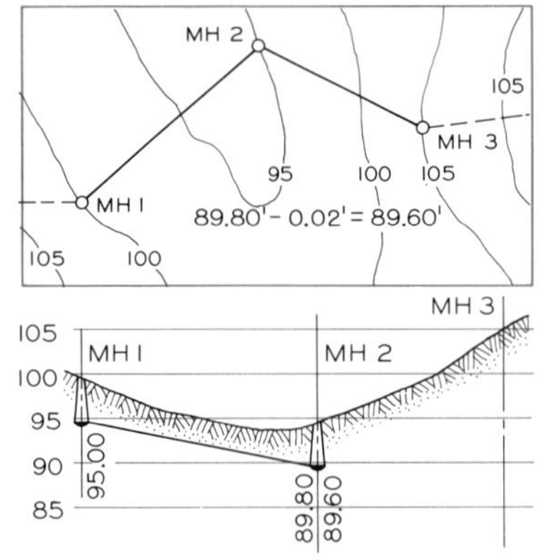

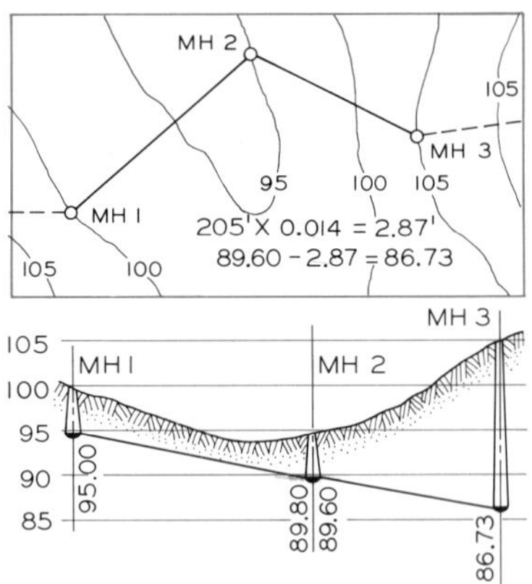

Given: The plan-profile and pipe specifications for a drainage system.

Required: Locate the manholes and give their elevations.

Step 1: The horizontal distances from the manholes and the percent grade of flow lines of the pipes are given on the plan. The elevation of the bottom of manhole 2 is calculated by subtracting from the given elevation of manhole 1.

Step 2: The lower side of manhole 2 is 0.20′ lower than the inlet side to compensate for loss of head (pressure) due to the turn in the pipeline. The lower side is found to be 89.60′ and is labeled.

Step 3: The elevation of manhole 3 is calculated to be 86.73′ since the grade is 1.40% from manhole 2 to manhole 3. The flow line of the pipeline is drawn from manhole to manhole and the elevations are labeled. This profile shows the relationship of the pipe to the surface above it.

FIGURE 5–33. EDGE VIEW OF A PLANE

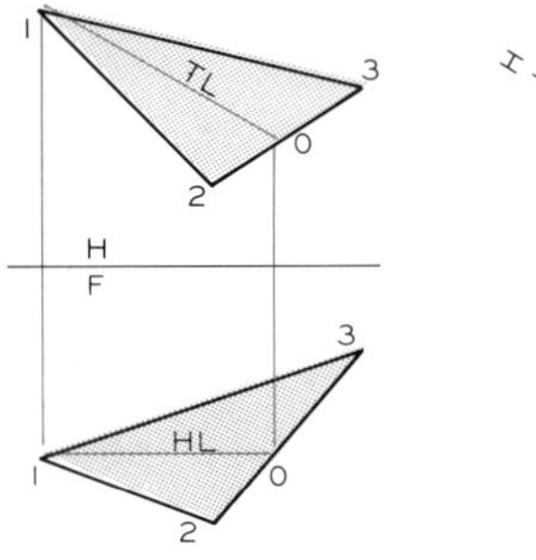

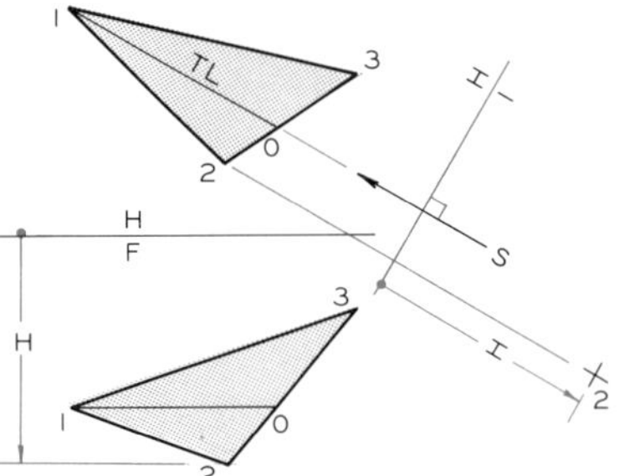

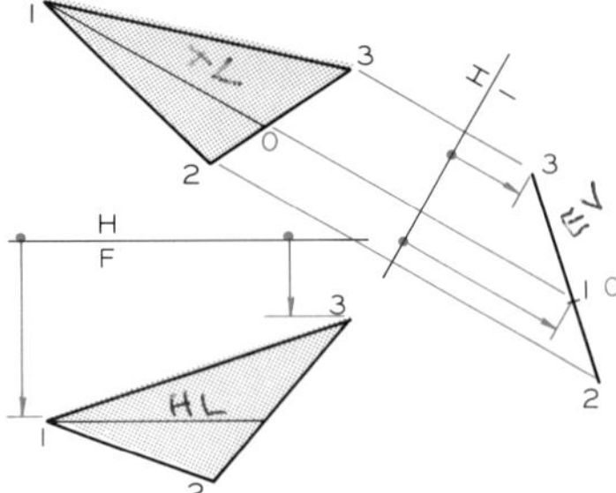

Step 1: To find plane 1–2–3 as an edge, horizontal line 1–0 is drawn on the plane in the front view. Line 1–0 is projected to the top view where it is true length.

Step 2: A line of sight is constructed parallel to the true-length line 1–0. Reference line H–1 is drawn perpendicular to the line of sight. Point 2 is found in the auxiliary view by transferring dimension *H* from the front view.

Step 3: Points 1 and 3 are found by transferring their height dimensions from the front view. These points will lie in a straight line which is the edge of the plane. Line 1–0 will appear as a point in this view.

the manhole to compensate for the loss of pressure (head) through the manhole. The elevations on both sides of the manhole are specified as shown in step 2 of Fig. 5–32. This system of calculation is used successively from manhole to manhole.

The true lengths of the pipes cannot be measured graphically in the profile section when the vertical scale is different from the horizontal scale. True length can be found by trigonometry.

5–13 EDGE VIEW OF A PLANE

The edge view of a plane can be found in any primary auxiliary view by applying previously covered principles. A plane will appear as an edge in any view where any line on the plane appears as a point (Fig. 5–33).

A true-length line can be constructed on any plane by drawing the line parallel to one of the principal planes and projecting it to the adjacent view, as shown in step 1, where a horizontal line is drawn. Since line 1–*O* is true length in the top view, its point view may be found as shown in Fig. 5–10. The remainder of the plane will appear as an edge in this view.

5–14 ANGLE BETWEEN TWO PLANES

It is often necessary for the angle between two planes, which is called a *dihedral angle*, to be found in order for a design to be refined. Perhaps a connecting bracket must be designed to assemble planes in a desired position based on the angle between them, as in Fig. 5–34, where the angular planes of a control tower must be joined with an acceptable bracket.

The angle between two planes can be measured in the view where their line of intersection appears as a point.

The angle between planes 1–2–3 and 1–2–4 in Fig. 5–35 can be found in a primary auxiliary view, since the line of intersection 1–2 is true length in the top view. Auxiliary line H–1 is drawn so that it is perpendicular to the direction of the top view of 1–2. A view is

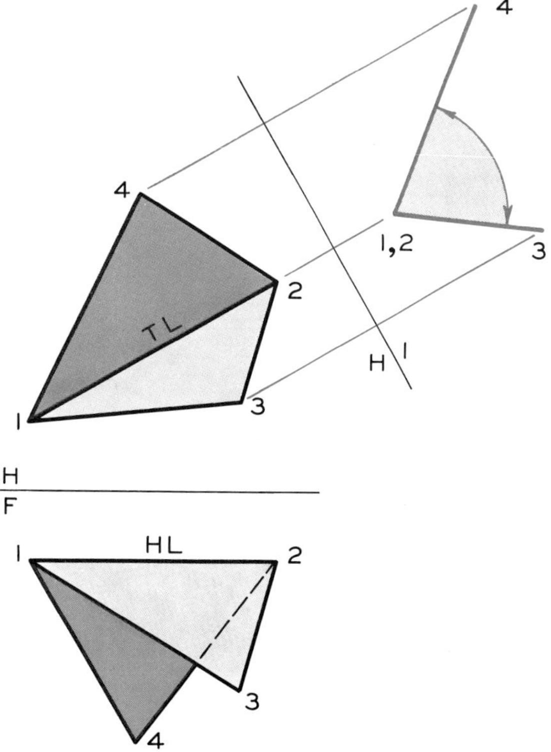

Fig. 5–35. The determination of the angle between two planes. This is called a *dihedral angle*.

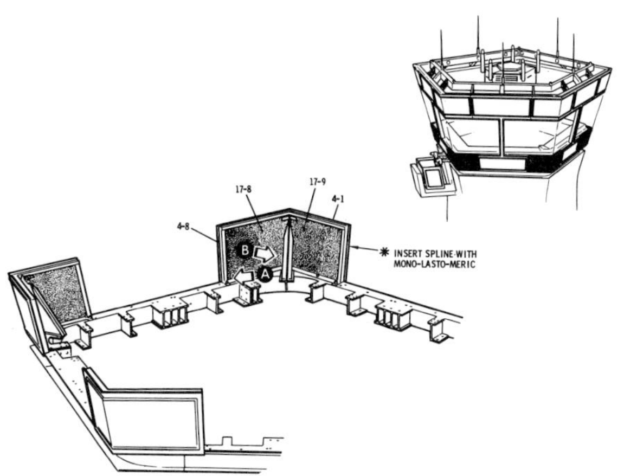

Fig. 5–34. The angles between the corner planes of this control tower had to be determined in order to design a connecting bracket. (Courtesy of the Federal Aviation Administration.)

Fig. 5–36. The angles between planes of the basic structure of this Comsat satellite were determined prior to the design of a system for fabricating the joints. (Courtesy of TRW Systems.)

FIGURE 5–37. PIERCING POINT OF A LINE THROUGH A PLANE—PROJECTION METHOD

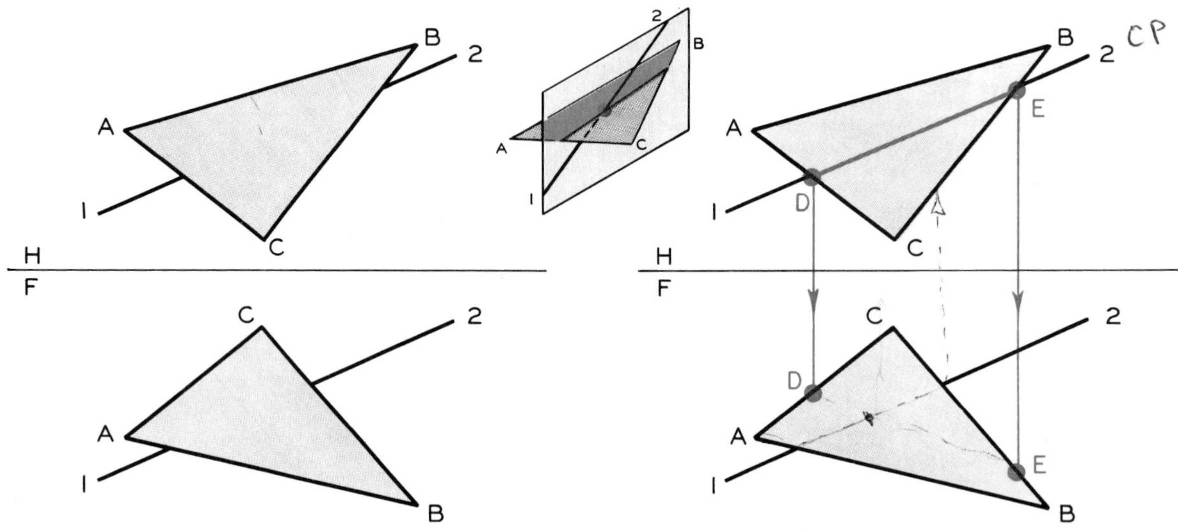

Given: The top and front views of plane *ABC* and line 1–2.

Required: Determine the piercing point of line 1–2 on the plane and the visibility of both views by the projection method.

References: Articles 4–12 and 5–15.

Step 1: Assume that a vertical cutting plane is passed through the top view of line 1–2. The plane intersects *AC* and *BC* at points *D* and *E*. Project points *D* and *E* to the front view.

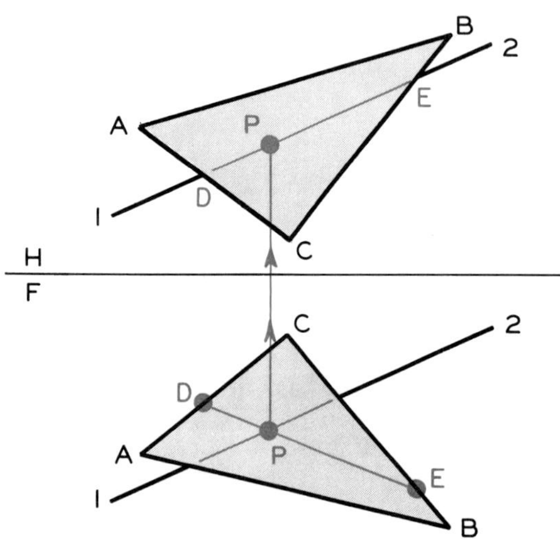

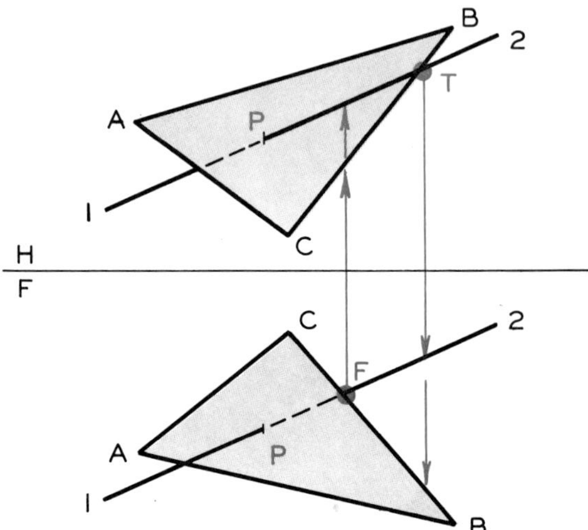

Step 2: Line *DE* represents the trace of the line of intersection between the imaginary vertical cutting plane and plane *ABC*. Any line that lies in the cutting plane and intersects plane *ABC* will intersect along line *DE*. Line 1–2 lies in the plane; therefore it intersects *ABC* at point *P* in the front view. Project point *P* to the top view.

Step 3: The visibility of line 1–2 in the front view is determined by analyzing point *F* where *P–2* and *BC* cross. By projecting this point to the top view, we see that *BC* is in front of *P–2;* therefore *BC* is visible in the front view. The top-view visibility is determined by analyzing point *T* in the same manner; we find that *P–2* is higher than *BC* in the front view and is therefore visible in the top view.

FIGURE 5–38. PIERCING POINT OF A LINE THROUGH A PLANE—AUXILIARY METHOD

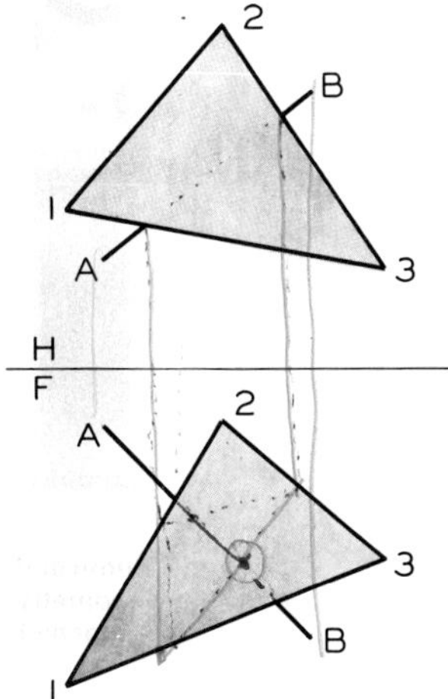

Given: The top and front views of plane 1–2–3 and line *AB*.

Required: Find the piercing point of line *AB* on plane 1–2–3 and the visibility in both views by the auxiliary-view method.

Reference: Article 5–16.

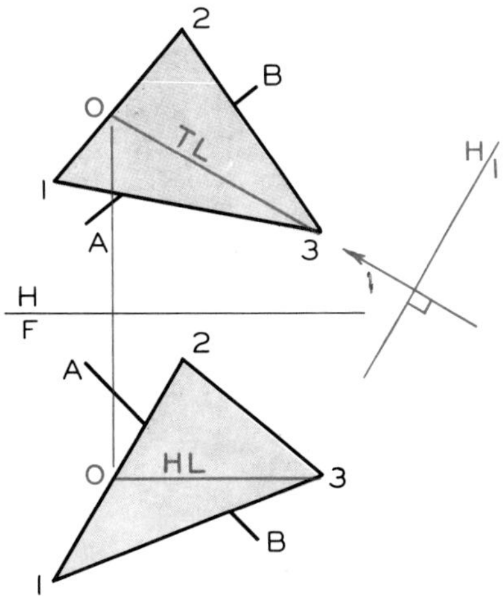

Step 1: Draw horizontal line *O*–3 in the front view and project it to the top view. Line *O*–3 projects true length in the top view. Establish the line of sight for the primary auxiliary view parallel to *O*–3 in the top view. Reference line H–1 is drawn perpendicular to the line of sight and *O*–3.

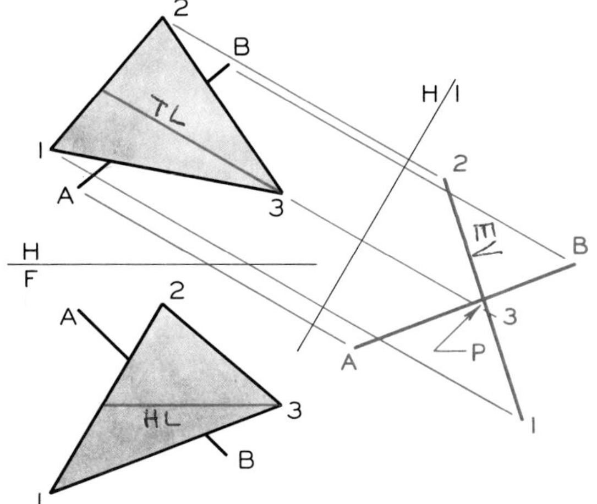

Step 2: Find the edge view of plane 1–2–3 by finding the point view of line *O*–3. Project line *AB* also. Point *P* in the auxiliary view is the piercing point of line *AB* on plane 1–2–3.

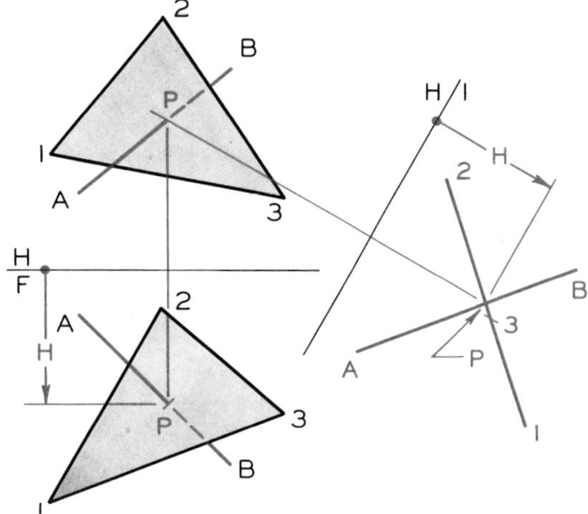

Step 3: Point *P* is projected to the top and front views in sequence. The front view of *P* can be checked by transferring distance *H* from the auxiliary view to the front view. Point *A* is closer to the H–1 line in the auxiliary view; therefore, line *AP* is higher than the plane and is visible in the top view. Visibility in the front view is found by the method used in Fig. 5–37.

projected to the auxiliary plane where the point view of line 1–2 is found. The edge views of both planes are found in this view, since line 1–2 is a line common to both planes. The dihedral angle is measured in this view.

The satellite pictured in Fig. 5–36 illustrates intersecting planes for which the determination of the dihedral angles was required. These angles affect the inner structural members and the methods of connecting the planes.

5–15 PIERCING POINT OF A LINE WITH A PLANE BY PROJECTION

The location of a point where a line intersects a plane is necessary to the design of many engineering projects. The line may represent a structural member that must be attached to an oblique plane, or a cable that must have clearance through a plane of an enclosure.

Figure 5–37 gives the sequential steps necessary for the determination of the piercing point of a line passing through a plane by the application of projection principles similar to those covered in Chapter 4. The visibility of the line can be found by the application of the principles covered in Article 4–12. If a line intersects a plane it will be visible on one side of the piercing point and hidden on the other side.

5–16 PIERCING POINT OF A LINE WITH A PLANE BY AUXILIARY VIEW

An alternative method for finding the piercing point of a line intersecting a plane is the auxiliary view method illustrated in Fig. 5–38. The edge view of the plane is found by projecting from either view into the primary auxiliary view. The piercing point is found to be the point where the line and edge view of the plane intersect. Piercing point P is found in the principal views by projecting from the auxiliary view back to line AB in the top view and then to the front view. The front view of point P will also lie on line AB. However, in many cases it is helpful to check the location for greater accuracy by transferring the distance H from the auxiliary view to the front view. This is especially necessary when the front view of the piercing line approaches being a vertical line. Visibility is easily determined in the view from which the primary auxiliary was projected. The portion of the line on the upper side of the plane in the primary auxiliary is visible in the principal view—line AP in this case.

Fig. 5–39. The piercing point between the steering column and the firewall could be designed using descriptive geometry principles. (Courtesy of LeTourneau-Westinghouse Company.)

The piercing point between the steering column and the inclined fireball of the truck shown in Fig. 5–39 can be found by applying this principle of descriptive geometry. The accurate location of this point is necessary to the function of the steering linkage.

5–17 LINE PERPENDICULAR TO A PLANE

Economy of design dictates that materials be reduced to a minimum. If a structural member were to be attached to an oblique plane, it would be more economical if the member were designed to be perpendicular to the plane, thus giving the shortest distance and requiring the minimum of materials.

Figure 5–40 is an example requiring that the shortest distance from a point to a plane be found. The edge view of the plane is found by

FIGURE 5–40. LINE PERPENDICULAR TO A PLANE

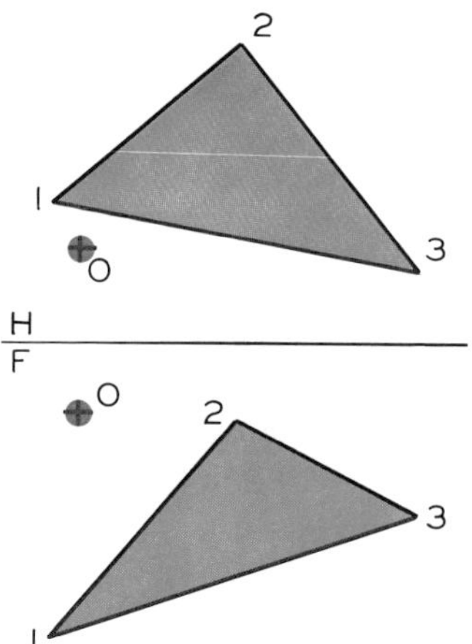

Given: Plane 1–2–3 and point *O* in two views.
Required: Find the shortest distance from point *O* to the plane 1–2–3 and show it in all views.
References: Articles 4–20 and 5–17.

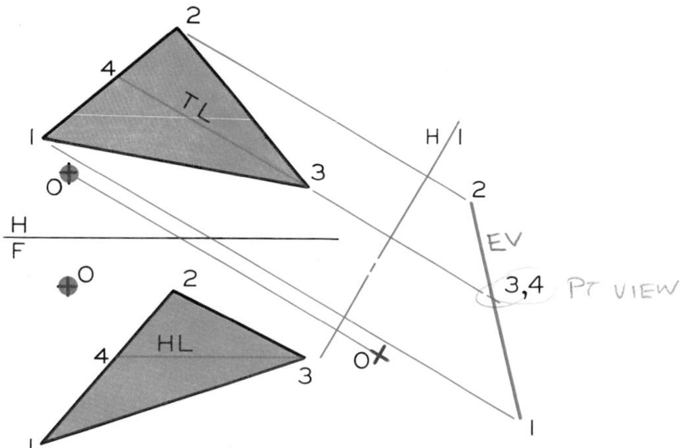

Step 1: Draw horizontal line 3–4 in the front view of the plane 1–2–3. This line will be projected true length in the top view. Plane 1–2–3 will be projected as an edge in the auxiliary view where 3–4 appears as a point. Project point *O* to this view also.

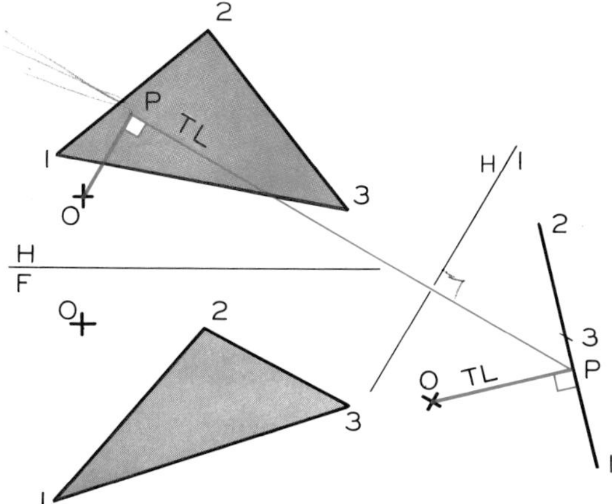

Step 2: Line *O–P* is drawn perpendicular to the edge view of plane 1–2–3, since the shortest distance is perpendicular to the plane. Because line *O–P* is true length in the auxiliary view, the top view of the line must be parallel to the H–1 reference line. It is also perpendicular to the direction of a true-length line in the top view.

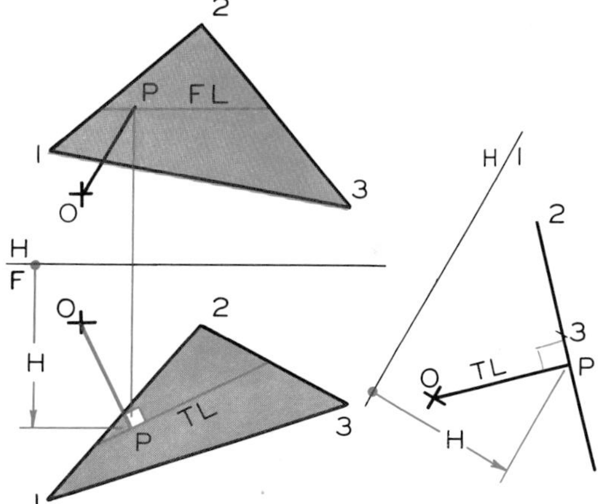

Step 3: The front view of line *O–P* is found by projecting point *P* to the front and locating it *H* distance from the H–F reference line by transferring the *H* distance from the auxiliary view. Line *O–P* is visible in all views. Line *O–P* is also perpendicular to a true-length frontal line in the front view.

an auxiliary view. The shortest distance (perpendicular distance) will be shown true length in this view when drawn from point *O* perpendicular to the edge view of the plane. The piercing point is projected to the top view to intersect with a line drawn from point *O* parallel to the H–1 reference line. Line *O–P* must be parallel to the H–1 line in the top view, since it is true length in the auxiliary view. Line *O–P* is also perpendicular to the true-length lines in the top view, since perpendicular lines will project as perpendicular when one or both of two perpendiculars are true length, as covered in Article 4–20.

Fig. 5–41. The solar panels of this satellite are designed to remain perpendicular to the sun's rays to take full advantage of its solar energy. (Courtesy of Ryan Aeronautics, Inc.)

A problem involving this principle is shown in Fig. 5–41. The solar panels are attached to the satellite in such a manner that they can rotate to take full advantage of the available solar energy.

5–18 INTERSECTION BETWEEN PLANES BY PROJECTION

The line of intersection between two intersecting planes can be found by applying the principles covered in Article 5–15. The line of intersection is found by locating the piercing points of two lines in a plane on the other plane. This procedure is shown by steps in Fig. 5–42. It can be determined by observation that lines 1–2, 2-3, and *AB* do not pierce either of the planes, since they fall outside of the planes in the given views. Therefore, lines *AB* and *BC* are selected to be lines which have a probability of intersecting plane 1–2–3. Line *AC* is analyzed as though it were a single line rather than a line on a plane. A vertical cutting plane is passed through the line in the top view. The piercing point *P* is found and projected to both views, and line *BC* is analyzed in the same manner for piercing point *T*. *Note*: It is necessary to work with one plane, instead of finding the piercing point of one line of a plane and then skipping to a line on the other plane.

If the piercing points of two lines on a plane are found, then all lines in the plane that pierce the other plane must intersect along a line connecting the two piercing points. Points *P* and *T* are connected to form the line of intersection. The visibility is determined by analyzing the points where the lines of each cross. Visibility analysis is covered in Article 4–12.

5–19 INTERSECTION BETWEEN PLANES—AUXILIARY VIEW METHOD

An alternative method of finding the line of intersection between two intersecting planes is the auxiliary view method, illustrated in Fig. 5–43. An edge view of either of the planes is found in step 1 by a primary auxiliary view, with the other plane appearing foreshortened. Piercing points *L* and *M* are projected from the auxiliary view to their respective lines, 5–6 and 4–6, in the top view of step 2. The visibility of plane 4–5–6 in the top view is apparent in step 3 by inspection of the auxiliary view, where sight line S_1 has an unobstructed view of the 4–5–*L*–*M* portion of the plane. Plane 4–5–*L*–*M* is visible in the front view, since sight line S_2 has an unobstructed view of the top view of this portion of the plane.

5–20 SLOPE OF A PLANE

Planes can be established in space by verbal specification of the *slope* and the *direction of slope* of the plane, as defined below.

Slope. The slope of a plane is the angle it makes with the horizontal plane.

Direction of Slope. The direction of slope is the compass bearing of a line which is perpen-

FIGURE 5–42. INTERSECTION OF PLANES BY PROJECTION

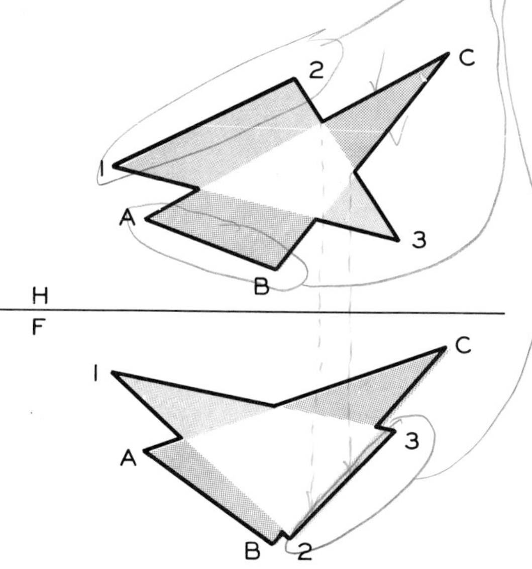

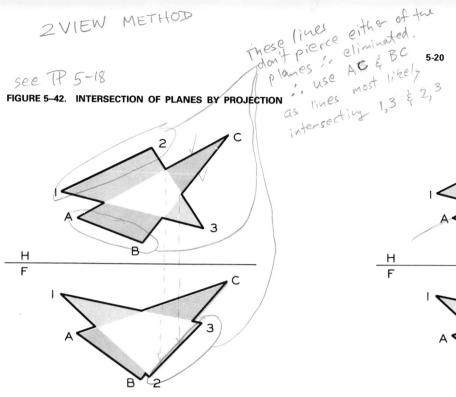

Given: The top and front views of planes 1–2–3 and *ABC*.

Required: Find the line of intersection between the planes and determine the visibility in both views by projection.

References: Articles 5–15 and 5–18.

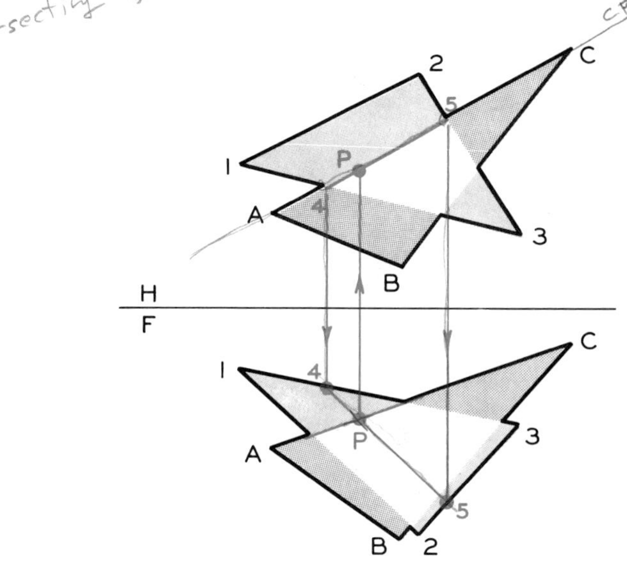

Step 1: Pass a vertical cutting plane through line *AC* in the top view to establish points 4 and 5. Project points 4 and 5 to lines 1–3 and 2–3 in the front view. Line *AC* pierces plane 1–2–3 where it crosses line 4–5. Project point *P* to *AC* in the top view.

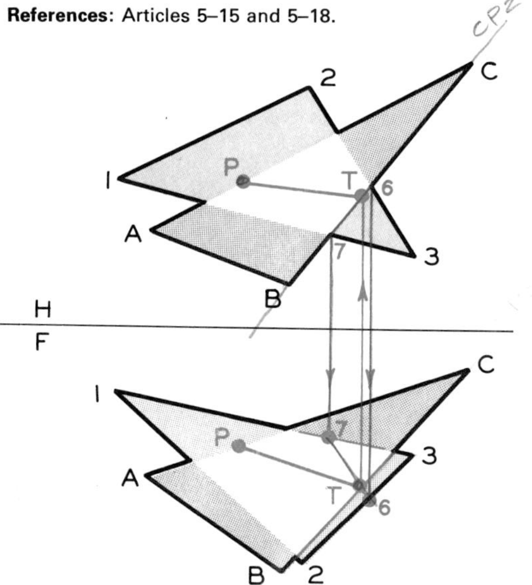

Step 2: Pass a vertical cutting plane through line *BC* in the top view to establish points 6 and 7. Project line 6–7 to the front view. Point *T* is the piercing point of line *BC* in plane 1–2–3. Line *PT* is the line of intersection between the planes. The piercing points of *AC* and *BC* are found as though they were independent lines rather than lines of a plane.

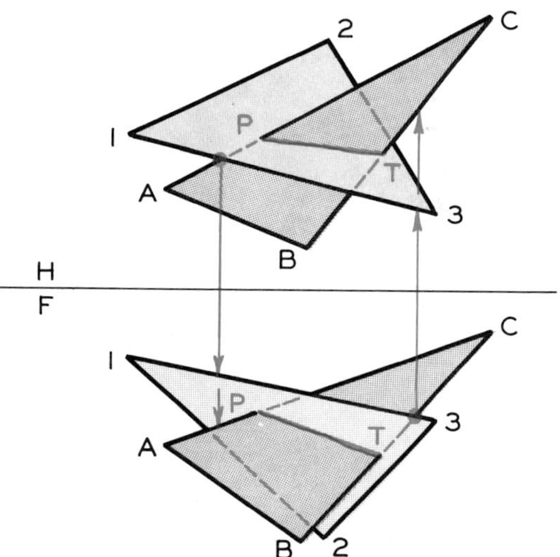

Step 3: Analyze the intersection of *AP* and 1–3 in the top view for visibility by projecting to the front view, where 1–3 is found to be higher and, consequently, visible in the top view. Line *PC* is also visible in the top view; therefore, *PCT* is visible. Frontal visibility is found by the analysis of the intersection of *CT* and 1–3, where 1–3 is in front and visible.

FIGURE 5–43. INTERSECTION OF TWO PLANES BY AUXILIARY VIEW

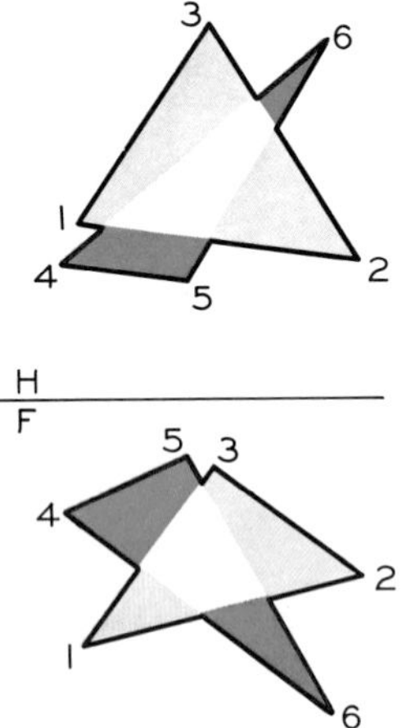

Given: The top and front views of planes 1–2–3 and 4–5–6.

Required: Find the line of intersection between the planes and determine the visibility in both views by the auxiliary view method.

References: Articles 4–12, 5–16, and 5–19.

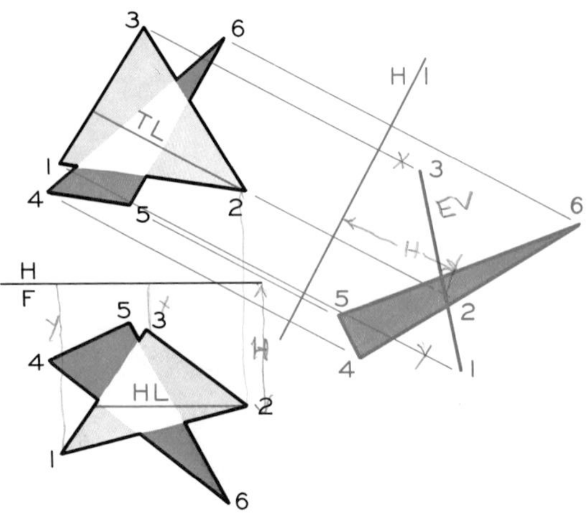

Step 1: Draw a horizontal line in plane 1–2–3 and project it to the top view where the line is true length. Find the edge view of plane 1–2–3 by finding the point view of the true-length line. Project plane 4–5–6 to the auxiliary view also.

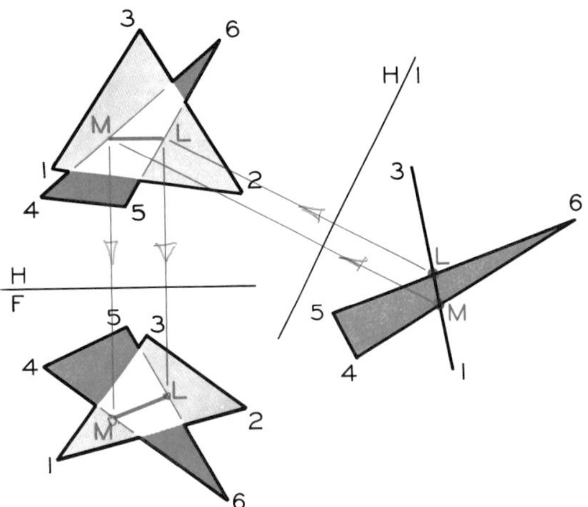

Step 2: Points *L* and *M* in the auxiliary view are the points of intersection of lines 5–6 and 4–6. These points are projected to the top and front views to give the line of intersection *LM*.

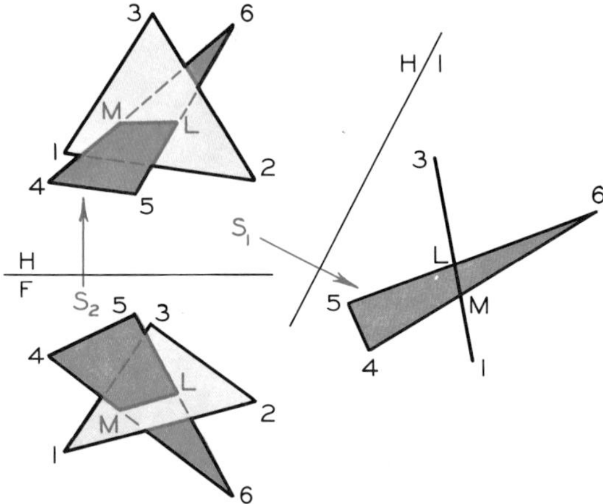

Step 3: Visibility in the top view is found by viewing the auxiliary view in the direction of S_1, where plane 4–5–*L*–*M* is seen to be above plane 1–2–3 and is visible in the top view. Frontal visibility is found by viewing the top view in the direction of S_2, where 4–5 is in front of 1–3 and is therefore visible in the front view.

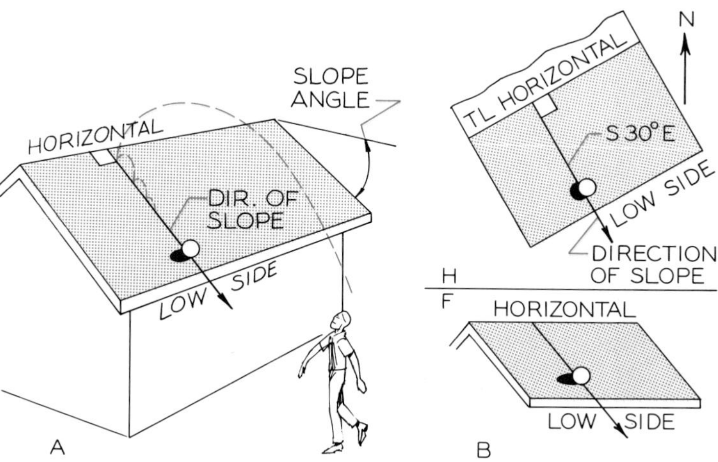

Fig. 5–44. The direction of slope of a plane is the bearing of a line perpendicular to a true-length line in the plane in the top view. This is shown pictorially in A and orthographically in B. Bearing is expressed as a compass direction toward the low side.

FIGURE 5–45. DETERMINATION OF SLOPE AND DIRECTION OF SLOPE OF A PLANE

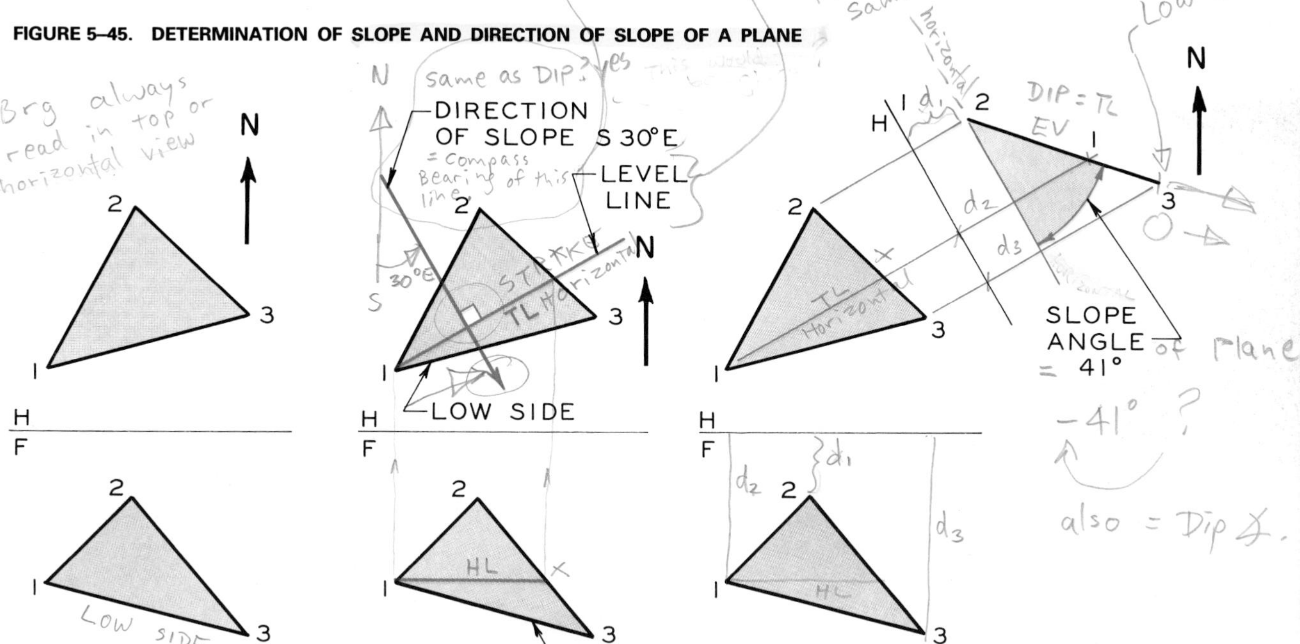

Required: Find the slope and the direction of slope of plane 1–2–3.

Step 1: A horizontal line is drawn in the front view and projected to the top view where it is a true-length line. The direction of slope is the compass bearing of a line perpendicular to the level line in the top view. The arrowhead is placed on the low side toward line 1–3.

Step 2: The slope is found by measuring the angle between the edge view of the plane and the horizontal reference plane in the auxiliary view.

Fig. 5–46. The slopes of inclined surfaces of a dam are constructed from written specifications. (Courtesy of Kaiser Engineers.)

dicular to a true-length line in the top view of a plane taken toward its low side. This is the direction in which a ball would roll on the plane.

These terms are illustrated pictorially in Fig. 5–44. Note that the true angle of the slope is seen when it is viewed parallel to the ridge of the roof (the line of intersection). Since the ridge line is horizontal, a ball will roll perpendicular to it thereby establishing the direction of slope, which is given as a compass bearing.

Figure 5–45 gives the steps involved in determining the slope and direction of slope of the oblique plane 1–2–3.

Following verbal specifications, we can locate plane 1–2–3 in space as a plane passing through point 1 with a slope of 41° and a slope direction of S 30° E. This information is not sufficient to determine the limits of the plane, but merely establishes an infinite plane of which plane 1–2–3 is a part. The slope of the dam in Fig. 5–46 is an example of a plane that can be established verbally in written specifications by the engineer.

5–21 CUT AND FILL OF A LEVEL ROADWAY

A level roadway routed through irregular terrain, such as the one pictured in Fig. 5–47, must cut through existing embankments in many locations. Also, volumes of fill must be provided to support the road at low points. It is more economical in most cases if the amount of fill is about equal to the cut volume in order that the earth removed by the cut can be transferred to the low area and used as fill. This problem

Fig. 5–47. This mountain road was constructed by cutting and filling volumes of the irregular hillside. (Courtesy of the Colorado Department of Highways.)

FIGURE 5–48. CUT AND FILL OF A LEVEL ROADWAY

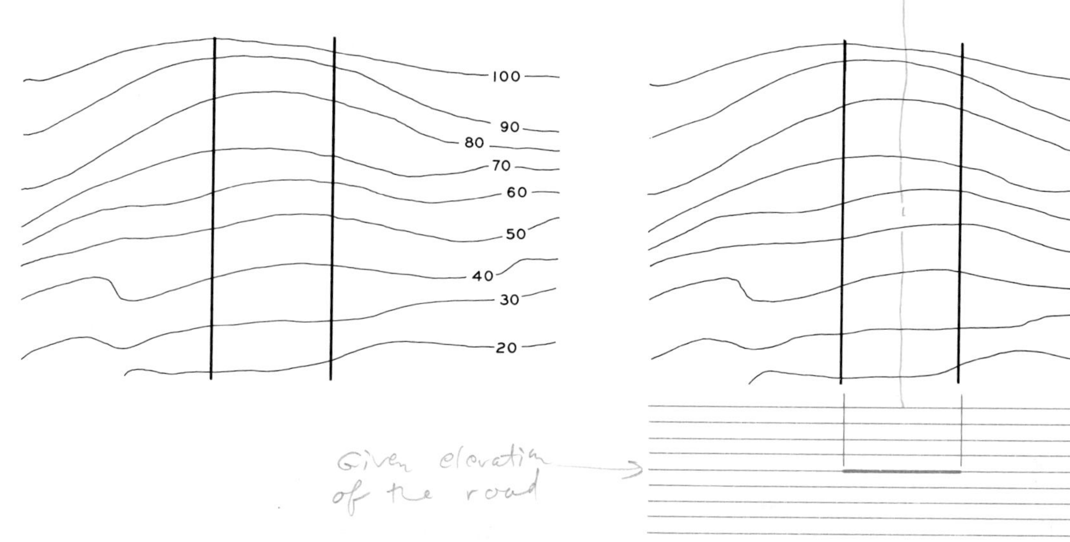

Given: A contour map with a level roadway given at an elevation of 60′.

Required: Find the top view given that the roadway is to have a 45° fill angle and a 30° cut angle.

Reference: Article 5–21.

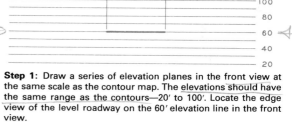

Step 1: Draw a series of elevation planes in the front view at the same scale as the contour map. The elevations should have the same range as the contours—20′ to 100′. Locate the edge view of the level roadway on the 60′ elevation line in the front view.

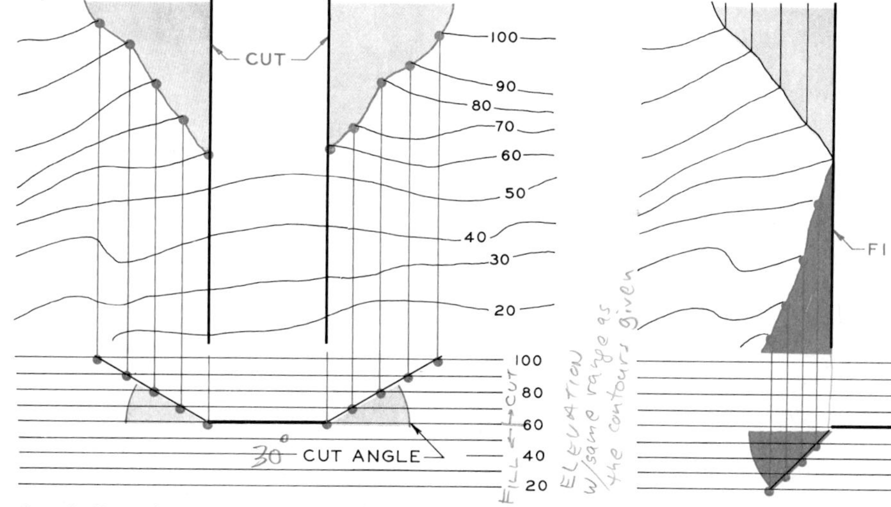

Step 2: Draw the cut angle of 30° with the horizontal on each side of the road in the upper portion of the front view. Project the points where the cut planes intersect the elevation lines to their respective contour lines in the top view. (*Example:* The points on the 100′ elevations in the front view are projected to the 100′ contour lines in the top view.) Connect these successive points to determine the cut area.

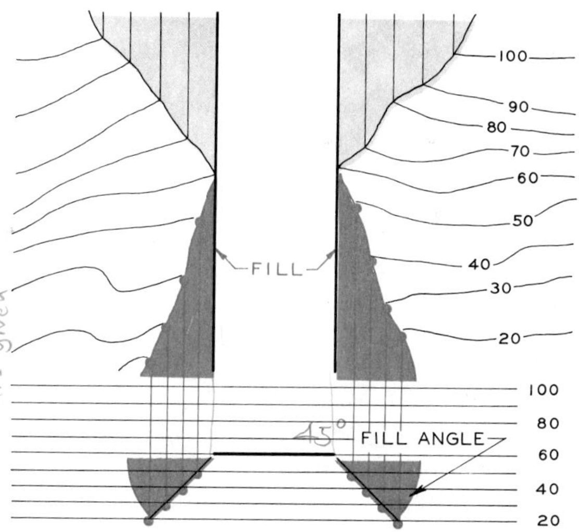

Step 3: The fill angles of 45° are drawn on each side of the roadway in the lower portion of the front view. The points on the plane of the fill are projected to the top view in the same manner that was used for the cut planes. The fill area is indicated by connecting the points. Note that the contour lines have been changed in the cut-and-fill areas to indicate the new contour of the land following construction.

lends itself to graphical solution by application of primary auxiliary views.

The steps for solution of a cut-and-fill problem are given in Fig. 5–48. A contour map, the route of the level roadway to be constructed, and the angles of cut and fill are given. The contour lines located in the front view of step 1 are spaced 10′ apart, using the same scale that was used on the contour map.

The angle of cut is drawn on the upper side of the roadway, toward the higher elevations, by measuring the angle of cut with the horizontal plane on each side of the road. The fill angle is on the low side, or toward the contour lines of the smaller elevations. The top views of these planes are found in steps 2 and 3 by projecting points that lie on the cut-and-fill planes at each elevation line in the profile view to their respective contour lines in the top view.

The volume of earth involved in each area can be approximated by passing a series of vertical cutting planes through the top view to determine several profiles where the cut-and-fill planes will project as edges in the front view. The areas of cut and fill can be averaged in these views and multiplied by the linear distance in the top view to give the volume of cut and fill.

The cut-and-fill areas found in the map view should be crosshatched or shaded by some method to indicate the solution. The shading or crosshatching used to represent the cut should be different from that used to indicate the fill area, and each symbol should be identified on the drawing.

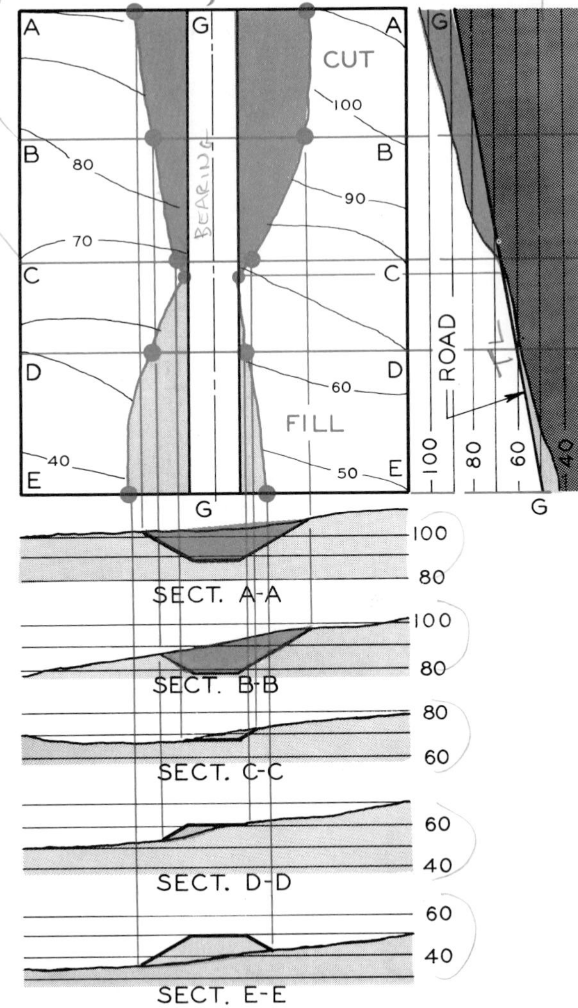

Fig. 5–49. Determining the cut and fill of a roadway on a grade.

5–22 CUT AND FILL OF A ROADWAY ON A GRADE

A highway cannot always be level; in many cases it will be constructed on a grade through irregular terrain. The problem given in Fig. 5–49 is an example of a road which required cut and fill during its construction on a grade. The following steps explain how the completed contour map view can be found and how an estimate can be made of the earth that must be cut and filled.

Step 1. A profile section, G–G, is taken through the center of the highway in order to show the true length and grade of the highway and its relationship to the terrain. It can be seen in profile G–G that a portion of the highway is beneath the terrain, therefore requiring that a cut be made to route the road at the specified grade. Fill is necessary where the level of the highway is above the earth's surface.

Step 2. The top view of the cut-and-fill areas can be approximated by constructing a series of vertical sections through the map view (top view of the terrain). These cutting planes, A–A, B–B, C–C, D–D, and E–E, are drawn perpendicular to the center line of the highway. The sections cut by these planes appear as profiles in the front view.

Step 3. Each of the sections formed by the cutting planes is drawn as a profile in the front view by constructing horizontal elevation planes in these views to correspond to the contour lines in the map view.

Step 4. Each of the profiles corresponding to the cutting planes is found by projecting to the profile section the intersection of of the cutting plane with each contour line in the top view. For example, Section B–B is drawn by projecting points on the 80′, 90′, and 100′ elevation contours to the profile to find the surface of the earth. The elevation of the highway at this section can be found in the profile by transferring the highway elevation found in profile G–G to profile B–B. The angles of cut and fill are constructed in each profile section by measuring their angle with the horizontal plane.

Step 5. The top view of the area of cut and fill is found by projecting the extreme points where the cut planes pierce the surface of the earth in each profile to the top view of the cutting plane used to form the profile. When these points are connected, they establish the limits of the cut and fill, as labeled in the map view.

The spacing of the cutting planes in the top view must be determined by judgment and is based on the characteristics of the terrain. The closer the sections are to each other, the more accurate the succeeding constructions and estimations of the volume of cut and fill will be. Note that the volume of fill can be determined by averaging the fill areas in profiles can be determined by averaging the fill areas in profiles C–C, D–D, and E–E, and multiplying this average cross section by the length of the road in the top view from section C–C to section E–E.

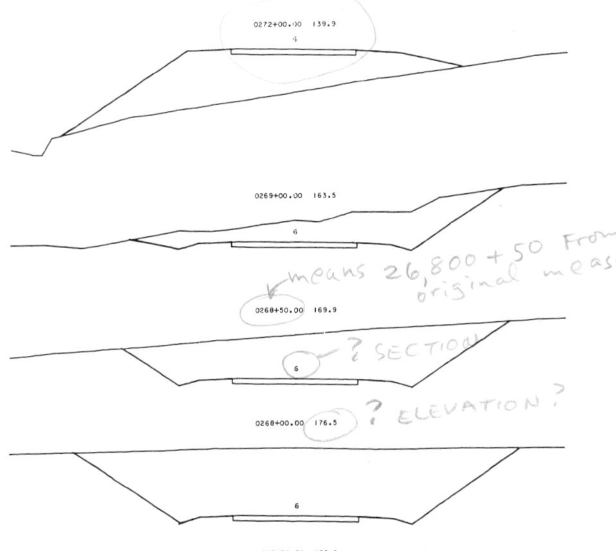

Fig. 5–50. The cut-and-fill areas of this roadway were plotted by a computer. Note that station points are given as 0268 + 50.00, for example. This is the equivalent to 26,850 feet from the origin of measurement. This is the civil engineer's technique of locating measurements by locating stations at 100-foot intervals. Intermediate measurements are made from the nearest 100-foot station. (Courtesy of EAI.)

Fig. 5–51. Field data are transcribed into the language of a computer prior to plotting. (Courtesy of IBM.)

An example solution to this type of problem as plotted by a computer is shown in Fig. 5–50. A series of cross sections are plotted to indicate cut and fill from field data that were fed into a system similar to the one shown in Fig. 5–51. A basic knowledge of the graphical process must be understood to permit the drawings to be interpreted or programmed.

NOT COVERERED

5–23 GRAPHICAL DESIGN OF A DAM

The design of a dam involves many of the same principles used in cut-and-fill problems. The basic definitions of terms associated with dams appear in Fig. 5–52. These are: (1) *crest* (the top of the dam), (2) *water level* (the level when the dam reaches capacity), and (3) *freeboard* (the height of the crest above the water level).

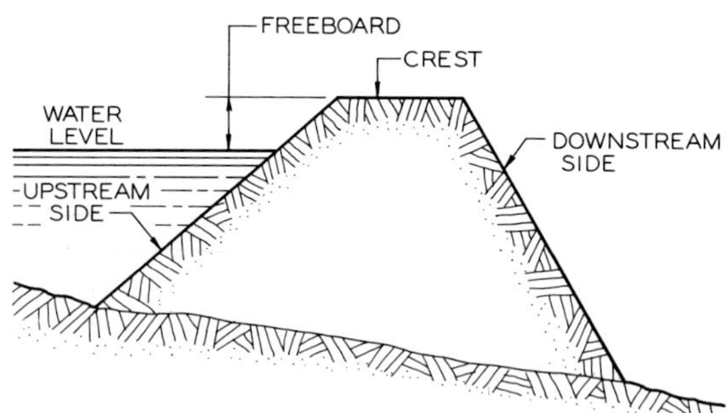

Fig. 5–52. The terms used in the construction of a dam.

FIGURE 5–53. GRAPHICAL DESIGN OF A DAM

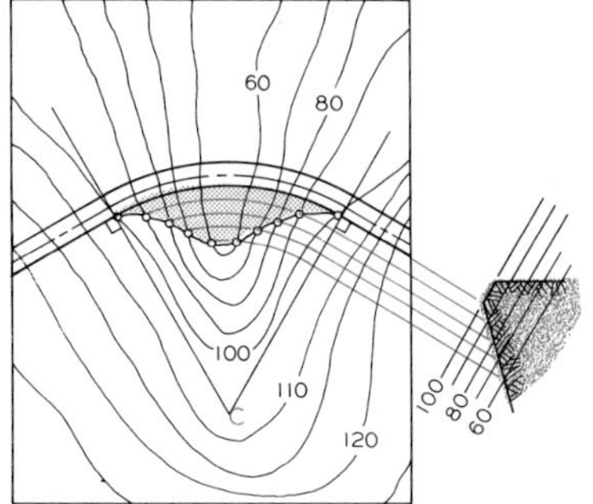

Step 1: A roadway is given at an elevation of 100 feet that is to pass over a dam that is an arc. Points of tangency are found in the top view, and a section is passed through the line from *C* to the point of tangency. The section of the dam is drawn using the given specifications. The upstream view of the dam is found by projecting to the respective contour lines from the section.

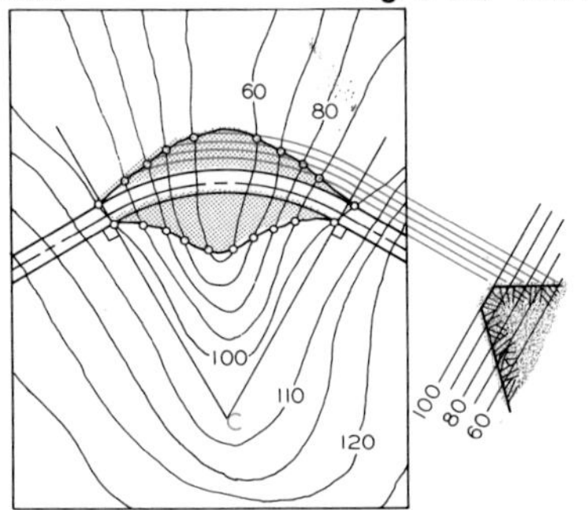

Step 2: The downstream side of the dam is found by projecting to the respective contour lines from the section. The projectors become arcs with center *C* at the curved portion of the dam. The contours in the section must be spaced using the same scale that was used to draw the plan view.

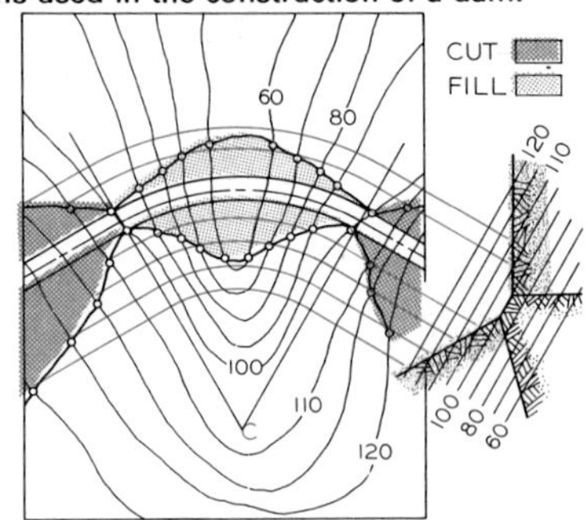

Step 3: The angles of cut are drawn in the section from given specifications. These points are projected from the section to the respective contour lines in the plan view. The projectors are parallel to the sides of the road with a portion of the projectors as arcs of a circle. The cut-and-fill areas are indicated by crosshatching or coloring.

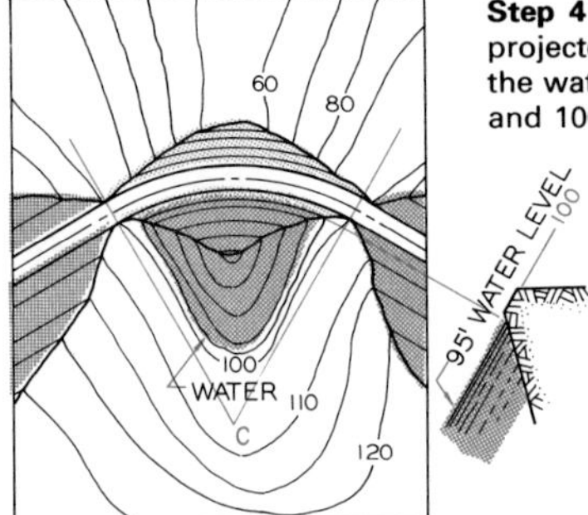

Step 4: If the water level is to be 95 feet, this can be drawn in the section and projected to the plan view to form an arc below the crest of the dam. The limits of the water can be found by assuming the 95-foot contour to be between the 90-foot and 100-foot contours.

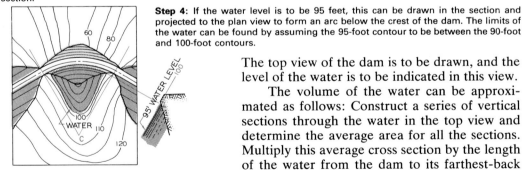

An earthen dam is located on the contour map shown in Fig. 5–53. The angle on each side of the dam and its radius of curvature from center point *C* are given. The top of the dam is to be level to provide a roadway on the surface. The top view of the dam is to be drawn, and the level of the water is to be indicated in this view.

The volume of the water can be approximated as follows: Construct a series of vertical sections through the water in the top view and determine the average area for all the sections. Multiply this average cross section by the length of the water from the dam to its farthest-back point. This method is very similar to the method suggested in the preceding article for estimating volumes of cut and fill.

An application of these principles can easily be related to the Hoover Dam (Fig. 5–54), which was built for water control and power generation during the period from 1931 to

Fig. 5–54. An aerial view of Hoover Dam and Lake Mead, which were built during the period 1931 to 1935. (Courtesy of the Bureau of Reclamation, U.S. Department of the Interior.)

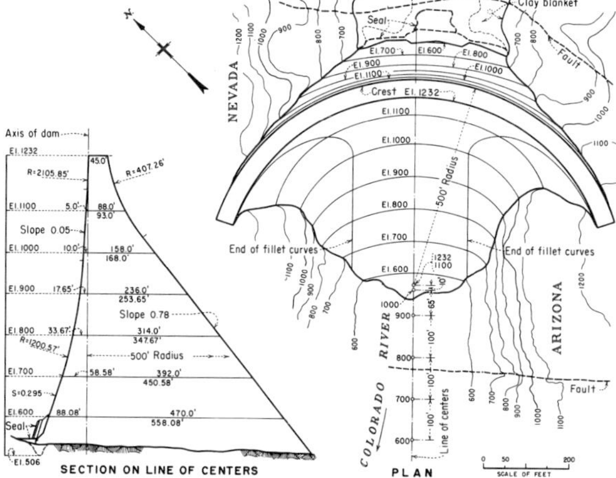

Fig. 5–55. The top view and a sectional view of Hoover Dam. (Courtesy of the Bureau of Reclamation, U.S. Department of the Interior.)

1935. Lake Mead, which is formed by this dam, originally had a capacity of 32,471,000 acre feet, the entire two-year flow of the Colorado River, making it the largest reservoir in the world.

The top view of the dam and a view of a section taken through the center of the dam are shown in Fig. 5–55, which is closely related to the example given in Fig. 5–53. The top view of the dam is built in the shape of an arch to take advantage of the compressive strength of concrete. The section shows that the dam is progressively thicker toward the bottom; this thickness is necessary to withstand the increased pressure at lower depths. The 726′-high structure required a total of 3,250,000 cu yd of concrete for its construction. Hoover Dam remains as one of engineering's wonders of the world, and Lake Mead is the world's largest man-made reservoir.

5-24 STRIKE AND DIP OF A PLANE

Strike and *dip* are terms used in geological engineering and mining to refer to strata of ore under the surface of the earth. It is important in these applications to locate the orientation of the strata by verbal terms that are somewhat similar to *slope* and *direction of slope.*

Strike. Strike is defined as the compass direction of a level line in the top view of a plane. All level lines in a plane are parallel and have the same compass bearing.

Dip. Dip is defined as the angle the edge view of a plane makes with the horizontal plane plus its general compass direction, such as NW or SW. The dip angle is found in the primary auxiliary view that is projected from the top view, and its general direction is measured in the top view. Dip direction is measured perpendicular to a level line in a plane in the top view toward the low inside.

Figure 5–56 illustrates the steps of finding the strike and dip of a given plane *ABC*.

5-25 DISTANCES FROM A POINT TO AN ORE VEIN

In the mining operation shown in Fig. 5–57, coal is being removed from its vein. Descriptive geometry principles can be used to find the most economical distances from a point on the surface of the earth to a productive vein of ore.

Three points are located on the top plane of a stratum of ore which lies under the surface of the earth (Fig. 5–58). Point *O* is a point on the surface from which tunnels will be drilled to the vein of ore for mining purposes. Point 4 is a point on the lower plane of the stratum of the

FIGURE 5–56. STRIKE AND DIP OF A PLANE (see page 87)

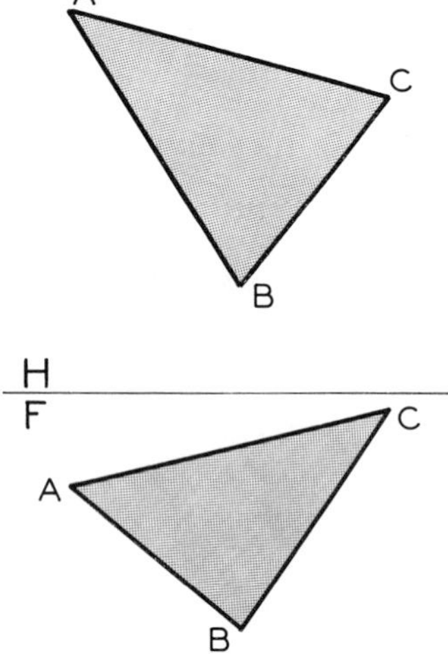

Given: The top and front views of plane *ABC*.
Required: The strike and dip of plane *ABC*.
Reference: Article 5–24.

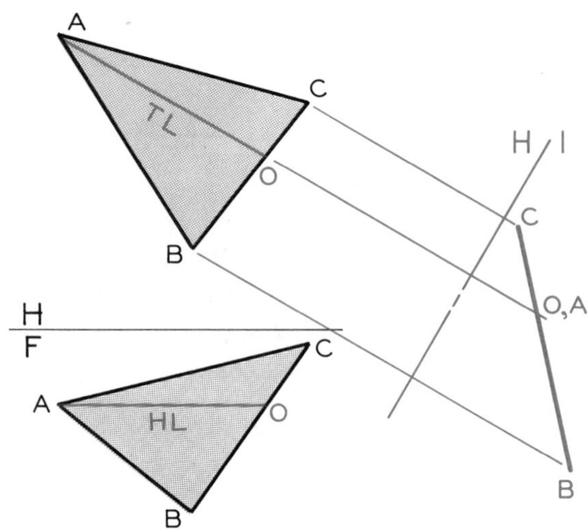

Step 1: Draw a horizontal line *AO* in the front view of plane *ABC*. This line will project true length in the top view. Project plane *ABC* as an edge in the auxiliary view where *AO* appears as a point. Project only from the top view in order to find the edge view of the horizontal plane.

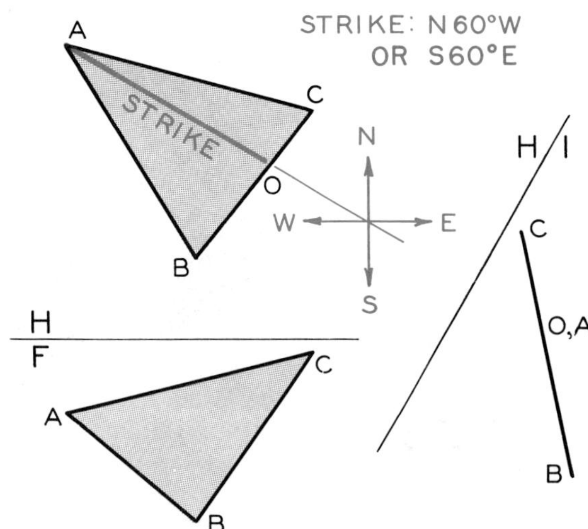

Step 2: The strike of a plane is the compass direction of a horizontal line in the plane. Line *AO* is the strike of *ABC* since it is a horizontal line. Its compass direction is measured in the top view as either N 60° W or S 60° E. The line has no slope; therefore either compass direction is correct. ✱

but when North is not specified on a plan, assume to be at top of paper.

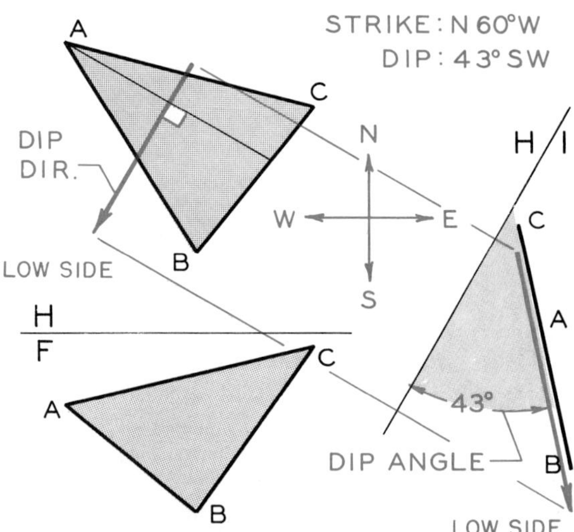

Step 3: The dip of a plane is its angle with the horizontal plane plus its general compass direction. This angle can be measured in the auxiliary view. The dip direction is perpendicular to the strike line toward the low side in the top view. The dip of *ABC* is 43° SW.

Fig. 5–57. Descriptive geometry principles have many applications to mining problems. (Courtesy of Joy Manufacturing Corporation.)

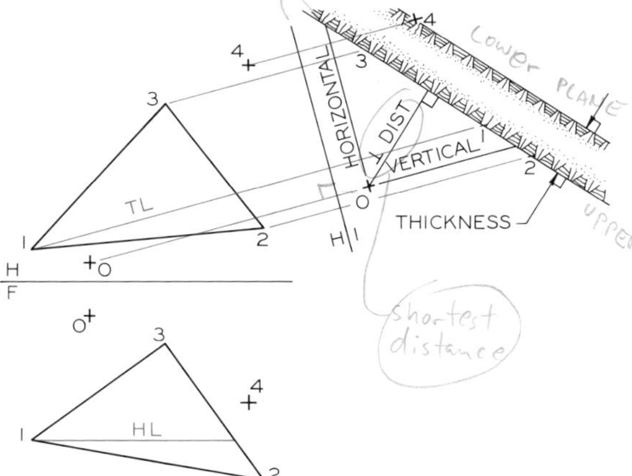

Fig. 5–58. Distances from a point to an ore vein.

vein. We are required to determine the lengths of the following tunnels: (1) the shortest distance to the ore, (2) the vertical distance to the ore, and (3) the shortest horizontal distance to the ore. The edge view of the ore vein is found by projecting from the top view. The lower plane is drawn parallel to the upper plane through point 4. The vertical distance is perpendicular to horizontal plane H–1, and the horizontal tunnel is parallel to the H–1 plane. The shortest tunnel is perpendicular to the plane. The vein thickness can be approximated by measurement in the auxiliary view.

5–26 OUTCROP OF AN ORE VEIN

Strata of ore or rock formations usually approximate planes of a somewhat uniform thickness. This assumption is employed in analyzing known data concerning the orientation of ore veins that are underground and are consequently difficult to study. A vein of ore may be inclined to the surface to the earth and may actually outcrop on its surface in some cases. Outcrops on the surface of the earth can permit open-surface mining operations at a minimum of expense (Fig. 5–59).

Figure 5–60 is an example of a problem in which an inclined ore vein is analyzed graphically to determine its area of outcrop, assuming that the plane is continuous to the surface of the earth. The locations of sample drillings, *A*, *B*, and *C*, are shown in the contour map and their elevations are plotted on the surface of the upper plane of the ore vein in the front view. The edge view of plane *ABC* is found by auxiliary view, where the elevation planes are used as datum planes in step 1. The lower plane of the vein is drawn parallel to the upper vein through point *D* to indicate the thickness of the vein. Points on the upper plane where the edge view intersects the elevation planes are projected to their respective contour lines in the contour map, as shown in step 2. The lower plane is also projected to the top view in the

Fig. 5–59. Open mines are located on sites where ore veins outcrop to the surface of the earth. (Courtesy of LeTourneau-Westinghouse Company.)

FIGURE 5–60. ORE VEIN OUTCROP

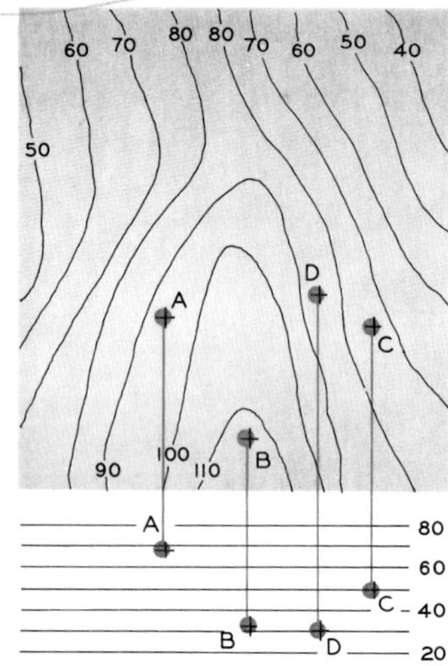

Given: Two views of points *ABC* on the upper plane of a stratum of ore and point *D* on the lower plane. The top view is a contour map and the front view is a series of elevation planes.

Required: Find the area where the vein outcrops on the surface, assuming that the vein in continuous.

Reference: Article 5–27.

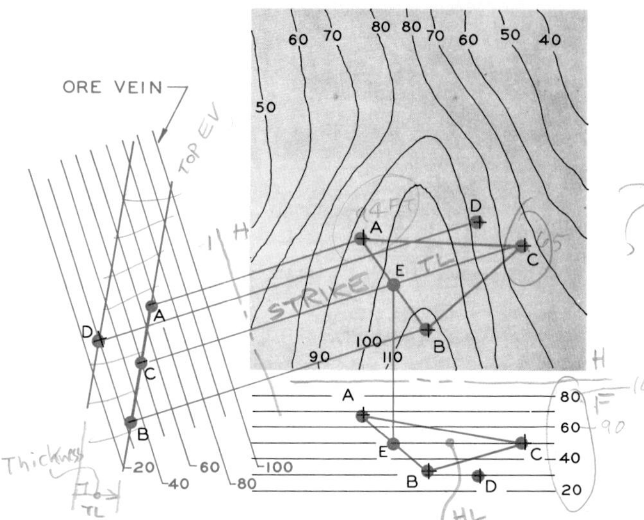

Step 1: Connect points *ABC* to form a plane in each view. Find the edge view of plane *ABC* by projecting from the top view. Locate point *D* in the auxiliary view. Construct the lower surface of the vein parallel to the upper plane in the auxiliary view with the same position as in the front view.

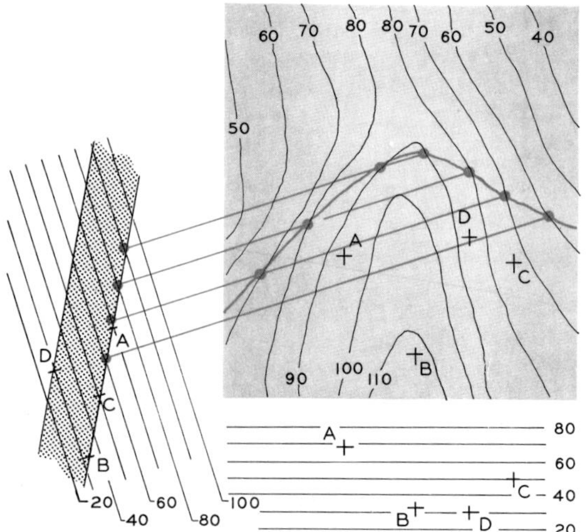

Step 2: Project the points where the upper plane of the vein crosses the elevation lines in the auxiliary view to their respective contours in the top view. For example, the point on the upper plane crossing the 90′ elevation line is projected to the 90′ contour line in the top view. Connect all points between contours in the top view.

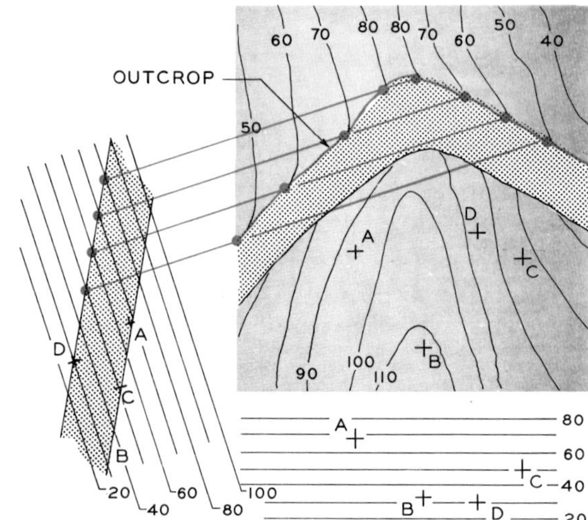

Step 3: Project the points where the lower plane crosses the elevation lines to the top view in the same manner as the upper plane. Connect these points with a line. The area between the two lines in the top view is the area where the stratum would outcrop on the surface if the plane were continuous.

same manner in step 3 to find its line of outcrop. The space between these lines is cross-hatched to indicate the area where the plane would outcrop through the surface of the earth provided that the ore vein were continuous in this direction.

5–27 INTERSECTION BETWEEN TWO PLANES–CUTTING PLANE METHOD

The intersection between strata or planes is important to the exploration for minerals, which are usually contained in strata that approximate planes. For example, if a stratum of oil-bearing sand were located beneath the surface, it would be significant to approximate the general location of the intersection of the stratum with the ocean floor if it were continuous to a point of intersection. This information would influence the location of offshore explorations such as that shown in Fig. 5–61.

Fig. 5–61. The position of a stratum of oil-bearing sand with respect to the ocean floor would influence the location of offshore exploration. (Courtesy of Exxon Corporation.)

FIGURE 5–62. INTERSECTION OF TWO INFINITE PLANES BY THE CUTTING-PLANE METHOD

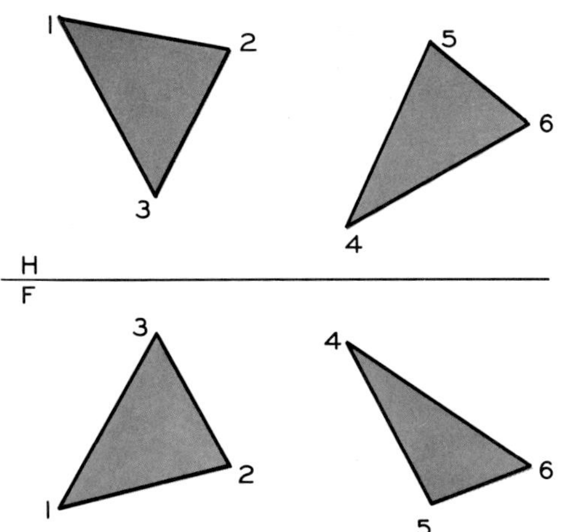

Given; The top and front views of two planes, 1–2–3 and 4–5–6.

Required: Find the line of intersection between these planes by projection, given that they are infinite in size.

Reference: Article 5–26.

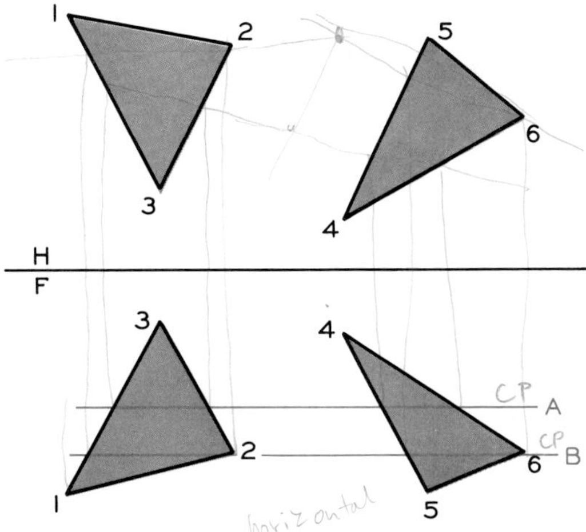

Step 1: Construct two cutting planes, *A* and *B,* in the front view. They are drawn parallel and horizontal in this case for convenience only. They could have been constructed in any direction and nonparallel.

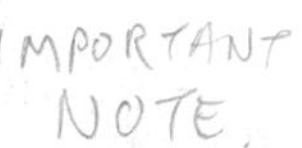

Planes 1–2–3 and 4–5–6 are segments of infinite planes in Fig. 5–62. We are required to find the line of intersection between them. Cutting planes are passed through either view at any angle and projected to the adjacent view. The two points where the lines formed by the cutting plane intersect in the top view establish the direction of the line of intersection. The compass direction of this line can be used to describe its orientation in space. The front view of the line of intersection is found by projecting the points from the top view to their respective planes in the front view.

5–28 INTERSECTION BETWEEN MINERAL VEINS–AUXILIARY METHOD

The locations of two mineral veins are given in a combination of verbal information and graphical representation in Fig. 5–63. We are required to locate the line of intersection, assuming that the two planes are continuous to a line of intersection.

The strike and dip of planes *C* and *D* are given in each view. Since the strike lines are true-length level lines in the top view, the edge view of the planes can be found in the view where the strike appears as a point. The plane can be drawn by the application of the dip angles given in specifications. Horizontal datum planes, H–F and H′–F′, are used to find lines on each plane that will intersect when projected from the auxiliary views to the top views. Points *A* and *B* are connected to determine the line of intersection between the two planes in the top view. These points are projected to the front view, where line *AB* is found.

Core samples taken from well sites are helpful in establishing information about the orientation of a mineral vein under the ground (Fig. 5–64). This information is used in the evaluation of the prospects for additional exploration in the immediate area.

FIGURE 5–62 (continued)

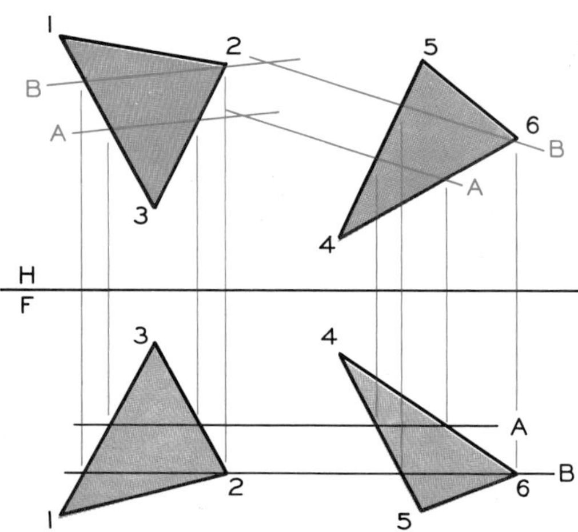

Step 2: The lines of intersection between the cutting planes and the given planes are projected to the top view. These lines are extended to cross their respective projections. Care should be taken to ensure that the lines determined by the *B* cutting plane intersect and that those determined by the *A* cutting plane intersect.

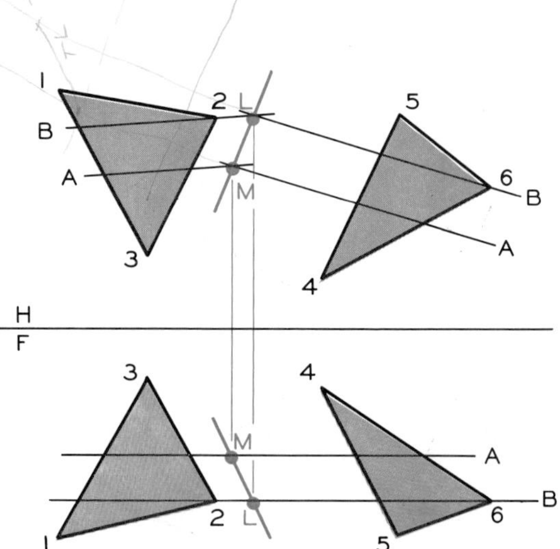

Step 3: Points *L* and *M* in the top view are the points where lines in a common horizontal plane intersect to form a line of intersection. Point *L* is projected to the front view of plane *B* and point *M* to the front view of plane *A*. The line of intersection is *LM*.

FIGURE 5–63. INTERSECTION BETWEEN ORE VEINS BY AUXILIARY VIEW

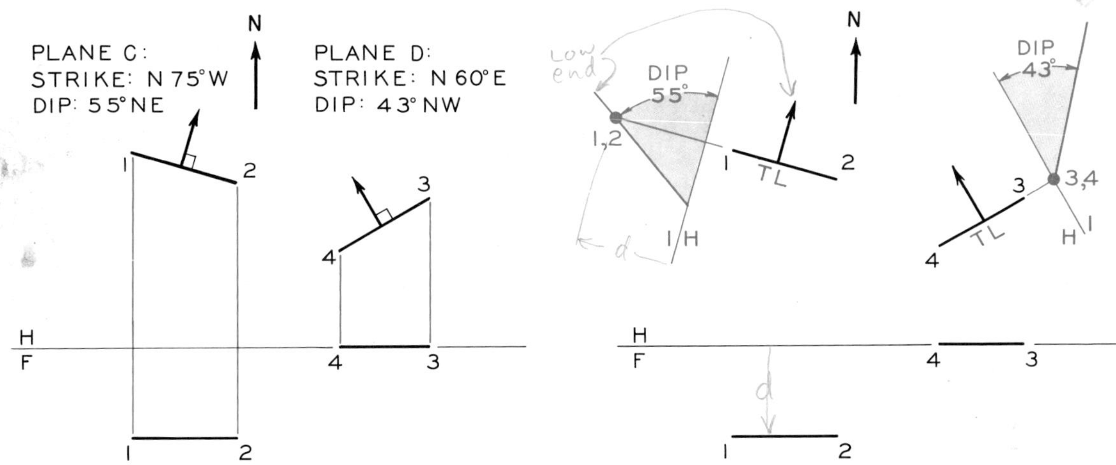

Given: The strike and dip of two ore veins, plane *C* and plane *D*.

Required: Find the line of intersection between the ore veins, assuming that each is continuous.

Reference: Article 5–28.

Step 1: Lines 1–2 and 3–4 are strike lines and are true length in the horizontal view. The point view of each strike line is found by an auxiliary view, using a common reference plane. The edge view of the ore veins can be found by constructing the dip angles with the H–1 line through the point views. The low side is the side of the dip arrow.

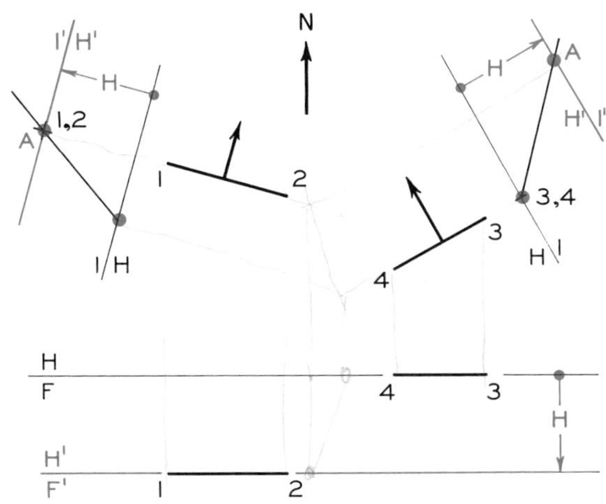

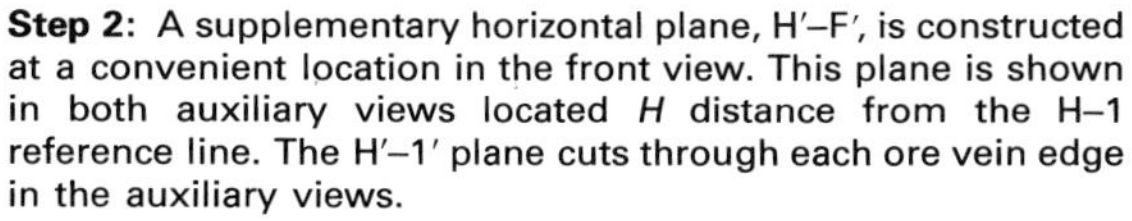

Step 2: A supplementary horizontal plane, H′–F′, is constructed at a convenient location in the front view. This plane is shown in both auxiliary views located *H* distance from the H–1 reference line. The H′–1′ plane cuts through each ore vein edge in the auxiliary views.

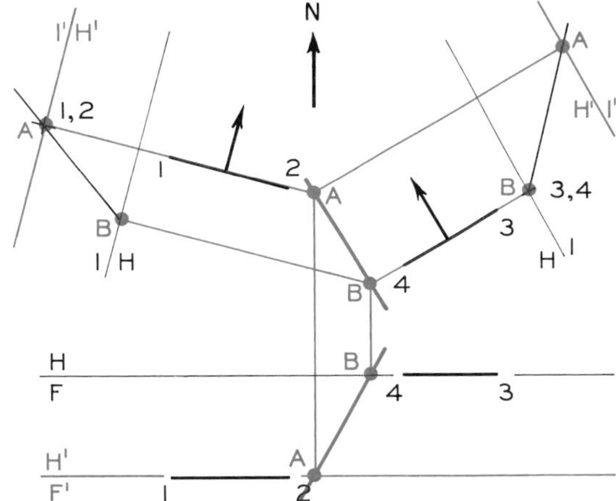

Step 3: Points *A*, which were established on each auxiliary view by the H′–1′ plane, are projected to the top view, and they intersect at point *A*. Points *B* on the H–1 plane are projected to their intersection in the top view at point *B*. Points *A* and *B* are projected to their respective planes in the front view. Line *AB* is the line of intersection between the two planes.

Fig. 5–64. Core samples are helpful in determining information about the orientation of a mineral vein under the ground. (Courtesy of Exxon Corporation.)

5–29 SUMMARY

The primary auxiliary view, which is projected from one of the principal views, has many applications similar to those covered in this chapter. Practically all engineering problems consist of a series of points, lines, and planes that represent most components of a design or project. Only a portion of the many applications of these principles have been covered in this chapter.

An understanding of the construction of auxiliary views of points, lines, and planes will enable the designer to solve many practical problems graphically when analytical solutions are impractical. Other applications of primary auxiliary views will become apparent when the principles are thoroughly understood.

Major emphasis has been placed on the theoretical concepts and principles throughout this chapter, as is the case in most chapters of this volume. However, the appearance of all drawings should be carefully considered for clarity; the drawings should clearly communicate ideas and specifications. A drawing presented with the minimum of notes and explanation may be very costly if insufficient information is given. The designer himself may be unable to interpret his own drawings after a few days unless he records each step of his solution as clearly as possible. The solution of a problem by the correct graphical procedures is insufficient unless the results, measurements, angles, and other findings are presented in an understandable form. The student can realize the importance of this if he attempts to interpret a problem unfamiliar to him that has been solved by a classmate.

PROBLEMS

Problems for this chapter can be constructed and solved on $8\frac{1}{2}'' \times 11''$ sheets, with instruments as illustrated by the accompanying figures. Each grid represents $\frac{1}{4}''$. References planes and points should be labeled in all cases using $\frac{1}{8}''$ letters with guidelines.

1. (A through D) In Fig. 5–65, find the true-size views of the inclined planes. Label all points and construction.

2. (A) In Fig. 5–66A find auxiliary views of point 2 as indicated by the lines of sight. Label all construction. (B) In part B of the figure, find auxiliary views of the lines 1–2 as indicated by the lines of sight.

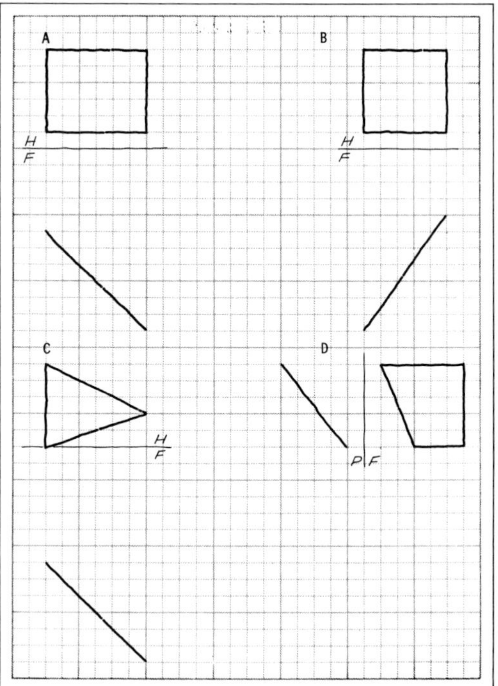

Fig. 5–65. Edge view of inclined planes.

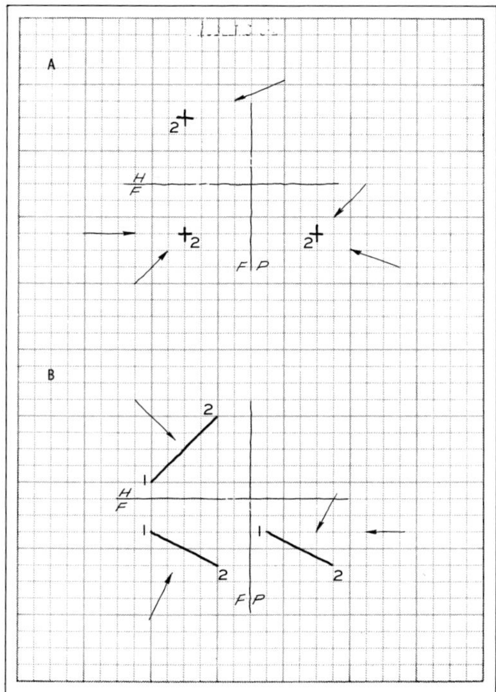

Fig. 5–66. Primary auxiliary views of a point and line.

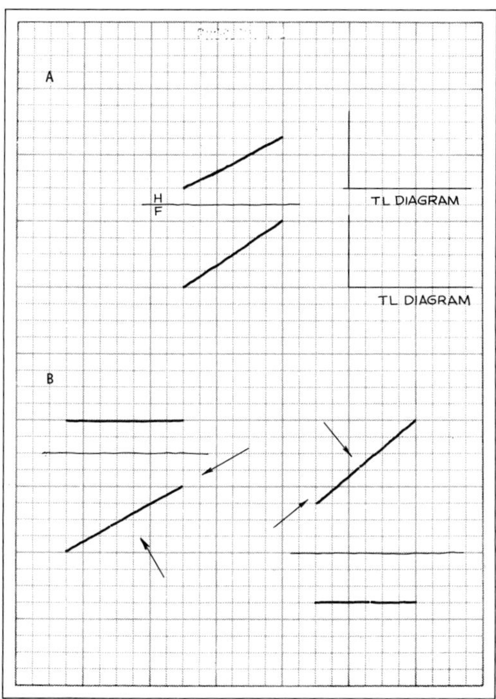

Fig. 5–67. True length of a line by auxiliary view and by true-length diagram.

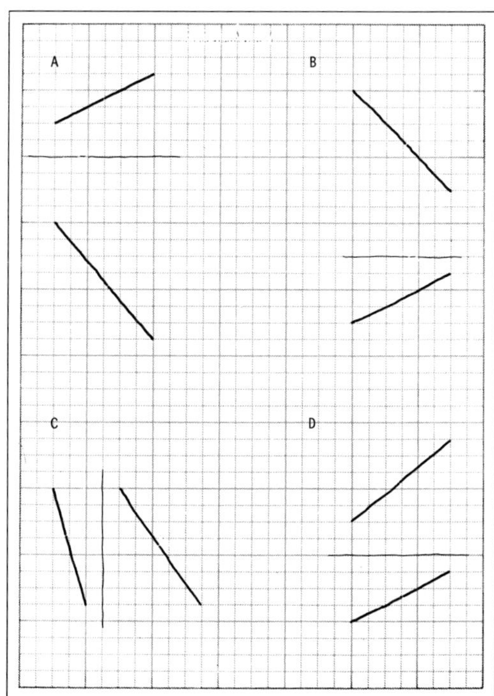

Fig. 5–68. Angle made by a line with principal planes.

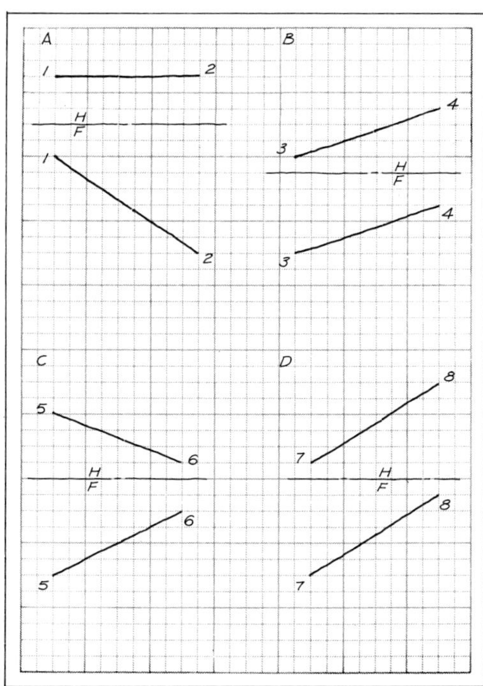

Fig. 5–69. Slope and percent grades of lines.

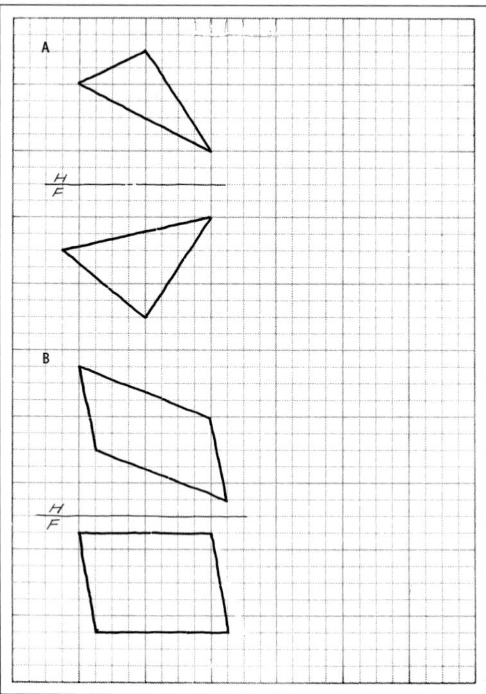

Fig. 5–70. Edge view of planes.

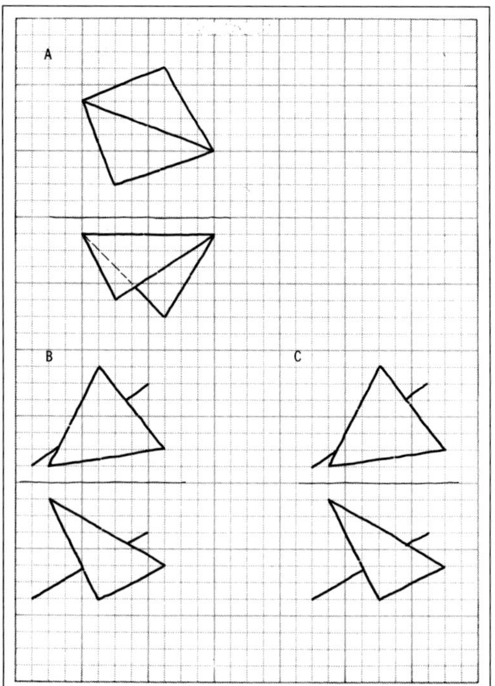

Fig. 5–71. Angle between planes and intersection of a line and plane.

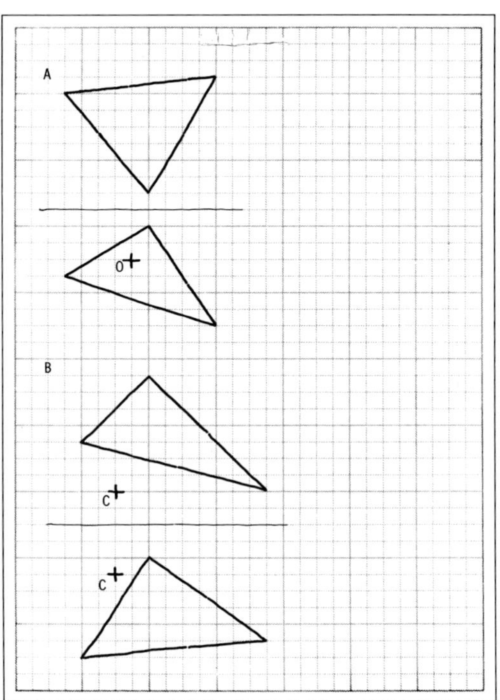

Fig. 5–72. Line perpendicular to a plane.

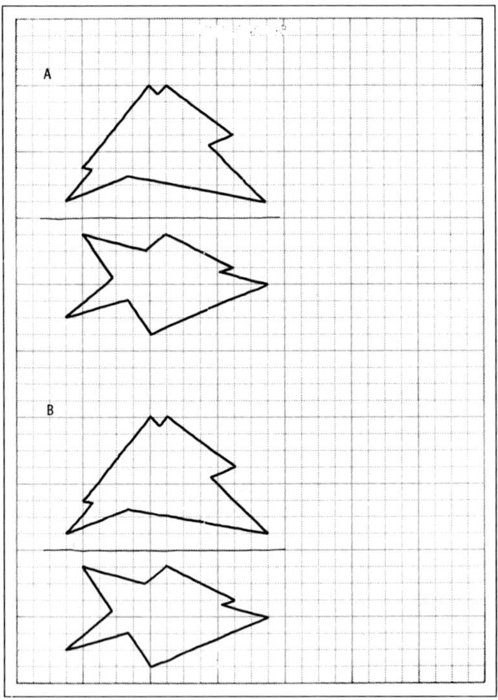

Fig. 5–73. Intersection between planes by projection and auxiliary views.

3. (A) In Fig. 5–67A, find the true length of the line by projecting auxiliary views from the top and front views. Find the true length of the line by true-length diagrams projected from the top and front views. (B) In part B of the figure, find auxiliary views of the lines as indicated by the lines of sight.

4. Use Fig. 5–68 for all parts of this problem. (A) Find the angle the line makes with the horizontal plane. (B) Find the angle the line makes with the frontal reference plane. (C) Find the angle between the line and the profile reference plane. (D) Find the slope of the line in the given top and front views.

5. Find the slope angle, tangent of the slope angle, and the percent grade of the four lines shown in Fig. 5–69. Determine the direction of the slope from the smaller numbered end to the larger numbered end of each line.

6. (A and B) In Fig. 5–70, find the edge views of the two planes.

7. Use Fig. 5–71 for all parts of this problem. (A) Find the angle between the two intersecting planes. (B) Find the piercing point and determine the visibility of the plane and line by projection methods. (C) Find the piercing point and determine the visibility of the plane and line by the auxiliary view method.

8. (A) In Fig. 5–72A, construct a line $\frac{1}{2}''$ long on the upper side of the plane through point *O* on the plane. (B) In part B of the figure construct a line from point *C* that will be perpendicular to the plane. Indicate the piercing point and visibility.

9. (A) In Fig. 5–73A, determine the line of intersection and the visibility between the two planes by projection. (B) In part B of the figure, find the line of intersection and the visibility of the two planes by the auxiliary-view method.

10. Use Fig. 5–74 for all parts of this problem. (A and B) Find the slope and direction of slope of each of the planes. Indicate all measure-

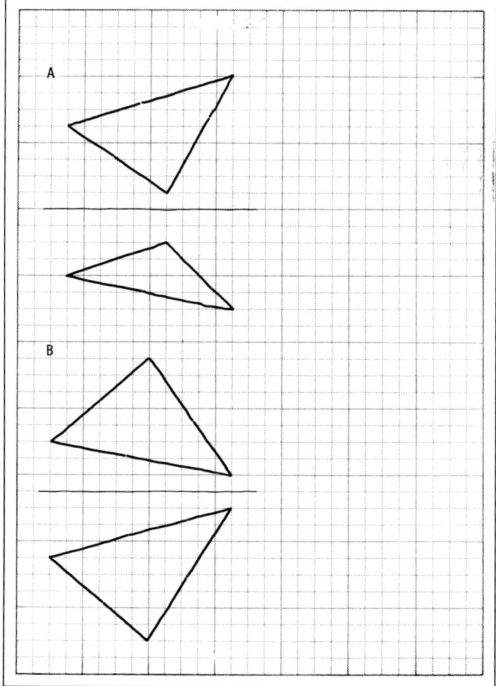

Fig. 5–74. Slope, strike, and dip of a plane.

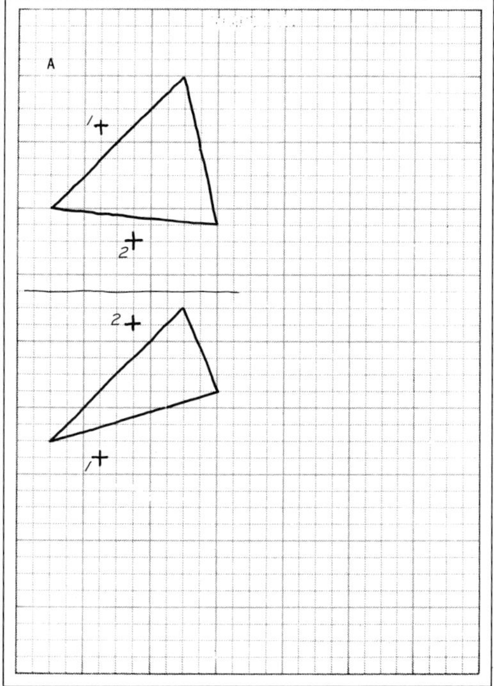

Fig. 5–75. Distance to an ore vein.

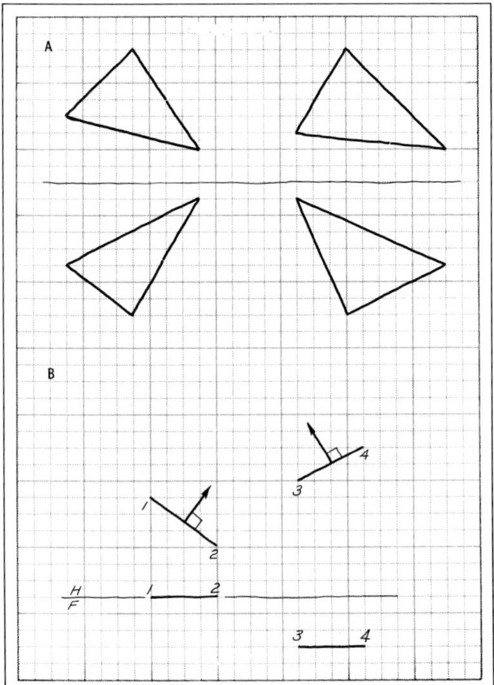

Fig. 5–76. Intersections between planes.

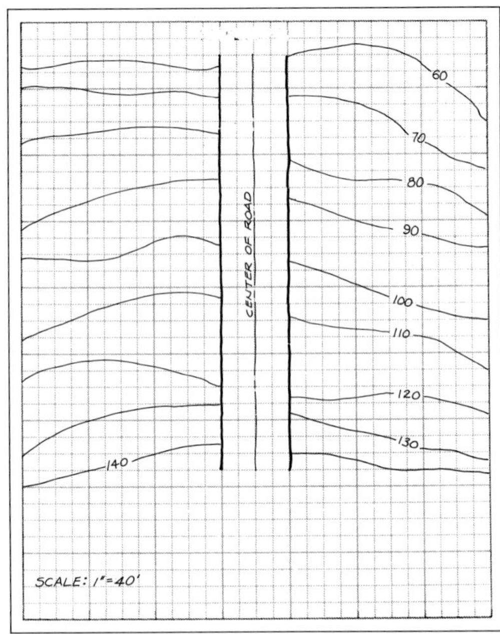

Fig. 5–78. Cut and fill of a level roadway.

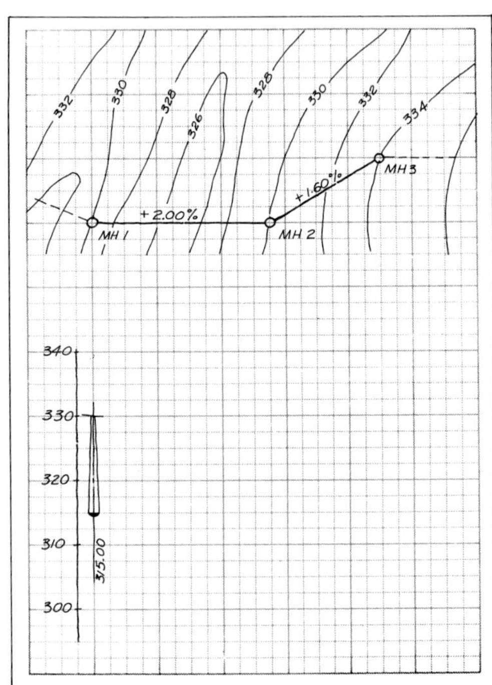

Fig. 5–77. A plan-profile problem.

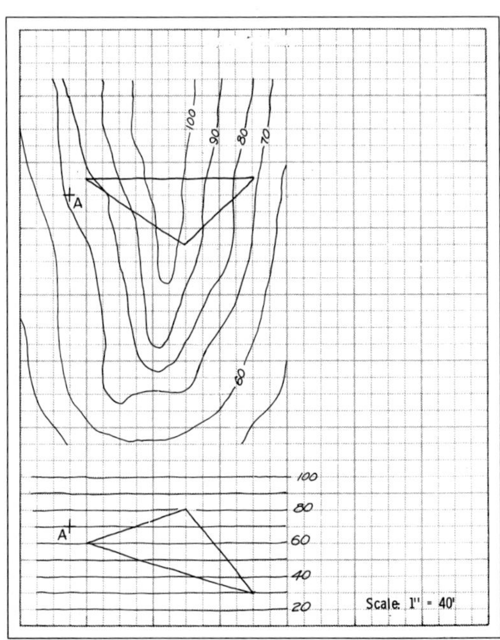

Fig. 5–79. Outcrop of an ore vein.

ments. (C and D) Find the strike and dip of each of the planes in parts A and B on a separate sheet of paper.

11. Use Fig. 5–75 for this problem. The plane represents the location of points on the upper plane of a stratum of ore under the ground. Point 2 is a point on top of the surface; point 1 is a point on the bottom plane of the stratum of ore. Find the shortest distance, the shortest horizontal distance, and the shortest vertical distance from point 2. What is the thickness of the stratum?

12. In Fig. 5–76A, find the line of intersection between the two planes by cutting planes, assuming that the two are continuous planes. In part B of the figure, two strike lines of an ore vein are given: 1–2 and 3–4. The plane with strike 1–2 dips 30° northeast, and the plane with strike 3–4 dips 45° northwest. Find the line of intersection between the two views in both views.

13. A drainage system is to be laid from manhole 1 to manhole 3 using the specifications given in Fig. 8–77. Complete the plan-profile drawing beginning with manhole 1, using the vertical scale of 1″ = 10′. Allow a drop of 0.20′ across each manhole to compensate for loss of head. The scale of the plan view is 1″ = 100′. Refer to Article 5–12.

14. (A) In Fig. 5–78, the level road has an elevation of 100′. Find the cut and fill in the top view given that the cut angle is 30° with the horizontal and the fill angle is 35° with the horizontal. Label all construction. (B) Assume that the roadway is sloping at a 10-percent grade with its low end toward the bottom of the sheet. Determine the cut and fill areas for this situation, using the same angles as in the previous problem. Use a separate sheet. Estimate the amount of earth to be cut and filled.

15. In Fig. 5–79, the plane represents two views of the top plane of a stratum of ore under the ground. Point *A* is in the bottom plane of the stratum of ore. Find the area of outcrop where the stratum pierces through the surface of the earth in the top view. Scale: 1″ = 40′.

6

Successive Auxiliary Views

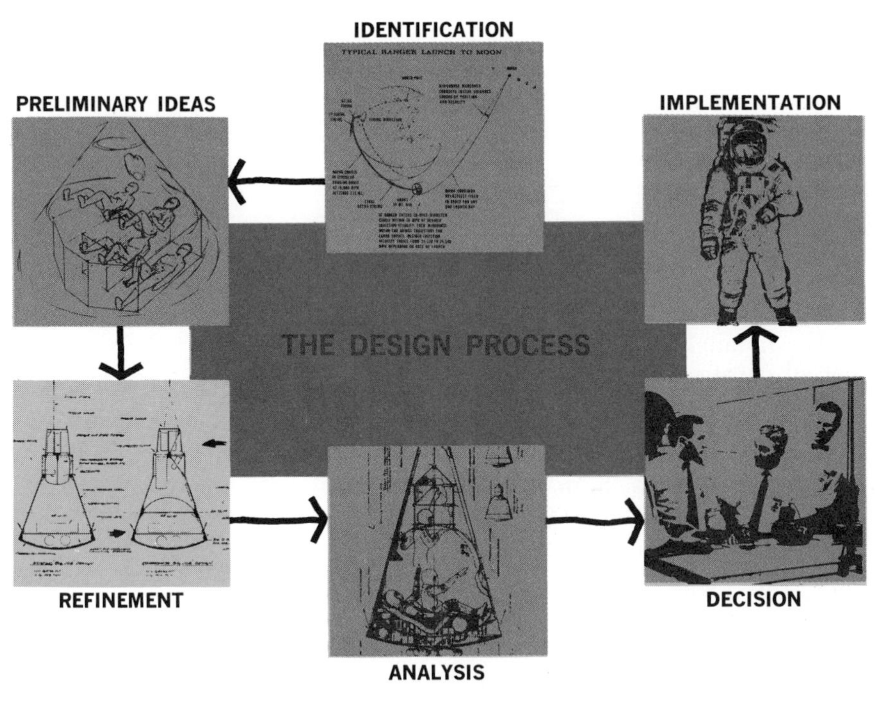

6–1 INTRODUCTION

Many industrial problems cannot be solved by primary auxiliary views, but instead they require that secondary auxiliary projections be made before the desired information can be obtained. The designer is concerned with much the same type of design information as that covered in Chapter 5, namely, physical dimensions and shapes. For example, a basic requirement of a design involving intersecting planes of irregular shapes is the true shape and size of the planes (Fig. 6–1). A design cannot be detailed with the complete specifications necessary for construction unless all details of fabrication have been determined; these details include true shapes of planes, angles between planes, distances from points to lines, and angles between lines and planes. An example of a typical project that required the solution of many descriptive geometry problems is a structural frame for a 65-ton truck (Fig. 6–2). Prior to its fabrication, drawings and specifications were prepared to describe completely each component of this system through the application of graphic principles. It should be rather obvious from observation of the problems covered in this chapter that many of the solutions obtained graphically would be quite complicated if attempted from an analytical approach or through the application of mathematical principles. The best method for solving complicated spatial problems is a mixture of graphical and analytical procedures which combines the advantages of each. The graphical analysis required to solve a spatial problem aids in the application of analytical methods.

Primary auxiliary views are supplementary views projected from primary orthographic views—the horizontal, frontal, or profile views. A *secondary auxiliary view* is a view projected from a primary auxiliary view. The reference line between the principal plane and the auxiliary view is labeled F–1, H–1 or P–1, but the reference line between a primary auxiliary view and a secondary auxiliary view is labeled 1–2, regardless of the primary view from which it is projected (Fig. 6–3). A *successive auxiliary view* is a view projected from a secondary auxiliary view or from another successive auxiliary view. In other words, an infinite sequence of auxiliary views can be produced by continuing to project successively from auxiliary view to auxiliary view.

Fig. 6–1. The many facets of the U.S. Pavilion dome at Expo 67 provide an example of the interrelationships between lines, points, and planes that require spatial analysis. (Courtesy of Rohm and Haas Company.)

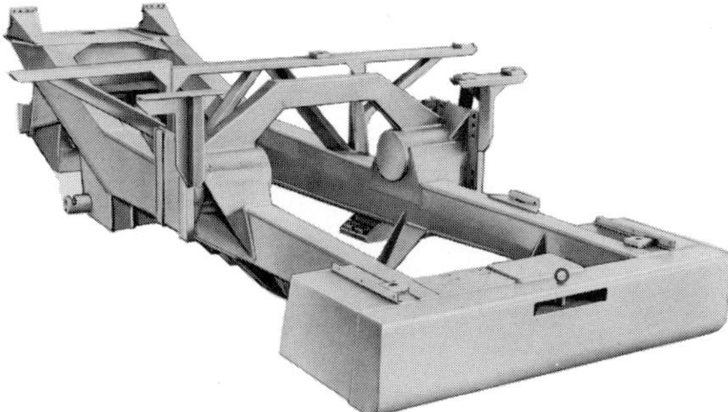

Fig. 6–2. This frame for a 65-ton Haulpak truck illustrates the many design problems that require the application of descriptive geometry. (Courtesy of LeTourneau-Westinghouse Company.)

Auxiliary views have the same relationship between their adjacent views as the principal views have with one another. A secondary auxiliary plane is perpendicular to the primary auxiliary plane, and the plane of a successive auxiliary view projected from a secondary auxiliary view is perpendicular to the secondary auxiliary plane. It should be remembered that all sequential auxiliary planes are perpendicular to the preceding plane from which the projection was made.

FIGURE 6–3. SECONDARY AUXILIARY VIEW OF A SOLID

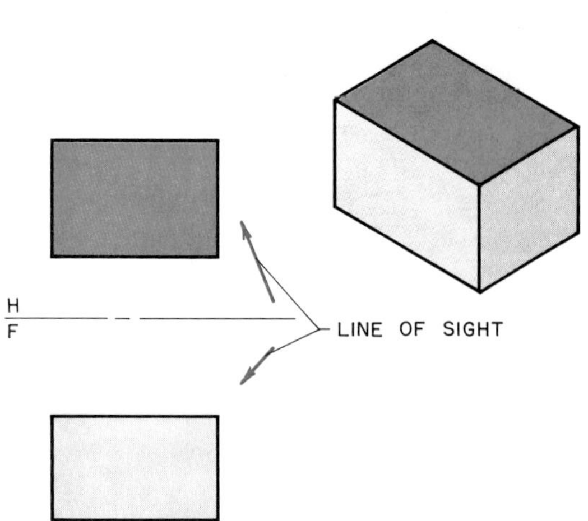

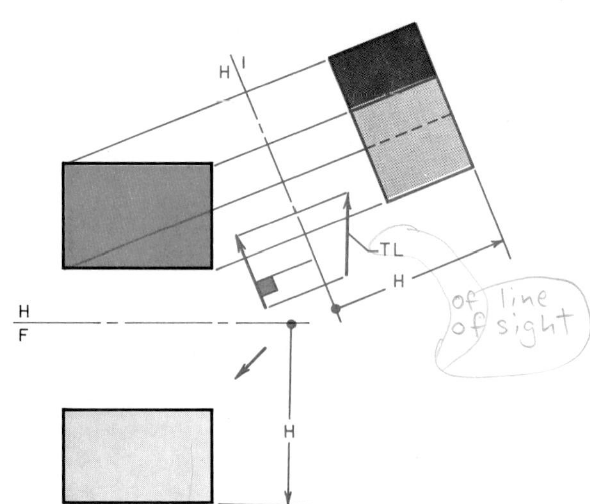

Given: The top and front views of a solid and a line of sight.

Required: Find the view of the solid indicated by the line of sight.

References: Articles 5–4 and 6–2.

Step 1: Project a primary auxiliary view from one of the given views so that it is perpendicular to the line of sight. The primary auxiliary view will give the true length of the line of sight since it is viewed perpendicularly. Project the solid to this view in the same manner. Transfer dimension *H* from the front view to establish points in the primary auxiliary view.

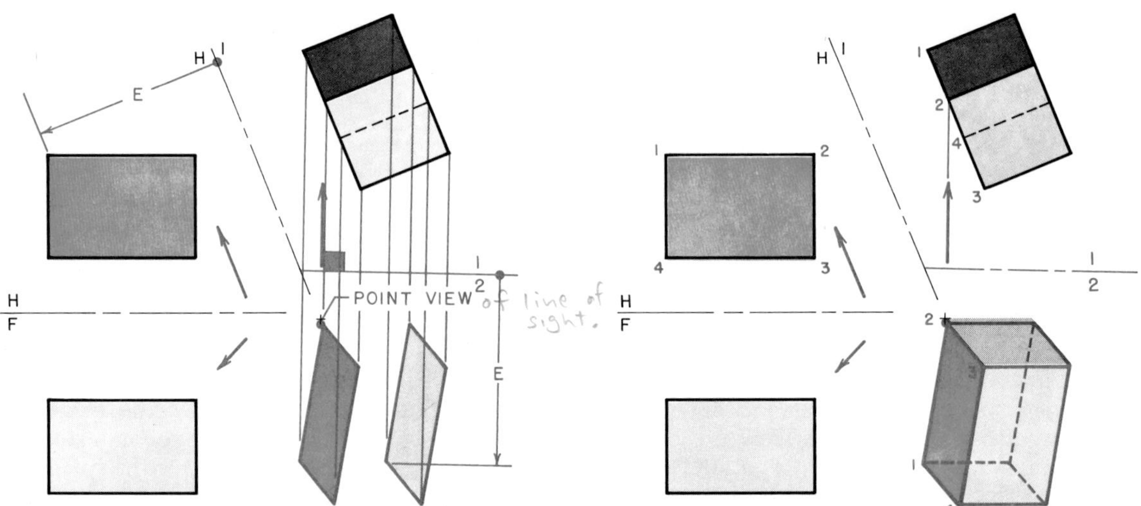

Step 2: Since the line of sight is true length in the primary auxiliary view, a point view of the line can be found in a secondary projection plane, which is drawn perpendicular to the line. This will give the required view of the object. Project the upper and lower planes to the secondary auxiliary view by transferring all dimensions from the H–1 line in the manner of measurement *E*.

Step 3: Complete the object by connecting the respective corners with the missing lines. Plane 1–2–3–4 will appear visible in the secondary auxiliary view since the line of sight gives an unobstructed view of the plane in the primary auxiliary view. Any line crossing this plane must be behind it and therefore is hidden, as shown above. The outlines of a solid are always visible.

6–2 SECONDARY AUXILIARY VIEW OF A SOLID

A simple rectangular prism is used to introduce secondary auxiliary-view construction in Fig. 6–3. A line of sight is arbitrarily chosen in the top and front views. Since we are required to view the prism in the direction of the line of sight, we must construct a view in which the line of sight will appear as a point. In step 1, a primary auxiliary view is drawn to find the true length of the line of sight; the object is also projected to this view. The secondary auxiliary plane, 1–2, is located perpendicular to the true-length line of sight in step 2. The two edge views of the prism are projected to the secondary auxiliary view as independent planes to simplify the formation of the solid and the determination of visibility.

It can be seen in step 3 that plane 1–2–3–4 is a visible plane in the secondary auxiliary view; that is, the line of sight from the secondary auxiliary view gives an unobstructed view of the plane in the primary auxiliary view. Since plane 1–2–3–4 is visible in the secondary auxiliary view, all lines crossing the plane must be behind it and consequently are hidden. All outlines of a solid are visible.

6–3 POINT VIEW OF A LINE

The principle involved in the determination of the point view of a line is a basic one that must be utilized to solve many problems in spatial geometry. Before the point view of a line can be found, a true-length view of the line must be found. This is illustrated with a primary view that is projected from the true-length view of line 3–4 in Fig. 6–4.

The line in Fig. 6–5 is not true length in either view, which requires that the line be found true length by a primary auxiliary view. The primary auxiliary view is projected from the front view in this case. The secondary auxiliary view gives line 1–2 as a point, since it is projected from a true-length view of the line.

When two adjacent views are given, the point view of a line can be found by projecting a secondary auxiliary view from either view.

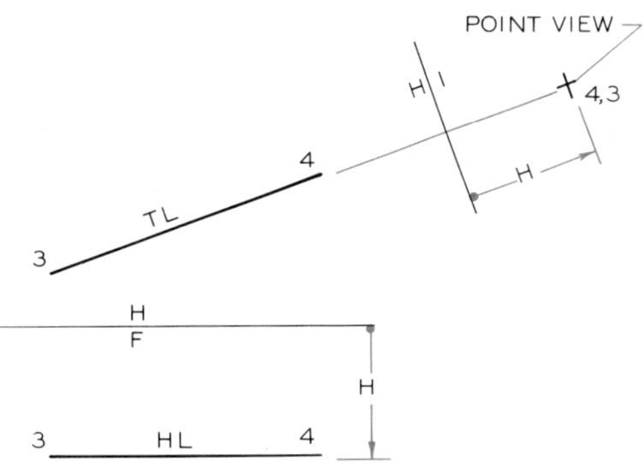

Fig. 6–4. The point view of a line can be found by projecting from a true-length view of the line.

For example, we could have found the true length of line 1–2 in Fig. 6–5 by projecting it from the top view and then developing a secondary auxiliary view from this projected view. Similarly, the primary auxiliary could have been projected from the profile view.

6–4 ANGLE BETWEEN TWO PLANES

Nearly all designs involve the intersection of planes at many unusual angles that must be specified in detail by the designer before fabrication. These angles must be known so that a means for connecting the two planes may be devised, or perhaps so that a form may be designed for casting the design in concrete, metal, or even glass. The nuclear detection satellite shown in Fig. 6–6 is an example of an assembly for which angles need to be determined. The dihedral angles between the planes of the exterior surface are very critical. These angles must be formed within a high degree of tolerance to permit a highly accurate joint, which is necessary for the successful function of the satellite in outer space.

The two intersecting planes in Fig. 6–7 represent a special case where the line of intersection, 1–2, is true length in the top view. This permits the point view of the line of intersection and the angle between the planes to be found.

The line of intersection between the two planes in Fig. 6–8 is oblique in the given views. Since this line of intersection does not appear true length in either view, we must develop a

FIGURE 6–5. POINT VIEW OF AN OBLIQUE LINE

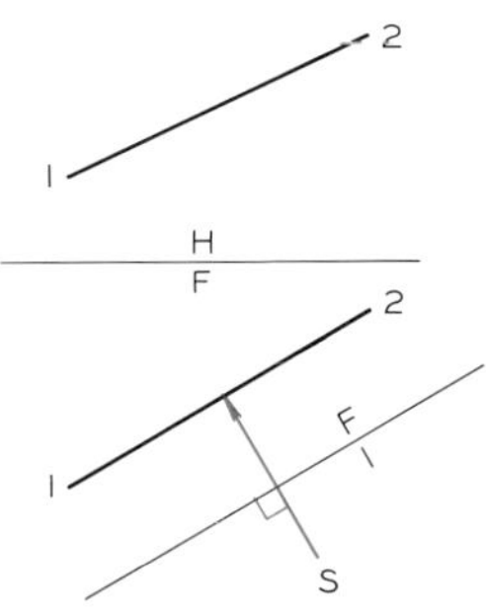

Step 1: Project an auxiliary view from one of the principal views. In this case, the reference line is established parallel to the front view of line 1–2. The projectors will be parallel to the given line of sight, which is perpendicular to the F–1 line.

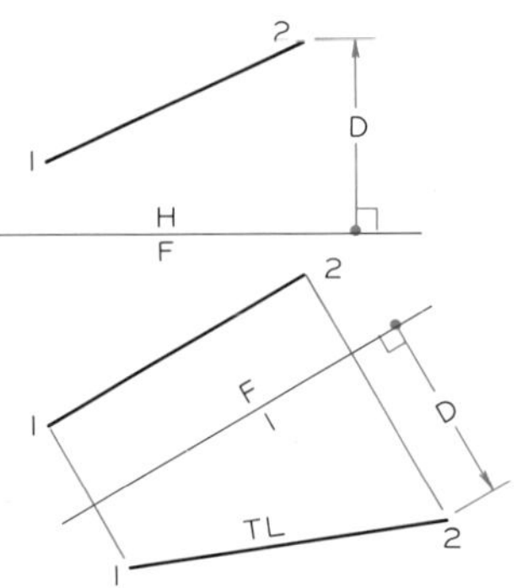

Step 2: Find the primary auxiliary view by projecting as specified in step 1. Line 1–2 is true length in this view, since the line is parallel to the reference plane in the preceding view. The primary auxiliary view could also have been projected from the top view.

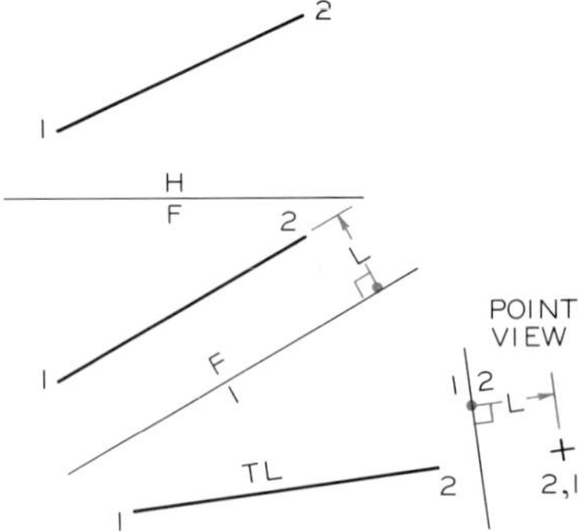

Step 3: Draw a secondary line, 1–2, perpendicular to the true-length view of line 1–2. Transfer measurement *L*, from the F–1 line to the front view of the line, to the secondary auxiliary view to find the point view. Measurement *L* will appear true length in these positions, since it is perpendicular to the primary auxiliary plane.

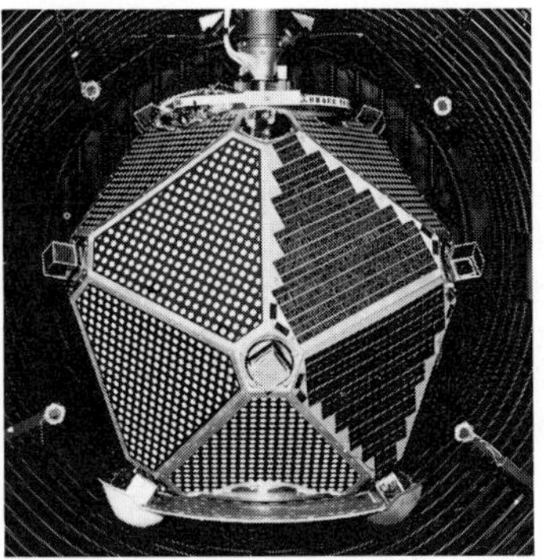

Fig. 6–6. A nuclear detection satellite is composed of planes and angles that can be determined with successive auxiliary views. (Courtesy of TRW Space Technology Laboratories.)

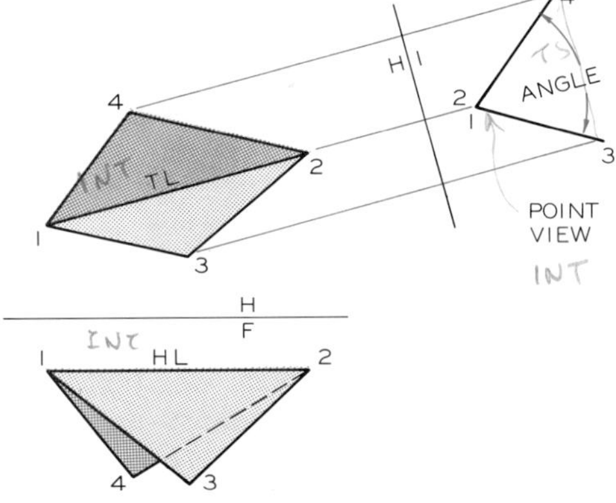

Fig. 6–7. The angle between two planes can be found in the view where the line of intersection projects as a point.

secondary auxiliary view to solve the problem. The true angle between two planes can be measured in the view in which the line of intersection appears as a point and both planes project as edges. The true length of the line of intersection is found in step 1, and the line is found as a point in step 2 by applying the procedures outlined in Fig. 6–5. The true angle is measured in step 3, where the plane of the angle, which is perpendicular to the line of intersection appears true size. This sequence of auxiliary views could have also been projected from the front view.

6–5 TRUE SIZE OF A PLANE

Most products and engineering designs contain many oblique surfaces and planes whose true size must be found in order that appropriate working drawings may be prepared for the con-

FIGURE 6–8. ANGLE BETWEEN TWO PLANES

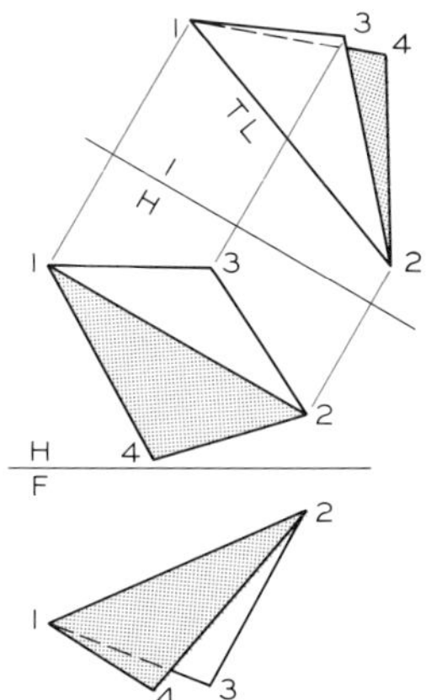

Step 1: The angle between two planes can be seen in a view where the line of intersection appears as a point. Project a primary auxiliary view perpendicularly from a principal view of the line of intersection. In this case, a view is projected from the top view. Line 1–2 will appear true length in the primary auxiliary view.

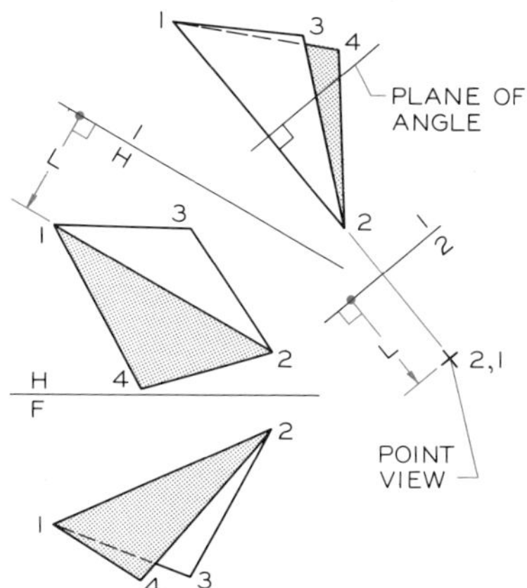

Step 2: The point view of the line of intersection 1–2 is found in the secondary auxiliary view. Locate this view by transferring measurement *L* from the edge view of the primary projection plane as indicated. The plane of the angle appears as an edge perpendicular to the true-length view of the line of intersection in the primary auxiliary view.

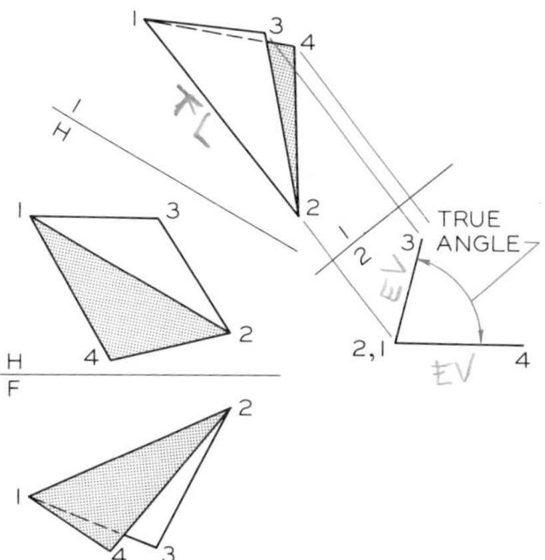

Step 3: The edge views of the planes are completed in the secondary auxiliary view by locating points 3 and 4 in the same manner as in step 2. The angle between the planes can be measured in this view since the line of intersection appears as a point and the planes appear as edges.

struction of the finished design. The nuclear detection satellite in Fig. 6–6 illustrates an assembly in which there is a need for finding the true size and shape of each of the oblique surface planes. These surfaces are precisely assembled to close tolerances.

The true size of plane 1–2–3 is found in sequential steps in Fig. 6–9. The edge view of the plane is found in step 1 by a primary auxiliary view. The secondary auxiliary reference line, 1–2, is constructed parallel to the edge view of plane 1–2–3. The true size of the plane is found in the secondary auxiliary view by projecting perpendicular to the 1–2 line and transferring the measurements from the H–1 line in the top view to the secondary auxiliary view, as shown in step 3.

It should be noted at this point that the representation of a plane, as defined in descriptive geometry, can take a variety of forms, including two intersecting lines, two parallel lines, three points, or a line and a point. This allows many applications of the principle for finding the true size of a plane. For instance, Fig. 6–10 illustrates the fuel system for a gas turbine engine; the tubing for the system must be bent to fit the contours of the engine properly. The determination of the lengths of tubing and the angular bends is an application of the principle for finding the true size of a plane. A problem similar to this is shown in Fig. 6–11, where the top and front views of points on the center line of a fuel line are given. The true angle, 1–2–3, can be found in the view where plane 1–2–3 appears true size. The primary auxiliary view is found by projecting in a direction parallel to line 1–2 in the top view, which is true length, to find the edge view of the plane. A secondary auxiliary view projected perpendicular to the edge view gives the plane true size, where both lines are true length and the angle between them is true size. A given bend radius can be used to construct the curvature arc at point 2 with point *C* as the center. The straight lengths can be scaled directly, while the arc distance can be found mathematically by application of the formula for finding the circumference of a circle. The solution to this problem would be quite complex if it had to be solved entirely by mathematical methods.

Another application of the angle between two lines is illustrated in Fig. 6–12, which shows a connecting joint designed to support the structural members of the 1965 New York

FIGURE 6–9. TRUE SIZE OF A PLANE

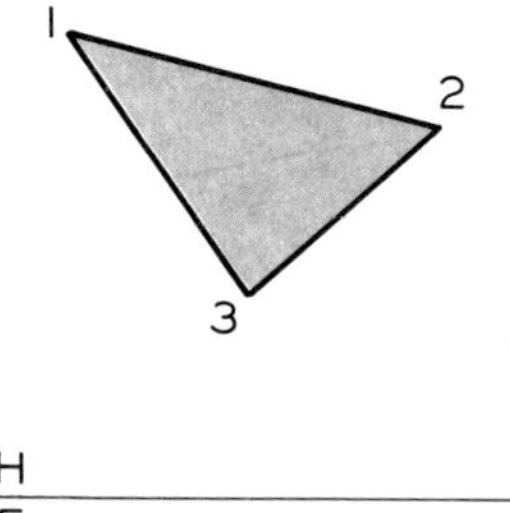

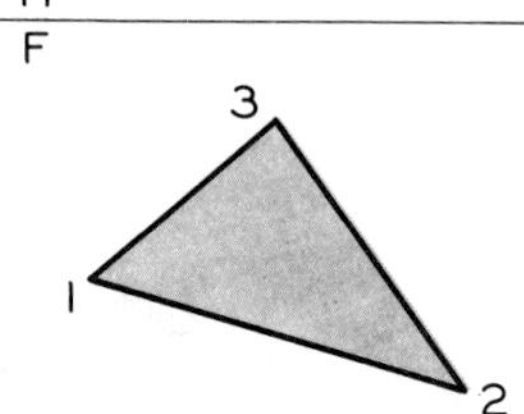

Given: The top and front views of a plane.

Required: Find the true size of the plane.

References: Articles 5–4 and 6–5.

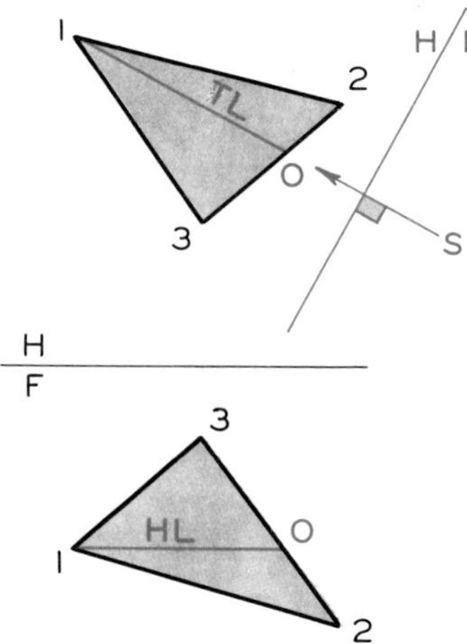

Step 1: Draw horizontal line 1–*O* in the front view of plane 1–2–3 and project it to the top view, where the line appears true length. Project a primary auxiliary view from the top view parallel to the direction of line 1–*O*. The H–1 reference line is perpendicular to 1–*O* and the line of sight.

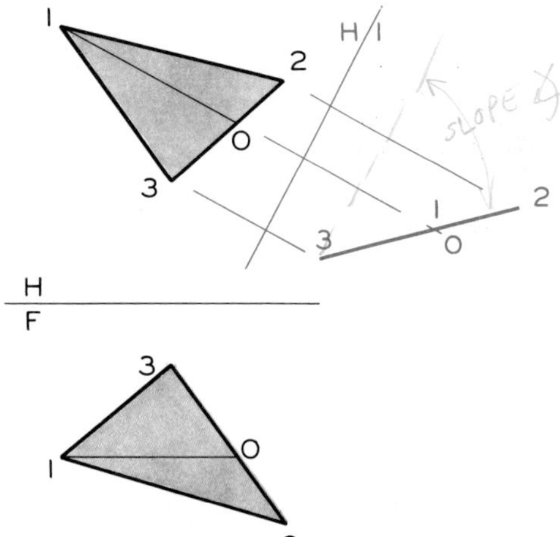

Step 2: The point view of line 1–*O* is found in the primary auxiliary view. Project points 2 and 3 to this view where the plane will appear as an edge.

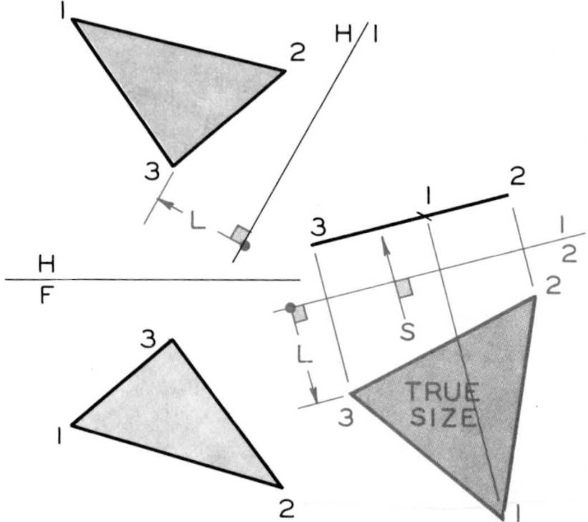

Step 3: A secondary auxiliary view of line 1–2 is drawn parallel to the edge view of plane 1–2–3. The line of sight is drawn perpendicular to the 1–2 line. The true-size view of the plane is found by locating each point with measurements taken perpendicularly from the edge view of the primary auxiliary plane, as indicated.

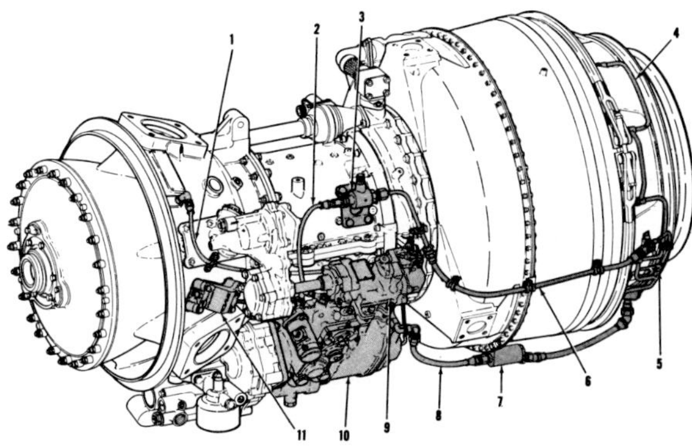

Fig. 6–10. The determination of the bends in a fuel line is an application of the principle of finding the angle between two lines. (Courtesy of Avco Lycoming.)

Fig. 6–13. The elliptical paths of satellites are shown in the partially completed Unisphere®. (Courtesy of U.S. Steel Corporation.)

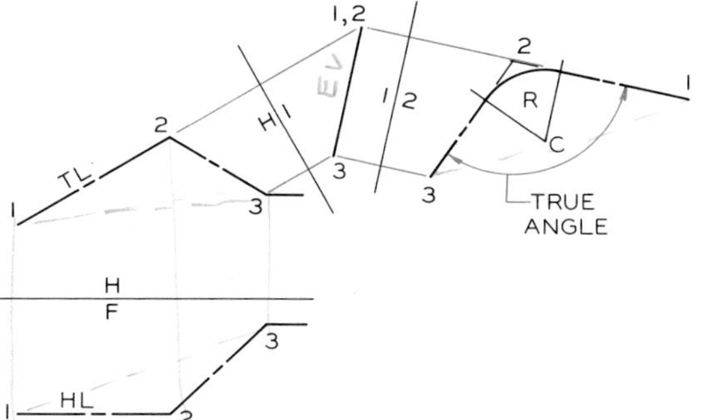

Fig. 6–11. The angle between two intersecting lines can be found by finding the plane of the lines true size.

Fig. 6–12. The base of the Unisphere®, symbol of the New York World's Fair, is shown under construction. This is an application of the principle of finding the angle between two lines. (Courtesy of U.S. Steel Corporation.)

World's Fair Unisphere®. It was necessary to construct a view in which the angle between the chordal member and the support element appeared true size. It was also necessary to find the angles between the intersecting planes at this point of support so that the details could be prepared for fabrication and erection.

6–6 ELLIPTICAL VIEWS OF A CIRCLE

Circular and cylindrical shapes are commonly used in most designs. The orbital paths of satellites will project as ellipses in most views, whether in actual space or as depicted symbolically in Fig. 6–13, which shows the Unisphere® in the final stages of construction. Circles appear true size and shape when the observer's line of sight is perpendicular to the plane of the circle. However, there are many instances when the line of sight is oblique to the plane of the circle; in the resulting foreshortened views the circles will appear as ellipses. The representation of circular features requires an understanding of the principles of ellipse construction.

The following definitions are given to explain terminology associated with ellipses. Refer to Fig. 6–14.

Ellipse. A view of a circle in which the line of sight is oblique to the plane of the circle.

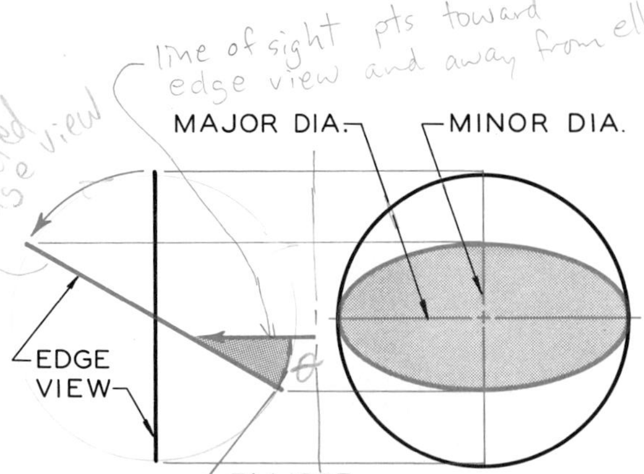

Fig. 6–14. The relationship of an ellipse to a circle and the selection of ellipse templates.

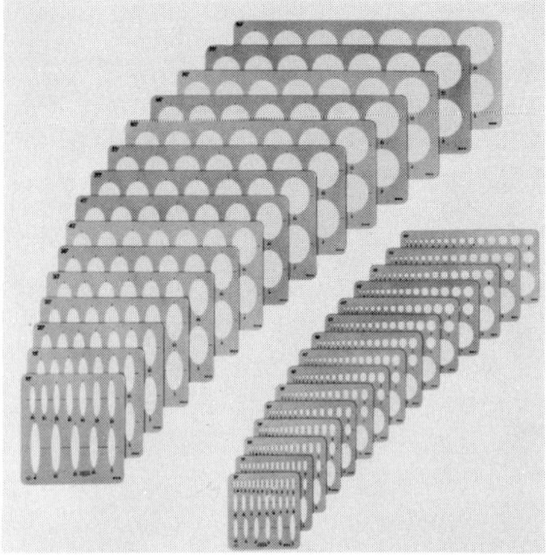

Fig. 6–15. Typical ellipse templates used for ellipse representation. (Courtesy of The A. Lietz Company.)

Major Diameter. The greatest possible diameter that can be measured across an ellipse. By definition, a diameter passes through the center of the ellipse. The major diameter is always true length in any view of a circle.

Minor Diameter. The shortest possible diameter that can be measured across an ellipse. This diameter is perpendicular to the major diameter at its midpoint in all views.

Ellipse Angle. The angle between the line of sight and the edge view of the plane of the circle, usually found in a primary auxiliary view.

Cylindrical Axis. In a right circular cylinder, an imaginary line connecting the centers of all right sections and perpendicular to them.

Ellipse Template. A template composed of a series of various sizes of ellipses, used for drawing the ellipses when the major and minor diameters are known. Ellipse guides are graduated in 5° intervals (ellipse angles) in most cases. A set of ellipse guides is illustrated in Fig. 6–15.

Suppose that we are required to construct a circle passing through points 1, 2, and 3 as shown in all views in Fig. 6–16. The true size of plane 1–2–3 must be found in order to construct the circle in true shape. The circle is found in step 1 by locating the center, where the three perpendicular bisectors of each line of the plane intersect, and by selecting a radius that will pass through each point. In step 2, the major and minor diameters are drawn in the secondary auxiliary view parallel and perpendicular to the 1–2 reference line. Next, the diameters of the circle are projected to the edge view in the primary auxiliary view; then they are projected to the top view where major diameter *CD* is parallel to a true-length line on the plane. The major-diameter length is found by transferring the measurements from the secondary auxiliary view, as shown in step 2. The minor diameter is drawn perpendicular to the major diameter through point *O*. These diameters will be used to position the ellipse template that will be used to draw the ellipse in the top view. The ellipse template angle is found in the primary auxiliary view by measuring the angle between the line of sight and the edge view of the plane.

The construction of the ellipse in the front view is found in much the same manner as in step 2; however, the true-size view is unnecessary since the center of the circle has been found in step 1. Point *O* is projected to the front view of plane 1–2–3. The edge view of the plane is found by projecting from the front view, where the ellipse guide angle can be

FIGURE 6–16. ELLIPTICAL VIEWS OF A CIRCLE

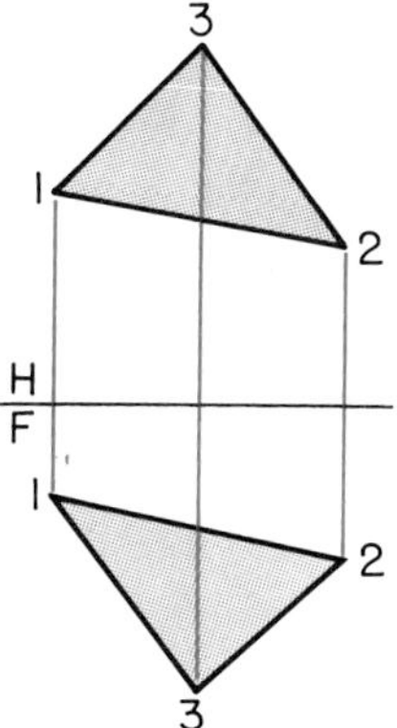

Given: The top and front views of plane 1–2–3.

Required: Construct a circle that will pass through each vertex of the plane. Show the circle in all views.

References: Articles 6–5 and 6–6.

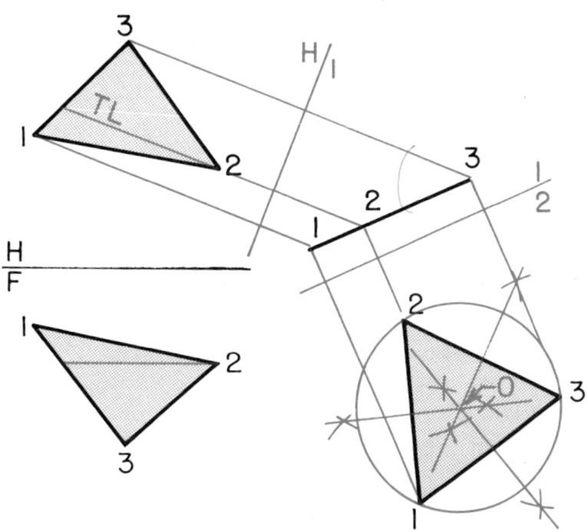

Step 1: Determine the true size of plane 1–2–3 in the manner illustrated in Article 6–5. Draw a circle through the vertexes in the true-size view. The center of the circle, *O*, is found at the intersection of the perpendicular bisectors of each of the triangle's sides.

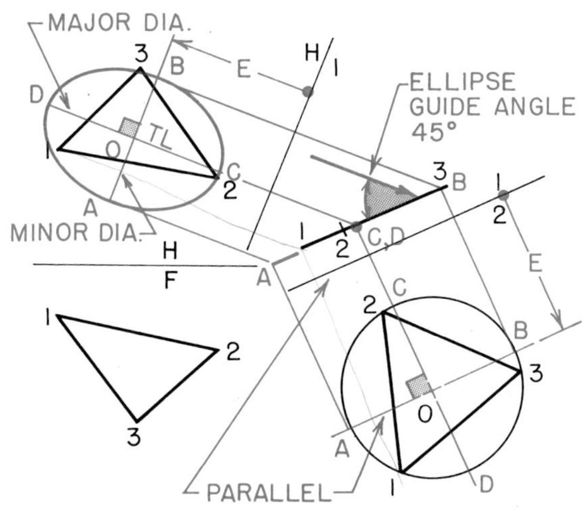

Step 2: Draw the diameters, *AB* and *CD*, parallel and perpendicular to the 1–2 line, respectively, in the secondary auxiliary view. Project these lines to the primary auxiliary and top views, where they will represent the major and minor diameters of an ellipse. Select the ellipse template for drawing the top view by measuring the angle between the line of sight and the edge view of the plane.

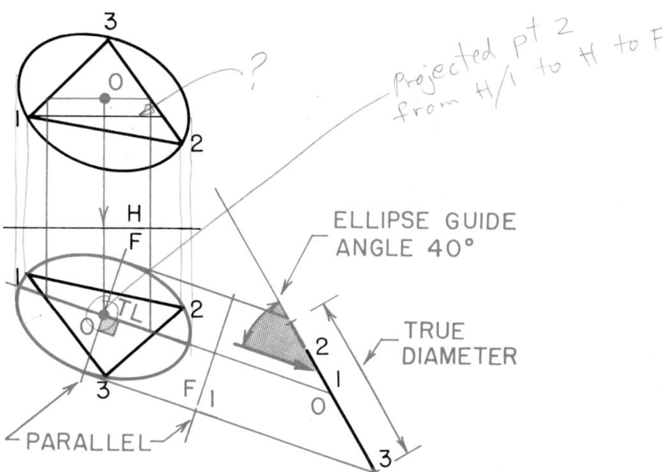

Step 3· Determine the particular ellipse template for drawing the ellipse in the front view by locating the edge view of the plane in an auxiliary view which is projected from the front view. The ellipse angle is measured in the auxiliary view as shown. Note that the major diameter is true length and that it is parallel to a true-length line on the plane in the front view. The minor diameter is perpendicular to it.

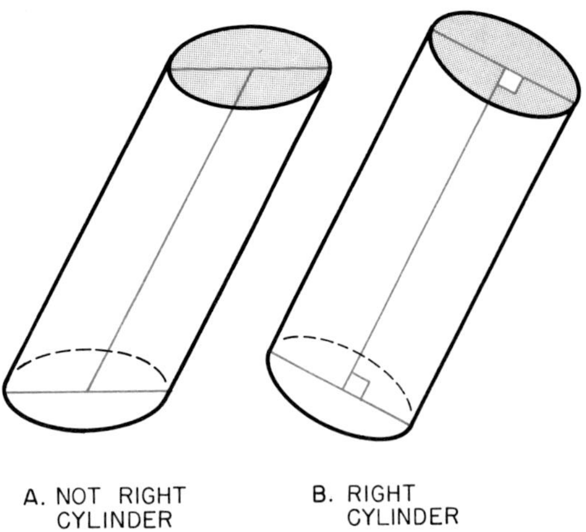

Fig. 6–17. The axis of a right cylinder is perpendicular to the major diameter of its right section.

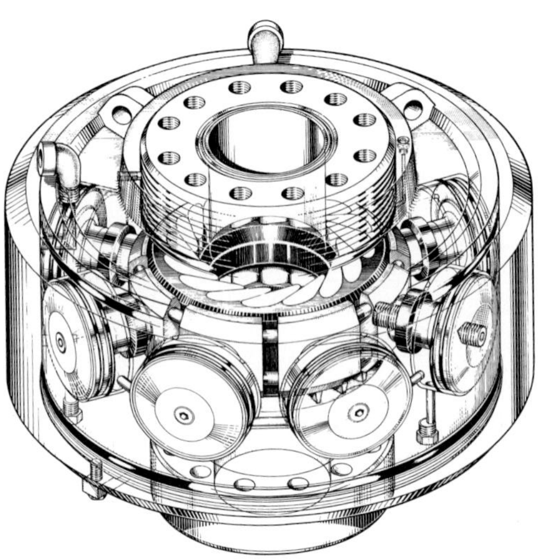

Fig. 6–18. This diaphanous view of a blowout preventer was drawn using an ellipse template to illustrate the circular features. (Courtesy of Cameron Iron Works, Inc., and L. G. Whitfield.)

found, as shown in step 3. The major diameter is drawn true length in the front view through point *O* parallel to a true-length line in the plane. The minor diameter is perpendicular to the major diameter and its length is found by projecting its extreme points from the edge view of the circle. The ellipse template can be used to construct the completed elliptical view as well as to find the top view.

Right circular cylinders are closely related to ellipses in that they are composed of a series of circles that may project as ellipses in conventional views. The right sectional ends of a cylinder will be perpendicular to the axis of a cylinder, as shown in Fig. 6–17B. It is rather obvious even to the untrained eye that the ends of the cylinder in part A of the figure are not perpendicular to the cylindrical axis. This is illustrated in Fig. 6–17B. An example of the application of this principle is shown in Fig. 6–18, a pictorial of a blowout preventer. Note that there are elliptical holes and features in the internal portion of the device; the drawing of these required the application of the previously covered principles.

6–7 SHORTEST DISTANCE FROM A POINT TO A LINE

The shortest distance from a given point to a line must be known in order to make the most economical use of material, whether it is pipe, structural members or power conductors. The steps required to find this shortest distance are given in Fig. 6–19.

The true length of the line is found in step 1 and its point view is found in step 2. The perpendicular distance from the point to the line can be seen true length in the view in which the line appears as a point. This line is projected to the primary auxiliary view, where it will be perpendicular to the line that is true length. Since line 3–*O* is true length in the secondary auxiliary view, it must be parallel to the 1–2 reference plane in the preceding view, as shown in step 3. Line 3–*O* is projected back to the other views in sequence.

The shortest distance from a point to a line can also be found by an alternative method covered in Article 6–5. The true size of plane 1–2–3 can be found where the perpendicular

FIGURE 6–19. SHORTEST DISTANCE FROM A POINT TO A LINE

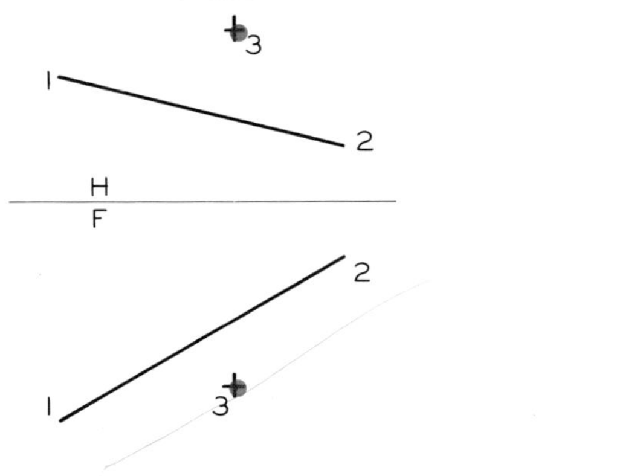

Given: The top and front views of line 1–2 and point 3.

Required: Find the shortest distance from point 3 to line 1–2 and show it in all views.

References: Articles 6–3 and 6–7.

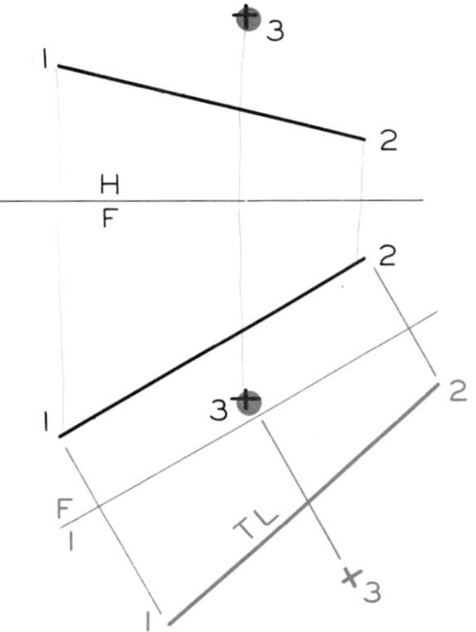

Step 1: Find the true length of line 1–2 by projecting a primary auxiliary view from the front view. Draw the reference line, F–1, parallel to the front view of line 1–2 and make all projections perpendicular to the F–1 line. Project point 3 to this view also.

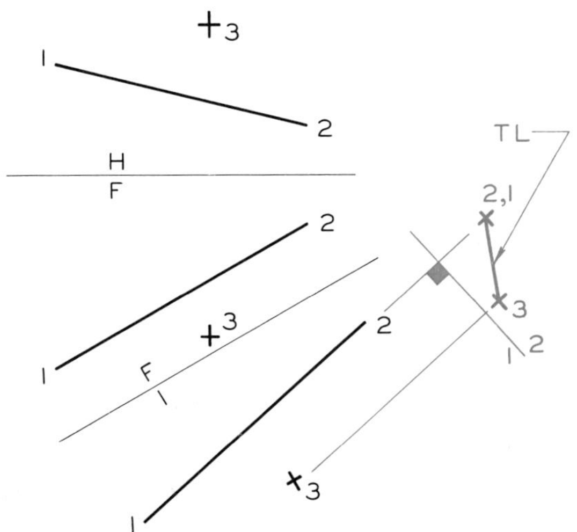

Step 2: Draw a secondary reference line, 1–2, perpendicular to line 1–2, in order to find the point view of line 1–2. The perpendicular distance from point 3 to line 1–2 can be seen true length in this view.

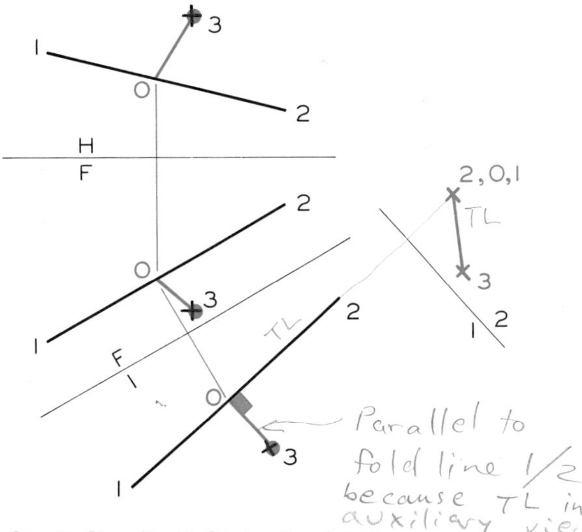

Step 3: Since line 3–*O* is true length in the secondary auxiliary view, it must be parallel to the 1–2 line in the primary auxiliary view and perpendicular to line 1–2. When one or both perpendicular lines are true length, they will project as perpendicular. Determine the front and top views of line 3–*O* by projecting from the primary auxiliary view in sequence.

distance can be drawn perpendicular to line 1–2 and measured true length in the same view.

It can be seen in Fig. 6–20 that in industry the determining of the shortest distances from points to lines is a frequent necessity to conserve expensive materials and labor.

6–8 SHORTEST DISTANCE BETWEEN SKEWED LINES—LINE METHOD

The determination of the shortest clearance between two lines is applicable to a number of industrial situations encountered by the engineer, technologist, and technician. The high-voltage power lines shown in Fig. 6–21 must have a minimum clearance, which is specified by regulations. The design of the support towers will be affected by this specified clearance, as will the safety factors related to the clearance.

Fig. 6–20. This top structure of a blast furnace involves application of finding the shortest distance from a point to a line. (Courtesy of Jones & Laughlin Steel Corporation.)

Two methods—the line method and the plane method—are used to find the shortest distance between two lines. The line method is presented in Fig. 6–22 in sequential steps. Since the shortest distance between two lines will be a line perpendicular to both, the line will appear true length in the secondary auxiliary view, where line 3–4 projects as a point. To establish point *P*, line *OP* is projected to the primary auxiliary view, where it is drawn perpendicular to line 3–4, which is true length in this view. Points *O* and *P* are projected to the front and top views to represent the shortest distance between the two lines.

6–9 SHORTEST DISTANCE BETWEEN SKEWED LINES—PLANE METHOD

The problem covered in Article 6–8 can be solved by the application of the plane method. This technique requires that a plane be constructed through one of the lines parallel to the other. This is illustrated in Fig. 6–23 where line *O*–2 is drawn parallel to line 3–4 in both views. Both lines will project as if they were parallel in a view where the plane appears as an edge.

In Fig. 6–24, a line is constructed through a point (point 4 in this case) parallel to line 1–2

Fig. 6–21. The shortest distance between two crossing high-voltage power lines can be found by descriptive geometry theory. (Courtesy of the Tennessee Valley Authority.)

FIGURE 6–22. SHORTEST DISTANCE BETWEEN SKEWED LINES—LINE METHOD

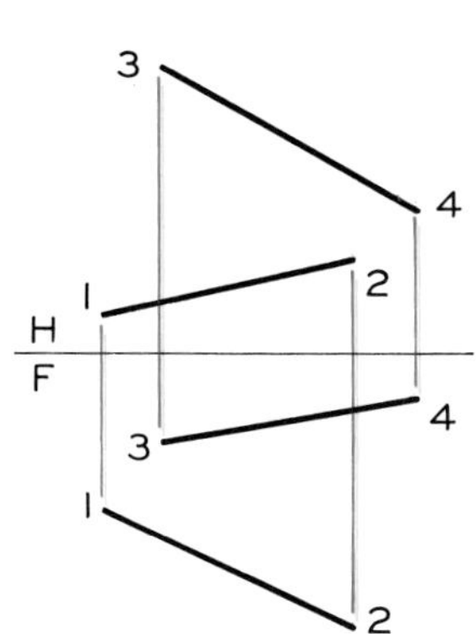

Given: The front and top views of lines 1–2 and 3–4.

Required: Find the shortest distance between the two lines by the line method and show it in all views.

References: Articles 6–3 and 6–8.

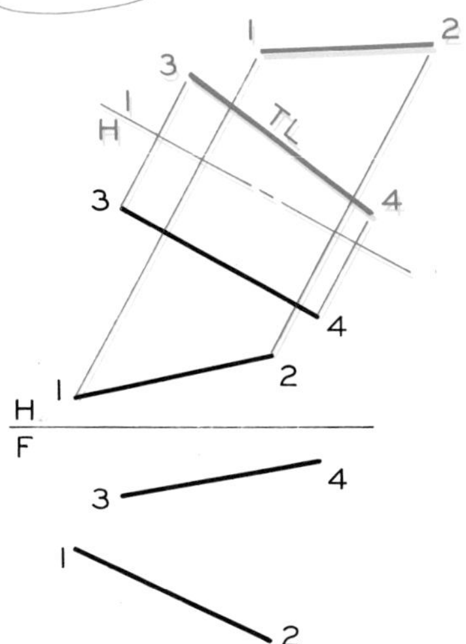

Step 1: Find line 3–4 true length in a primary auxiliary view projected from the horizontal view. Project line 1–2 to this view also. The primary auxiliary view could have been projected from the front view equally well.

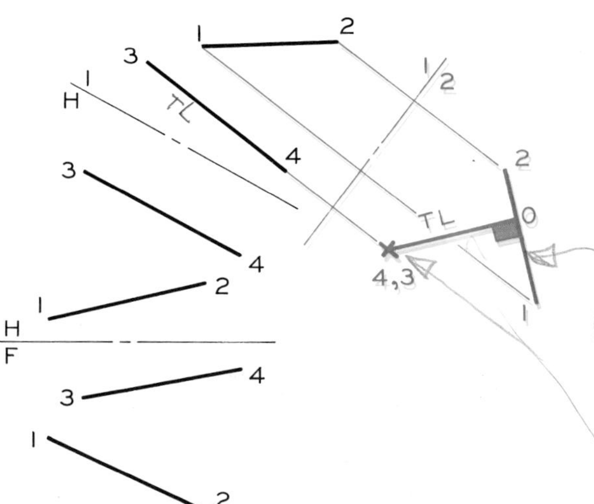

Step 2: Draw a secondary auxiliary view to find line 3–4 as a point. The shortest distance between the two lines is a line perpendicular to both. This line will appear true length in the secondary auxiliary view, where it is drawn perpendicular to line 1–2.

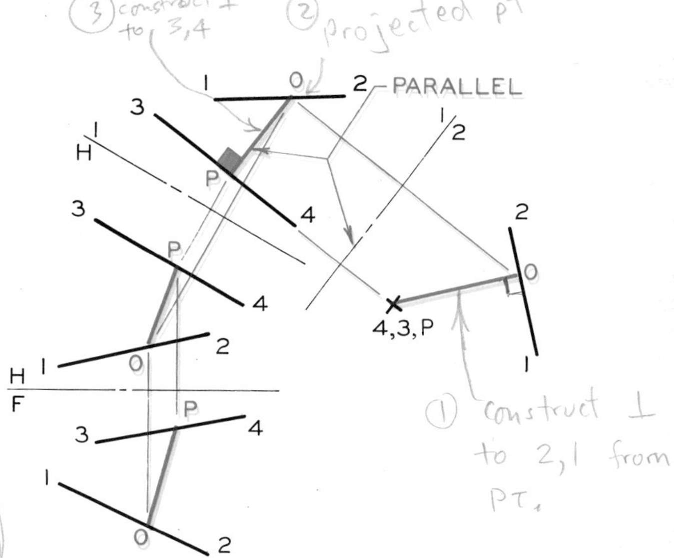

Step 3: Locate point *O* in the primary auxiliary view by projection. Locate point *P* on line 3–4 by constructing line *O–P* through point *O* perpendicular to line 3–4. These points are projected back to the top and front views to represent the line. Note that line *O–P* is parallel to the 1–2 reference line in the primary auxiliary view since it is true length in the secondary view.

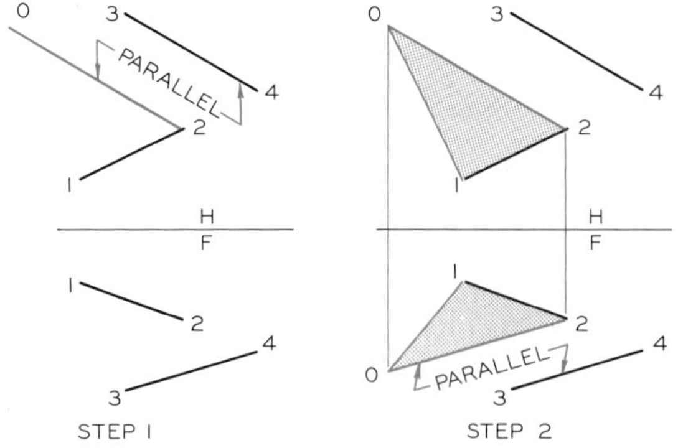

Fig. 6–23. A plane can be constructed through a line parallel to another by construction. In step 1, line *O*–2 is drawn parallel to the top view of line 3–4. The front view of *O*–2 is drawn parallel to the front view of line 3–4. Plane 1–2–*O* is parallel to line 3–4.

in both views. Two intersecting lines, 4–*O* and 3–4, form a plane. A number of planes could be constructed in this manner, but all would lie on a common infinite plane. When an edge view of the plane is found in a primary auxiliary view, the two lines project as parallel, as shown in step 1, where the plane is found as an edge. The plane has served its purpose when the auxiliary view is found and can be ignored in the remaining steps of the solution.

The shortest distance between the two lines will be projected as perpendicular to each of the lines in the primary auxiliary view, where the lines are parallel (step 2). Although a number of lines can be drawn apparently perpendicular to the lines, only one will be truly perpendicular—the one that appears true length in the primary auxiliary view. This line will appear as a point in the secondary auxiliary view, which is projected perpendicularly from the primary auxiliary view, as shown in step 3. Both lines, 1–2 and 3–4, appear true length in this view; consequently, the shortest distance between them is at the point of crossing, where line *LM* projects as a point. The true-length view of line *LM* is found in the primary auxiliary view. The line is projected to the two principal views to complete the requirements of the problem.

This method of solving for the shortest distance between two lines is the general case that can be used in the solution of problems covered in Articles 6–10 and 6–11. Complicated traffic systems, such as that shown in Fig. 6–25, must be analyzed to determine the clearances between the center lines of the crossing highways. Also, the vertical clearances are critical to the design of the overpasses. Vertical distances between skewed lines appear true length in the front view directly beneath the point where the two lines cross in the top view.

6–10 SHORTEST LEVEL DISTANCE BETWEEN TWO SKEWED LINES

The shortest level, or horizontal, distance between two lines is found by using the plane method in the initial steps, as illustrated in Article 6–9. A plane is constructed through one of the lines to be parallel to the other given line in step 1, Fig. 6–26. An edge view of the plane must be projected from the *top view* in order that the horizontal reference line may appear as an edge in the primary auxiliary view, where the level distance can be drawn parallel to the horizontal plane. This problem *cannot* be solved by projecting from the front view, because in that case the frontal plane, rather than the horizontal plane, will appear as an edge. The direction of the shortest level line will appear true length in step 2 and parallel to the H–1 reference line. Only the *shortest* level line will appear true length in this view. A secondary auxiliary view is required to locate its position. The point view of line *LM* is found in the secondary auxiliary view and is projected to the primary auxiliary view, where it appears true length. As a check on the accuracy of the construction, line *LM* should appear level in the front view.

An application of this problem is the connection of two roadways with a level tunnel (Fig. 6–27). This principle is also used to connect mining shafts with tunnels which must remain level, but which also must be as short as possible for reasons of economy.

FIGURE 6–24. SHORTEST DISTANCE BETWEEN SKEWED LINES—PLANE METHOD

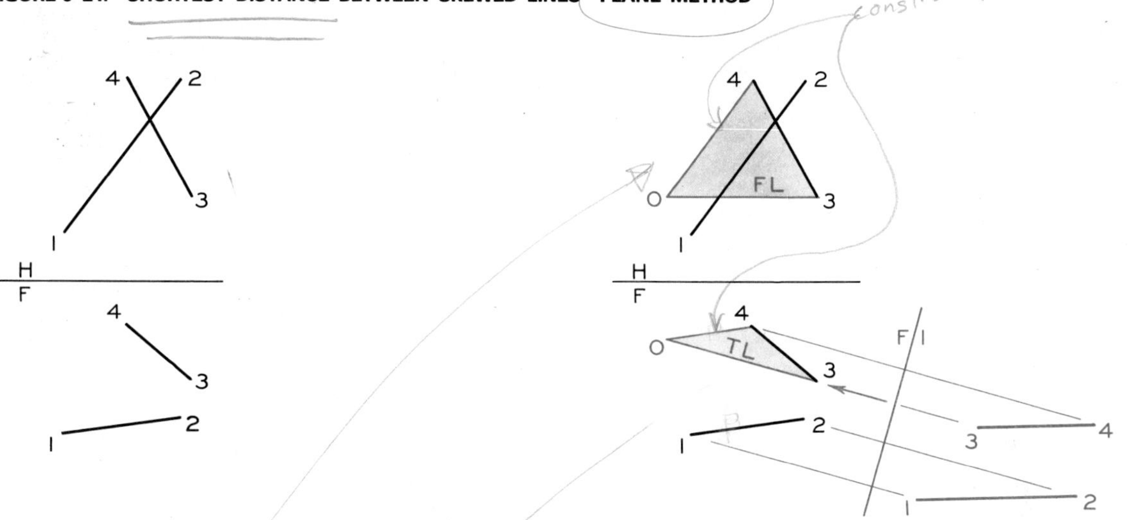

Given: The top and front views of lines 1–2 and 3–4.

Required: Find the shortest distance between the two lines by the plane method. Show this distance in all views.

References: Articles 4–19 and 6–9.

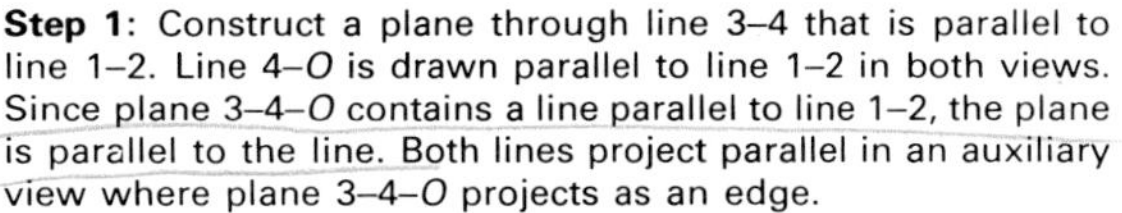

Step 1: Construct a plane through line 3–4 that is parallel to line 1–2. Line 4–*O* is drawn parallel to line 1–2 in both views. Since plane 3–4–*O* contains a line parallel to line 1–2, the plane is parallel to the line. Both lines project parallel in an auxiliary view where plane 3–4–*O* projects as an edge.

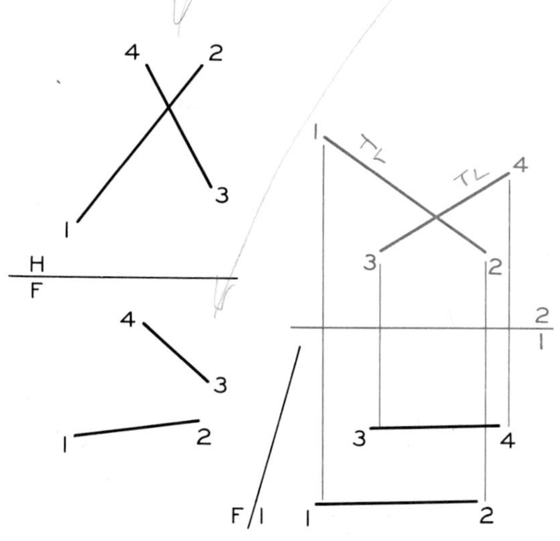

Step 2: The shortest distance will appear true length in the primary auxiliary view, where it will be perpendicular to both lines. Draw a secondary auxiliary view by projecting perpendiculary from the lines in the primary auxiliary view. Lines 1–2 and 3–4 cross in this view.

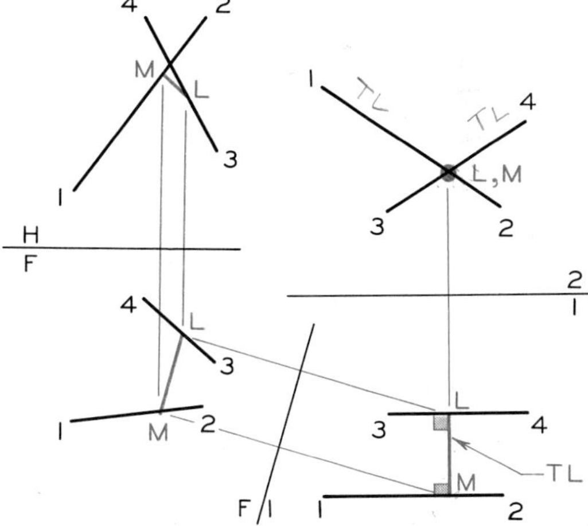

Step 3: The crossing point of lines 1–2 and 3–4 establishes the point view of the perpendicular distance, *LM*, between the lines. This distance is projected to the primary auxiliary view, where it is true length. The line is found in the front and top views by projecting points *L* and *M* to their respective lines in these views.

Fig. 6–25. The interchange of Harbor and Santa Monica Freeways in downtown Los Angeles illustrates a variety of skewed-line applications. (Courtesy of the California Division of Highways.)

Fig. 6–27. Construction of a three-mile tunnel to be used for a rapid transit system in the Berkeley Hills area of San Francisco. (Courtesy of Kaiser Engineers.)

6–11 SHORTEST GRADE DISTANCE BETWEEN SKEWED LINES

The plane method introduced in Article 6–9 must be employed in solving for the shortest grade line between two skewed lines. It can be seen from this series of problems that the plane method is a general approach to solving all skewed-line problems, whereas the line method is applicable only to the perpendicular distance between two skewed lines.

The lines are projected as parallel in the primary auxiliary view by finding the edge view of a plane constructed parallel to one of the lines through the other, as shown in step 1 of Fig. 6–28. This primary auxiliary view *must* be projected from the *top view* in order for the horizontal plane to appear as an edge, from which the percent grade of a line can be drawn. Recall that grade is the ratio of the vertical rise to the horizontal run of a line, expressed as a percentage (step 2). The grade could be drawn in two directions with respect to the H–1 reference line. However, the shortest grade between two lines will be the line drawn in the direction that is most nearly perpendicular to the lines. The secondary auxiliary line, 1–2, is drawn perpendicular to the grade line that has been constructed, and the view is projected parallel to the direction of the grade line. The lines in this example do not cross in the secondary auxiliary view. Since lines 1–2 and 3–4 are only segments of longer, continuous lines, they can be extended to their point of intersection, as shown in step 3. This locates the point view of the shortest line that can be drawn at a 50-percent grade. This line, *LM*, can be projected back to the horizontal and frontal views, as illustrated.

Figure 6–29 illustrates a multitude of pipes that had to be designed to conform to grade specifications in order for the system to function under design conditions. Figure 6–30 shows a complex traffic interchange where highways connect with intermediate arteries on a grade. Drainage problems and sewer systems must also be critically analyzed with respect to grade distances between drainage channels and culverts.

6-27

FIGURE 6–26. SHORTEST LEVEL DISTANCE BETWEEN SKEWED LINES

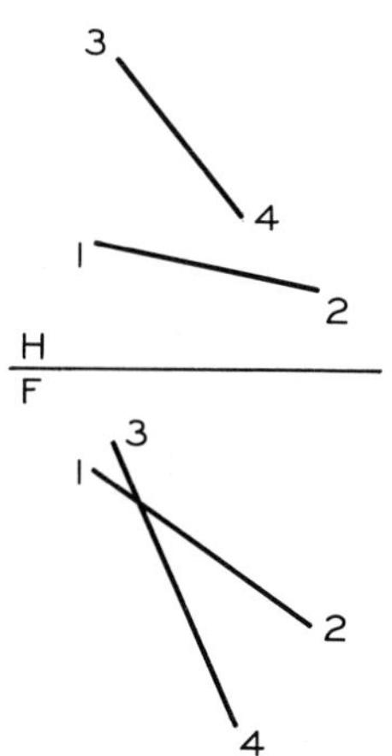

Given: The top and front views of lines 1–2 and 3–4.

Required: Find the shortest level distance between the lines and project it to all views.

References: Articles 4–19, 6–9, and 6–10.

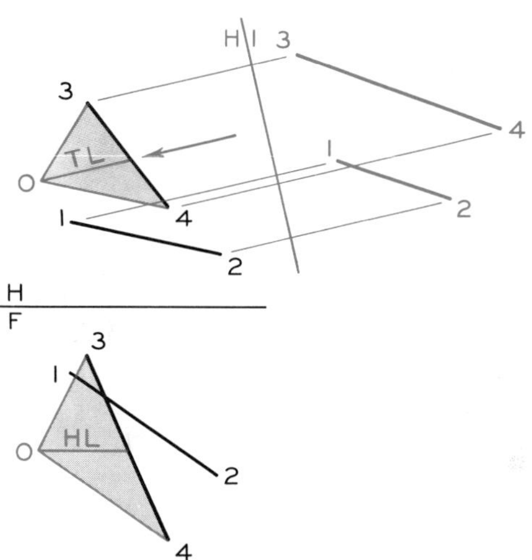

Step 1: Construct plane 3–4–*O* parallel to line 1–2 by drawing line 4–*O* parallel to line 1–2 in the top and front views. Plane 3–4–*O* is found as an edge in the primary auxiliary view where the lines are projected as parallel. *Note:* The primary auxiliary *must* be projected from the *top view* to find the horizontal plane as an edge.

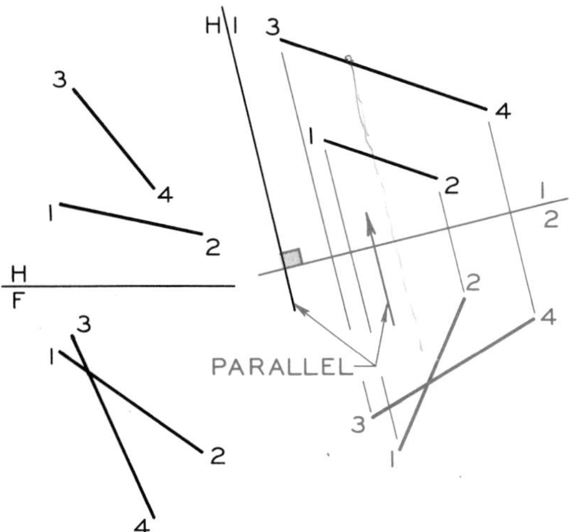

Step 2: An infinite number of horizontal (level) lines can be drawn parallel to H–1, between the lines in the primary auxiliary view, but only the shortest level line will project true length in the primary auxiliary view. Draw the secondary auxiliary plane, 1–2, perpendicular to the H–1 line, and project lines 1–2 and 3–4 to this view.

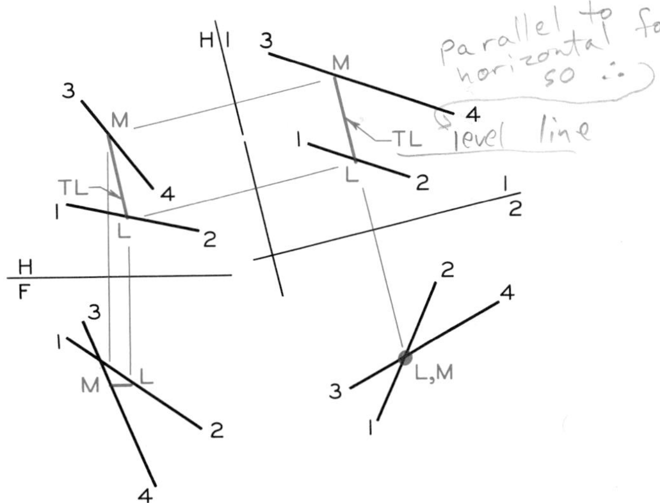

Step 3: The point where lines 1–2 and 3–4 cross in the secondary auxiliary view establishes the point view of line *LM* that will appear true length in the primary auxiliary view. Project line *LM* to the top view and front views. Line *LM* is parallel to the H-plane in the front view, which verifies that it is a level line.

FIGURE 6–28. GRADE DISTANCE BETWEEN SKEWED LINES

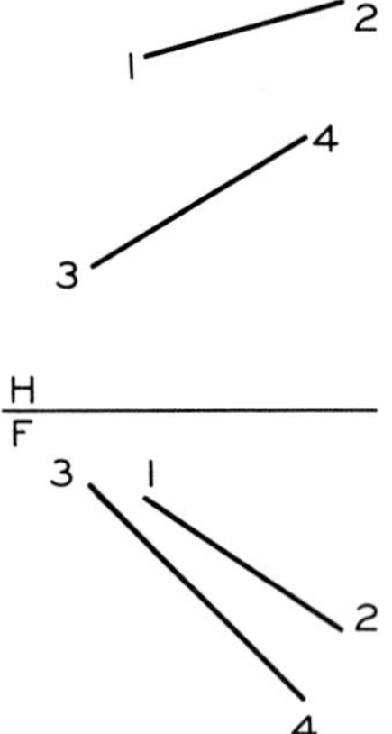

Given: The top and front views of lines 1–2 and 3–4.

Required: Find the shortest line having a 50-percent grade between the two lines.

References: Articles 4–19 and 6–11.

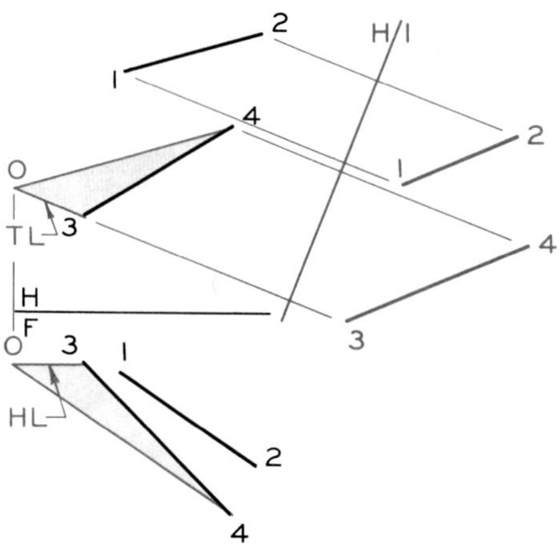

Step 1: Draw plane 3–4–*O* parallel to line 1–2 by drawing line 4–*O* parallel to line 1–2 in both views. The edge view of the plane is found where both lines project as parallel. *Note:* The primary auxiliary *must* be projected from the *top view* in order that the horizontal plane may appear as an edge in the primary auxiliary view.

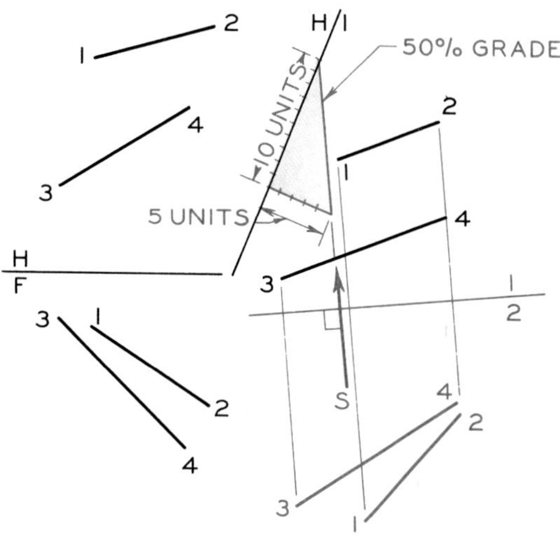

Step 2: Construct a 50-percent grade line with respect to the edge view of the H–1 line in the primary auxiliary view. Draw this line as nearly perpendicular to the lines as possible for the shortest connector. Project a secondary auxiliary parallel to the direction of the grade line. The shortest grade distance will appear true length in the primary auxiliary view.

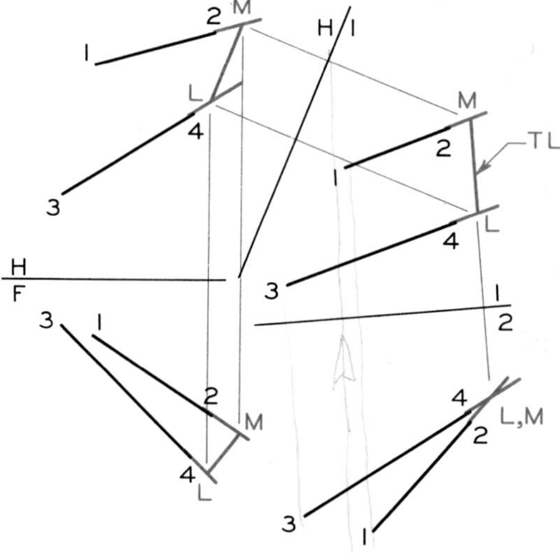

Step 3: Extend the lines in the secondary view to establish their point of intersection, where the point view of *LM* is located. Line *LM* will appear true length at a 50-percent grade in the primary auxiliary view. Line *LM* is projected back to the top and front views. Lines 1–2 and 3–4 must be extended in each view.

Fig. 6–29. Clearances between interrelated pipes must be evaluated to reduce cost of materials and installation. (Courtesy of Standard Oil Corporation of New Jersey.)

Fig. 6–30. Highways often present skewed-line problems requiring graphical solutions. (Courtesy of the California Division of Highways.)

6–12 A LINE THROUGH A POINT AT A GIVEN ANGLE TO A LINE

Standard connectors are available for joining pipes, structural beams, and other engineering forms encountered in industrial projects. Of course, the standard connectors have been designed for only the most commonly used angles, since it would be economically impossible to provide connectors that varied from 0° to 90° at 1° intervals. Consequently, it is important to know how to design connections corresponding to a given common angle.

The example problem in Fig. 6–31 illustrates the procedure for constructing a line from a given point to another line such that a standard 45° angle will be formed between the lines. This procedure could be utilized practically to design a connection that would allow a standard connector to be used. Symbols for piping drawings as well as the dimensions for standard connections are given in Appendixes 10, 12, and 13.

This problem specifies that the line from point *O* slopes downward to line 1–2. Observation of the front view tells you that the point of intersection will be located closer to point 2 than to point 1, since point 2 is the low end of the line. Line *OP* is drawn from point *O* to intersect line 1–2 at 45°, the standard angle, in step 3. Line *OP* is projected back to the given views in sequence.

Fig. 6–32. The design of a processing plant such as this involves problems which lend themselves to solution by successive auxiliary views. (Courtesy of Standard Oil Corporation of New Jersey.)

Fig. 6–33. In complicated systems, clearances and optimum distances are often checked by analysis of three-dimensional models. (Courtesy of Exxon Research of New Jersey.)

FIGURE 6–31. LINE THROUGH A POINT WITH A GIVEN ANGLE TO A LINE

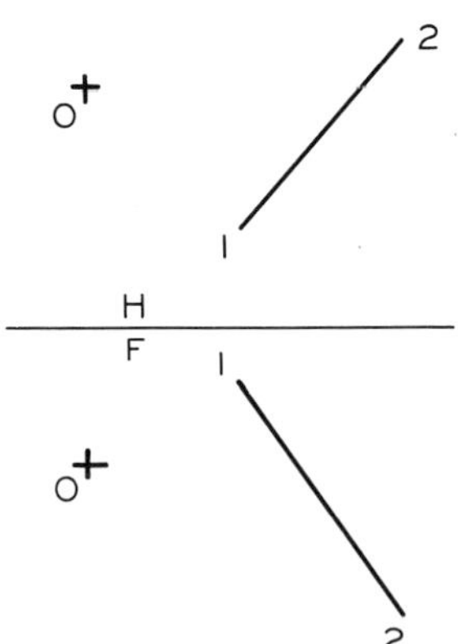

Given: The top and front views of line 1–2 and point *O*.

Required: Construct a line sloping downward from point *O* that will make an angle of 45° with line 1–2.

References: Articles 6–5 and 6–12.

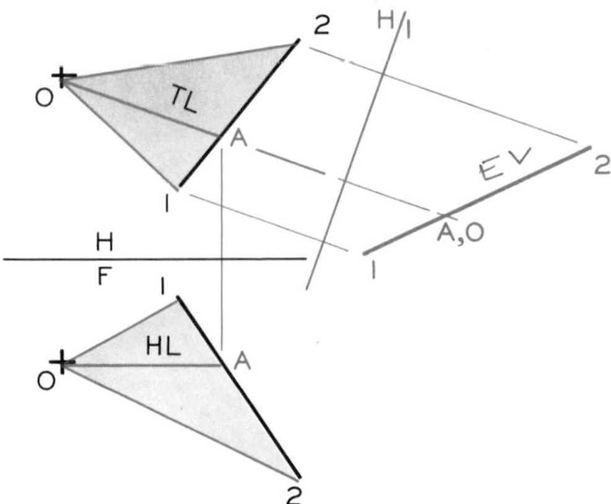

Step 1: Connect point *O* to each end of the line to form plane 1–2–*O* in both views. Draw a horizontal line in the front view of the plane and project it to the top view, where it is true length. Determine the edge view of the plane by finding the point view of line *OA*.

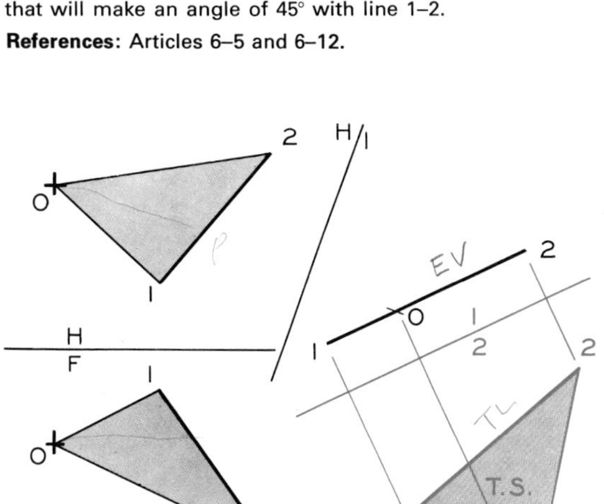

Step 2: Determine the true size of plane 1–2–*O* by perpendicularly projecting an auxiliary view from the edge view of the plane in the primary auxiliary view. The plane need not be drawn in the secondary auxiliary view since it will not be used further.

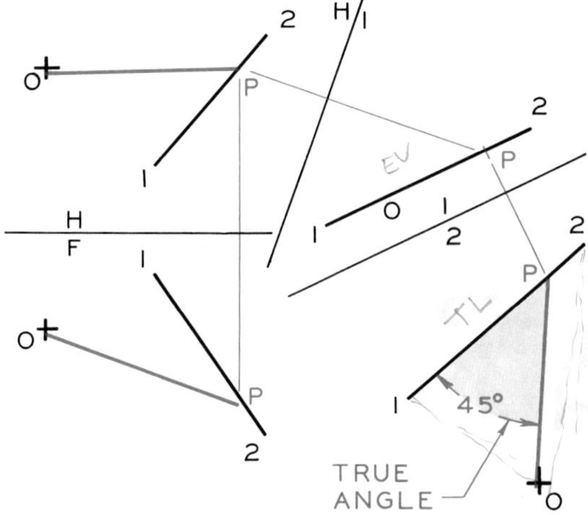

Step 3: Line *OP* can be constructed at the specified angle of 45° with line 1–2 in the secondary auxiliary view toward the low end, since point *O* and line 1–2 lie in the same plane. Project point *P* back to the primary auxiliary, top, and front views, and connect it with point *O*. This problem also could have been solved by projecting from the front view.

Designing for standard connectors is common practice in the complex chemical and petroleum industry, as illustrated in Fig. 6–32. The details of construction of a refinery or processing plant are so complicated that often three-dimensional models are used to assist in the solution of the design problems (see Fig. 6–33).

FIGURE 6–34. ANGLE BETWEEN A LINE AND A PLANE—PLANE METHOD

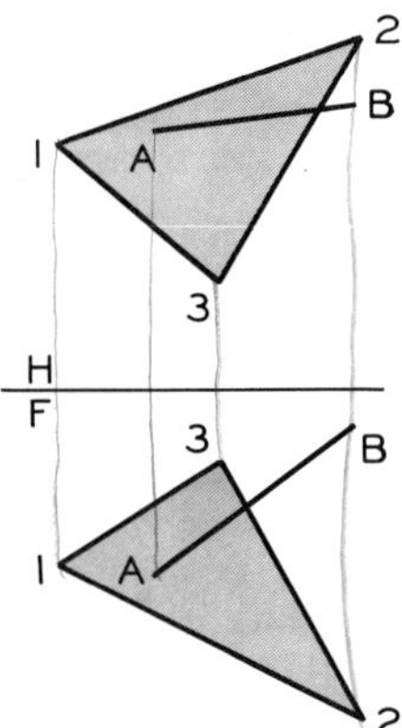

Given: The top and front views of plane 1–2–3 and line *AB*.

Required: Find the angle between the line and plane and determine its visibility in all views.

References: Articles 6–5 and 6–13.

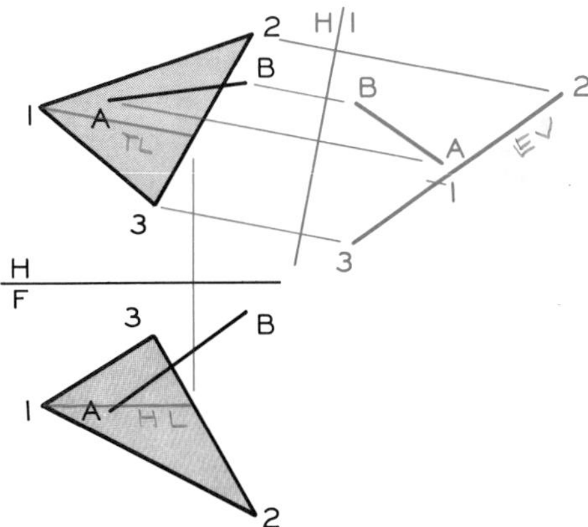

Step 1: Determine the edge view of plane 1–2–3 by projecting from either the front or top view. Find the edge view by projecting from the top view in this example; project line *AB* also. The angle cannot be measured in this view since the line is not true length.

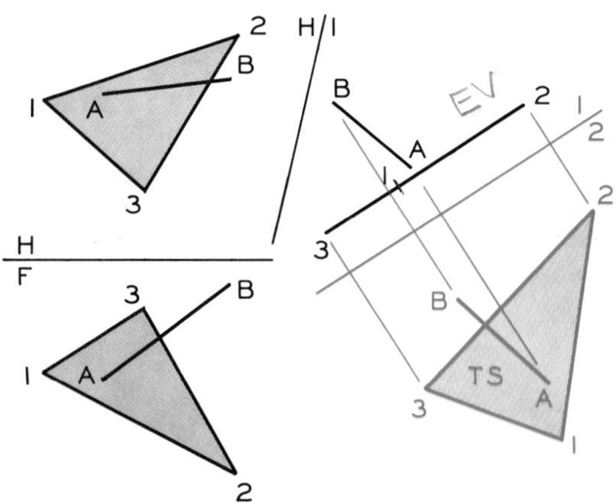

Step 2: Determine the true size of plane 1–2–3 in a secondary auxiliary view projected perpendicularly from the edge view of the plane. Line *AB* is not true length in this view. Draw line *AB* lightly since visibility must be determined.

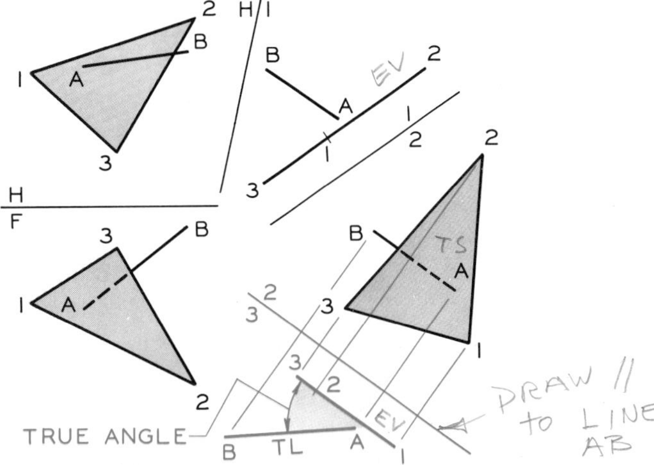

Step 3: A view projected in any direction from a true-size view of a plane will result in an edge view of the plane. Since line *AB* must be true length at the same time, project a third auxiliary view perpendicularly from line *AB*. The line appears true length and the plane appears as an edge, thus satisfying the conditions for measuring the angle between them. Visibility is shown in all views.

6–13 ANGLE BETWEEN A LINE AND A PLANE—PLANE METHOD

Although standard connectors and hardware components should be considered for connecting structural members to a plane, there will be cases where a particular nonstandard angle is unavoidable, so the designer must be able to determine the angle between a line and a plane to design the special connector. This principle has many applications; for instance, in space vehicles the angle of the observer's line of sight to the plane of the instrument panel must fall within the operational limitations previously determined.

These problems are solved by finding a view where the plane appears as an edge and the line appears true length (Fig. 6–34). The plane is found as an edge in the primary auxiliary view in step 1 and true size in step 2. Since the plane is true shape in this view, any auxiliary view projected from it will result in an edge view of the plane. The third auxiliary view plane, 2–3, is drawn parallel to line *AB* in step 3 in order to find the true length of line *AB* and an edge view of the plane. This condition exists in step 3; therefore, the true angle can be measured in the third auxiliary view.

Fig. 6–35. The Hydra 5 sea-test vehicle required application of the principle for finding the angle between a line and a plane

An application of this principle can be seen in Fig. 6–35 in the Hydra 5 sea-test vehicle. A series of tripods have been constructed at intervals along the body of the vehicle. Finding the angle which one leg of each tripod forms with the other two is the same as finding the angle between a line and a plane, since two intersecting lines form a plane. This information was necessary to design the tripods.

6–14 ANGLE BETWEEN A LINE AND A PLANE—LINE METHOD

An alternative method for finding the angle between a line and a plane is the line method illustrated in Fig. 6–36. The true length of line *AB* is found in step 1, and the point view of the line is found in step 2. Any view of a line projected from its point view will show the true length of the line. A line is constructed on the plane in the primary auxiliary view that is parallel to the 1–2 reference line. This line is projected to the plane in the secondary auxiliary view, where it appears true length on plane 1–2–3. Since line *AB* appears as a point in the secondary auxiliary view, any view projected from it will result in a true-length view of the line. Therefore the edge view of the plane can be found in the third auxiliary view by projecting a point view of the true-length line on the plane (step 3). The line appears true length and the plane appears as an edge, making possible the measurement in the final view. The point of intersection is located in each view and the visibility is determined.

6–15 SUMMARY

Successive auxiliary views can be used to great advantage to refine preliminary designs and to determine information that is needed for design finalization and analysis. Many of the solutions illustrated would be virtually impossible without using the principles of descriptive geometry and graphical methods. The engineer and technician should have command of these methods in order to recognize problems that lend them-

FIGURE 6–36. ANGLE BETWEEN A LINE AND A PLANE—LINE METHOD

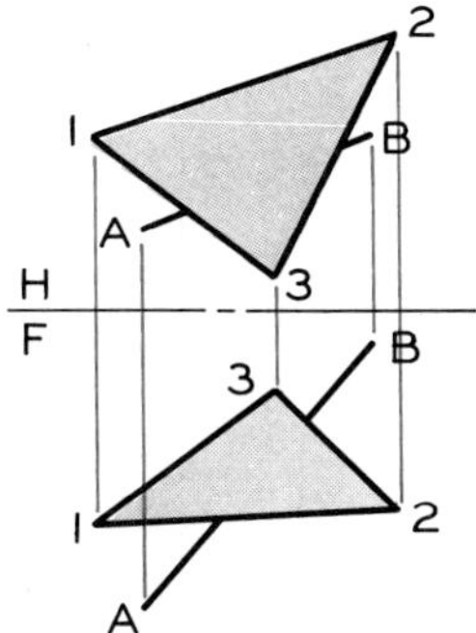

Given: The top and front views of plane 1–2–3 and line *AB*.

Required: Find the angle between line *AB* and plane 1–2–3 by the line method.

References: Articles 6–13 and 6–14.

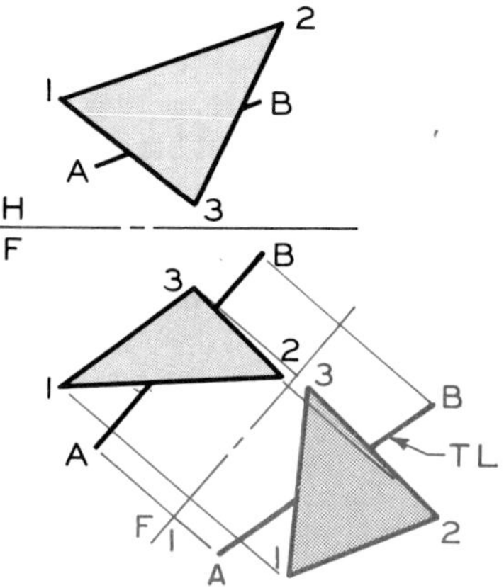

Step 1: Determine the true length of line *AB* in a primary auxiliary view by projecting from either principal view. Plane 1–2–3 is projected also; however, it does not appear true size in this view except in a special case.

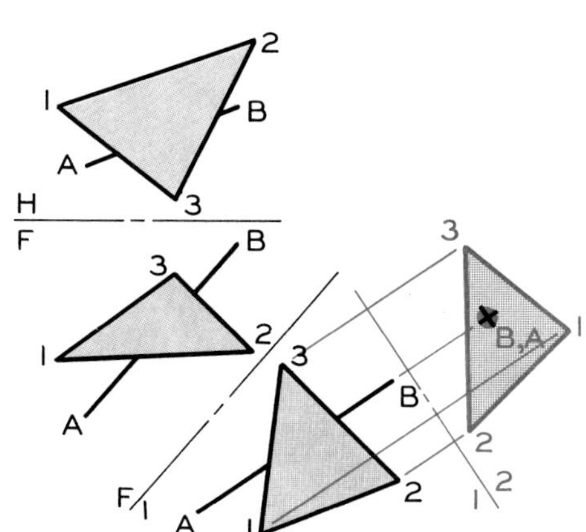

Step 2: Construct the point view on line *AB* in the secondary auxiliary view. Plane 1–2–3 does not appear true size in this view unless the line is perpendicular to the plane. The point view of the line in this view is also the piercing point on the plane.

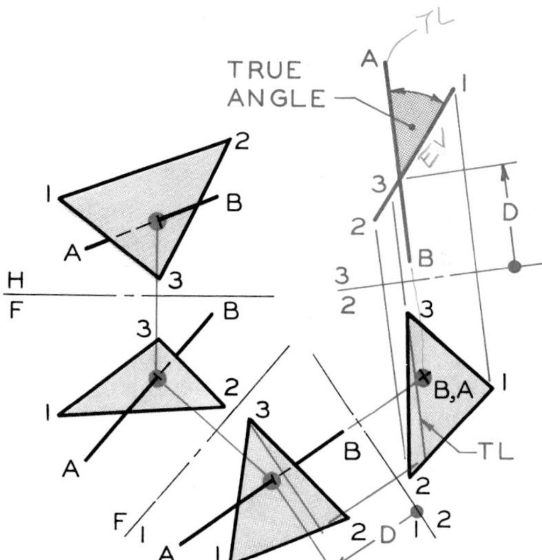

Step 3: Construct a true-length line on plane 1–2–3 in the secondary auxiliary view, from which the edge view of the plane can be found in the third auxiliary view. Line *AB* will be true length in this view, since it appeared as a point in the secondary auxiliary view. Measure the angle in the third auxiliary view and determine the piercing point and visibility in the previous views.

selves to graphical solutions but which would be difficult to solve by other methods.

It should be remembered that auxiliary views are merely orthographic projections that have the same relationship to each other as do principal views. (Fundamentals of orthographic projection can be reviewed in Chapter 4.) A thorough understanding of these basic principles is a prerequisite to the solution of problems by successive auxiliary views, since each construction step must be analyzed for spatial relationships before the next view is projected.

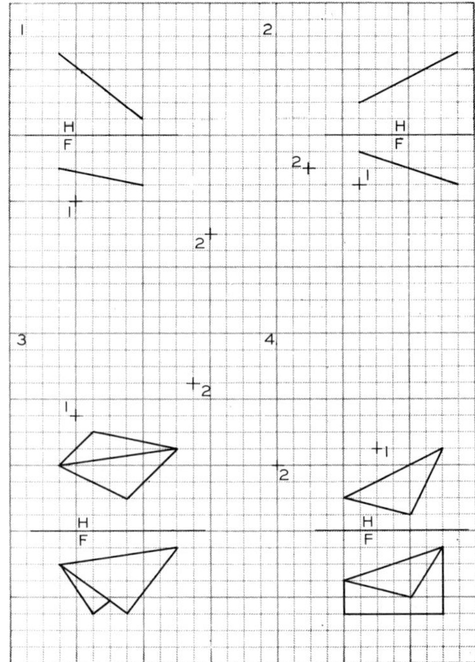

FIGURE 6–37

PROBLEMS

Problems for this chapter can be constructed and solved on $8\frac{1}{2}'' \times 11''$ sheets with instruments as illustrated by the accompanying figures. Each grid represents $\frac{1}{4}''$. Reference planes and points should be labeled in all cases using $\frac{1}{8}''$ letters with guidelines.

The crosses marked "1" and "2" are to be used for placing the primary and secondary reference lines. The primary reference line should pass through "1" and the secondary through "2." Refer to Article 3–6 for layout rules.

Point View of a Line

1 and 2. Find the point views of the lines in Fig. 6–37.

Angle between Two Planes

3 and 4. Find the angles between the intersecting lines in Fig. 6–37. Project from the top views in both problems.

True Size of a Plane

5 and 6. Find the true size views of the planes in Fig. 6–38. Project from the front view in problem 5 and from the top view in problem 6.

7 and 8. Find the angle between lines 1–2 and 2–3 in problem 7 (Fig. 6–39). Project from the top view. Find the angle that the line from point 7 makes with line 5–6 in Fig. 6–39. Project from the top view.

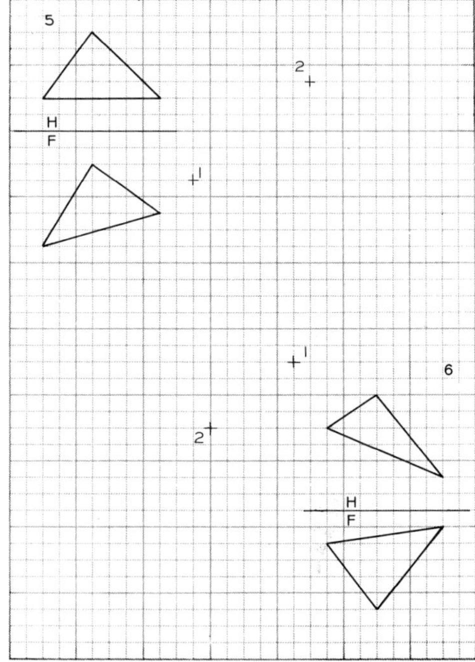

FIGURE 6–38

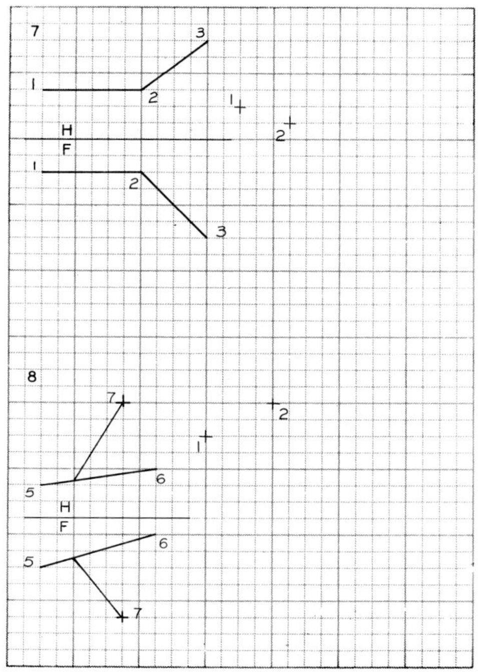

FIGURE 6–39

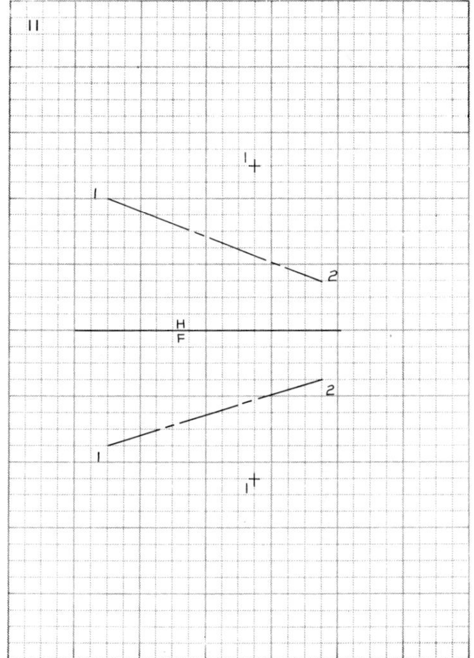

FIGURE 6–41

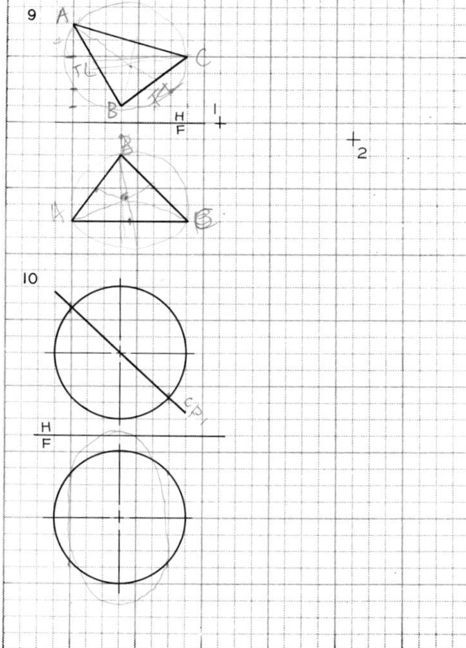

FIGURE 6–40

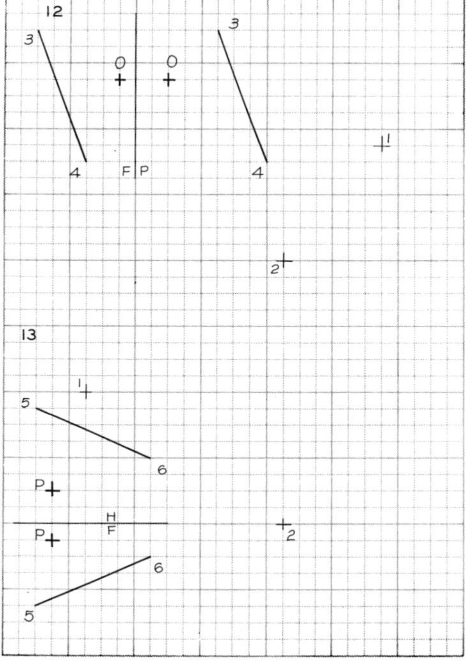

FIGURE 6–42

Elliptical View of a Circle

9 and 10. Construct a circle that will pass through each vertex of the triangle in problem 9 (Fig. 6–40). Show the elliptical views of the circle in the given views. Project from the top view. In problem 10, find the front view of the elliptical path of a satellite whose circular path appears as an edge in the top view.

11. Line 1–2 in Fig. 6–41 represents the center line of a right cylinder in which each circular end is perpendicular to the axis. Show the cylinder in all views with a 25 mm diameter.

Shortest Distance from a Point to a Line

12. Find the shortest distance from point *O* to line 3–4 (Fig. 6–42) and show the line in all views. Use the plane method and project from the front view. Scale: full size.

13. Find the shortest distance from point *P* to line 5–6 (Fig. 6–42) and show the line in all views. Use the line method and project from the top view. Scale: full size.

14. Find the distance between the two skewed lines in Fig. 6–43 and show this line in all views. Use the line method. Scale: full size.

15. Find the distance from the two skewed lines in Fig. 6–43 and show the line in all views. Use the line method by finding line 3–4 true length by projecting from the top view. Scale: full size.

16. Find the shortest distance between the two skewed lines in Fig. 6–44 by the plane method. Project from the top view. Scale: full size.

17. On a separate sheet of paper, redraw problem 16 (Fig. 6–44) and find the shortest horizontal distance between the two lines and show this line in all views. Project from the top view. Scale: full size.

18. Find the shortest 20-percent grade distance between the lines given in Fig. 6–45. Project from the top view. Scale: full size.

Angle from a Point to a Line

19. In Fig. 6–46, find the shortest distance from point *O* to line 1–2 that will intersect at 60°.

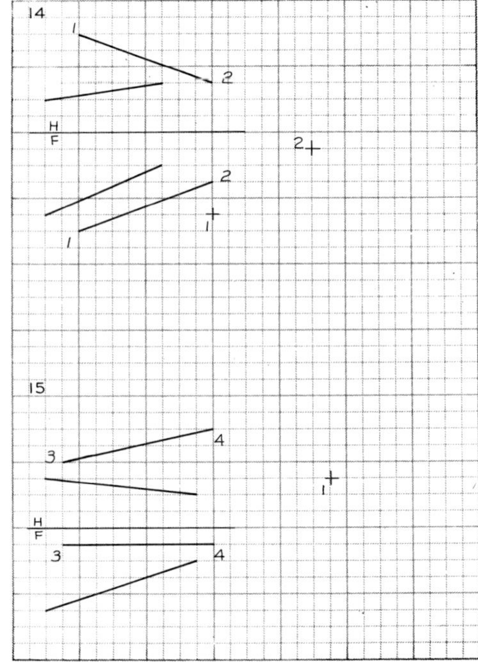

FIGURE 6–43

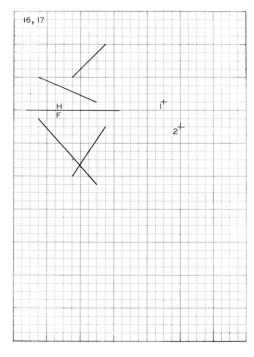

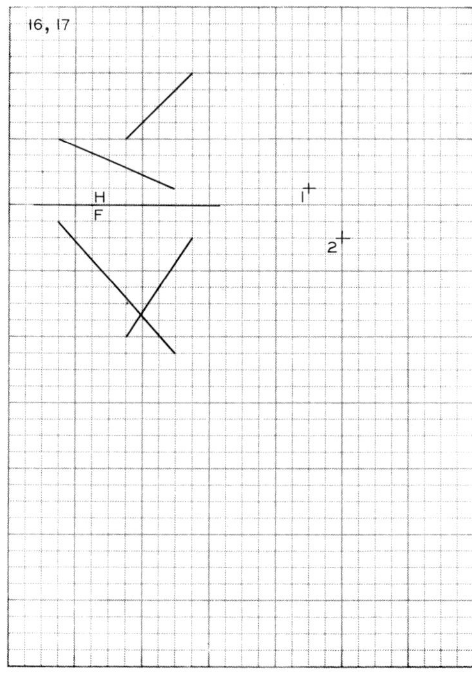

FIGURE 6–44

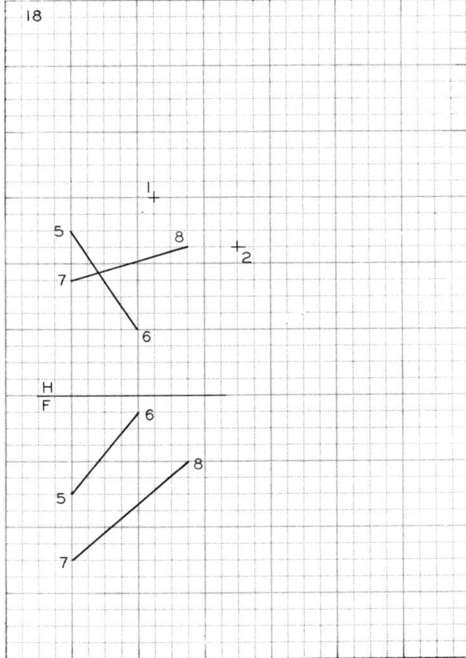

FIGURE 6–45

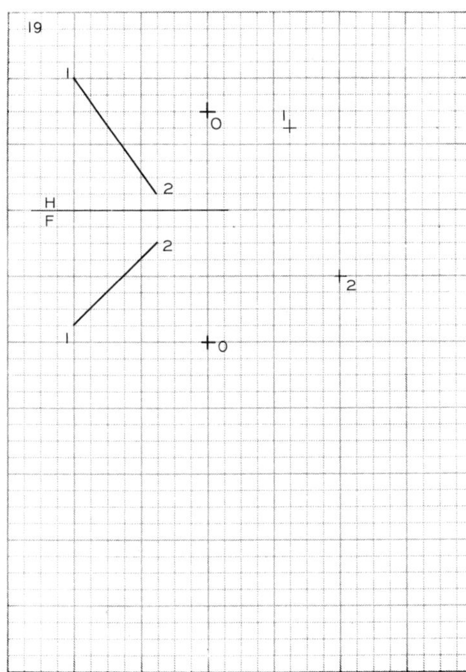

FIGURE 6–46

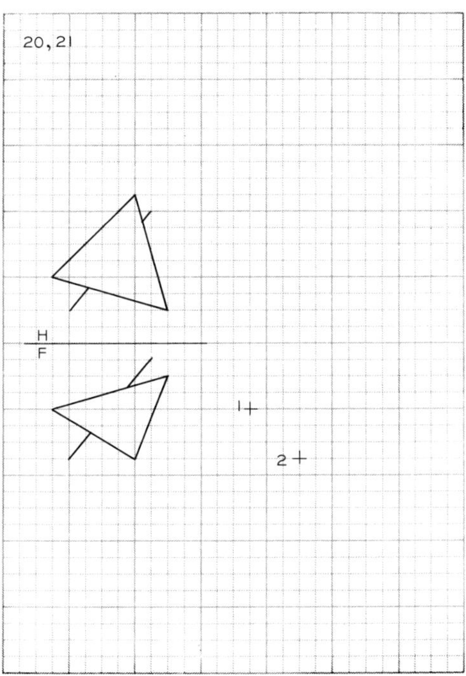

FIGURE 6–47

Show this line in all views. Scale: full size. Project from the top view.

Angle between a Line and a Plane

20. Find the angle between the line and the plane in Fig. 6–47 by projecting from the front view. Show visibility in all views. Use the plane method.

21. Same as problem 20, except use the line method.

Combination Problem

22. In Fig. 6–48, a line that is 2.3″ long is to be constructed through point *A* that will have a bearing of N 66° E and slope upward from point *A* with a 20° slope. Draw this line in all views. Find the angle between this line and the plane.

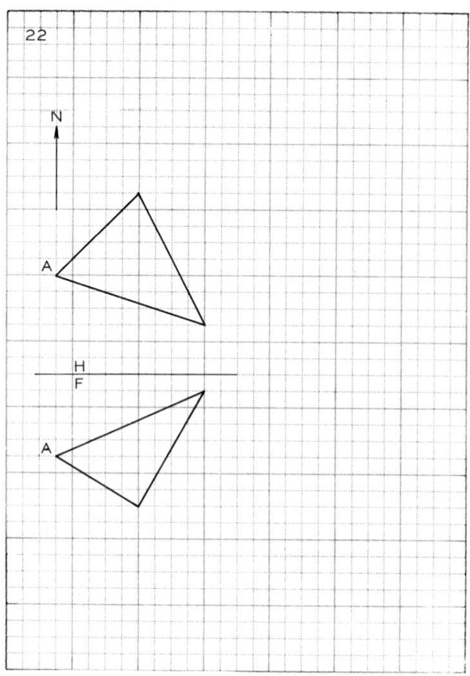

FIGURE 6–48

7

Revolution

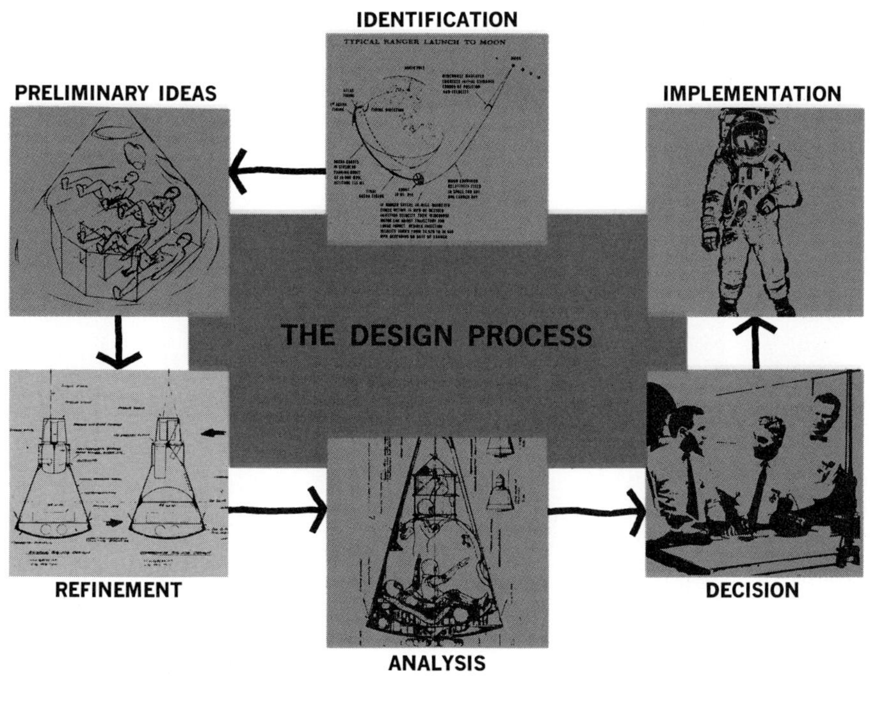

7–1 INTRODUCTION

The F–111, the world's first variable-geometry aircraft, is shown in Fig. 7–1 in a sequence of photographs which illustrate the full range of positions of its wings during flight. The wings are shown revolved from a 16° spread at takeoff to a fully swept 72.5° they assume for supersonic speed.

The development of an aircraft of this type involved many hours of testing, planning, and design. The wing system of the plane is an example of an application of revolution to the design of an aircraft that permits variable positions during flight.

Revolution is another method of solving problems that could also, in most cases, be solved by auxiliary views. It is sometimes more advantageous to use the revolution method than the auxiliary-view method covered in Chapters 5 and 6. An understanding of revolution will reinforce an understanding of auxiliary-view principles, which is necessary for the solution of spatial problems. Many engineering designs utilize rotating or revolving mechanisms that must be analyzed for critical information by means of the principles of revolution.

Fig. 7–2. The true length of the structural members of Saturn S–IVB could be found by revolution during the refinement process. (Courtesy of NASA.)

7–2 TRUE LENGTH OF A LINE IN THE FRONT VIEW BY REVOLUTION

The Saturn S–4B shown in Fig. 7–2 has a conical configuration composed of intersecting structural members. The lengths of these and the angles they make with the circular planes at each end can be determined by applying revolution principles as well as by applying auxiliary

Fig. 7–1. The F–111, the world's first variable-geometry aircraft designed as an operational plane, involves principles of revolution in the design of its wings. (Courtesy of General Dynamics Corporation.)

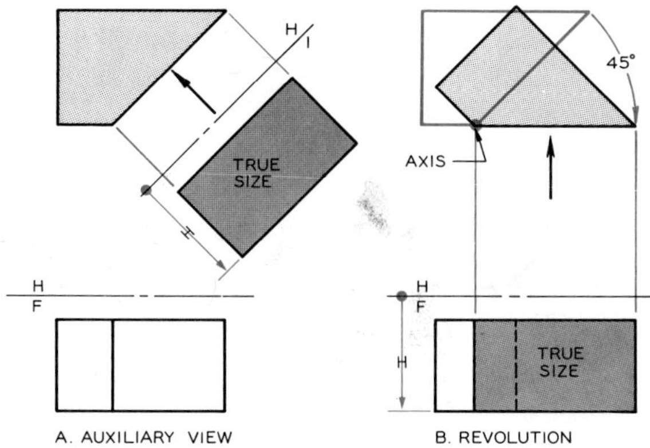

Fig. 7–3. A comparison of the auxiliary view with revolution to find the true size of a plane as it would appear on your drawing paper.

views. The procedure for this is developed in the following explanation.

A simple object (Fig. 7–3) is used to introduce the principles required to find a plane true size by revolution. The slanted surface that appears as an edge in the top view can be found true size by a primary auxiliary view or by a single revolution. Your position as an observer must be changed so that your line of sight is perpendicular to the slanted surface in the auxiliary view in A. However, it is possible to find this surface true size using the same line of sight used for finding the front view of the object and by revolving the object. Since the axis of the revolution is vertical, it appears as a point in the top view. The top view is revolved until the slanted surface is parallel to the frontal plane; the slanted surface is then projected to the front view. The height dimensions are shown to complete the true-size view.

The true length of line *AB* is found in the front view by revolution in Fig. 7–4. The top and front views of an oblique line, *AB*, are given. Point *A* is used as the apex of a cone, and the half view of the top view of the cone is drawn in step 1, using line *AB* as a radius. The front view of the cone is projected from the top view. The top view of line *AB* is revolved into the frontal plane of the cone and its projection found in the front view in step 2. Since line *AB′* has been revolved into the frontal plane, its true length is found in the front view where it is an extreme element of the cone. Point *B* traveled in the horizontal plane, so the vertical

FIGURE 7–4. TRUE LENGTH IN THE FRONT VIEW

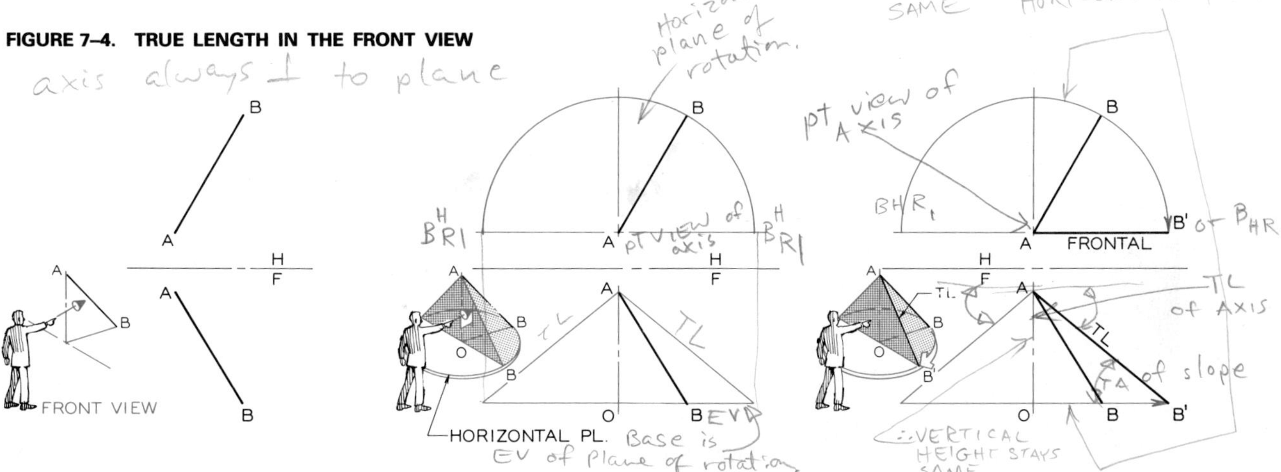

Given: The top and front views of line *AB*.

Required: Find the true-length view of line *AB* in the front view by revolution.

Step 1: The top view of line *AB* is used as a radius to draw the base of a cone with point *A* as the apex. The front view of the cone is drawn with a horizontal base through point *B*. Line *AO* is the axis of the cone.

Step 2: The top view of line *AB* is revolved to be parallel to the frontal plane, *AB′*. When projected to the front view, frontal line *AB′* is the outside element of the cone and is true length.

height between points A and B was not changed. Consequently, the front view of point B' is found by projecting horizontally from the front view of point B to the projector from the top view of point B'.

7–3 TRUE LENGTH OF A LINE IN THE HORIZONTAL VIEW BY REVOLUTION

The handle for operating the speed control on the lathe shown in Fig. 7–5 allows the operator to apply the principle of revolution of a line about an axis. The handle has been positioned to take into account the human factors involved in the operation of the lathe and the position of the operator.

A slanted surface that appears as an edge in the front view can be found true size by a primary auxiliary view or by a single revolution. This comparison is shown in Fig. 7–6. When revolution is used, you need not change your position, but the same line of sight can be used that gives the top view.

An axis is located in the front view as a point, and is a true-length line in the top view.

Fig. 7–5. The spindle speed lever on the lathe was designed through the use of principles of revolution and with consideration of human factors. (Courtesy of Jones and Lamson Corporation.)

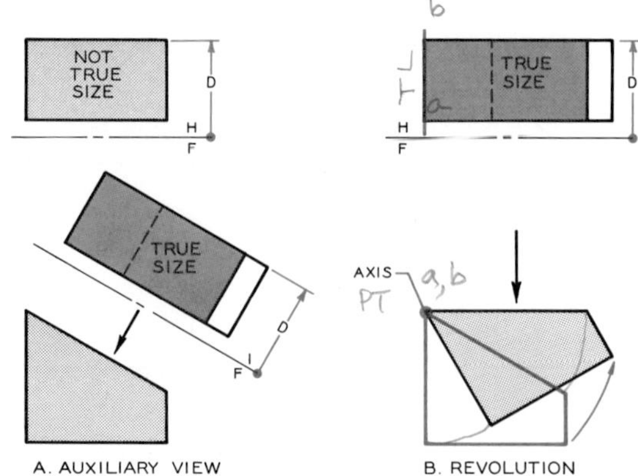

Fig. 7–6. Determination of the true size of a surface that appears as an edge in the front view by an auxiliary view and by revolution.

The front view is revolved about the axis until the slanted surface is a horizontal edge in the front view; from this it is possible to project a true-size view in the top view. As in the auxiliary view, the depth dimension (D) does not change.

These principles are applied to find a line true length by revolution in Fig. 7–7. The front view of line CD is used as a radius to draw the front view of the cone that is obtained when point D is revolved parallel to a frontal plane (step 1). The triangular view of the cone is constructed in the top view by projection, as shown in step 1. Line CD is revolved in the front view to position CD', where it is horizontal (step 2). It becomes the extreme, outside element of the cone in the top view, where it appears true length. Note that points D and D' are in the same frontal plane in the top view. Point D' is found by projecting parallel to the H–F reference line until it intersects the projector from the front view.

Figure 7–8 shows a crucible that revolves about an axis, pouring aluminum to form ingots. The design of this crucible and its operating system was analyzed through the use of the principles of revolution to establish its limits of operation.

FIGURE 7–7. TRUE LENGTH OF A LINE IN THE TOP VIEW

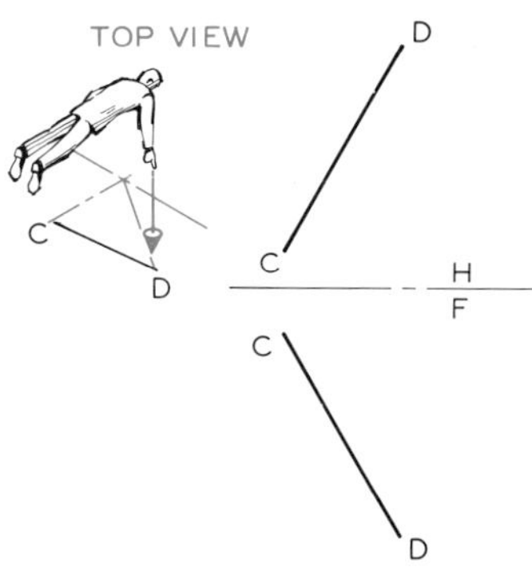

Given: The top and front views of line *CD*.

Required: Find the true-length view of line *CD* in the front view by revolution.

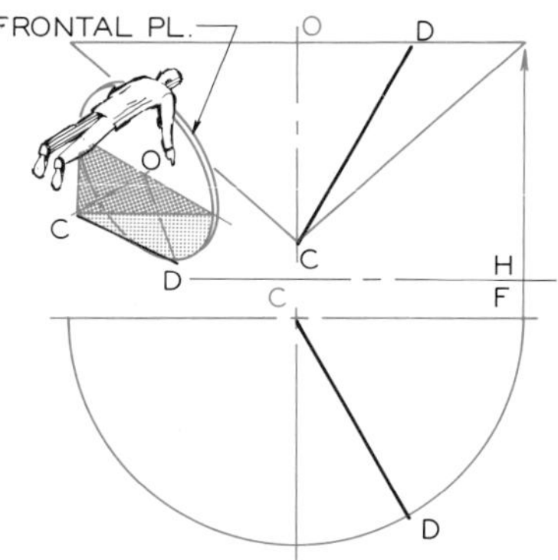

Step 1: The front view of line *CD* is used as a radius to draw the base of a cone with point *C* as the apex. The top view of the cone is drawn with the base shown as a frontal plane. The axis, *CO*, is perpendicular to the frontal base.

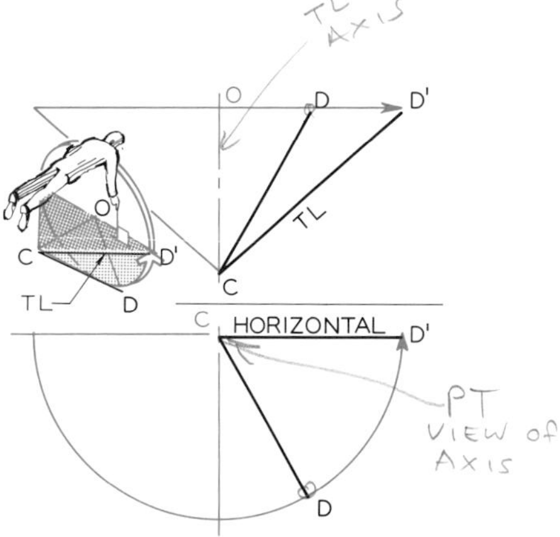

Step 2: The front view of line *CD* is revolved into position *CD'* where it is horizontal. When projected to the top view, *CD'* is the outside element of the cone and is true length.

Fig. 7–8. The crucible used to pour 700-pound ingots of aluminum was designed to revolve about an axis to the position required for efficient flow of metal. (Courtesy of ALCOA.)

7–4 TRUE LENGTH OF A LINE IN THE PROFILE VIEW BY REVOLUTION

The orthographic projections and revolutions of line *EF* are given in Fig. 7–9. In order for the line to be true length in the side view, it is revolved in the front view as though it were an element of a cone (step 1). The circular view of the cone is projected to the side view, where its triangular shape is seen. In step 2, point *F* is revolved to *F'* in the front view and projected to the side view, where the line represents the extreme element of the cone and is true length, since it is a profile line in this position.

It should be noted that the true length of any line can be found by revolution in either view when two adjacent views are given. The true length of line *EF* could have been found in the front view of Fig. 7–9 by revolving the line into a position that was parallel to the frontal plane instead of the profile plane.

The examples previously covered revolve each line about one of its given ends, because this is a simple way of introducing the principles of revolution. However, the line could be revolved about any point on its length equally as well (Fig. 7–10).

FIGURE 7–9. TRUE LENGTH OF A LINE IN THE SIDE VIEW

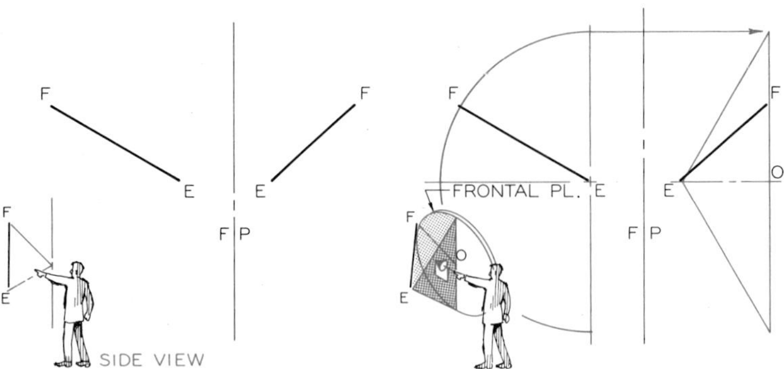

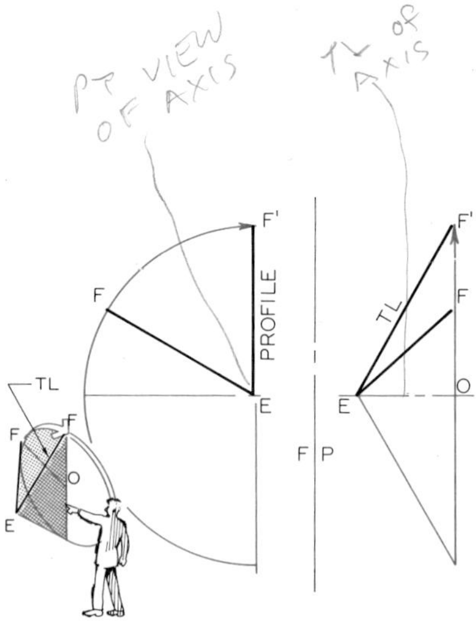

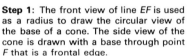

Given: The front and side views of line *EF*.

Required: Find the true-length view of line *EF* in the profile view by revolution.

Step 1: The front view of line *EF* is used as a radius to draw the circular view of the base of a cone. The side view of the cone is drawn with a base through point *F* that is a frontal edge.

Step 2: Line *EF* in the front view is revolved to position *EF'* where it is a profile line. Line *EF'* in the profile view is true length, since it is a profile line and the outside element of the cone.

Fig. 7–10. The axis of revolution used to find the true-length view of a line can be placed anywhere on the line to be revolved. In this case a vertical axis was placed through point *O* of line *GH*.

The portable well work-over equipment and the pump shown in Fig. 7–11 illustrate revolutions about an axis. The design of each was analyzed by revolution principles to refine and develop operational functions.

Fig. 7–11. The portable well work-over equipment and the pump are examples of mechanisms that were designed to revolve about an axis into a variety of positions. (Courtesy of Exxon Corporation.)

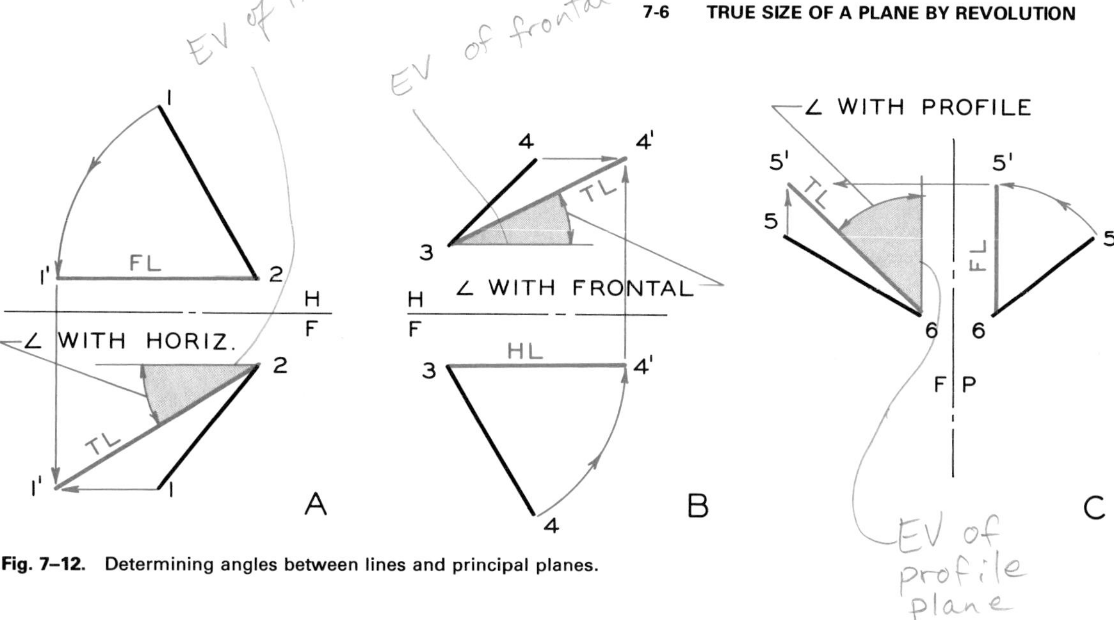

Fig. 7–12. Determining angles between lines and principal planes.

7–5 ANGLES BETWEEN A LINE AND PRINCIPAL PLANES BY REVOLUTION

The angle between a line and a plane will appear true size in the view where the plane is an edge and the line is true length. In all principal views, two principal planes appear as edges. Consequently, when a line appears true length in a principal view, the angle between the line and two principal planes can be measured.

A line may be revolved to find its true length in any principal view, as illustrated in Fig. 7–12. Since the horizontal plane appears as an edge in the front view, the true angle between the horizontal plane and an oblique line, 1–2, can be found by constructing the true length of the line in the front view by revolution, as shown in part A. The frontal plane projects as an edge in the top view (part B). The angle between line 3–4 and the frontal plane can be found in the top view by revolving point 4 to a horizontal position in the front view and then projecting to find the true length of 3–4 in the top view, as shown in part B.

The angle between line 5–6 and the profile plane can be found in the front view, where the profile plane is a vertical edge (part C). Line 5–6 is revolved in the side view until it becomes a frontal line and is projected to the front view. The angle line 5′–6 makes with the profile plane can be measured in the front view.

7–6 TRUE SIZE OF A PLANE BY REVOLUTION

A plane can be revolved about an axis until it becomes true size in much the same manner as a truck bed is revolved about an axis (Fig. 7–13). This principle of revolving a plane is

Fig. 7–13. The 45-ton Haulpak truck was designed to permit the bed to revolve about an axis, as required for functional operation. (Courtesy of LeTourneau-Westinghouse Corporation.)

FIGURE 7–14. TRUE SIZE OF A PLANE BY REVOLUTION

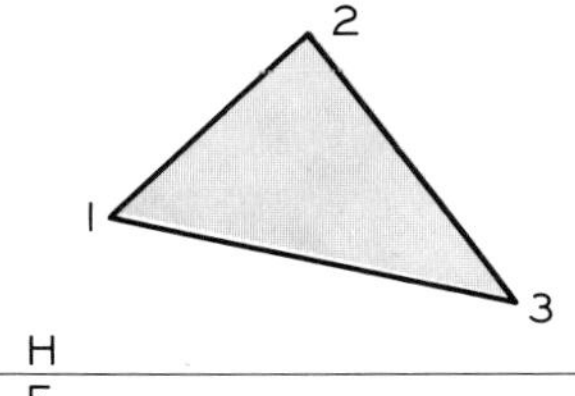

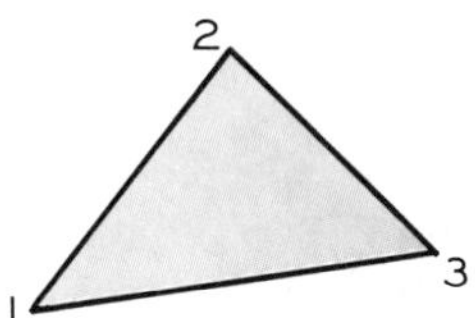

Given: The top and front views of plane 1–2–3.

Required: Find the true size of the plane by revolution.

References: Articles 5–13 and 7–6.

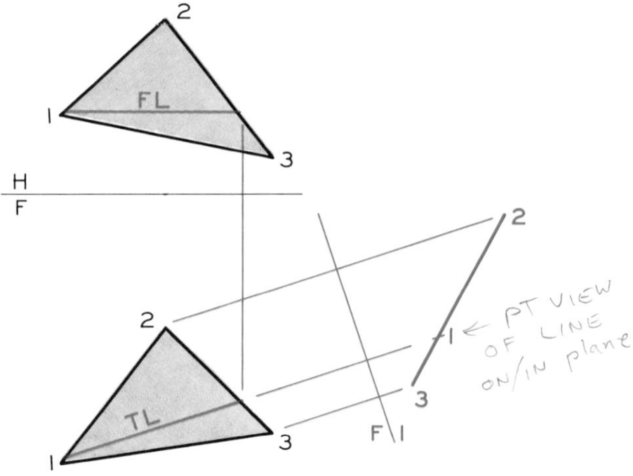

Step 1: Construct a true-length line in the front view of plane 1–2–3. Since the plane will appear as an edge in a view where the true-length line projects as a point, project a primary auxiliary view of the plane from the front view. The edge view could have been projected from the top view as well.

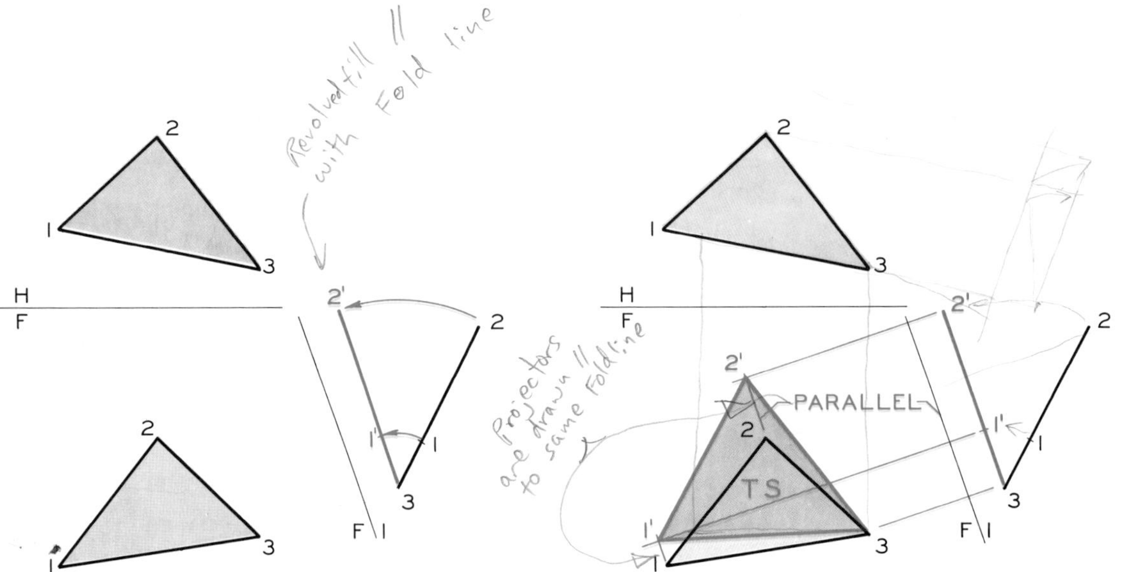

Step 2: Revolve the edge view of the plane about an axis through point 3 until the plane is parallel to the F–1 line. The plane, 1′–2′–3, will be a true-size projection when projected to the front view, since it has revolved into a plane which is parallel to the frontal projection plane.

Step 3: Projects points 1′ and 2′ from their revolved positions to the front view. Locate points 1 and 2 in the front view by extending projectors from the original points 1 and 2 parallel to the F–1 line, because the plane was revolved to a position parallel to the auxiliary plane in step 2.

closely related to the revolution of a line when the plane being revolved appears as an edge.

The steps required to find the true size of a plane by revolution are presented in Fig. 7–14 through a combination of auxiliary-view and revolution methods. The plane is found as an edge in the primary auxiliary view by locating the point view of a line on the plane (step 1). Because the edge view is oblique to the F–1 line in this projection, it has to be revolved in the primary auxiliary view until it is parallel to the F–1 line in step 2. Any point could have been selected for the axis of revolution. Since the plane was revolved parallel to the auxiliary plane, the true size of the plane is obtained by projecting the original points in the front view parallel to the F–1 line to intersect the projectors from 1′ and 2′. The true size of the plane could have also been found by projecting the edge view from the top view and revolving the plane in this auxiliary view.

7–7 EDGE VIEW OF A PLANE BY REVOLUTION

The edge view of a plane can also be found by revolution without using auxiliary views as was done in Fig. 7–14. The revolution method is illustrated in Fig. 7–15. In this case, plane 1–2–3 is given in the top and front views in part A, where a frontal line is drawn on the plane in the top view and projected to the front view. The plane is revolved until the true-length line becomes vertical in the front view (part B). The true-length line will project as a point in the top view. The edge view of the plane is found in part C by projecting original points, 2 and 3, in the top view parallel to the H–F reference line to intersect the projectors from the revolved points, 2′ and 3′, in the front view.

A second revolution, sometimes called a *double revolution*, can be made to revolve the edge view of the plane in the top view until it is parallel to the edge view of the frontal plane. This is illustrated in Fig. 7–16, where a plane is found true size by double revolution. In step 1, line 1–2 (which appears true length in the top

FIGURE 7–15. EDGE VIEW OF A PLANE BY REVOLUTION

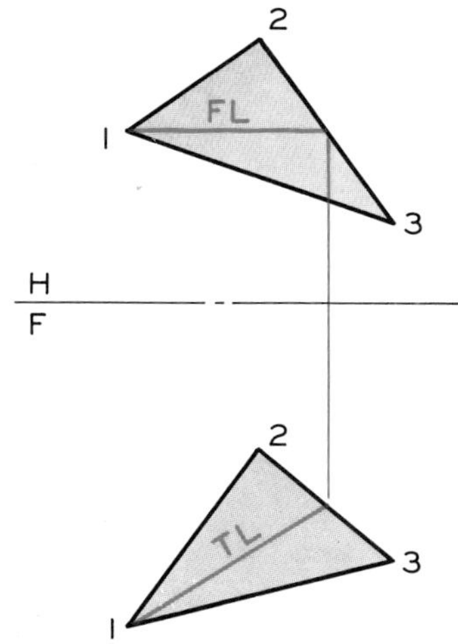

Step 1: It is required that we find the edge view of plane 1–2–3. A frontal line is found true length on the front view of the plane.

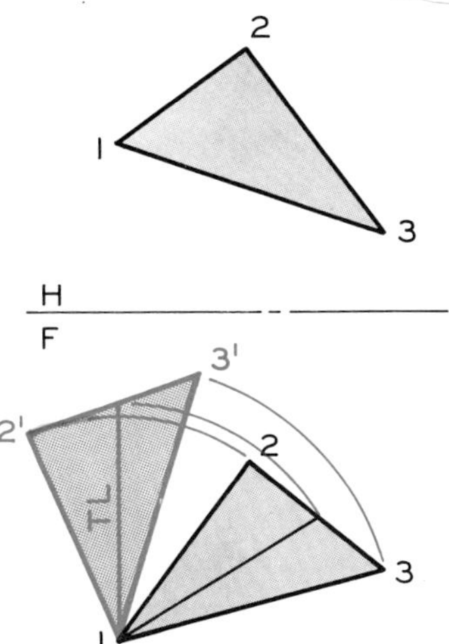

Step 2: The front view of the plane is revolved until the true-length line is vertical.

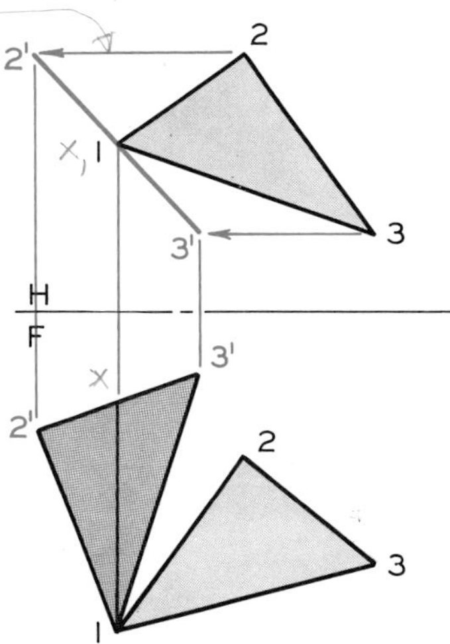

Step 3: Since the true-length line is vertical, it will appear as a point in the top view and the plane will appear as an edge, 1–2′–3′.

FIGURE 7–16. TRUE SIZE BY DOUBLE REVOLUTION

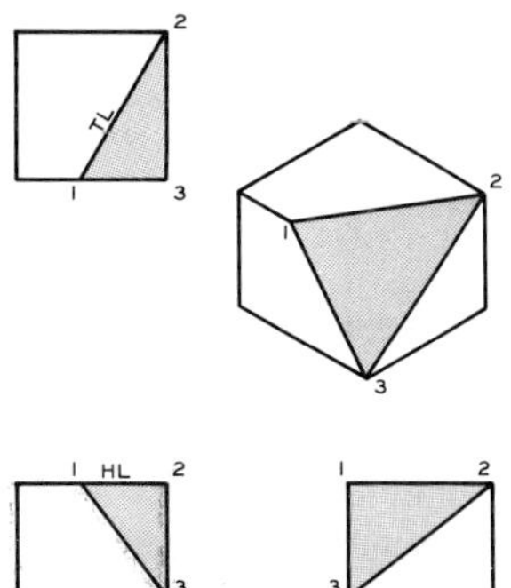

Given: Three views of a block with an oblique plane across one corner.
Required: Find the plane true size by revolution.

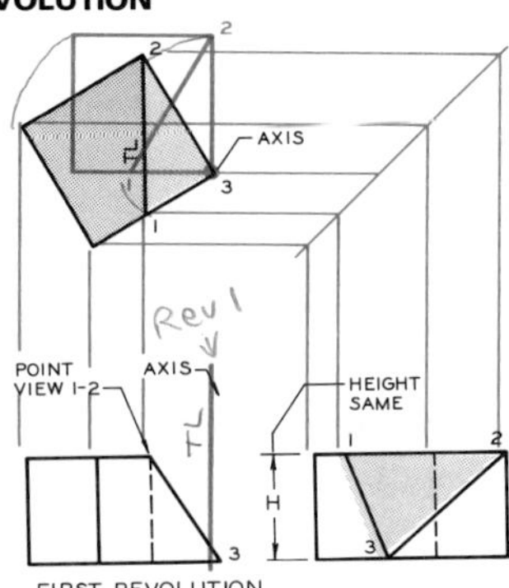

Step 1: Since line 1–2 is horizontal in the front view, it is true length in the top view. The top view is revolved into a position where line 1–2 can be seen as a point in the front view.

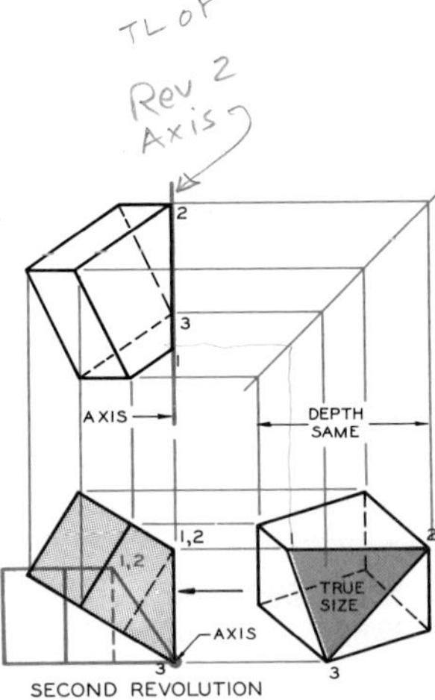

Step 2: Since plane 1–2–3 was found as an edge in step 1, this plane can be revolved into a vertical position in the front view, so that it will appear true size in the side view. The depth dimension does not change, since it is parallel to the axis of revolution.

view) is revolved in the top view until it is parallel to the projectors between the top and front views. The axis of revolution appears as a point in the top view and as a vertical axis in the front view. The height does not change.

Now that plane 1–2–3 appears as an edge in the front view, this view can be revolved until the edge view is vertical. This will give a true-size view of the plane in the side view. The depth will remain the same, since this dimension is parallel to the axis that appears as a point in the front view. The entire object is found in this view by projecting dimensions from the top and front views.

FIGURE 7–17. ANGLE BETWEEN TWO PLANES

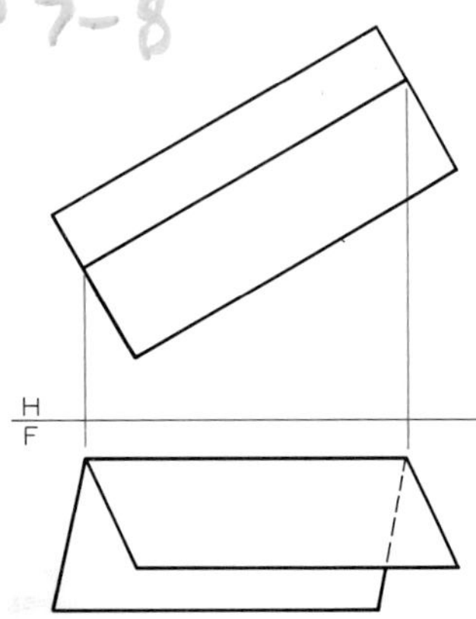

Given: The top and front views of two intersecting planes.

Required: Find the angle between the planes by revolution.

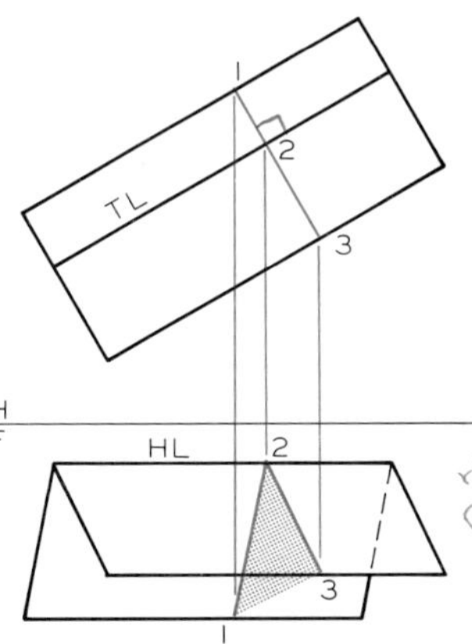

Step 1: A right section is drawn perpendicular to the true-length line of intersection between the planes in the top view and is projected to the front view. The section is not true size in the front view.

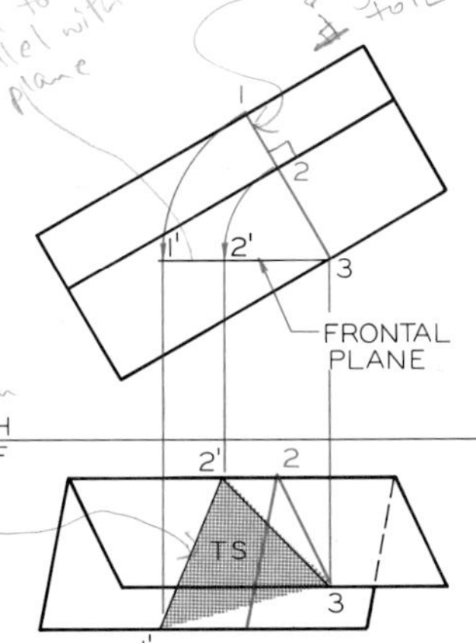

Step 2: The edge view of the right section is revolved to position 1′–2′–3 in the top view to be parallel to the frontal plane. This section is projected to the front view, where it is true size since it is a frontal plane.

7–8 ANGLE BETWEEN TWO PLANES BY REVOLUTION

The line of intersection of the two intersecting planes given in Fig. 7–17 is true length in the top view. The plane of the angle between the two planes will appear as an edge, in this view, since it is perpendicular to the true-length line of intersection. Points 1, 2, and 3 in the plane of the angle are projected to the front view where the plane of the angle appears foreshortened (step 1). The true size of this plane can be found in the front view by revolving the edge view of the plane in the top view until it becomes a frontal plane (step 2). Points 1′, 2′, and 3 are then projected from the top view to the front view and located in the same horizontal planes as the original points in the foreshortened view. Angle 1′–2′–3 appears true size in the front view.

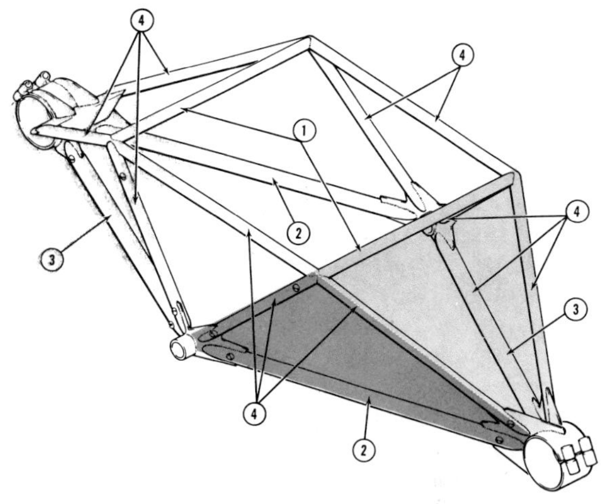

Fig. 7–18. The angular measurements between the two planes of the helicopter engine mount can be determined by revolution principles. (Courtesy of Bell Helicopter Corporation.)

FIGURE 7–19. ANGLE BETWEEN OBLIQUE PLANES

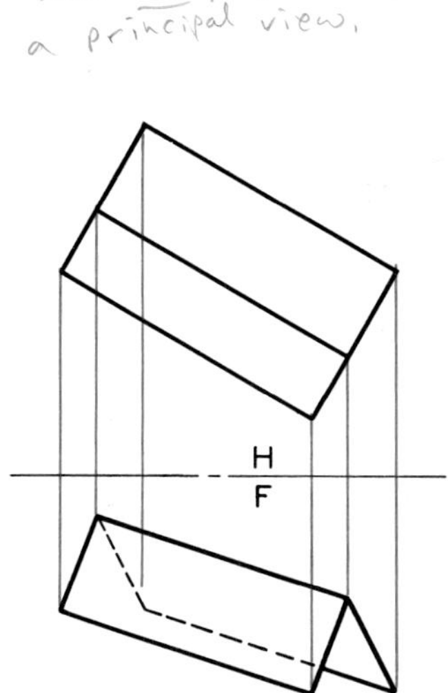

Given: The top and front views of two intersecting planes.

Required: Find the angle between the two planes.

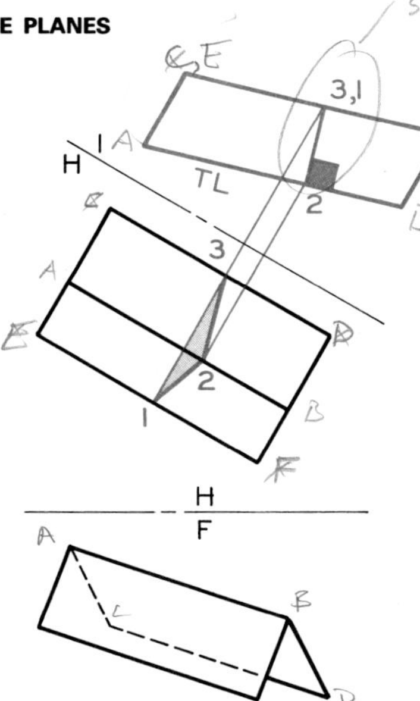

Step 1: A true-length view of the line of intersection is found in an auxiliary view projected from the top view. The right section is constructed perpendicular to the true length of the line of intersection and is projected to the top view.

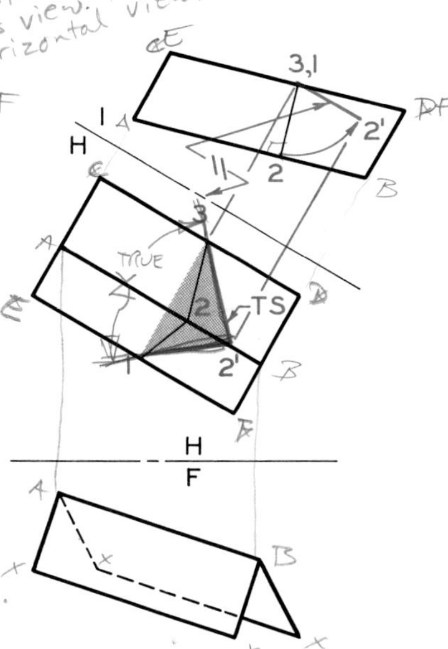

Step 2: The edge view of the right section is revolved to be parallel to the H–1 reference line so the plane will appear true size in the top view after being revolved. The angle between the planes can be found by measuring angle 1–2′–3.

Often the line of intersection between two intersecting lines will not project as true length in a principal view. Such is the case in Fig. 7–18, which shows an engine mount frame of a helicopter. The angle between these planes must be determined to design the joints and to analyze the clearances within the frame.

In the top and front views of two intersecting planes in part A of Fig. 7–19, the line of intersection does not appear true length in either view. The true length of the line of intersection is found in a primary auxiliary view which is projected perpendicularly from the line of intersection in the top view. The plane of the angle between the two planes projects as an edge that is perpendicular to the true-length view of the line of intersection (step 1). Plane 1–2–3 is projected as a foreshortened plane in the top view. The edge view of plane 1–2–3 is therefore revolved about the axis, 3–1, in the primary auxiliary view until it is parallel to the H–1 line (step 2). It is then projected back to the top view. Angle 1–2′–3 appears true size in the top view when point 2′ has been located by projecting from point 2 parallel to the H–1 line.

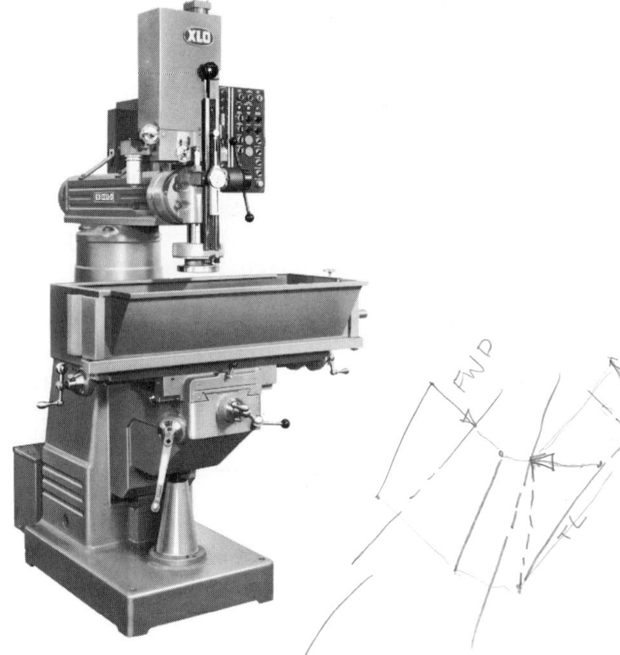

Fig. 7–20. The hand cranks on the mass-production machine were designed to permit adequate clearance when they are revolved about their axes. (Courtesy of Ex-Cell-O Corporation.)

7–9 REVOLUTION OF A POINT ABOUT AN OBLIQUE AXIS

Handwheels and hand cranks are mechanical means of adjusting all types of machines from common household appliances to mass-production equipment such as the machine shown in Fig. 7–20. Principles of revolution were applied to these hand adjustments in the early stages of their design. The same principles can be applied to the location of a power line with respect to another by revolving a point on one wire about the axis of the other to determine the required minimum clearance.

Often a point is revolved about an axis to a specified position such as the highest, lowest, most forward, or most backward position. This requires that an arrow be located in the two given views pointing in the specified direction, as illustrated in Fig. 7–21. This arrow is projected to the view where the path of rotation appears as a circle to find the specified position.

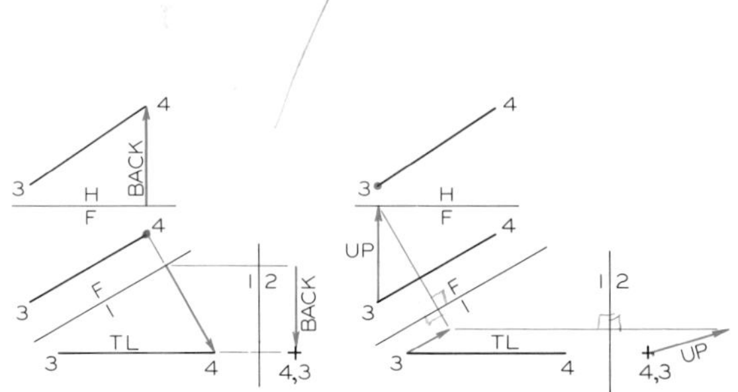

Fig. 7–21. To find the direction of back and up, arrows are drawn in the two given views. These arrows are projected to the auxiliary views in the same manner as any other line in the drawing.

The revolution of a point about a line is illustrated in Fig. 7–22. The top and front views of axis 1–2 and point O are given. We are to revolve point O into its highest position to determine its location in the given views and its relationship to adjacent components. The

FIGURE 7–22. REVOLUTION OF A POINT ABOUT AN AXIS

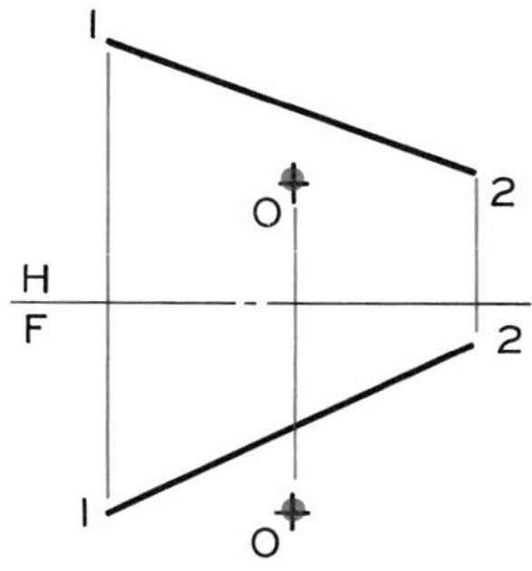

Given: The top and front views of axis 1–2 and point *O*.

Required: Revolve point *O* about the axis, locate its highest position, and show it in all views.

References: Articles 4–3, 6–3, 6–6, and 7–9.

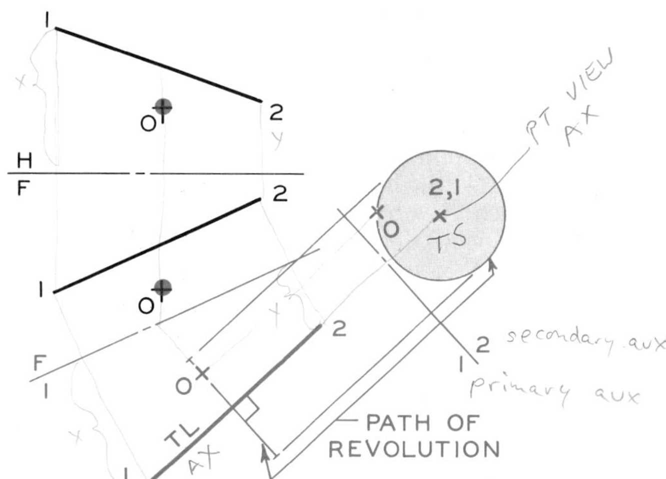

Step 1: Locate the true length of axis 1–2 in a primary auxiliary view, and construct its point view in the secondary auxiliary view. Project point *O* to these views also. Using as a radius the distance from the point view of axis 1–2 to point *O*, construct the circular path of revolution in the secondary auxiliary view. The path of revolution appears as an edge in the primary auxiliary view.

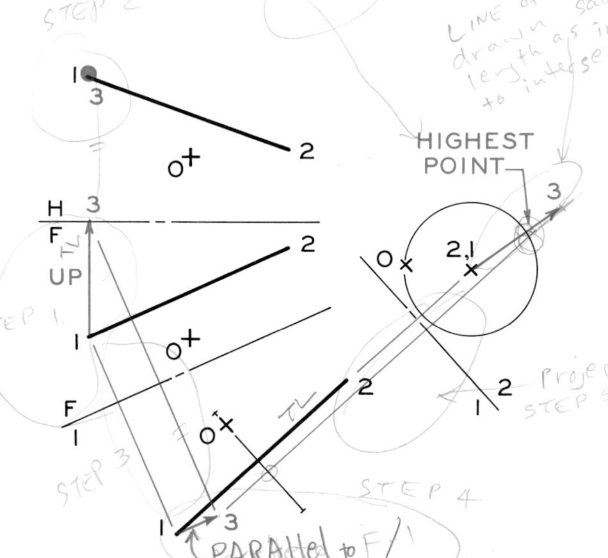

Step 2: Determine the highest point by constructing line 1–3 upward in the front view and projecting it as a point in the top view. This line is then projected back through the primary auxiliary view to the secondary auxiliary view. The point where this directional arrow crosses the circular path locates the highest point of the path of point *O*.

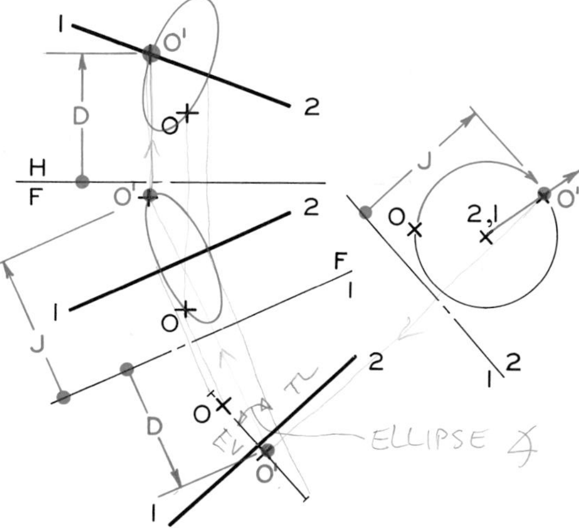

Step 3: Point *O′* is projected from its circular view back through the successive views. Note that the highest point, *O′*, lies on line 1–2 in the top view, which verifies that it is in its highest position. This problem could have been solved by projecting from the top view as well. The circular path appears elliptical in the front and top views.

circular path of revolution of point O about line 1–2 can be seen in the view where line 1–2 appears as a point (step 1). The highest position of point O is found in step 2 by constructing line 1–3 in an upward direction in the two given views. The directional line is then projected back to the secondary auxiliary view where the highest point is located on the circular path. Point O', the highest point on the path, is found in each view by projecting to the primary auxiliary view, the front view, and the top view (step 3). Note that the circular path of the point projects as an edge which is perpendicular to axis 1–2 in the primary auxiliary view.

The elliptical paths of point O can be constructed in principal views by application of the principles covered in Article 6–6. The ellipse guide angle for the front view is the angle formed by the line of sight from the front view and the edge view of the circular path of revolution. The ellipse angle for the top view must be found by an auxiliary view projected from the top view in which the axis appears true length and the circular path appears as a perpendicular edge. The angle formed by the line of sight from the top view and the edge view of this circular path establishes the ellipse angle that will be used in selecting the proper ellipse template for drawing the ellipse in the top view.

The point could have been located in any specified position, such as its highest, lowest, or most forward position, by constructing a line of the required direction in the principal views and projecting it into all views.

7–10 REVOLUTION OF A LINE ABOUT AN AXIS

A line can be revolved about another line, as shown in Fig. 7–23, if the line to be used as an axis is found as a point. The point view of line 1–2 is obtained in part A of the figure in the primary auxiliary view, since the top view of line 1–2 is true length. A circle is drawn tangent to line 3–4 with its center at the axis 1–2. Each end of line 3–4 is revolved the specified number of degrees and drawn in its new position as shown in part A. The top view of line 3′–4′ is found by projecting parallel to the H–1 line

FIGURE 7–23. REVOLUTION OF A LINE ABOUT AN AXIS

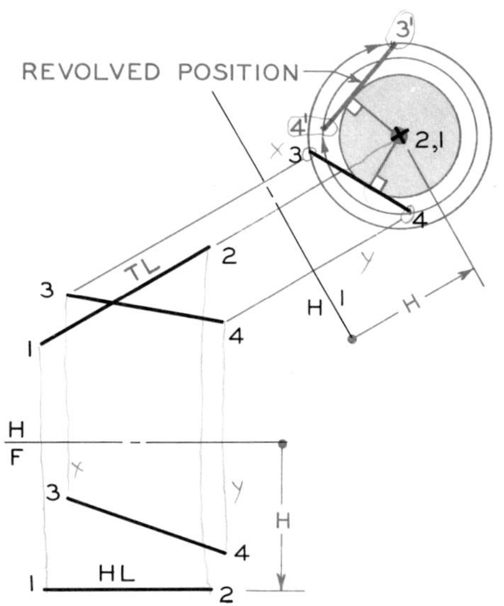

A. The axis, 1–2, is found as a point in the auxiliary view. Line 3–4 is revolved to its specified position in this view.

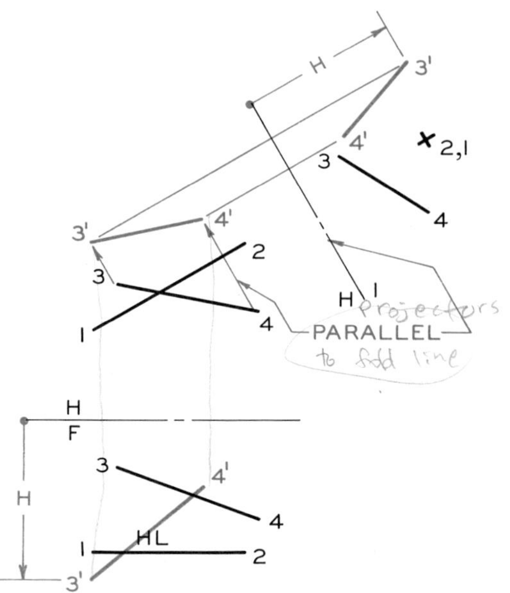

B. The new position of line 3–4 is projected back to the top view where projectors parallel to the top view of the H–1 reference line intersect those from the auxiliary view to find line 3′–4′. Line 3′–4′ is located in the front view.

Fig. 7–24. The cradle of this orthicon camera was designed to permit revolution to any position for tracking space vehicles and astronomical bodies. (Courtesy of ITT Industrial Laboratories.)

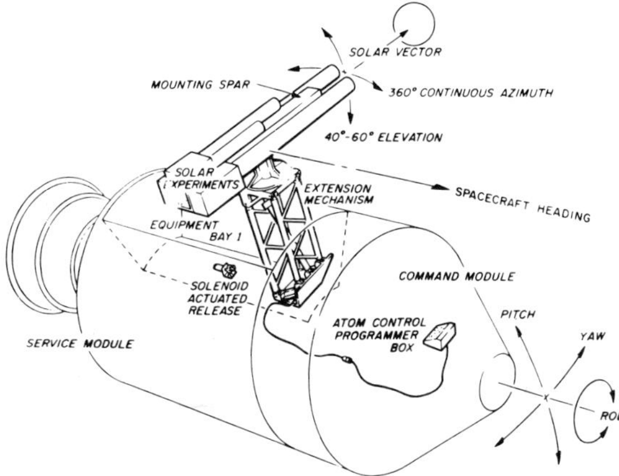

Fig. 7–25. The Apollo command service module was designed to permit the automatic revolution of a telescope to keep it directed toward the sun as the vehicle travels through space. (Courtesy of NASA.)

from the original points of 3 and 4 in the top view, as shown in part B. These projectors will intersect the projectors from the primary auxiliary view. The front view is obtained by projecting from the top view and transferring the height dimensions from the primary auxiliary view, as shown in part A.

Principles of revolution are also applied to the design of cameras used to track space vehicles (Fig. 7–24).

An artist's drawing of the atom system in the Apollo command service module is shown in Fig. 7–25. The maneuvers of pitch, roll, and yaw are illustrated with respect to the path of the spacecraft. Pitch is the up-and-down revolution with respect to the spacecraft heading, while yaw is the left-or-right rotation, and roll is the revolution of the spacecraft along the path of its heading. The commands given to a module will be for revolutions about these three axes, whether the commands come from the earth by radio waves or from the astronauts aboard. The telescopes are mounted on a spar that will extend outside the service module on a two-axis gimbal that can automatically correct for the yaw or the pitch of the spacecraft. This system will be used for observing the sun through telescopes.

7–11 REVOLUTION OF A RIGHT PRISM ABOUT AN AXIS

Many mass-production plants employ automatic tow lines located overhead or under the floor to transport parts and materials through the manufacturing process. An example of a portion of an overhead system is shown in Fig. 7–26. The track is an I-beam and the trolleys are designed to roll on its lower flange, as shown in Fig. 7–27. The tracks are designed to be suspended from the structural beams of the plant's interior structure. It is obvious that the trolley system will work effectively only when the track's right section is positioned so that the interior web of the beam is positioned in a vertical plane and the lower flange is positioned horizontally. If the web were not vertical, the trolley would bind and not roll properly. Conse-

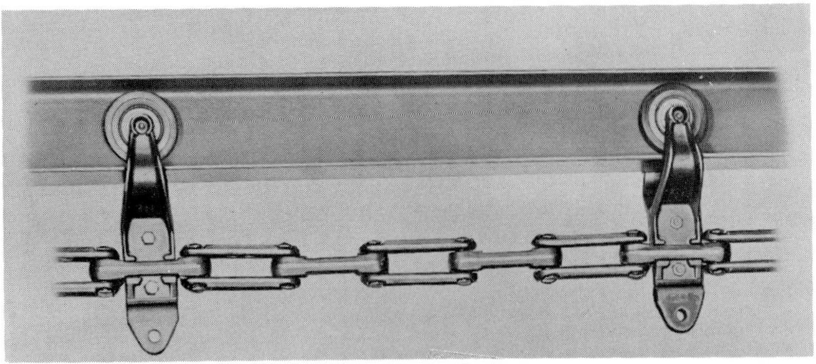

Fig. 7–26. The overhead track of this conveyor system must be designed in such a manner that the track web is positioned in a vertical plane at all times. (Courtesy of Mechanical Handling Systems, Inc.)

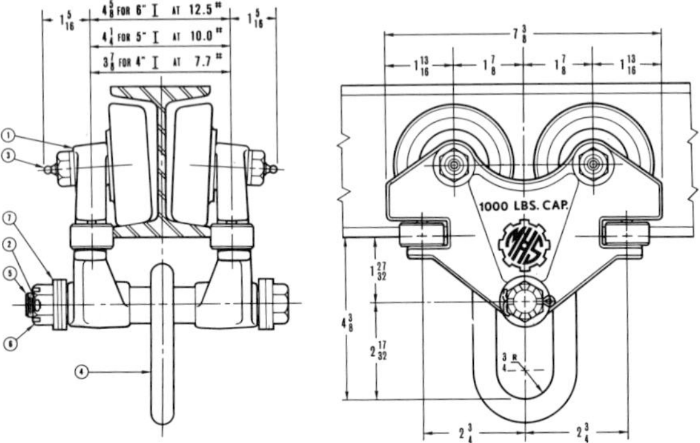

Fig. 7–27. A detail view of a trolley that will utilize a track like that shown in Fig. 7–26. (Courtesy of Mechanical Handling Systems, Inc.)

quently, it is necessary to design a method for suspending the track in such a way that the web is positioned in a vertical plane.

A problem of this type is illustrated in Fig. 7–28, in which a prism with a square right section is revolved about its axis until two of its planes are vertical. This prism could represent the I-beam mentioned above; for simplicity, the details of each flange are not drawn. The center line is found as a point in a secondary auxiliary view in step 1. The direction of vertical is found in this view by projecting a vertical directional arrow in the front view to the secondary auxiliary view. In step 2, the square right section is positioned about the point view of the center line so that two sides are parallel to the directional arrow. The sides of the prism are found by drawing the lateral sides parallel to the center line through the corner points of the right section in all views. The length of the prism is drawn from specifications in the primary auxiliary view; the ends will be perpendicular to the center line (step 3). They are found in the top and front views by projection.

Fig. 7–29. Each of the chutes for transporting iron ore in this installation was designed so that two edges of its right section are vertical and the other two are horizontal. These designs were developed by applying the principle of revolving a prism about its axis. (Courtesy of Kaiser Steel Corporation.)

These principles apply to structural members, hallways, and conveyor belts which, to function properly, must be designed so that their surfaces are positioned with respect to certain planes. A chute connecting two planes through which material will be conveyed must have two sides of its right section in the vertical plane and the other two sides in the horizontal plane (Fig. 7–29).

FIGURE 7–28. REVOLUTION OF A RIGHT PRISM ABOUT ITS AXIS

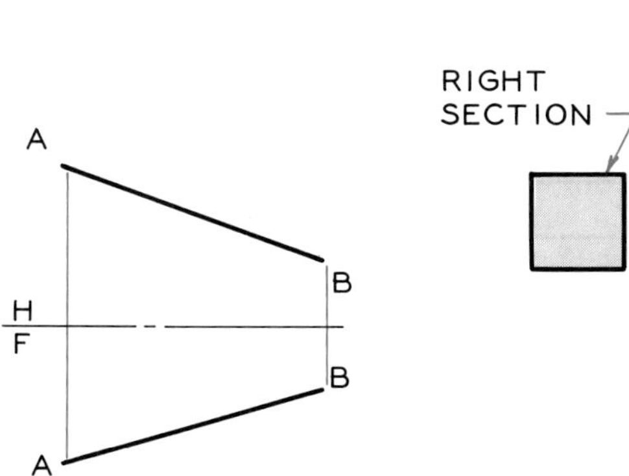

Given: The top and front views of line *AB*, the center line of a prism.

Required: Revolve the given right section about the axis to establish the prism with two surfaces in the vertical plane. Show the prism in all views.

References: Articles 4–3, 6–3, 7–9, and 7–11.

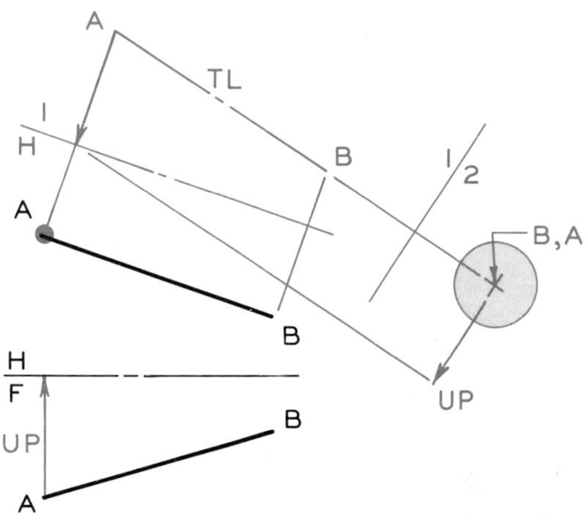

Step 1: Locate the point view of center line *AB* in the secondary auxiliary view by drawing a circle about the axis with a diameter equal to one side of the square right section. Draw a vertical arrow in the front and top views and project it to the secondary auxiliary view to indicate the direction of vertical in this view.

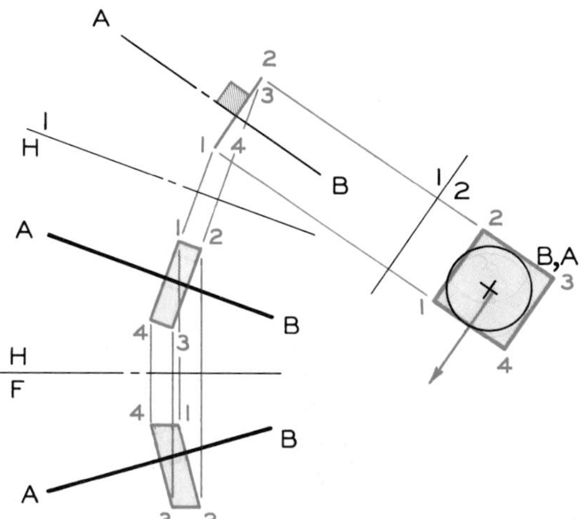

Step 2: Draw the right section, 1–2–3–4, in the secondary auxiliary view with two sides parallel to the vertical directional arrow. Project this section back to the successive views by transferring measurements with dividers. The edge view of the section could have been located in any position along center line *AB* in the primary auxiliary view, so long as it was perpendicular to the center line.

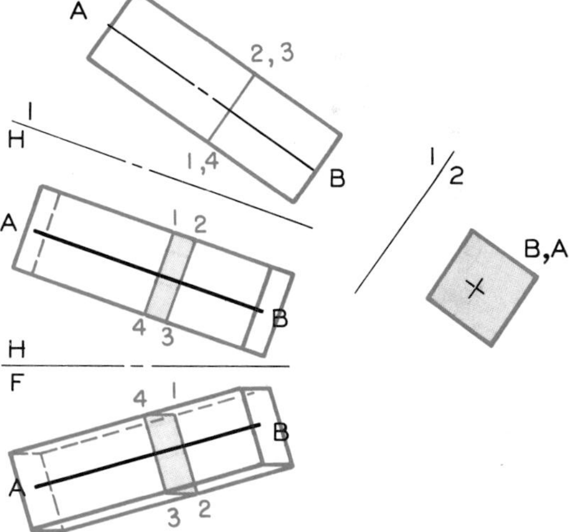

Step 3: Draw the lateral edges of the prism through the corners of the right section so that they are parallel to the center line in all views. Terminate the ends of the prism in the primary auxiliary view where they appear as edges that are perpendicular to the center line. Project the corner points of the ends to the top and front views to establish the ends in these views.

7–12 ANGLE BETWEEN A LINE AND A PLANE BY REVOLUTION

A third way of finding the angle between a line and a plane is by revolution, as shown in Fig. 7–30. This problem was solved by auxiliary view methods in Articles 6–13 and 6–14.

The true size of the plane is found in a secondary auxiliary view in step 1 of Fig. 7–30. The line is revolved in the secondary auxiliary view until it is parallel to the 1–2 reference line (step 2). Line 1–2′ will project true length in the primary auxiliary view because it was parallel to the edge view of the primary auxiliary line. Point 2′ is found by projecting point 2 parallel to the 1–2 line in the primary auxiliary view, as shown in step 3. Since the line appears true length and the plane appears as an edge, the true angle between the line and plane can be measured in this view.

The spacecraft shown in Fig. 7–31 is an example of planes formed by intersecting lines that connect with other lines. These angles must be determined during the refinement stages of developing the final design.

7–13 A LINE AT SPECIFIED ANGLES WITH TWO PRINCIPAL PLANES

The facet-eye camera shown in Fig. 7–32 can be revolved about three axes, giving it full mobility and flexibility for viewing any point in the sky. These television cameras are used for tracking satellites and bodies in space, and they receive excellent contrast even under poor visibility conditions. The design of the camera's cradle involves applications of revolution about several axes. These cameras can be positioned to make a required angle with the two adjacent principal planes. For instance, the direction of the cameras could be positioned to make an angle of 44° with the horizontal and 35° with the frontal plane. The example is illustrated in Fig. 7–33.

In step 1 of Fig. 10–33, cone *A* is drawn to contain all the lines making an angle of 35° with the frontal plane. These lines are the elements on the surface of the cone. Cone *A* will be triangular in the top view and circular in the front view. Cone *B* is drawn in step 2 to contain elements which make an angle of 44°

Fig. 7–31. Angles between the structural members and planes of the spacecraft can be determined by revolution. (Courtesy of NASA.)

Fig. 7–32. The cradle of these image orthicon cameras was designed to permit revolution to any position for tracking space vehicles and astronomical bodies. (Courtesy of ITT Industrial Laboratories.)

FIGURE 7–30. ANGLE BETWEEN A LINE AND A PLANE BY REVOLUTION

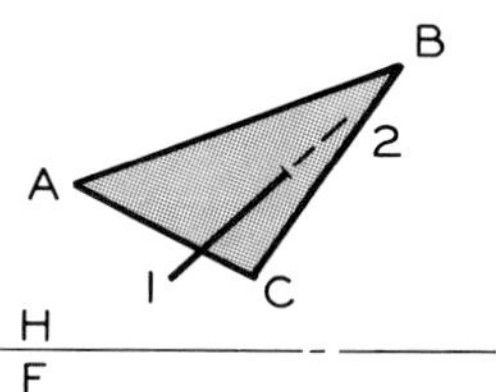

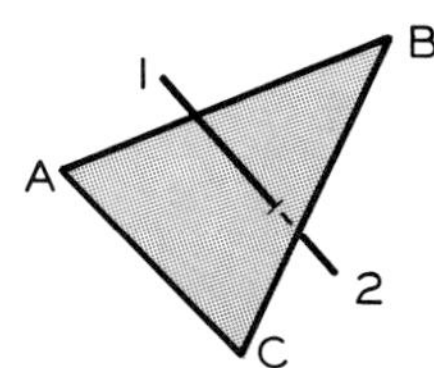

Given: The top and front views of plane *ABC* and line 1–2.

Required: Find the angle between the line and the plane by revolution.

References: Articles 6–13, 6–14, and 7–12.

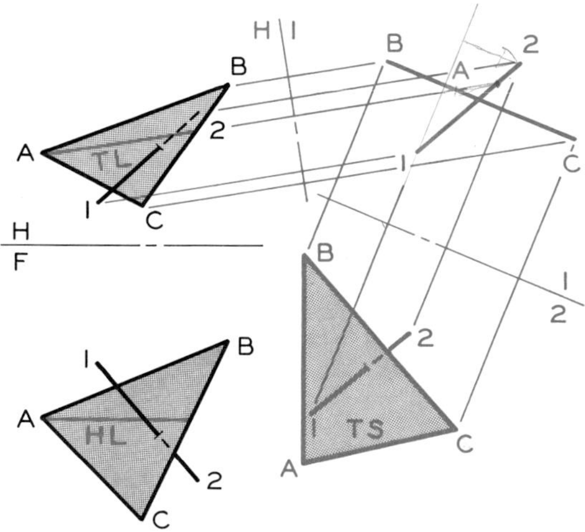

Step 1: Construct plane *ABC* as an edge in a primary auxiliary view, which can be projected from either view. Determine the true size of the plane in the secondary auxiliary view, and project line 1–2 to each view.

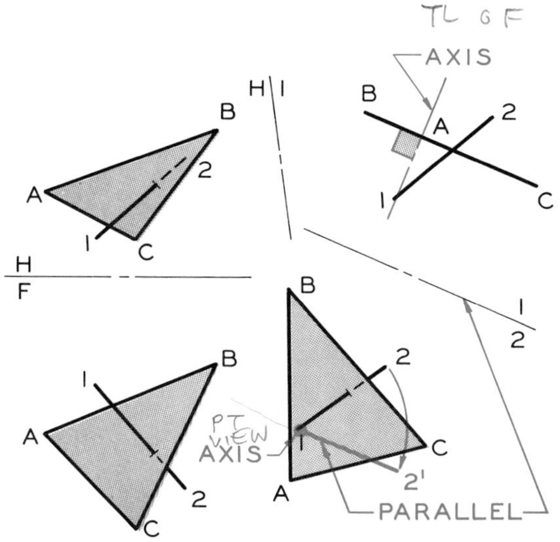

Step 2: Revolve the secondary auxiliary view of the line until it is parallel to the 1–2 reference line. The axis of revolution appears as a point through point 1 in the secondary auxiliary view. The axis appears true length and is perpendicular to the 1–2 line and plane *ABC* in the primary auxiliary view.

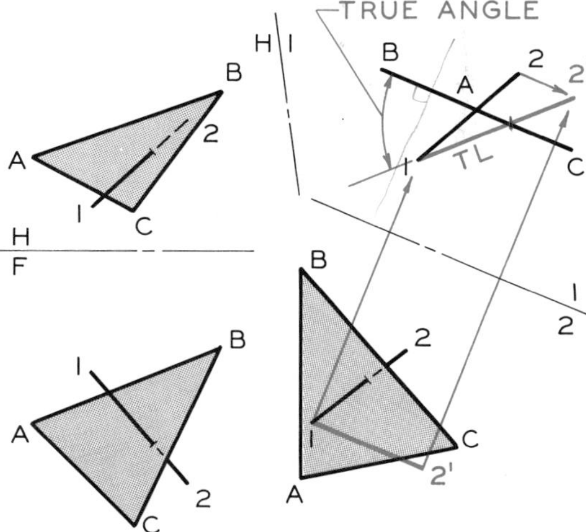

Step 3: Point 2′ is projected to the primary auxiliary view where the true length of line 1–2′ is found by projecting the primary auxiliary view of point 2 parallel to the 1–2 line as shown. Since the plane appears as an edge and the line appears true length in this view, the true angle between the line and the plane can be measured.

FIGURE 7–33. A LINE AT SPECIFIED ANGLES TO TWO PRINCIPAL PLANES

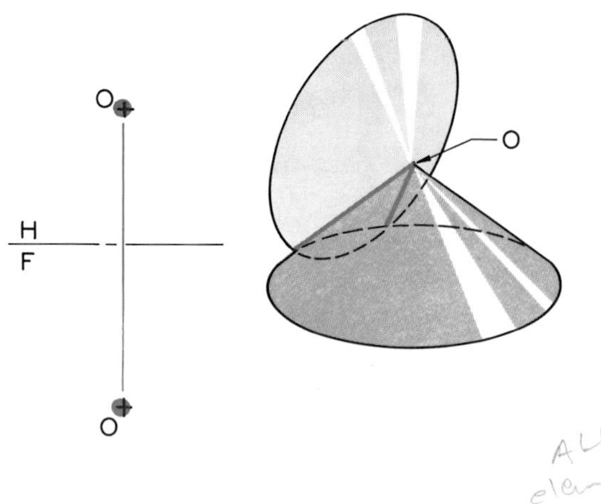

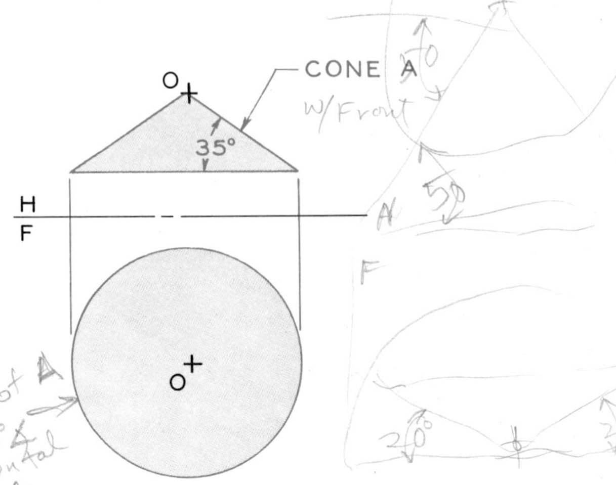

Given: The top and front views of point *O*.

Required: Construct a line through point *O* that will make angles of 35° with the frontal plane and 44° with the horizontal plane, sloping forward and downward.

Reference: Article 7–13.

Step 1: Draw a triangular view of a cone in the top view such that the extreme elements make an angle of 35° with the edge view of the frontal plane. Construct the circular view of the cone in the front view, using point *O* as the apex. All elements of this cone make an angle of 35° with the frontal plane.

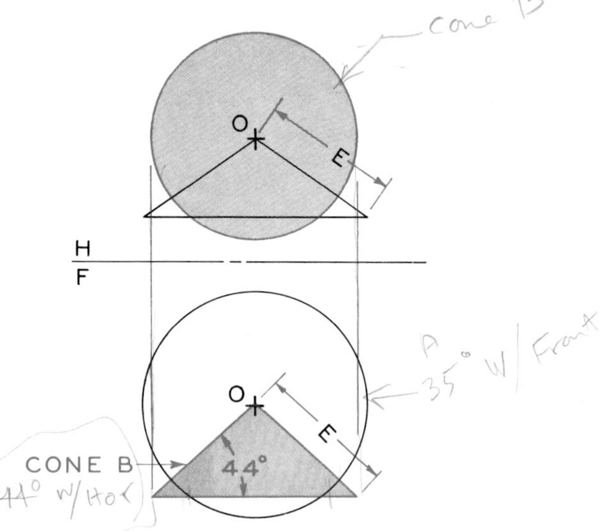

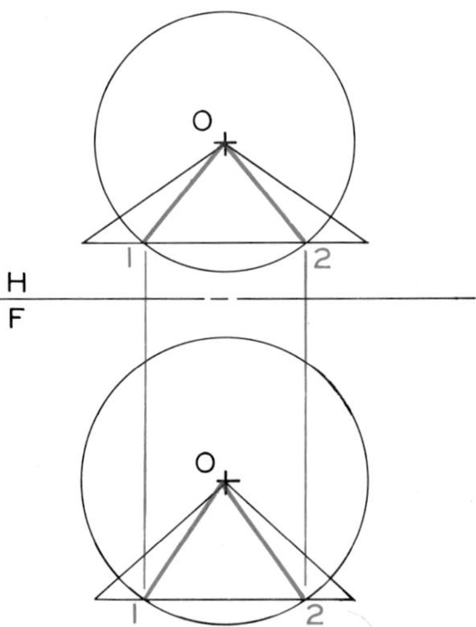

Step 2: Draw a triangular view of a cone in the front view such that the elements make an angle of 44° with the edge view of the horizontal plane. Draw the elements of this cone equal in length to element *E* of cone *A*. All elements of cone *B* make an angle of 44° with the horizontal plane.

Step 3: Since the elements of cones *A* and *B* are equal in length, there will be two common elements that lie on the surface of each cone, elements *O*–1 and *O*–2. Locate points 1 and 2 at the point where the bases of the cone intersect in both views. Either of these lines will satisfy the problem requirements.

with the horizontal plane. The elements are drawn equal in length to element E of the previously drawn cone A. The two cones will intersect with common elements, since the elements of each cone are equal in length. Two lines, O–1 and O–2, satisfy the requirements of the problem. If the requirements had specified that the line slope to the right or left, then only one of the lines would have satisfied the requirements.

These principles can also be applied to determining the intersections between piping systems that must be joined with standard connectors which are cast in standard angles.

7–14 SUMMARY

Principles of revolution are closely related to principles of auxiliary-view projections. In revolution, the observer maintains his position to view principal views in the conventional direction while the object is revolved into the desired position to give the required view. The auxiliary view method of projection moves the observer's position about the stationary object so that the object is viewed from auxiliary positions.

In many instances, the principles of revolution can be used to supplement those of auxiliary views, allowing the designer to find the true sizes and shapes of geometric figures with greater ease than would be possible with auxiliary views alone. Spatial problems should always be analyzed to determine the most appropriate method of solution available. Once the preliminary designs have been scaled and drawn in preliminary form, the configurations can be refined through the application of revolution principles and auxiliary views. Angles, true lengths, true sizes, and other physical properties must be found to permit further analysis of the final design, as will be discussed in the succeeding chapters.

PROBLEMS

Problems for this chapter should be constructed and solved on $8\frac{1}{2}'' \times 11''$ sheets, as illustrated by the accompanying figures, in accordance with

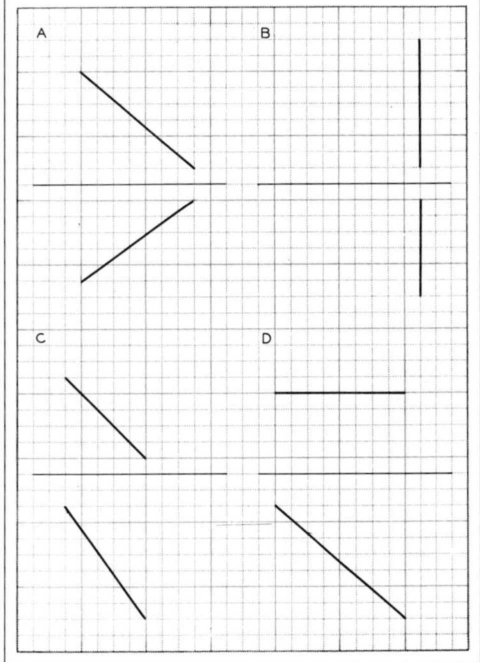

Fig. 7–34. Revolution of lines.

the practices outlined in Article 3–6. Each grid represents $\frac{1}{4}''$. All reference planes and points should be labeled using $\frac{1}{8}''$ letters with guidelines.

1. Use Fig. 7–34 for all parts of this problem. (A) Find the true length of the line in the front view by revolution. Indicate the angle this line makes with the horizontal plane. (B) Find the true length of the line in the front view by revolution. Indicate the angle this line makes with the horizontal plane. (C) Find the true length of the line in the horizontal view by revolution. Indicate the angle this line makes with the frontal plane. (D) Find the true length of the line in the horizontal view by revolution. Indicate the angle this line makes with the frontal plane.

2. Use Fig. 7–35 for all parts of this problem. (A) Find the true length of the line in the profile view by revolution. Indicate the angle this line makes with the frontal plane. (B) Find the true length of the line in the profile view by

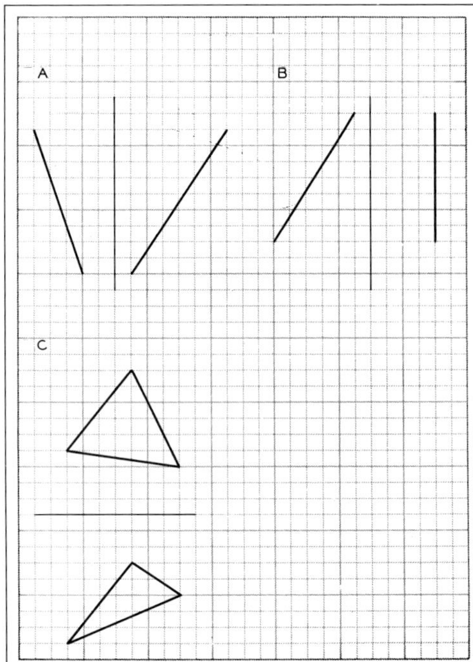

Fig. 7–35. Revolution of lines and planes.

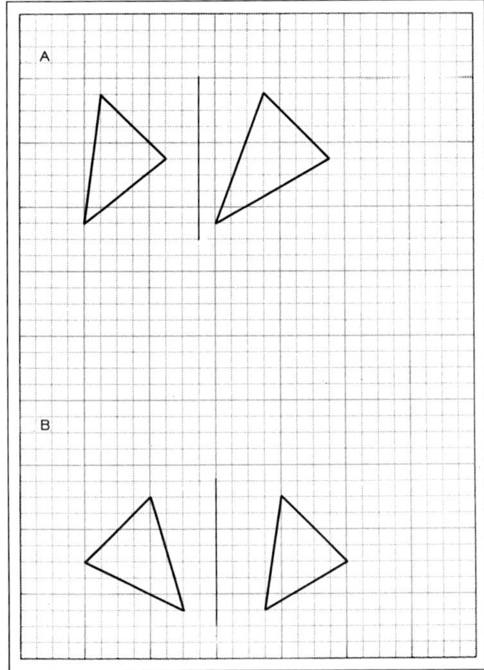

Fig. 7–36. Revolution of a plane.

revolution. Indicate the angle this line makes with the frontal plane. (C) Find the true size of the plane by a primary auxiliary view and a single revolution.

3. (A) In Fig. 7–36A, find the edge view of the plane by revolution. (B) Find the true size of the plane by double revolution in part B of the figure.

4. (A) In Fig. 7–37A, find the angle between the two intersecting planes by revolution. (B) Find the angle between the intersecting planes in part B of the figure by revolution.

5. (A) In Fig. 7–38A, revolve the point about the line and locate its highest and lowest positions. (B) Revolve the point about the oblique line in part B of the figure. Locate its highest and most forward positions. Indicate it in all views.

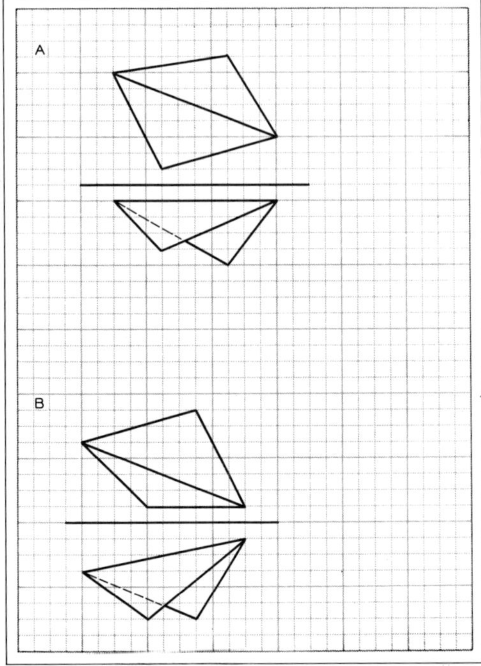

Fig. 7–37. Determining the angle between planes by revolution.

▶

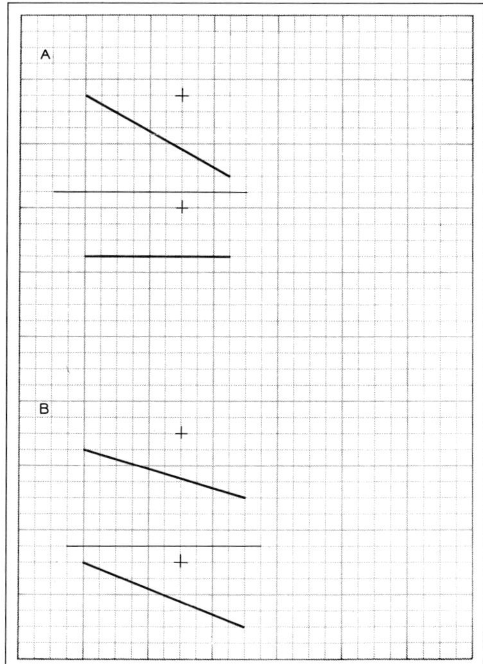

Fig. 7–38. Revolution of a point about a line.

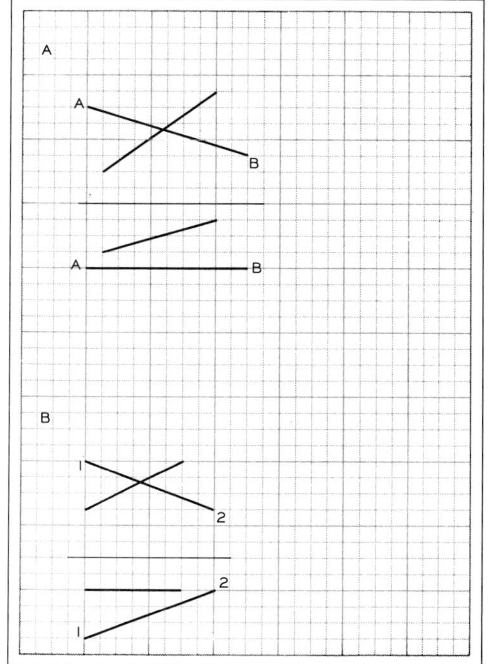

Fig. 7–39. Revolution of a line about an axis.

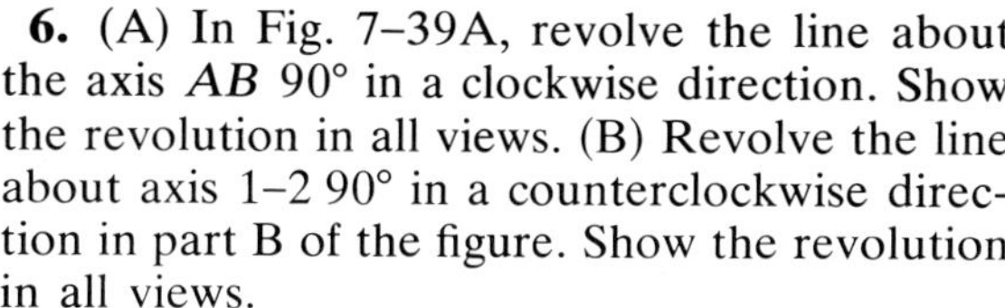

6. (A) In Fig. 7–39A, revolve the line about the axis *AB* 90° in a clockwise direction. Show the revolution in all views. (B) Revolve the line about axis 1–2 90° in a counterclockwise direction in part B of the figure. Show the revolution in all views.

7. Line 1–2 in Fig. 7–40 is the center line of a conveyor chute, such as that shown in Fig. 10–41, which has a 10-foot-square cross section. Construct the necessary views to revolve the 10-foot square into a position where two sides of the right section will be vertical planes. Show the chute in all views. Scale: 1″ = 10′.

8. In Fig. 7–42A, find the angle between the line and plane by two auxiliary views and one revolution. Show all construction. (B) Find the angle between the line and plane in part B of the figure by the auxiliary-view method. Compare the solutions obtained in both parts.

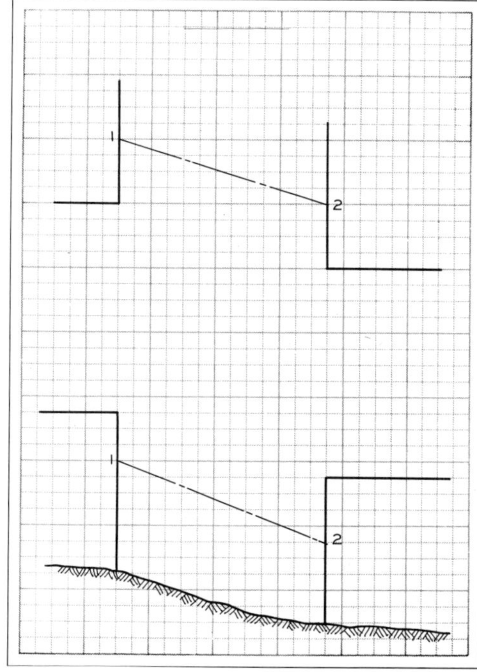

Fig. 7–40. Revolution of a prism about an axis. ▶

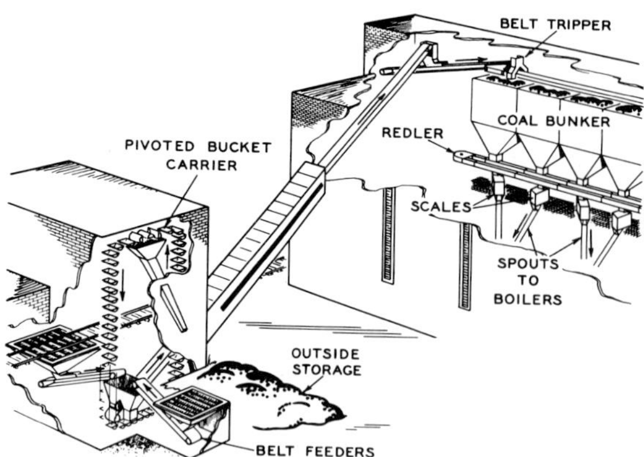

Fig. 7–41. An example of a coal chute between two buildings which is similar to that represented by Fig. 7–40. (Courtesy of Stephens-Adamson Manufacturing Company.)

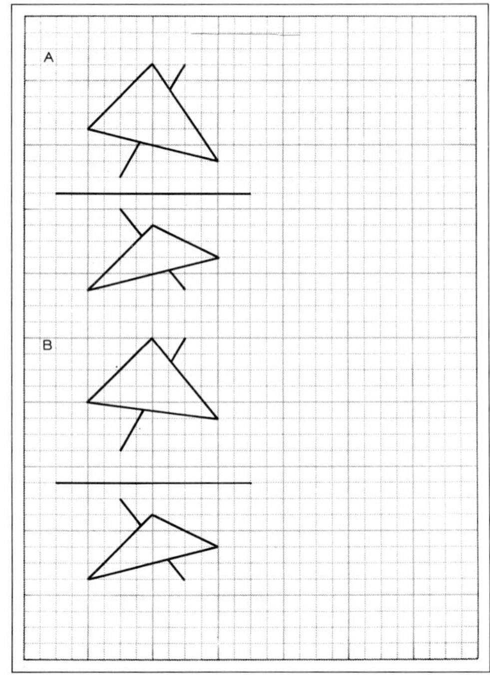

Fig. 7–42. Determining the angle between a line and a plane by revolution and by auxiliary view.

9. Locate two views of a point $3\frac{1}{2}''$ apart on an $8\frac{1}{2}'' \times 11''$ sheet. Using a conical element of 2″, find the direction of a line that slopes forward and makes an angle of 30° with the frontal plane and slopes downward and makes an angle of 50° with the horizontal plane. Show all construction.

10. A fixture block (Fig. 7–43) must have a hole drilled perpendicular to the inclined surface with its center at point *B*, which lies on the plane. Using principles of revolution, determine the angles of revolution necessary to position the block under a vertically mounted drill for drilling. This information will be needed to design a jig for holding the block during this operation. Show the new positions of the block after revolution in all views.

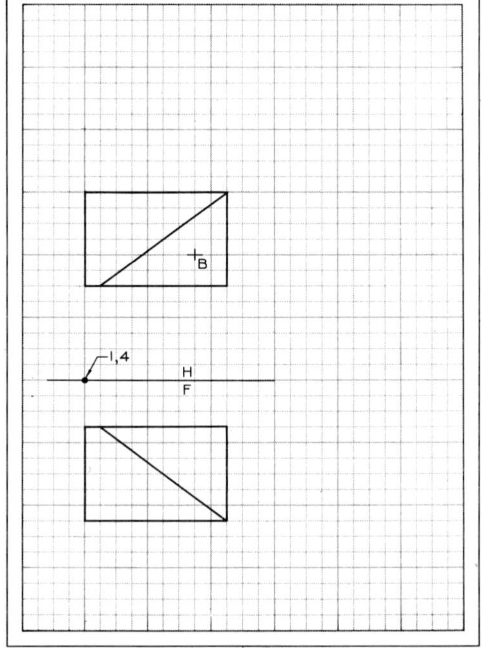

Fig. 7–43. Fixture block. ▶

8

Intersections

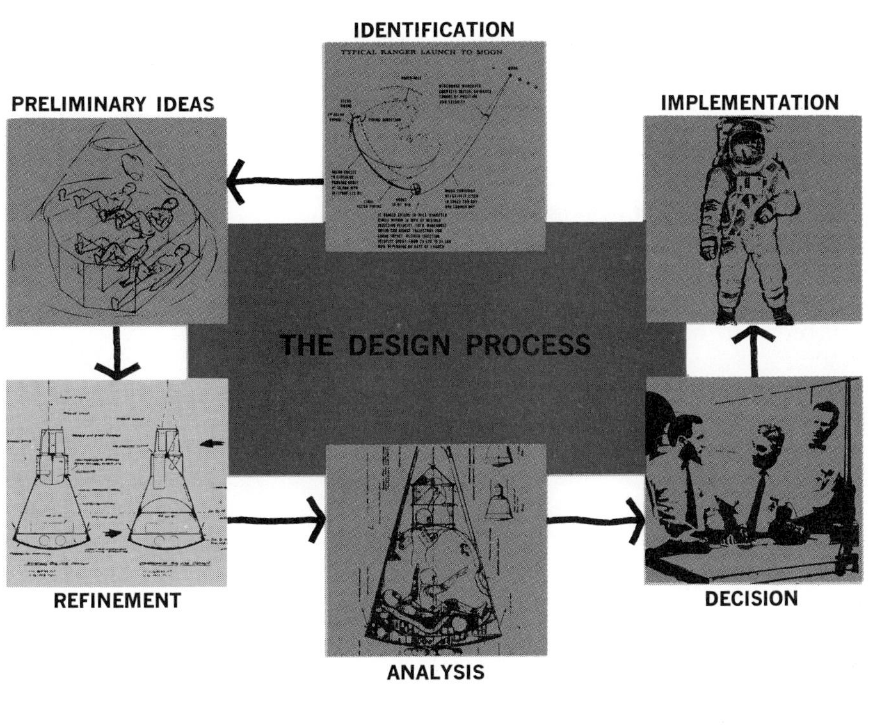

8–1 INTRODUCTION

Practically every product or engineering project designed is composed of planes, lines, or solids that intersect, often at unusual angles. The simple attachment of a rear-view mirror on the exterior body of an automobile is an intersection problem that must be solved prior to production. Massive concrete structures often involve the intersection of geometric forms with each other. The engineer must understand the principles of intersections in order to present his designs and to supervise the construction of the concrete forms into which the concrete will be poured.

The design of the dashboard of an automobile requires the solution of many intersection problems. The forms that intersect vary from lines to contoured shapes. All intersections must be developed graphically in the early stages of the design refinement, since the intersections will influence the final configuration and appearance of the instrument panel and the included accessories.

Many of the principles of intersection covered in this chapter are of a conventional nature and involve planes and regular geometric shapes. An understanding of the principles given will be sufficient for practically any problem encountered, since all intersection applications will involve variations of fundamental examples. As often as possible in our discussion, we shall include industrial examples to illustrate applications of the principles being covered.

It is advantageous to letter the significant points and lines that are used in the solution of intersection problems, as shown in the examples that follow. It is unnecessary to letter each and every point, but key points should be labeled to clarify construction and projection. Guidelines and constructions should be drawn very lightly to avoid the need for erasure upon completion of the problems.

8–2 INTERSECTIONS OF LINES AND PLANES

The basic step of finding an intersection between geometric shapes is the determination of the intersection between a line and a plane. This is illustrated in Fig. 8–1, where the plane is inclined at a 45° angle and the intersecting line is *AB*. This is a special case, since the plane appears as an edge in the side view, where the point of intersection can be found and projected to the front view. Visibility is found in step 2.

If you can understand the simple principle of this problem, it will be easy for you to solve more complex problems, since all shapes are composed of lines. This principle is used in Fig. 8–2 to find the line of intersection between two planes. Since plane *EFGH* appears as an edge in the side view, points of intersection 1

FIGURE 8–1. INTERSECTION OF A LINE AND A PLANE

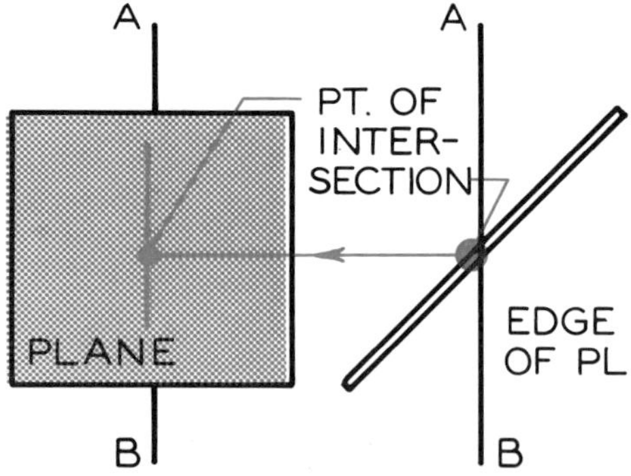

Step 1: The point of intersection can be found in the view where the plane appears as an edge, the side view in this example.

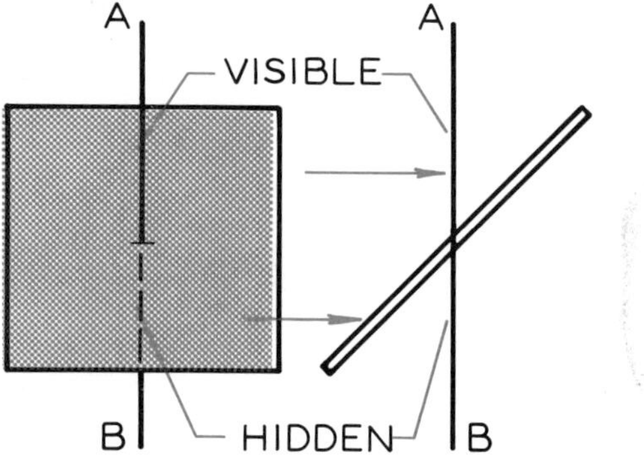

Step 2: Visibility in the front view is determined by looking from the front view to the right side view.

FIGURE 8–2. INTERSECTION BETWEEN PLANES

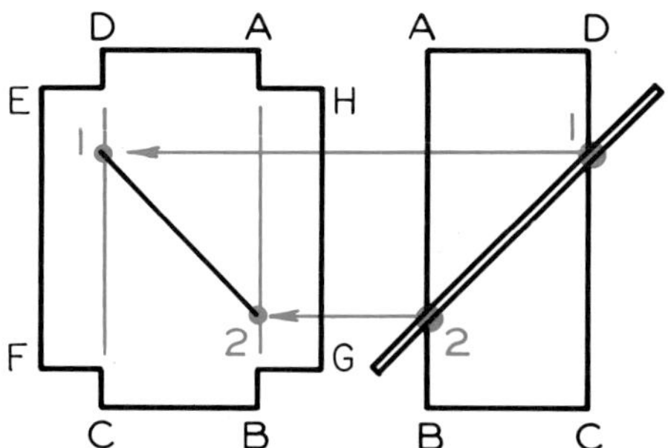

Step 1: The points where plane *EFGH* intersects lines *AB* and *DC* are found in the view where the plane appears as an edge. These points are projected to the front view.

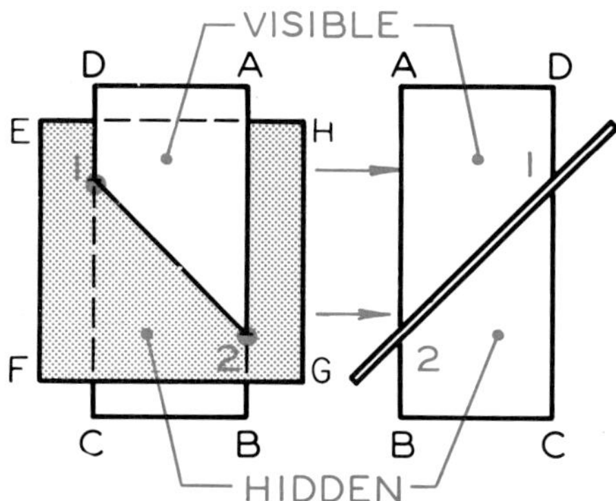

Step 2: Line 1–2 is the line of intersection. Visibility is determined by looking from the front view to the right side view.

and 2 can be found and projected to the front view. The problem is completed by determining visibility in step 2. Note that the problem was solved by finding the points of intersection (piercing points) of lines *AB* and *DC*.

The intersection of a plane and a corner of a prism results in a line of intersection that will bend around the corner (Fig. 8–3). This requires the application of the previously covered principles plus finding the location of the bend point. This point lies on the corner line where the latter pierces the intersecting plane. The visibility changes in the front view in this example, where the line of intersection bends around the corner.

FIGURE 8–3. INTERSECTION OF A PLANE AT A CORNER

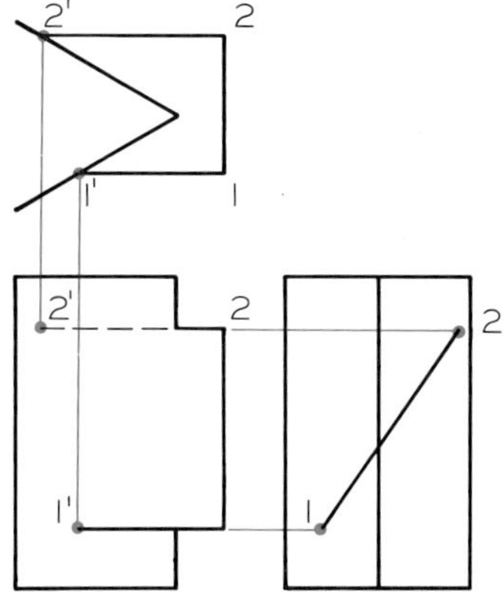

Step 1: The intersecting plane appears as an edge in the side view. Intersection points 1′ and 2′ are projected from the top and side views to the front view.

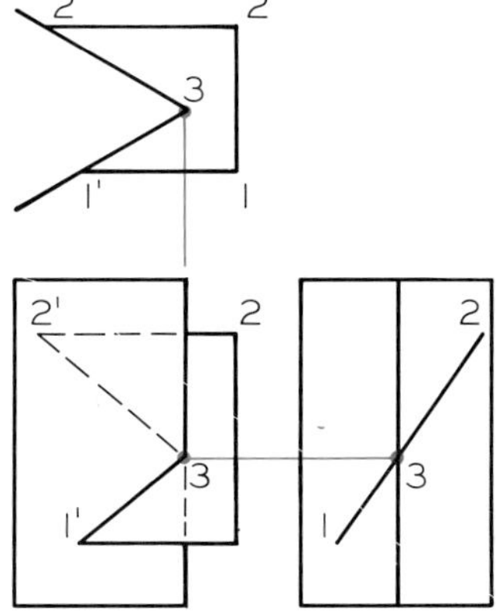

Step 2: The line of intersection from 1′ to 2′ must bend around the vertical corner at point 3 in the top and side views. This point is projected to the front view to locate line 1′–3′–2′.

The same principle is used to find the intersection of a plane and a prism in Fig. 8–4, where the plane appears as an edge. A prism is composed of planes, which are in turn composed of lines; hence, it is possible for the principle introduced in Fig. 8–1 to be used again. The points of intersection are found for each line and are connected; visibility is determined to complete the solution.

FIGURE 8–4. INTERSECTION OF A PLANE AND A PRISM

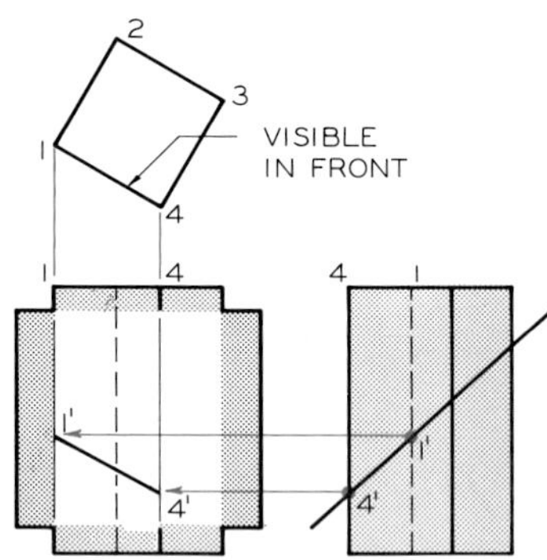

Step 1: Vertical corners 1 and 4 intersect the edge view of the plane in the side view at 1′ and 4′. These points are projected to the front view and are connected with a visible line.

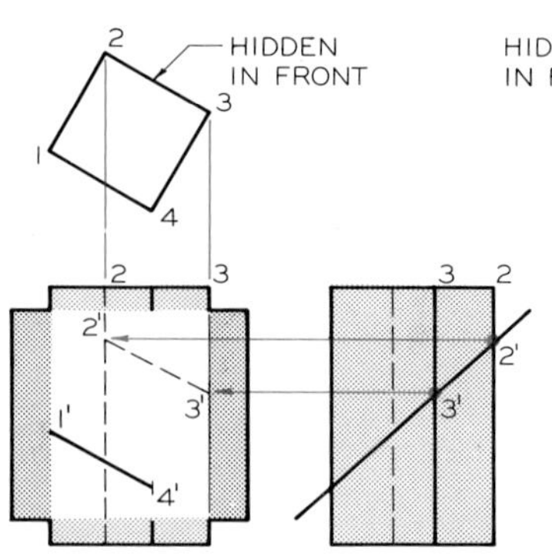

Step 2: Vertical corners 2 and 3 intersect the edge of the inclined plane at 2′ and 3′ in the side view. These points are connected in the front view with a hidden line. Inspection of the top view tells us that this line is hidden.

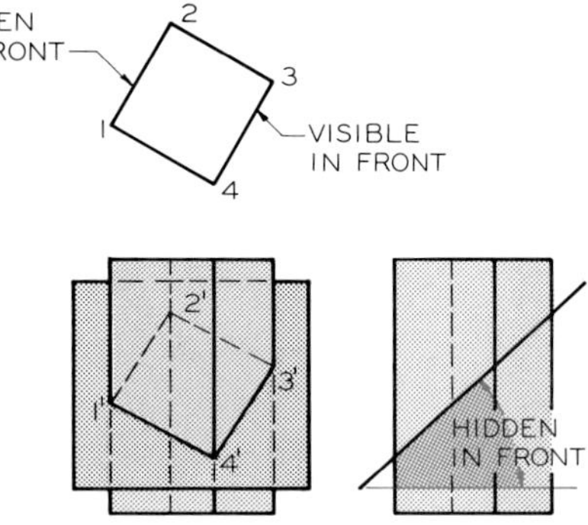

Step 3: Lines 1′–2′ and 3′–4′ are drawn as hidden and visible lines, respectively. Visibility is determined by inspection of the top and side views and by projection to the front view.

A general case of intersection between a plane and a prism is shown in Fig. 8–5. The vertical corners of the prism project true length in the front view and the plane appears foreshortened in both views. Imaginary cutting planes are passed vertically through the planes of the prism in the top view to find piercing points of the corners of these planes in the front view. The points are connected and the visibility is determined to complete the solution.

The general case of the intersection between an oblique plane and an oblique prism is

FIGURE 8–5. INTERSECTION OF AN OBLIQUE PLANE AND A PRISM

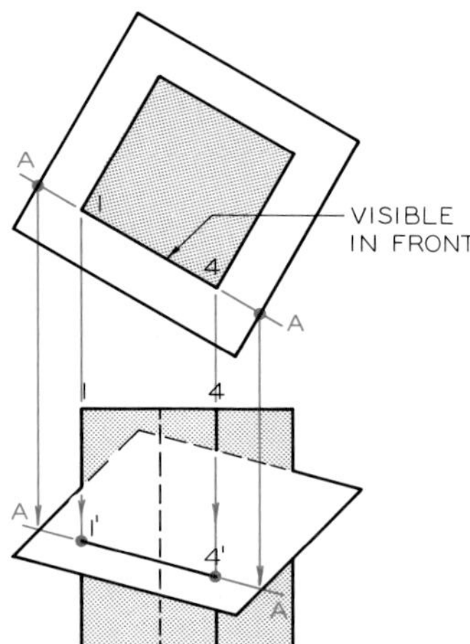

Step 1: Vertical cutting plane *A–A* is passed through the vertical plane, 1–4, in the top view and is projected to the front view. Piercing points 1′ and 4′ are found in this view.

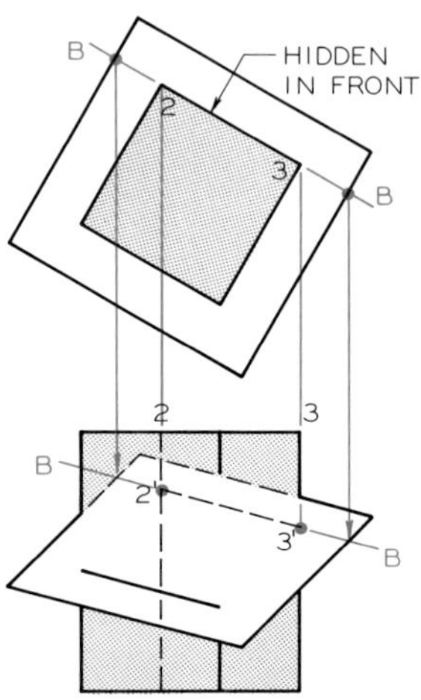

Step 2: Vertical plane *B–B* is passed through the top view of plane 2–3 and is projected to the front view where piercing points 2′ and 3′ are found. Line 2′–3′ is a hidden line.

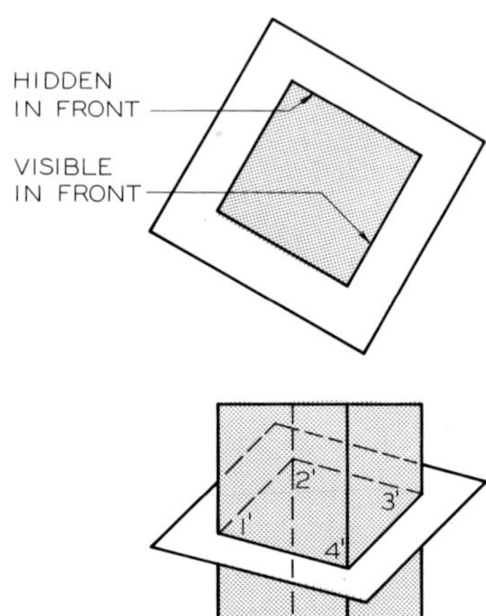

Step 3: The line of intersection is completed by connecting the four points in the front view. Visibility in the front view is found by inspection of the top view.

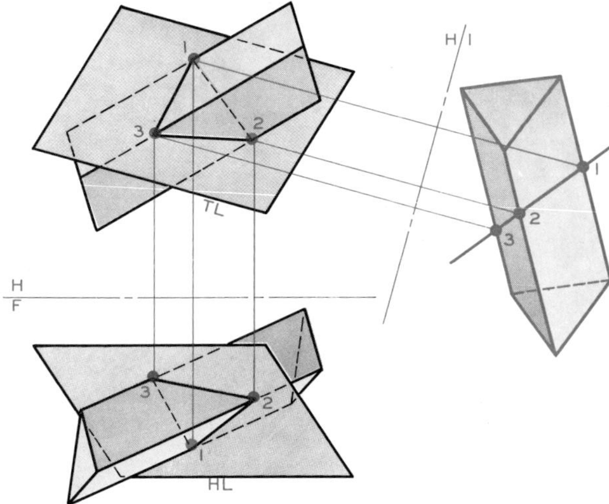

Fig. 8–6. Intersection of an oblique plane and an oblique prism.

shown in Fig. 8–6. Since both the plane and the prism are oblique in the given views, neither the plane nor any of the planes of the prism appears as an edge. This problem can be solved in the same manner as the example in Fig. 8–4, if a view can be found where the plane projects as an edge. This view can be found in a primary auxiliary view by taking the point view of a line on the plane by projection from either view. The lateral edges of the prism do not appear true length in the auxiliary view, but this does not complicate the problem. In the figure, points 1, 2, and 3 are located in the auxiliary view where the corner edges of the prism intersect the plane. These points are projected to the top and front views as shown. Visibility is determined in both views by inspection. Principles of visibility can be reviewed in Article 4–12.

8–3 INTERSECTION BETWEEN PRISMS

The intersection between two prisms can be found by applying the principles that were used to find the intersection between a single plane and a prism in the preceding article. Prisms are composed of planes; consequently, it is possible to work with each plane individually until all lines of intersection are found. An example is solved by steps in Fig. 8–7.

The top and front views of two intersecting prisms are given, and we are required to find

FIGURE 8–7. INTERSECTION BETWEEN TWO PRISMS

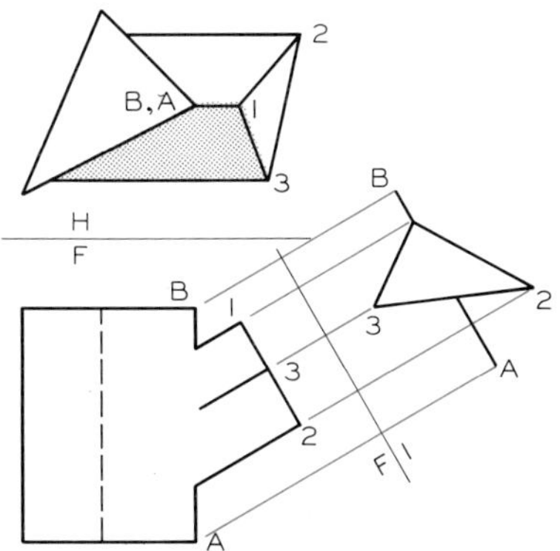

Step 1: Construct the end view of the inclined prism by projecting an auxiliary view from the front view. Show only line *AB*, the corner line of the vertical prism, in the auxiliary view because this is the only critical line. Letter the points.

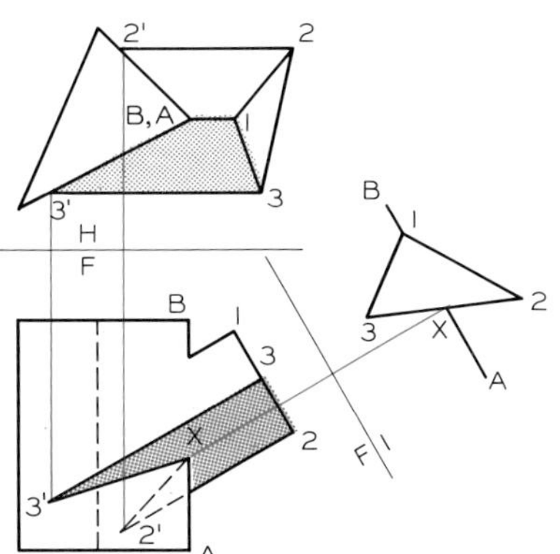

Step 2: Locate the piercing points of lines 2–2′ and 3–3′ in the top view and project them to the front view to the extension of the corner lines. It is rather obvious that a line connecting points 2′ and 3′ will not be a straight line, but will bend around the corner line *AB*. The point where this line intersects the corner is found to be point *X* in the primary auxiliary view. Project it back to the front view.

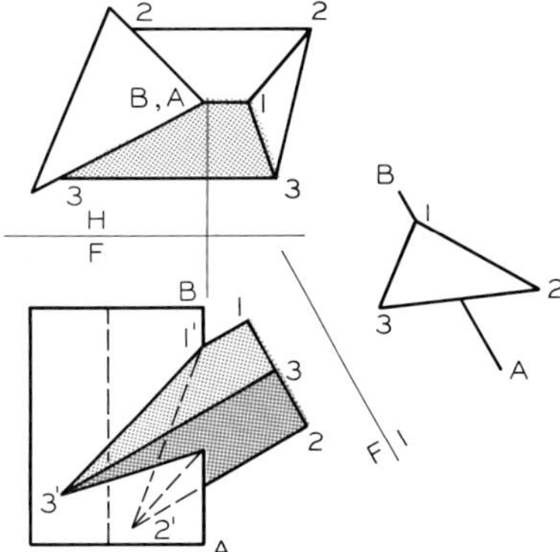

Step 3: It can be seen in the primary auxiliary view that lines 1′–3′ and 1′–2′ do not bend around corner *AB*. Draw lines 1′–3′ and 1′–2′ in the front view. Line 1′–2′ is found to be invisible in the front view by inspection of the primary auxiliary view. Draw line 1′–3′ as a visible straight line.

their lines of intersection. The surfaces of one prism appear as edges in the top view, which makes it possible to see where the corner edges of the other prism intersect. It is necessary to draw an auxiliary view (step 1) in order to find the relationship between line AB and the end view of the prism, since one of the lines of intersection must bend around this corner line. The points of intersection, 3′ and 2′, are found in the top view in step 2, and then are projected to the front view. Point X, the point where line 2′–3′ bends around line AB, is found in the auxiliary view and projected to the front view. Line 2′–X–3′ is completed to the front view and visibility is indicated. The remaining lines of intersection are found, as explained in step 3, by repeating this procedure.

An alternative method of finding the line of intersection between two prisms is illustrated in Fig. 8–8. Vertical cutting plane A is passed through the point view of corner line 3–4 in the top view, and points 1 and 2 are established on the second prism. Since the cutting plane is parallel to the corner edges of the upper surface of the oblique prism, line 1–2 will be drawn parallel to these sides in the front view, where it is found to intersect line 3–4 at point 2. Point 2 is the point where the line of intersection bends around corner line 3–4. The other piercing points are found in the manner described in Fig. 8–7.

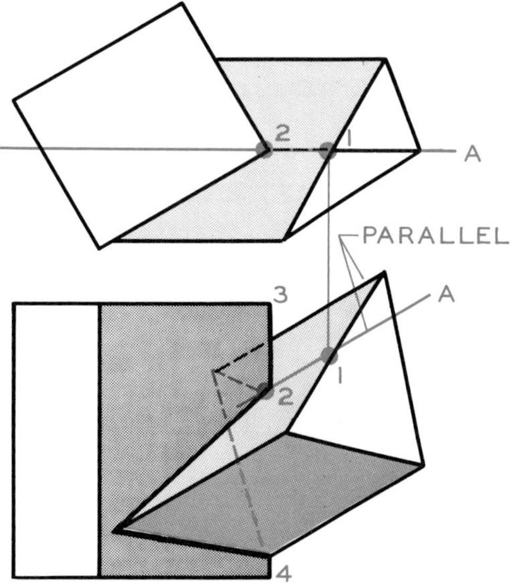

Fig. 8–8. Intersection between prisms by projection.

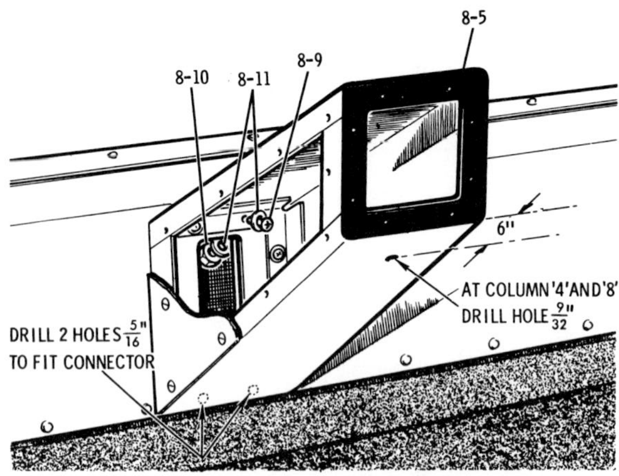

CONDUIT CONNECTOR

Fig. 8–9. This conduit connector was designed through the use of the principles of intersection of a plane and a prism. (Courtesy of the Federal Aviation Administration.)

The conduit connector shown in Fig. 8–9 is an example of the intersection of planes and prisms. This connector is designed to intersect an oblique wall at an angle.

8–4 INTERSECTION OF A PLANE AND A CYLINDER

The catalytic cracking unit shown in Fig. 8–10 illustrates many intersections of cylindrical shapes with geometric forms. Cylinders are fundamental shapes that are used extensively in engineering plants of this type and have many engineering applications in all other industries as well.

A special case of a plane intersecting a cylinder is shown in Fig. 8–11. The plane appears as an edge in the side view.

The line of intersection is found by passing vertical cutting planes through the vertical cylinder in the top view to establish elements on it and their points of intersection. These points can be located in the side view, where the plane appears as an edge, by the principle introduced in Fig. 8–1.

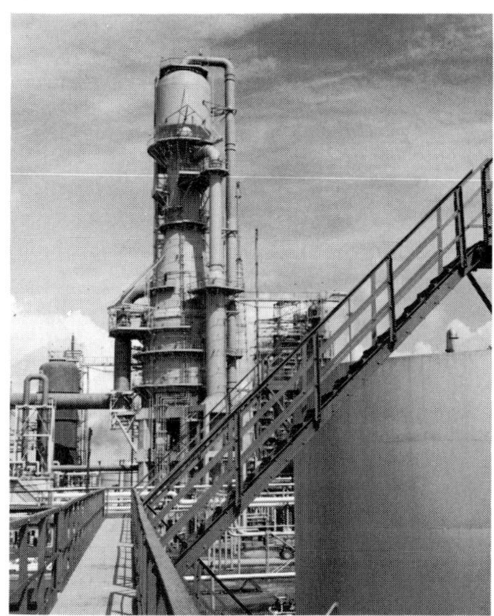

Fig. 8–10. Many intersections of cylinders and planes can be seen in this refinery. (Courtesy of Exxon Corporation.)

The more cutting planes that are used, the more accurate will be your solution. The line of intersection will be an ellipse.

A number of planes intersecting cylinders are shown in Fig. 8–12. This is a model of a refinery that has been constructed to analyze the relationships of the components and to serve as a guide during construction.

A similar problem is solved in Fig. 8–13, in which the plane is oblique in the principal views.

Vertical cutting planes are passed through the vertical cylinder and the oblique plane in the top view to find points of intersection that can be projected to the front view. This process is repeated until a sufficient number of points are found to draw the elliptical line of intersection in the front view.

The general case for the intersection between an oblique plane and an oblique cylinder is shown in Fig. 8–14, in which the cylinder does not appear true length, and the plane does

FIGURE 8–11. INTERSECTION BETWEEN A CYLINDER AND A PLANE

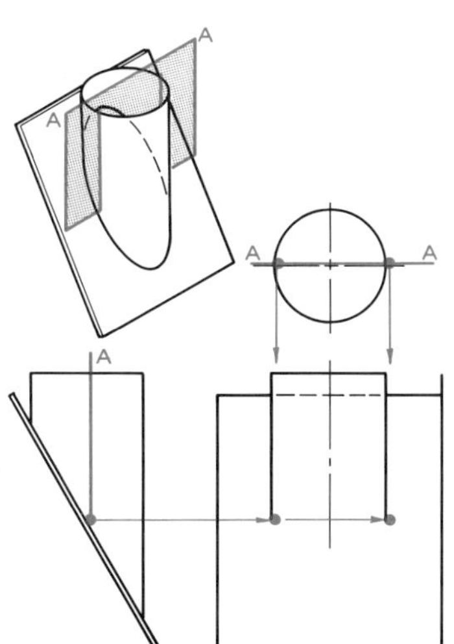

Step 1: A vertical cutting plane, *A–A*, is passed through the cylinder parallel to its axis to find two points of intersection.

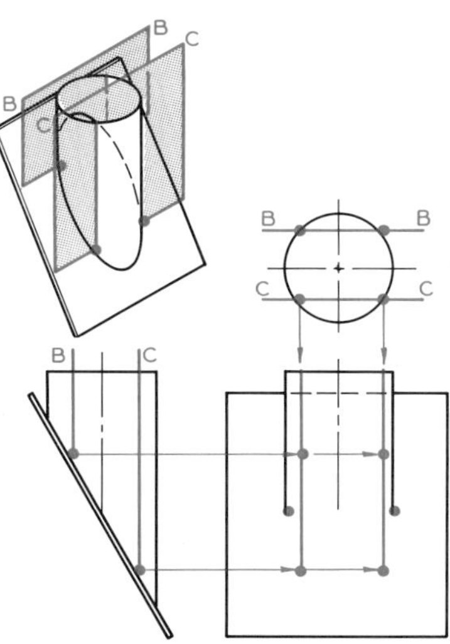

Step 2: Two more cutting planes, *B–B* and *C–C*, are used to find four additional points in the top and left side views; these points are projected to the front view.

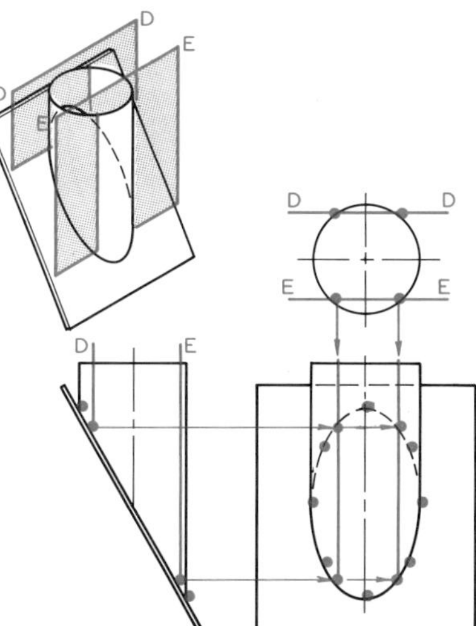

Step 3: Additional cutting planes are used to find more points. These points are connected to give an elliptical line of intersection.

Fig. 8–12. Models are sometimes used to refine a design of a complicated installation. Cylindrical shapes are prominent in this design. (Courtesy of Standard Oil Company of New Jersey.)

not appear as an edge. The line of intersection between these forms is determined by finding an edge view of the plane in a primary auxiliary view, where the projection of the cylinder is foreshortened. Cutting planes are used to establish lines on the surface of the cylinder in the primary auxiliary view. These lines are found to intersect the edge view of the plane at points 1, 2, 3, 4, 5, and 6. The lines on the surface of the cylinder are projected from the auxiliary view to the profile view. Each of the points of intersection found in the auxiliary view is projected to its respective line in the profile view. These points are connected to establish the elliptical line of intersection. Visibility is shown. The line of intersection is found in the front view by transferring measurements of points from the primary auxiliary view to the front view using dividers.

Aircraft designs require the solution of many intersection problems because of the many intersecting geometric forms. Examples

FIGURE 8–13. INTERSECTION OF A CYLINDER AND AN OBLIQUE PLANE

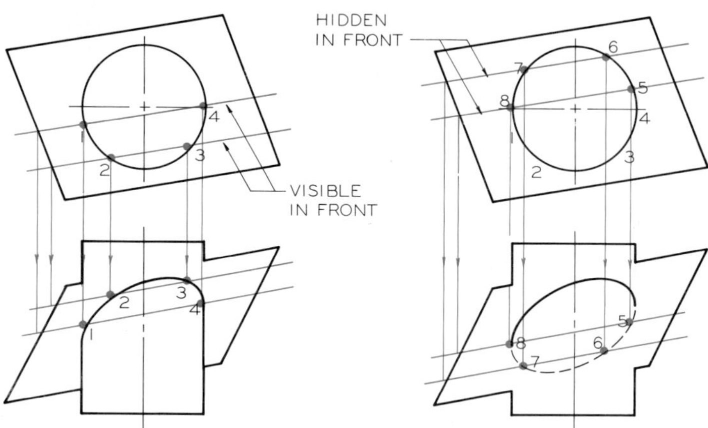

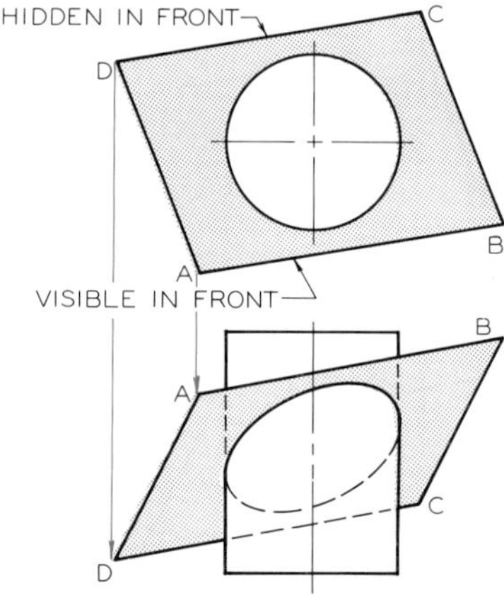

Step 1: Vertical cutting planes are passed through the cylinder in the top view to establish elements on its surface and lines on the oblique plane. Piercing points 1, 2, 3, and 4 are projected to the front view to their respective lines and are connected with a visible line.

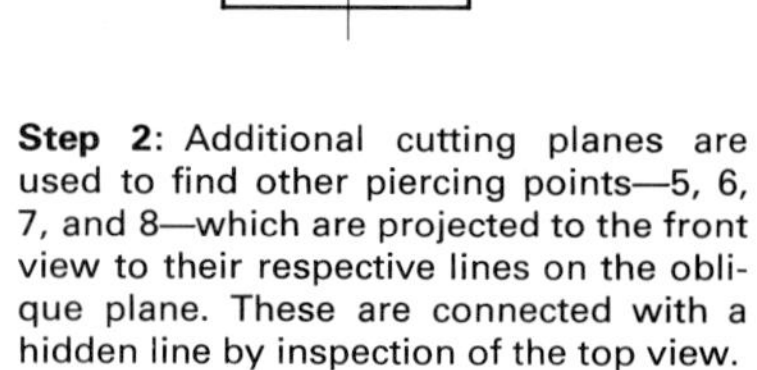

Step 2: Additional cutting planes are used to find other piercing points—5, 6, 7, and 8—which are projected to the front view to their respective lines on the oblique plane. These are connected with a hidden line by inspection of the top view.

Step 3: Visibility of the plane and cylinder is completed in the front view. Line *AB* is found to be visible by inspection of the top view, and line *CD* is found to be hidden.

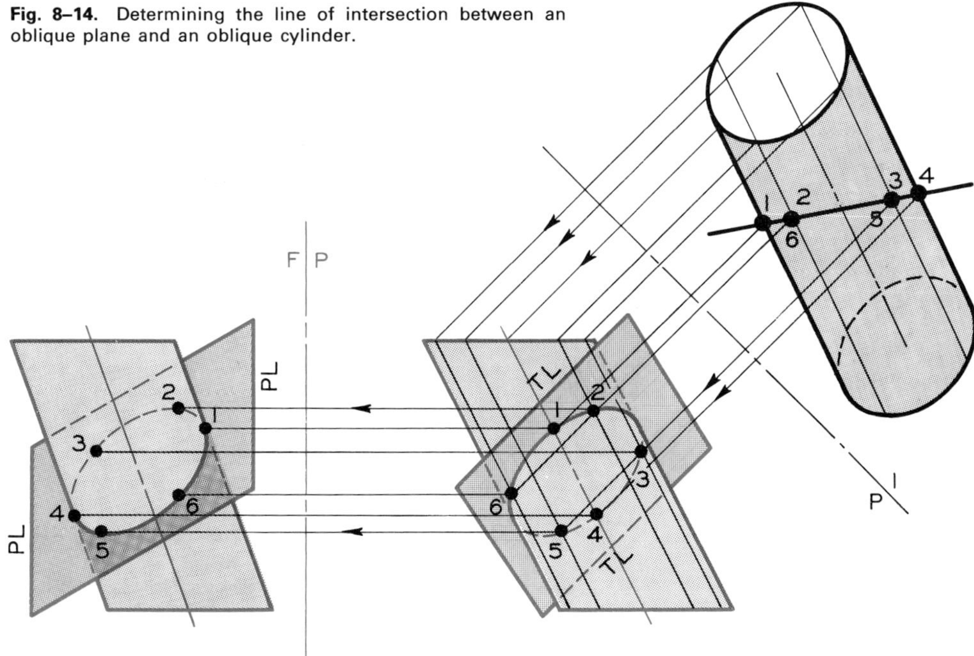

Fig. 8–14. Determining the line of intersection between an oblique plane and an oblique cylinder.

of intersections can be seen in the F–105 shown in Fig. 8–15. In addition to the intersections made by the tail assembly and the wings with the fuselage, the cockpit represents the principle of a cone intersecting an approximate cylinder.

Fig. 8–15. The F–105 fighter-bomber is composed of many intersections between a variety of geometric shapes. (Courtesy of Republic Aviation Corporation.)

8–5 INTERSECTION BETWEEN A CYLINDER AND A PRISM

A series of cutting planes are used in Fig. 8–16 to establish lines that lie on the surfaces of the cylinder and the prism. Since these lines lie in a common cutting plane, they will intersect where they cross in the views of projection. A primary auxiliary view is necessary to locate the lines on the surface of the prism in the front view (step 1). Rather than attempting to find the lines of intersection of two or more planes simultaneously, we shall analyze each plane independently. The line of intersection is projected from the edge view of plane 1–3 in the auxiliary view to the front view in step 2. A plane intersects a curved surface with a curved line. Note that the change in visibility of a line passing around a cylinder in the front is found to be point X in the top view. Step 3 is a continuation of this system of locating points until the final line of intersection is found with the visibility shown.

A practical example of intersections such as we have been discussing can be seen in an electric coffee pot. The lines of intersection made by the spout and handle with the cylindrical body of the container can be determined by

FIGURE 8–16. INTERSECTION BETWEEN A CYLINDER AND A PRISM

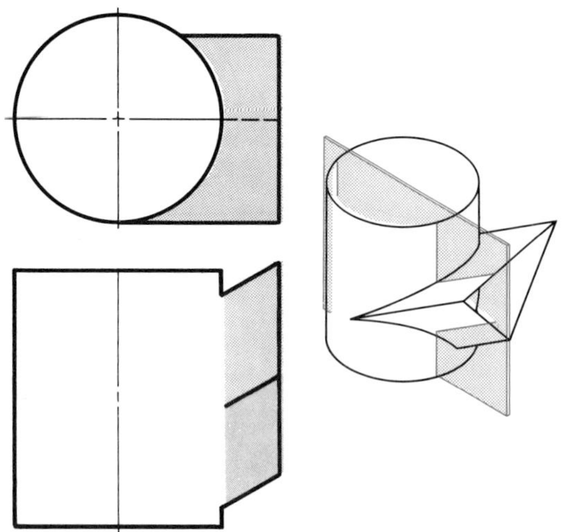

Given: The top and front views of an intersecting cylinder and prism.

Required: Find the line of intersection between the cylinder and the prism.

Reference: Article 8–5.

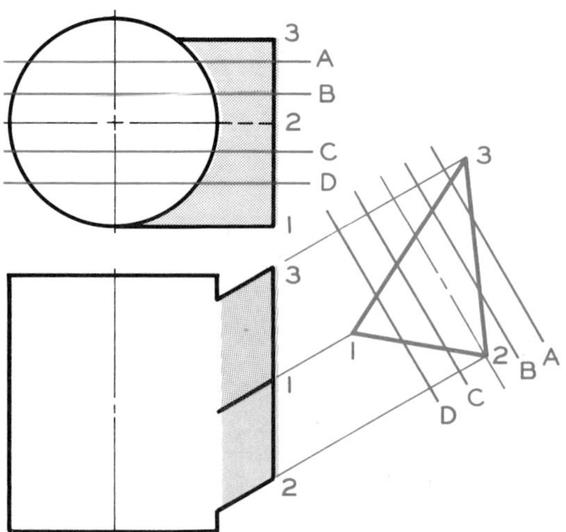

Step 1: Project an auxiliary view of the triangular prism from the front view to show three of its surfaces as edges. Pass frontal cutting planes through the top view of the cylinder and project them to the auxiliary view. The spacing between the planes is equal in both views.

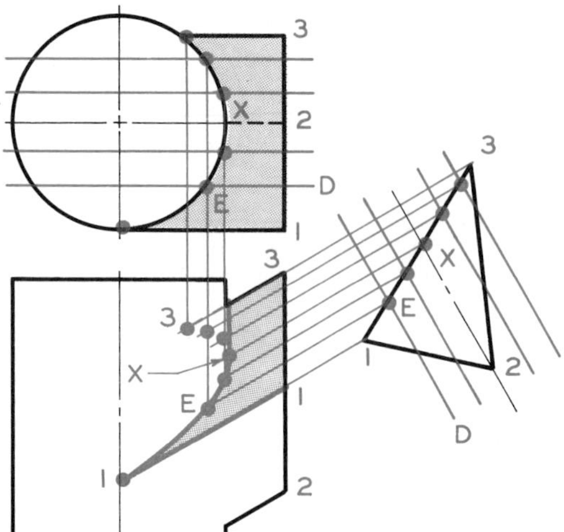

Step 2: Locate points along the line of intersection of the cylinder and plane 1–3 in the top view and project them to the front view. The intersection of these projectors with those coming from the auxiliary view establishes points on the line of intersection in the front view. *Example:* Point *E* on cutting plane *D* is found in the top and primary auxiliary views and projected to the front view where the projectors intersect. Point *X* on the center line is the point where visibility changes from visible to hidden in the front view.

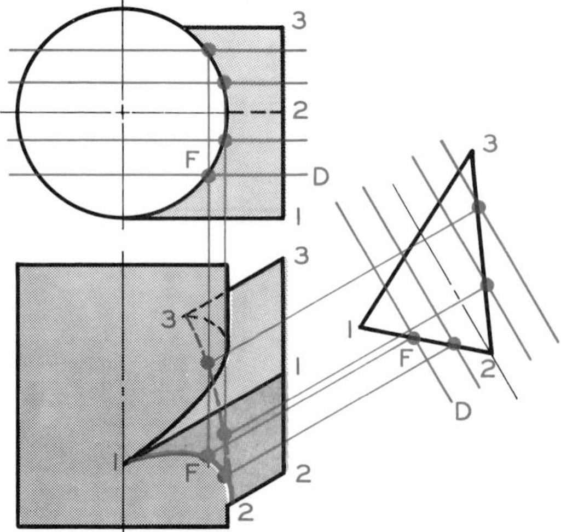

Step 3: Determine the remaining points of intersection by using the same cutting planes. Point *F* is shown in the top and primary auxiliary views and is projected to the front view on line of intersection 1–2. Connect the points and determine visibility. Judgment should be used in spacing the cutting planes so that they will produce the most accurate representation of the line of intersection.

constructing a series of horizontal cutting planes and projecting points of intersections to the other views. Determining these lines of intersection is a preliminary step in designing the components for accurate assembly.

8–6 INTERSECTION BETWEEN TWO CYLINDERS

The intersection between the two cylinders shown in Fig. 8–17 is found with the cutting plane method. Frontal cutting planes are drawn in the top view to establish points common to each cylinder. For example, the locations of points 1 and 2 in the front view are determined by the intersections of the projectors coming from the top and primary auxiliary views where plane *D* intersects the cylinders. These points are on the line of intersection in the front view. The trace of the cutting plane *D* is shown in the front view to illustrate the path of the cutting plane. Additional points are found and connected to find the complete line of intersection. Visibility is indicated in the front view.

A similar intersection between two cylinders is shown in Fig. 8–18. Cutting planes are used to find the line of intersection.

Many intersections between cylinders are shown in Fig. 8–19, in which a thermometer at an Oklahoma gas-processing plant is being checked. Intersections of this type must be found to permit the shapes to be formed for proper joining on the site.

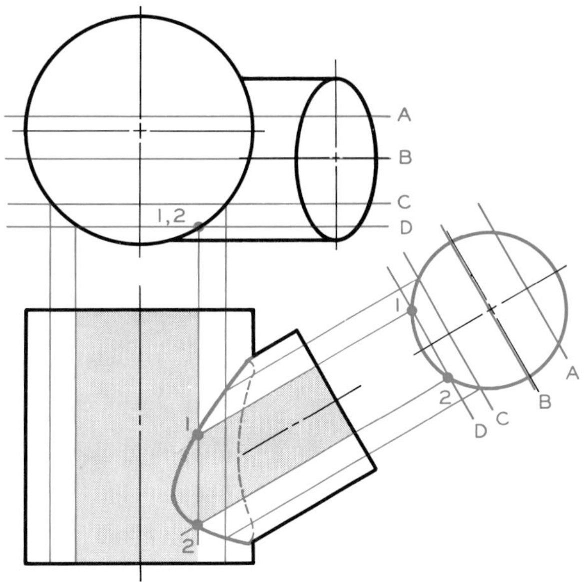

Fig. 8–17. Determining the intersection between cylinders.

FIGURE 8–18. INTERSECTION BETWEEN TWO CYLINDERS

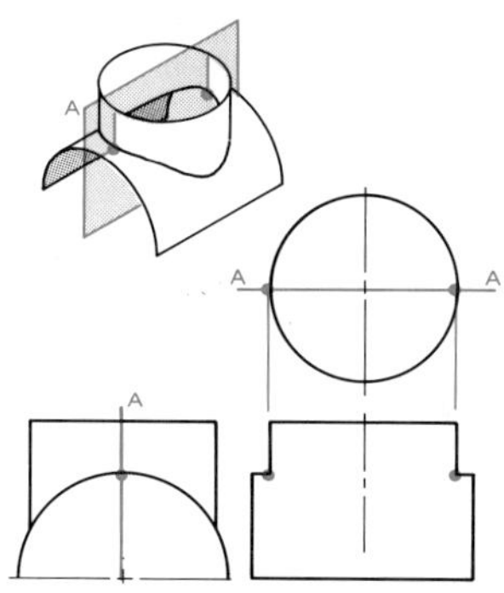

Step 1: A cutting plane, *A–A*, is passed through the cylinders parallel to the axes of both. Two points of intersection are found.

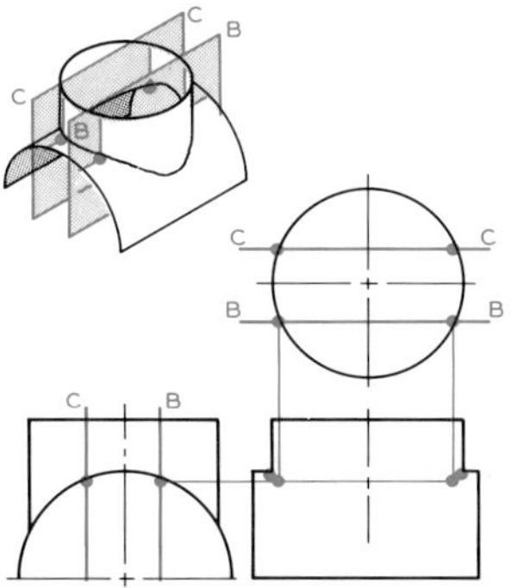

Step 2: Cutting planes *C–C* and *B–B* are used to find four additional points of intersection.

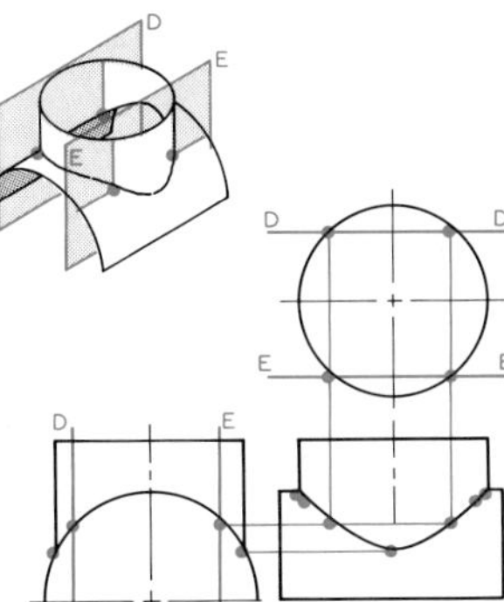

Step 3: Cutting planes *D–D* and *E–E* locate four more points. Points found in this manner are connected to give the line of intersection.

Fig. 8–19. Many intersections between cylinders can be seen in this gas processing plant. (Courtesy of Exxon Corporation.)

8–7 INTERSECTION BETWEEN A PLANE AND A CONE

A cone is a geometric shape which is used in many engineering designs in combination with other forms. The determination of the intersection of a plane with a cone is shown in Fig. 8–20.

In this case, a series of radial lines from the cone's apex to its base (elements) is used to find the line of intersection. These elements cross the edge view of the plane in the front view to locate piercing points that are projected to the top view of these elements. The points are connected to find the line of intersection.

A cone and an oblique plane are given in Fig. 8–21, in which the line of intersection is determined with a series of cutting planes.

FIGURE 8–20. INTERSECTION OF A PLANE AND A CONE

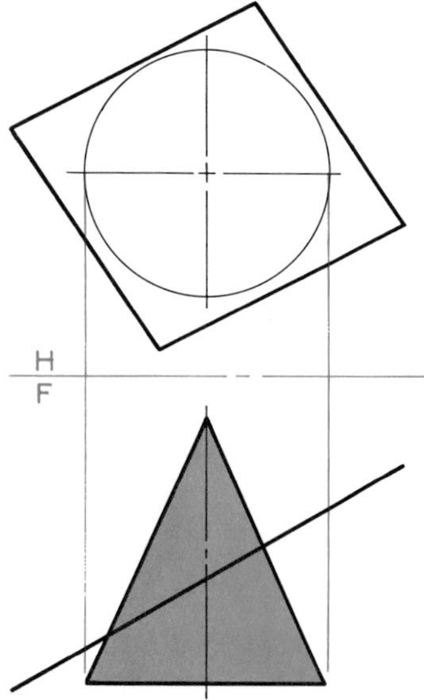

Given: The top and front views of a cone and an intersecting plane.

Required: Find the line of intersection of the two geometric shapes.

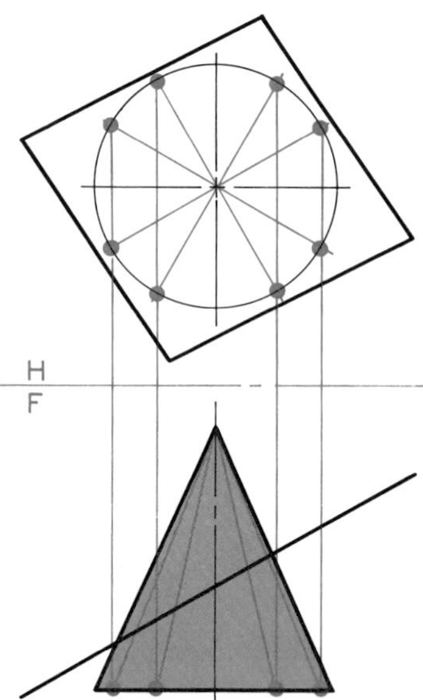

Step 1: Divide the base into even divisions in the top view and connect these points with the apex to establish elements on the cone. Project these to the front view.

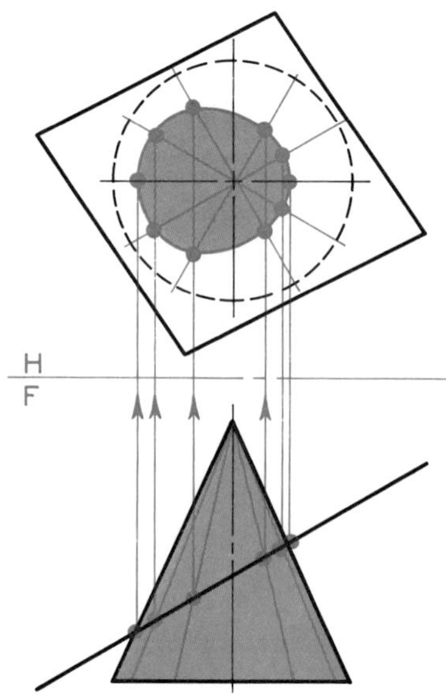

Step 2: The piercing point of each element on the edge view of the plane is projected to the top view to the same elements, where they are connected to form the line of intersection. Visibility is shown to complete the drawing.

FIGURE 8–21. INTERSECTION OF AN OBLIQUE PLANE AND A CONE

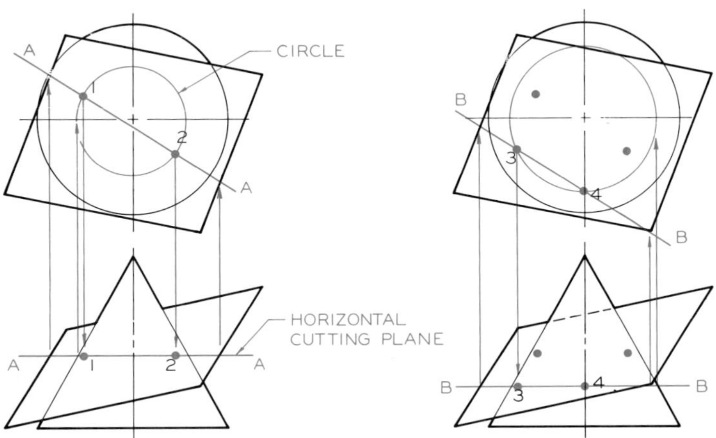

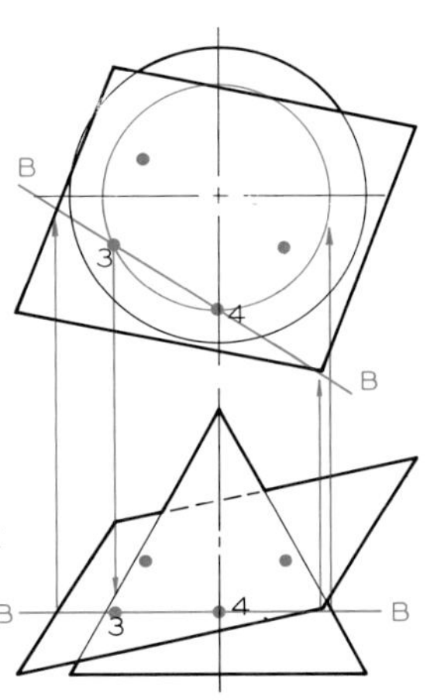

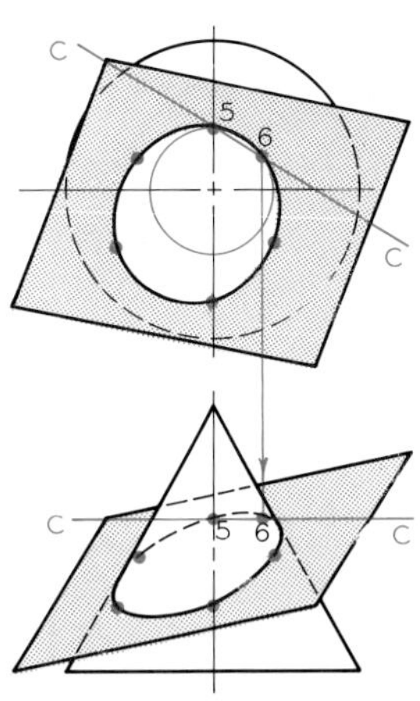

Step 1: A horizontal cutting plane is passed through the front view to establish a circular section on the cone and a line on the oblique plane in the top view. The piercing point of this line must lie on the circular section. Piercing points 1 and 2 are projected to the front view.

Step 2: Horizontal cutting plane *B–B* is passed through the front view in the same manner to locate piercing points 3 and 4 in the top view. These points are projected to the horizontal plane in the front view from the top view.

Step 3: Additional horizontal planes are used to find sufficient points to complete the line of intersection. Determination of the visibility completes the solution.

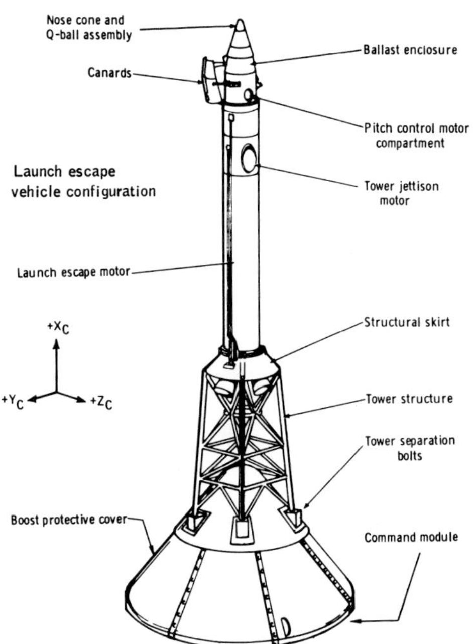

Fig. 8–22. Intersections with a conical shape were solved in designing the launch escape vehicle. (Courtesy of NASA.)

Horizontal cutting planes are used in the front view to give easy-to-draw circular sections in the top view. Also, these cutting planes will locate lines on the oblique plane that will intersect the same circular sections cut by each respective cutting plane. The points of intersection are projected to the adjacent view.

The cutting-plane method could have been used to solve the example shown in Fig. 8–20 as an alternative method. Most descriptive geometry problems have more than one method of solution.

Figure 8–22 illustrates the utilization of conical shapes in the configuration of a launch escape vehicle. Points of intersection, as well as lines of intersection, were determined in the final refinement of this spacecraft.

8–8 INTERSECTION BETWEEN A CONE AND A PRISM

A primary auxiliary view is drawn in Fig. 8–23 to show the lateral planes of the prism as edges so that the line of intersection between the prism and cone may be found. Cutting planes

FIGURE 8–23. INTERSECTION BETWEEN A CONE AND A PRISM

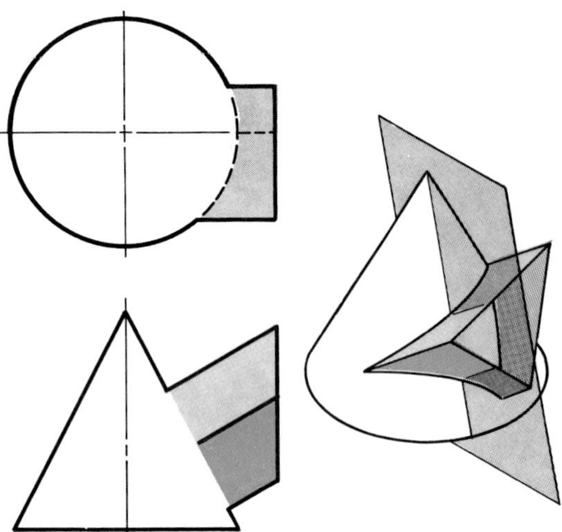

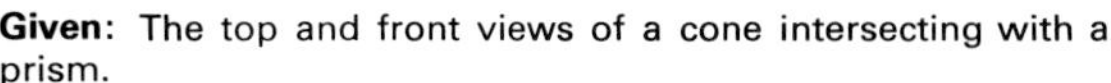

Given: The top and front views of a cone intersecting with a prism.

Required: Find the line of intersection between the cone and the prism and determine visibility.

Reference: Article 8–8.

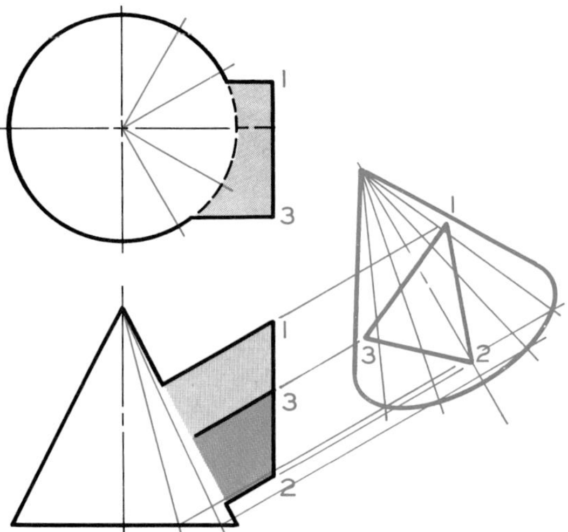

Step 1: Construct a primary auxiliary view to obtain the edge views of the lateral surfaces of the prism. In the auxiliary view, pass cutting planes through the cone which radiate from the apex to establish elements on the cone. Project the elements to the principal views.

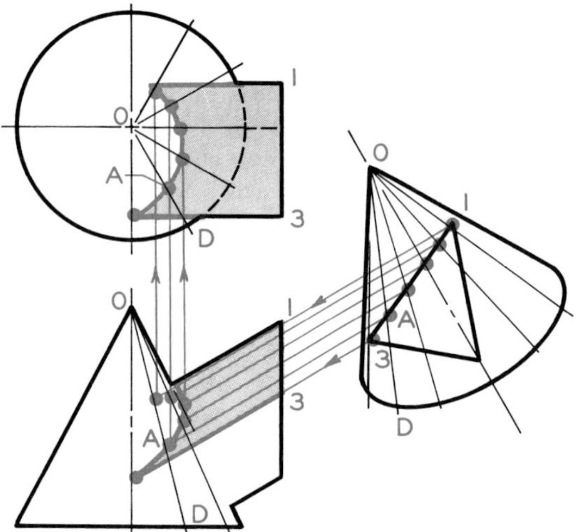

Step 2: Locate the piercing points of the cone's elements with the edge view of plane 1–3 in the primary view and project them to the front and top views. *Example:* Point *A* lies on element *OD* in the primary auxiliary view, so it is projected to the front and top views of element *OD*. Locate other points along the line of intersection in this manner.

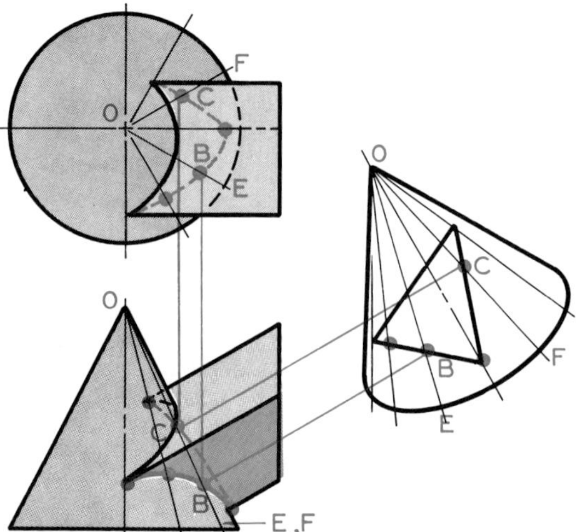

Step 3: Locate the piercing points where the conical elements intersect the edge views of the other planes of the prism in the primary auxiliary view. *Example:* Point *B* is found on *OE* in the primary auxiliary view and is projected to the front and top views of *OE*. Show visibility in each view after the location of a sufficient number of points.

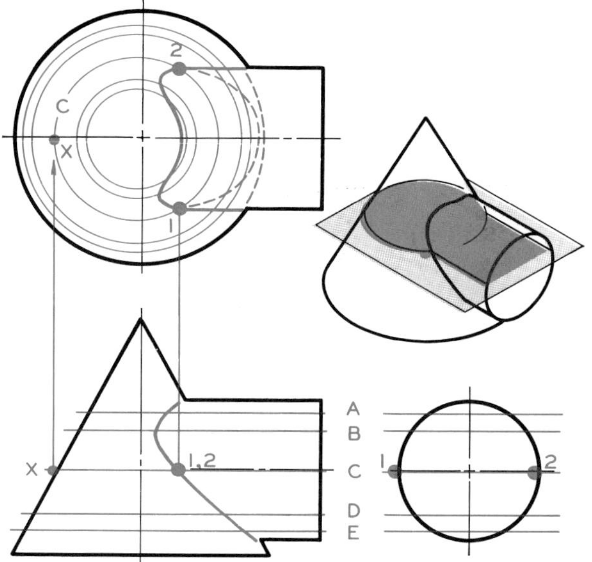

Fig. 8–24. Intersection between a cone and a cylinder using horizontal cutting planes.

Fig. 8–25. This electrically operated distributor illustrates intersections between a cone and a series of cylinders. (Courtesy of GATX.)

that radiate from the apex of the cone in the primary auxiliary view are used to establish lines lying on the surface of the cone and the surface of the prism, as shown in step 1. Inspection of the auxiliary view in step 2 shows that point *A* lies on cutting plane *OD*. Point *A* is projected to the front and top views of element *OD*, which was established on the surface of the cone by the cutting plane.

Other piercing points are found by repeating this system on the other two planes of the prism in step 3. All projections originate in the auxiliary view, where the cutting planes and the planes of the prism appear as edges.

The horizontal cutting-plane method for finding the line of intersection between a cylinder and a cone is shown in Fig. 8–24. This is not a feasible method if the axis of the intersecting cylinder is not horizontal and the axis of the cone not vertical, since the sections cut by the cutting planes would be irregular in shape and would require and the tedious plotting of many points.

An unusual example of cylinders intersecting a cone can be seen in Fig. 8–25. The lines of intersection between the cones of this distributor housing were determined through the use of the auxiliary view method, as illustrated in Fig. 8–23.

8–9 INTERSECTION BETWEEN A PYRAMID AND A PRISM

The intersection between a prism and pyramid can be found by a method similar to that used in Fig. 8–23. An auxiliary view is constructed to show the lateral planes of the prism as edges in step 1 of Fig. 8–26. Cutting planes are drawn to radiate from apex *O* through the corner edges of the prism in the auxiliary view (step 2). Lines *OA* and *OB* are found in the principal views by projection. The lateral edges of the prism are projected to these lines in steps 2 and 3 to find piercing points, which are connected to determine the line of intersection. A pictorial located adjacent to step 1 illustrates the cutting-plane principle used in solving this intersection problem.

The intersection between a horizontal prism and a pyramid is determined in Fig. 8–27 through the use of the horizontal cutting-plane method. An auxiliary view is projected from the top view to find the edge views

FIGURE 8–26. INTERSECTION BETWEEN A PRISM AND A PYRAMID

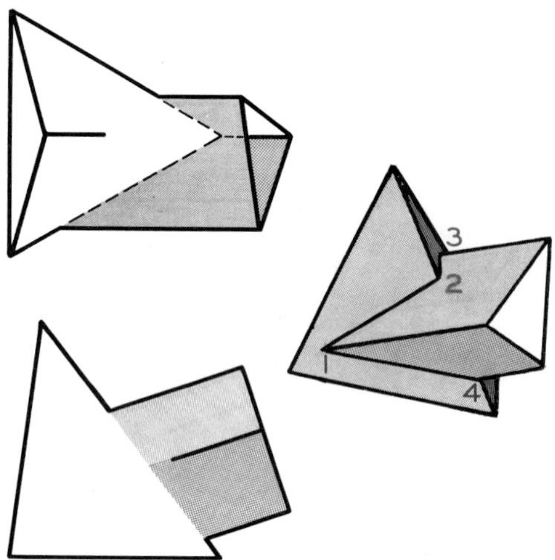

Given: Top and front views of a prism intersecting a pyramid.

Required: Find the line of intersection between the two geometric shapes.

Reference: Article 8–9.

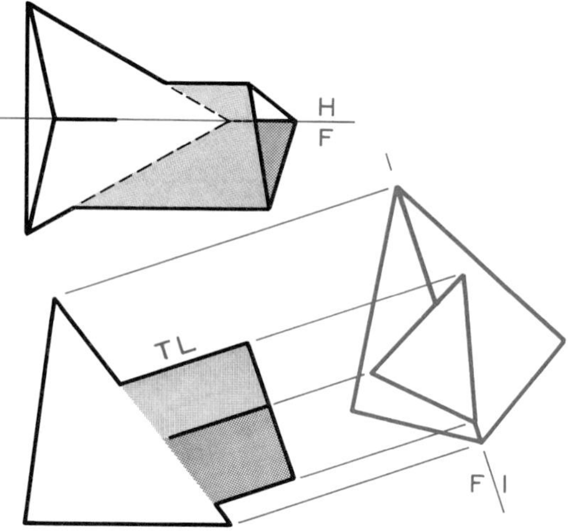

Step 1: Find the edge view of the surfaces of the prism by projecting an auxiliary view from the front view. Project the pyramid into this view also. Only the visible surfaces need be shown in this view.

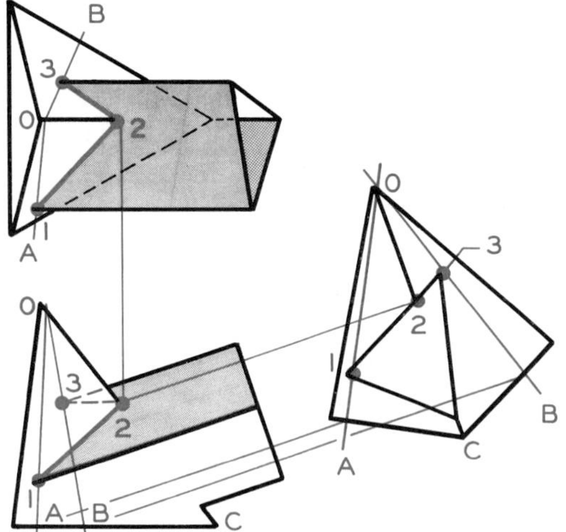

Step 2: Pass planes *A* and *B* through apex *O* and points 1 and 3 in the auxiliary view. Project the intersections of the planes *OA* and *OB* on the surfaces of the pyramid to the front and top views. Project points 1 and 3 to *OA* and *OB* in the principal views. Point 2 lies on line *OC*. Connect points 1, 2, and 3 to give the intersection of the upper plane of the prism with the pyramid.

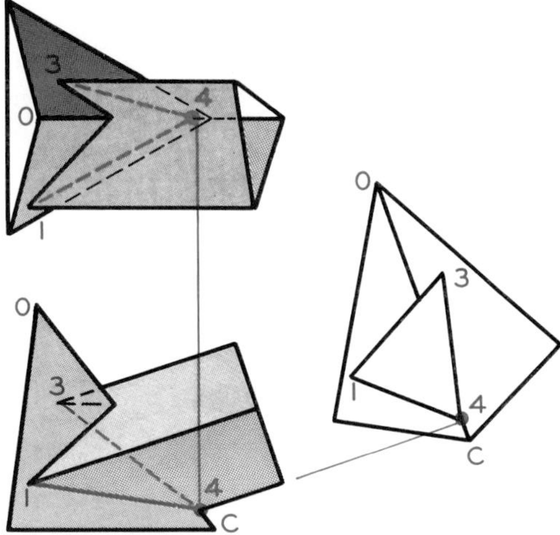

Step 3: Point 4 lies on line *OC* in the auxiliary view. Project this point to the principal views. Connect point 4 to points 3 and 1 to complete the intersections. Visibility is indicated. Note that these geometric shapes are assumed to be hollow as though constructed of sheet metal.

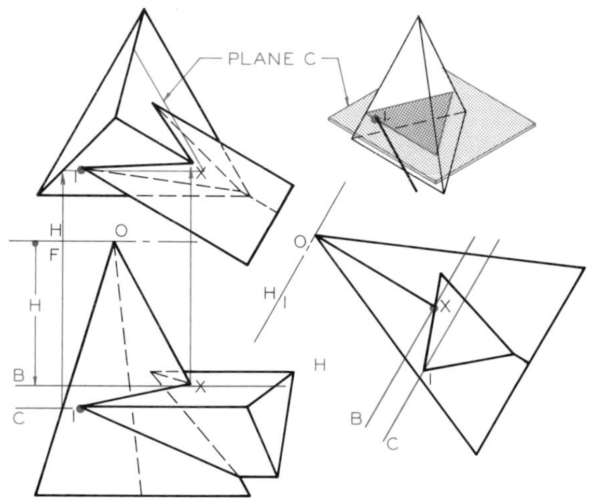

Fig. 8–27. Determining the line of intersection between a pyramid and a prism by horizontal cutting planes.

Fig. 8–28. Examples of intersections of a variety of geometric shapes can be seen in this compressor station installation. (Courtesy of Trunkline Gas Company.)

FIGURE 8–29. INTERSECTION OF A SPHERE AND A PLANE

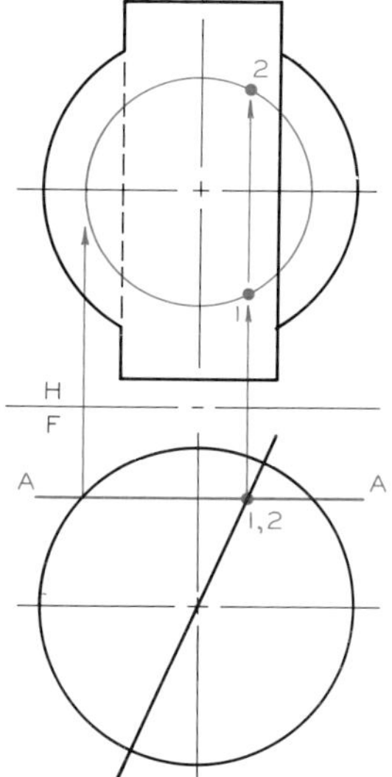

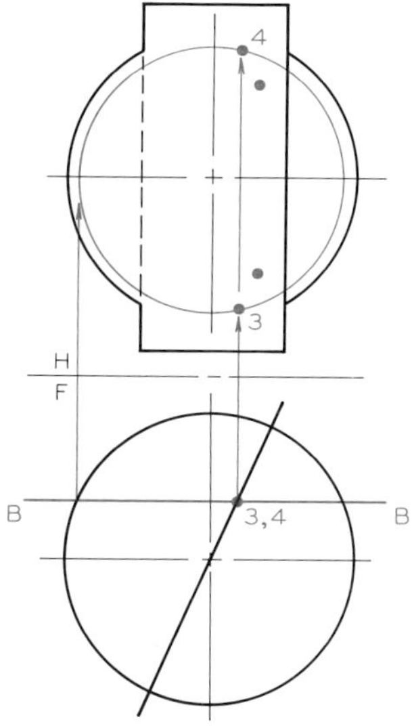

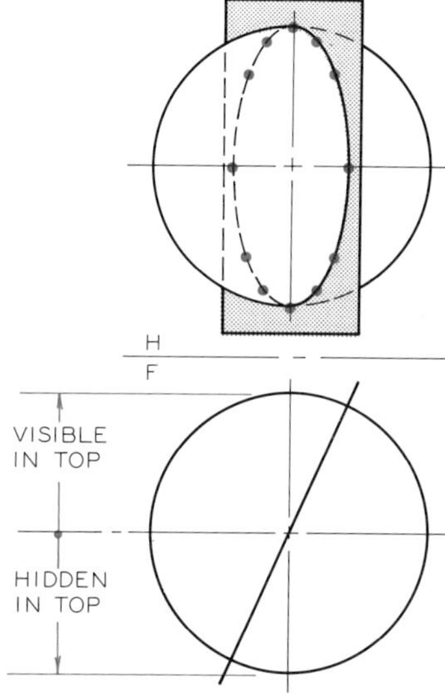

Step 1: Horizontal cutting plane *A–A* is passed through the front view of the sphere to establish a circular section in the top view. Piercing points 1 and 2 are projected from the front view to the top view, where they lie on the circular section.

Step 2: Horizontal cutting plane *B–B* is used to locate piercing points 3 and 4 in the top view by projecting to the circular section cut by the plane in the top view. Additional horizontal planes are used to find sufficient points in this manner.

Step 3: Visibility of the top view is found by inspection of the front view. The upper portion of the sphere will be visible in the top view and the lower portion will be hidden.

of the lateral surfaces of the prism. Horizontal cutting planes are passed through the front and auxiliary views, where they appear as edges that are parallel to the horizontal. These planes will cut triangular sections in the top view which have sides that are parallel to the base of the pyramid. Note that these planes are drawn to pass through the given corner edges of the prism. Each corner edge is extended in the top view to the point of intersection with the section of the pyramid formed by the cutting plane passed through that particular line. Plane *B* is used to locate point *X* in the auxiliary view, which is where the line of intersection 1-*X*-3 bends at line *O*-2. Visibility is determined in each view.

The intersection could have been found by using radial cutting planes, as in Fig. 8–26. As can be seen in these examples, the use of a systematic lettering procedure to plot each important point is helpful in intersection problems.

Figure 8–28 shows the interior of a compressor station where natural gas is compressed for transmission through pipelines over long distances. Many intersection problems are apparent in this complex facility. Complicated layouts of this type are presented in combinations of drawings and models to improve visualization and communication of spatial relationships.

8–10 INTERSECTION BETWEEN A SPHERE AND A PLANE

The sphere is a shape that has many engineering applications, from petroleum storage tanks to the plotting of the paths of satellites traveling in space. An example of the determination of the intersection between a plane and a sphere is shown in Fig. 8–29. The resulting line of intersection will be an ellipse in the top view, but a circle when the line of sight is perpendicular to the intersecting plane.

The ellipse could have been drawn with an ellipse template which was selected by measuring the angle between the edge view of the plane in the front view and the projections coming from the top view. The major diameter of the ellipse would be equal to the true diameter of the sphere, since the plane passes through the center of the sphere.

In the partially constructed Unisphere® shown in Fig. 8–30, the structural members represent intersections between imaginary cutting planes and the surface of the sphere. All the circles passing through the poles are equal in size, while those passing perpendicularly to the axis of the sphere vary in size. Straight members are used to approximate the spherical

Fig. 8–30. Structural members of this spherical shape represent the intersection between imaginary cutting planes and the sphere. (Courtesy of U.S. Steel Corporation.)

Fig. 8–31. The orbital paths depicted by metal rings which encircle the sphere can be projected to the surface of the sphere to locate support brackets. (Courtesy of U.S. Steel Corporation.)

shape in which it appeared in its finished form (Fig. 8–31). The paths of the satellites, depicted by metal rings, can be projected to the surface of the globe to form circular paths that would appear as ellipses in the view shown.

The sphere in Fig. 8–32 has three points, 1, 2, and 3, located on its surface. A circle is to be drawn through these points so that it will lie on the surface of the sphere. This problem is solved by drawing the plane of the circle, plane 1–2–3, in the top and front views. The plane is found as an edge in the primary auxiliary view when projected from the top and front views as shown. The circle on the sphere cut by plane 1–2–3 will appear as an ellipse in the top and front views. The ellipse-guide angle for each view is found by measuring the angle made by the projectors with the edge view of the plane in the primary auxiliary view. The major diameters of ellipse are drawn parallel to the true-length lines on plane 1–2–3 in the top and front views.

Satellites circling the earth are tracked by determining the lines of intersection made by the planes of their flight with the surface of the earth. Tracking stations (Fig. 8–33) receive

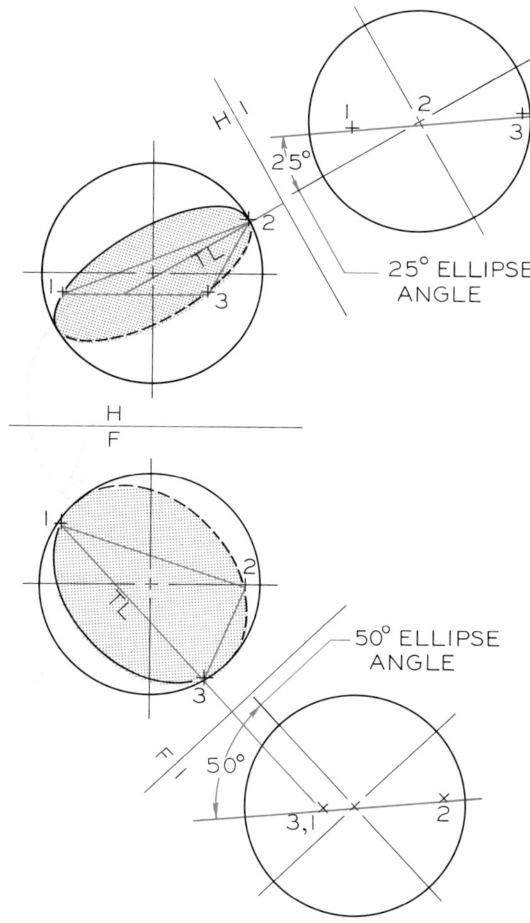

Fig. 8–32. Determining the location of an orbital path on a sphere.

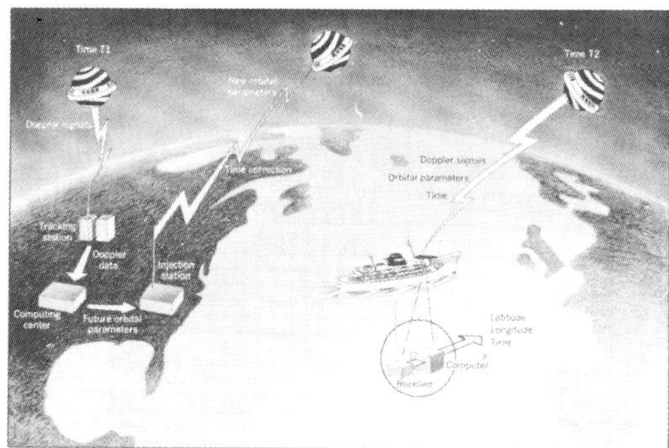

Fig. 8–33. Tracking stations are used to project the path of a satellite to the spherical surface of the earth. (Courtesy of the Coast and Geodetic Survey.)

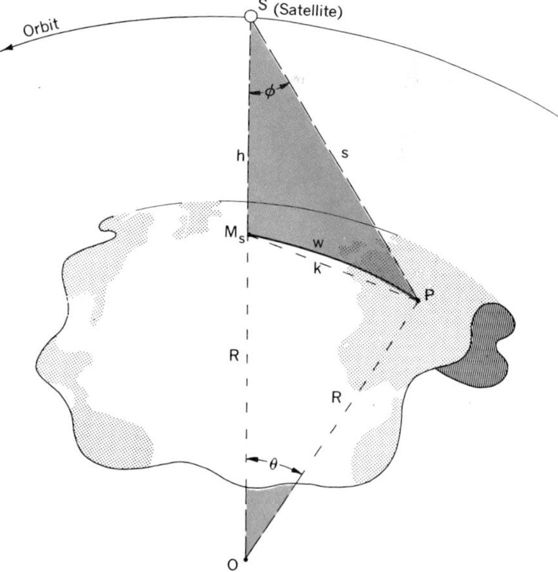

Fig. 8–34. The projection of a satellite's path to the surface of the earth. (Courtesy of the Coast and Geodetic Survey.)

signals from the satellite that give its location in space at a particular instant. Additional locations in space establish its plane of travel and, consequently, its projected path on the earth's surface. The intersection of the plane of a satellite's orbit with the surface of the earth is shown in Fig. 8–34. The path on the earth is found by projecting the orbital path toward the center of the earth to locate points *M* and *P*.

8–11 INTERSECTION BETWEEN A SPHERE AND A PRISM

A prism which intersects a sphere is shown in Fig. 8–35. Cutting planes are passed through the sphere in the top and side views so that they are parallel to the frontal plane; they appear as circles in the front view. In the side view, the intersections made by the cutting planes with the edges of the prism are projected to the front view, where they are found to intersect with their respective circles, i.e., those formed by the same cutting plane. *Example:* Points 1 and 2 are found to lie on cutting plane *A* in the side view. These are projected to the front view to circle *A*, which was established by cutting plane *A*. Point *X* in the side view locates the point where the visibility of the intersection in the front view changes. Point *Y* in the side view is the point where the visibility of the intersection changes in the top view. Note that both these points lie on center lines of the sphere in the side view.

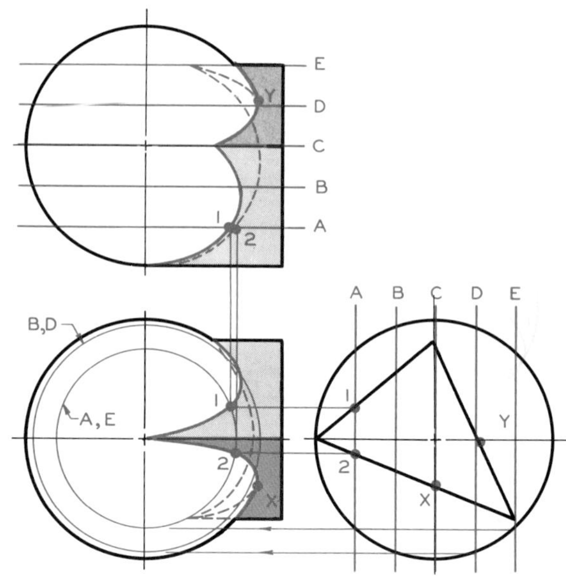

Fig. 8–35. Determining the intersection between a sphere and a prism.

8–12 INTERSECTION BETWEEN TWO OBLIQUE CYLINDERS

To determine the line of intersection of two cylinders, as shown in Fig. 8–36, a plane must be drawn in space such that it is parallel to both cylinders. This will be the case if the plane contains lines that are parallel to the axes of each cylinder, as shown in step 1. It is necessary that the planes be passed through the top views of the intersecting cylinders as cutting planes; consequently, a line is constructed in the plane that is parallel to the edge view of the base planes of the cylinder in the front view. This line is projected to the top view, where its direction will represent the line of intersection of the cutting planes on the circular bases of the cylinders in the top view. A series of planes is passed through the bases, as shown in step 1, where the elements are formed on the surface of the cylinders parallel to their axes, as shown in step 2. Elements that lie in common cutting planes establish points on the line of intersection when they cross in the top view. A systematic lettering procedure will assist in plotting the points as they are found. The points are connected in sequence to obtain the continuous lines of intersection shown in step 2 in the top view.

The elements on each cylinder are projected to the front view, where they will be parallel to the axes of both cylinders. The points on the lines of intersection lying on the cutting planes are projected to the front view and connected to form the desired lines of intersection. The visibility of each is determined by examining each pair of intersecting elements. If both elements are visible, their points of intersection will be visible. If only one element is visible, the point of intersection is hidden.

FIGURE 8–36. INTERSECTION BETWEEN OBLIQUE CYLINDERS

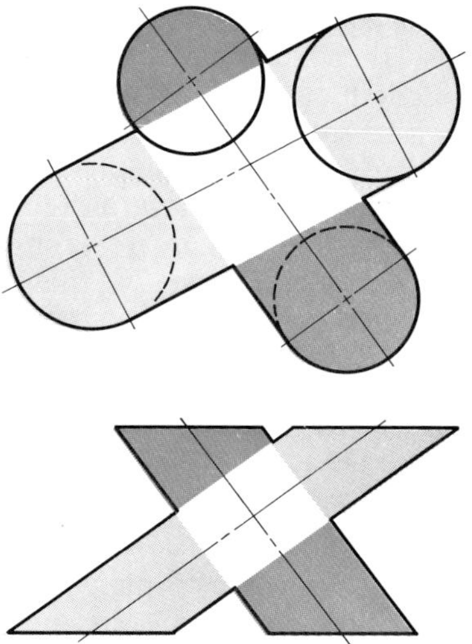

Given: The top and front views of two intersecting cylinders.

Required: Determine the line of intersection between the two cylinders in both views.

Reference: Article 8–12.

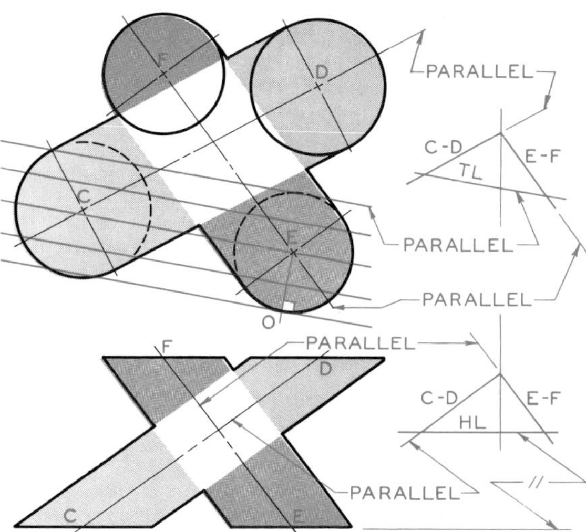

Step 1: Construct a triangular plane so that it will contain lines which are parallel to both axes of the cylinders. Draw a horizontal line in the front view of the triangular plane so that it lies in the base plane of the cylinders. Project this line to the triangular plane in the top view, where its direction is used as the direction for the cutting planes that will be drawn in the top view to pass parallel to the axes of the cylinders.

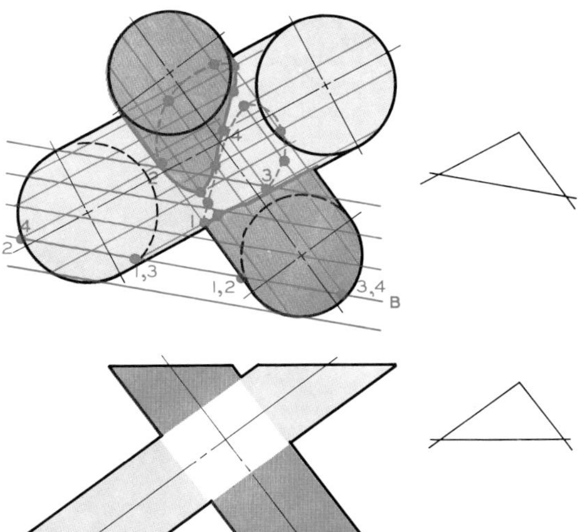

Step 2: Where the cutting planes intersect the bases of cylinders in the top view, elements are formed on the cylinders and they are parallel to the axes of each cylinder. Note that four elements are cut by each cutting plane. Where these common elements intersect, a point on the line of intersection is found. These points should be labeled, as are the four example points shown.

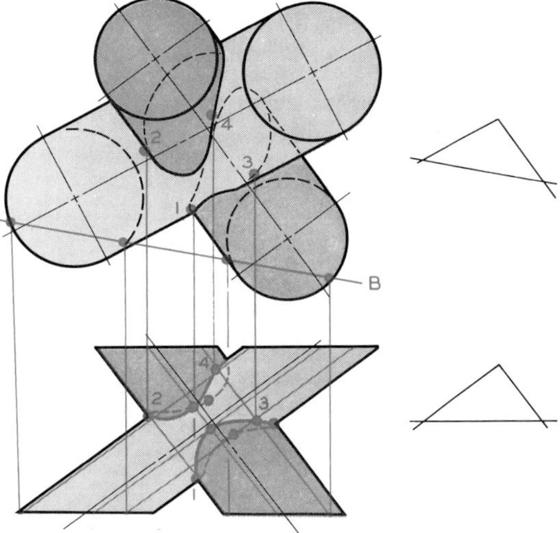

Step 3: Project the elements on the cylinders from the top view to the front view. The points found to lie on specific elements in the top view are projected to their respective elements in the front view. Several points have been projected as examples. Visibility in this view is determined by analysis of the top view.

The ducts which intersect at unusual angles in Fig. 8–37 require the same type of analysis as that covered in Fig. 8–36. The lines of intersection were used to design the joints of the intersecting ducts.

8–13 SUMMARY

It can be seen from the examples given in this chapter that principles of intersection have many applications to engineering, technology, and science. Use of the principles of intersections involves most of the previously covered techniques of descriptive geometry and orthographic projection. Piercing points and visibility analysis can be reviewed in Chapter 4 to assist in a better understanding of intersections. Auxiliary views—primary and secondary—are used to find intersections of geometric shapes, as was covered in Chapters 5 and 6.

The fundamental principles covered in this chapter are basic to practically any problem that involves intersections. An attempt should always be made to identify an intersection in terms of its geometric elements. Perhaps several shapes are joined in combination to form the configuration of a design. The intersections will be easier to find if the problem is treated as though it involved an intersection between two geometric elements, then two more, etc., in sequence, until the complete line is found.

The student must understand intersection principles before proceeding to Chapter 9, which deals with developing flat patterns that are used to fabricate products that are composed of geometric shapes. Many of these shapes will be intersected by other forms; consequently, the lines of intersection must be found before the patterns can be completed.

Fig. 8–37. The intersection between these cylindrical shapes is an example of the principle covered in Article 8–12. (Courtesy of Ryan Aeronautics, Inc.)

PROBLEMS

Problems for this chapter should be constructed from the given sketches with instruments on $8\frac{1}{2}'' \times 11''$ sheets, as illustrated in the accompanying figures. For laying out the problems on grid or plain paper, assume that each grid represents $\frac{1}{4}''$. All reference planes and figure points should be labeled, using $\frac{1}{8}''$ letters with

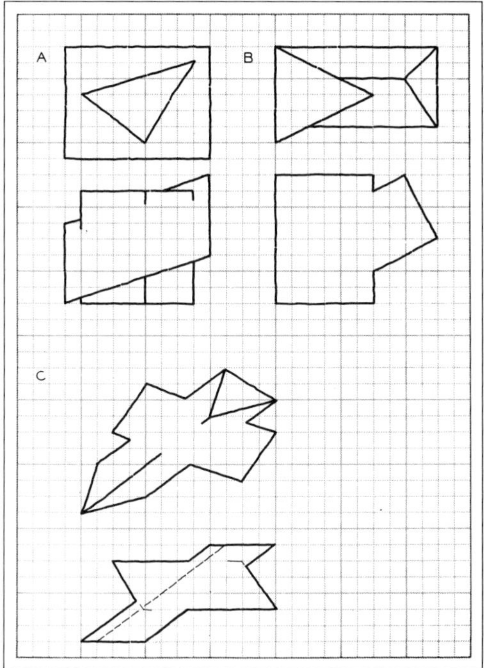

Fig. 8–38. Intersections of planes and prisms.

guidelines. Solutions should be sufficiently noted and labeled to explain all construction. Refer to Article 3–6.

1. (A) In Fig. 8–38A, find the intersection between the prism and the plane. (B) In part B of the figure, find the intersection between the two prisms using the projection method. (C) In part C of the figure, find the line of intersection by the projection method. Lay out the same problem on a separate sheet and solve it by the auxiliary-view method.

2. (A) In Fig. 8–39A, find the intersection between the cylinder and the plane. (B) In part B of the figure, find the intersection between the cylinder and the prism.

3. In Fig. 8–40A, find the intersection between the two cylinders. In part B of the figure, find the intersection between the two cylinders.

4. (A) In Fig. 8–41A, find the top view of the intersection formed by the cutting plane, and construct the auxiliary view of the section. What type of conic section is this? (B) In part B of the figure, find the intersections formed by the cutting planes in the front view. Construct the sections indicated by the cutting planes. Identify the types of conic sections in each case.

5. (A) Two views of a sphere are given in Fig. 8–42A, in which points *A* and *B* are located on the upper surface and point *C* is located at the sphere's center. Find the line of intersection formed by a plane that passes through these points and extends through the surface of the sphere. Show the line of intersection in all views. (B) In part B of the figure, find the intersection between the prism and the sphere.

6. In Fig. 8–43, determine the line of intersection of the two oblique cylinders in both views.

7. Construct the lines of intersection in parts A and B of Fig. 8–44. Determine visibility and show in all views.

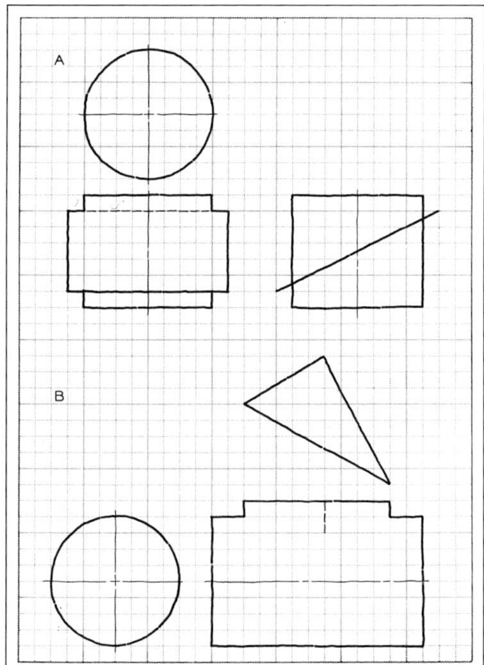

Fig. 8–39. Intersections of cylinders and planes.

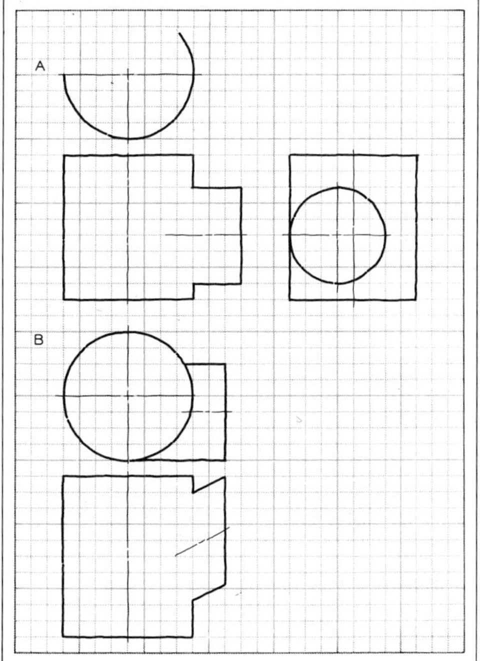

Fig. 8–40. Intersections of cylinders.

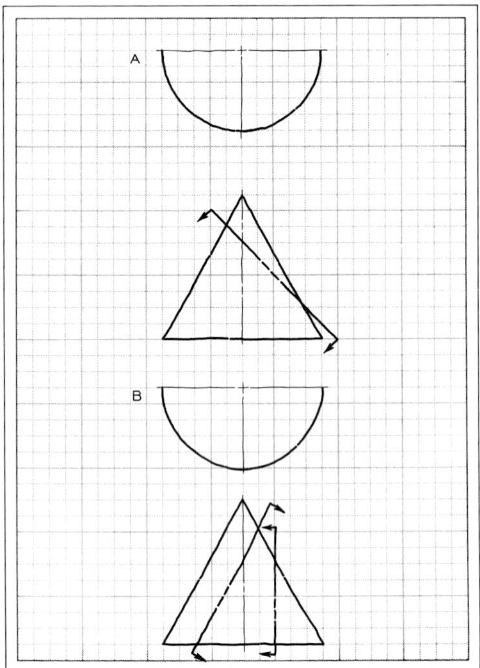

Fig. 8–41. Conic sections.

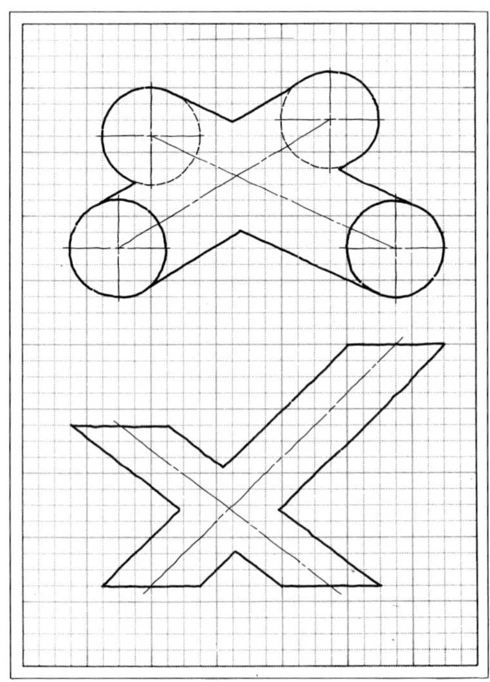

Fig. 8–43. Intersection between oblique cylinders.

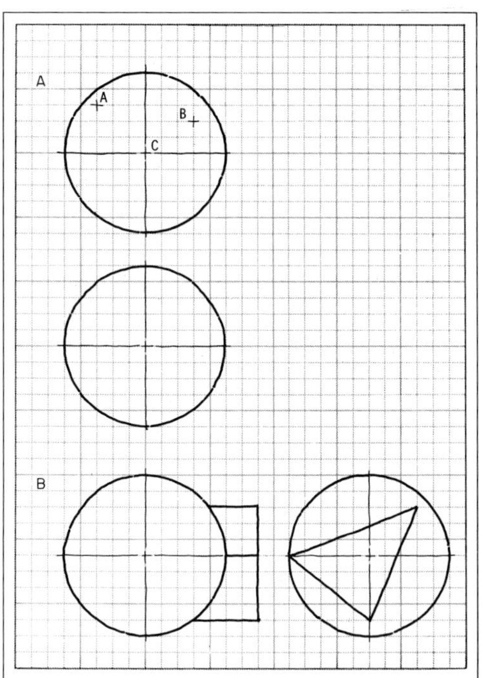

Fig. 8–42. Spherical intersections.

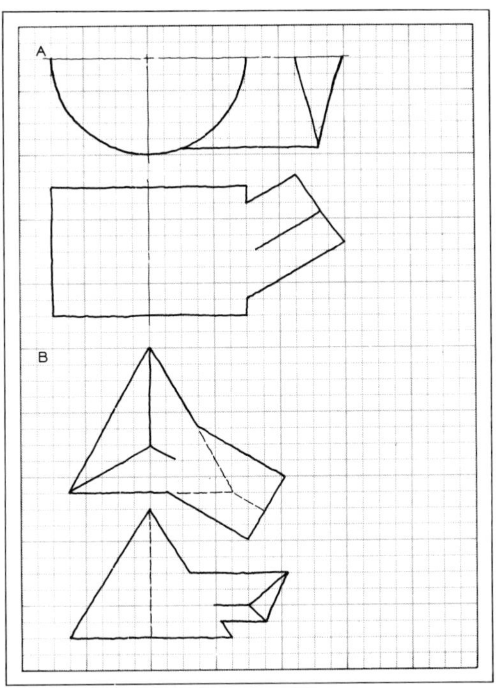

Fig. 8–44. Intersections.

9 Developments

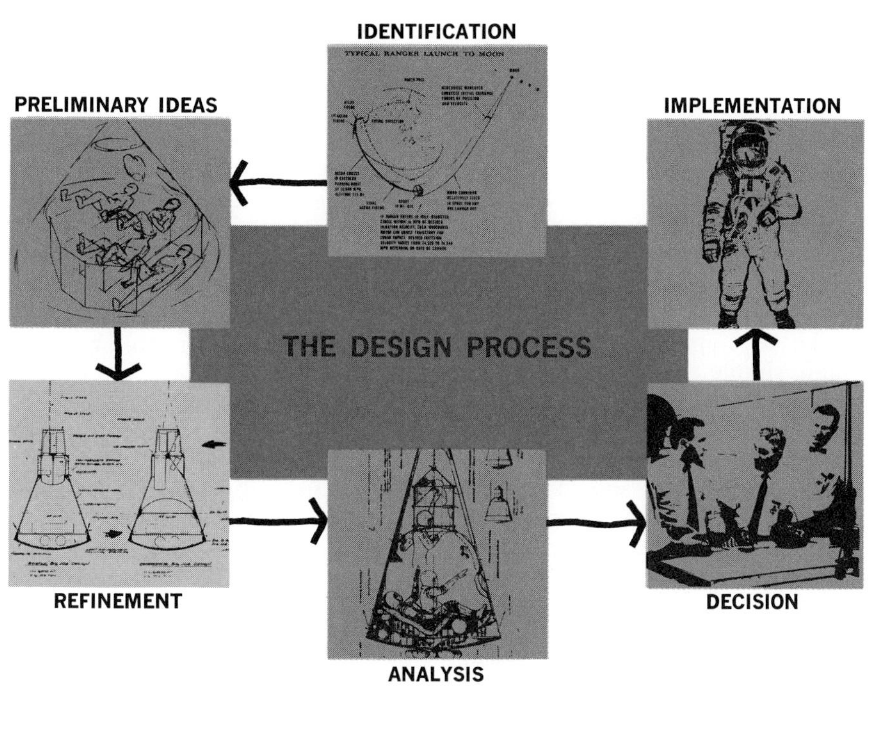

Fig. 9–1. The surface of this F–105 Thunderchief is an application of developed surfaces which were designed to conform to a specified shape. (Courtesy of Republic Aviation Corporation.)

Fig. 9–2. Many examples of intersections and developments can be seen in this model of a processing installation. (Courtesy of Bechtel Corporation.)

9–1 INTRODUCTION

The Supersonic F–105 Thunderchief shown in Fig. 9–1 is an example of a highly complicated shape that has been formed with sections of flat sheet metal. This chapter is concerned with the geometric principles and techniques used in the fabrication of such a shape from flat materials.

Creating a flat pattern for a three-dimensional object involves a *development* ("unfolding") of the object. Developments are closely related to the intersections we studied in Chapter 8, since provision for the joining of component parts must be made in the flat pattern of an object.

We may make developments for all applications from small, simple shapes made of thin sheet metal to sophisticated pieces of hardware such as space capsules, which must be fabricated within a high degree of accuracy. The model of the processing plant shown in Fig. 9–2 illustrates a wide variety of shapes that must be designed, developed, and specified by the designer. Whether the design is fabricated by bending flat metal or by casting a solid object, the designer must have a grasp of development principles. This chapter will cover the fundamentals of this area and will relate these principles to industrial applications where possible.

9–2 DEVELOPMENT OF A PRISM

A cylinder or a prism (Fig. 9–3) can be laid out to result in either an inside or an outside pattern. An inside development is more frequently desired because (1) most bending machines are designed to fold metal such that the markings are folded inward, and (2) markings and lines etched on the patterns will be hidden when the development is assembled into its finished form. Whether the pattern is an inside or an outside pattern will depend upon the material and the equipment being used. In any case, it is important that the pattern *always* be labeled as inside or outside when presented. The patterns in the following examples will be inside patterns, since these are the most common; however, the principles for finding inside patterns can be applied to outside patterns.

The following rules apply generally to cylinders and prisms. These should be reviewed as example problems are studied. The most important rule in developing patterns is that *all lines of a development must be true length.*

Rules for Developing Cylinders and Prisms

1. Find the view in which the right section appears as an edge.
2. Lay out the stretch-out line of the develop-

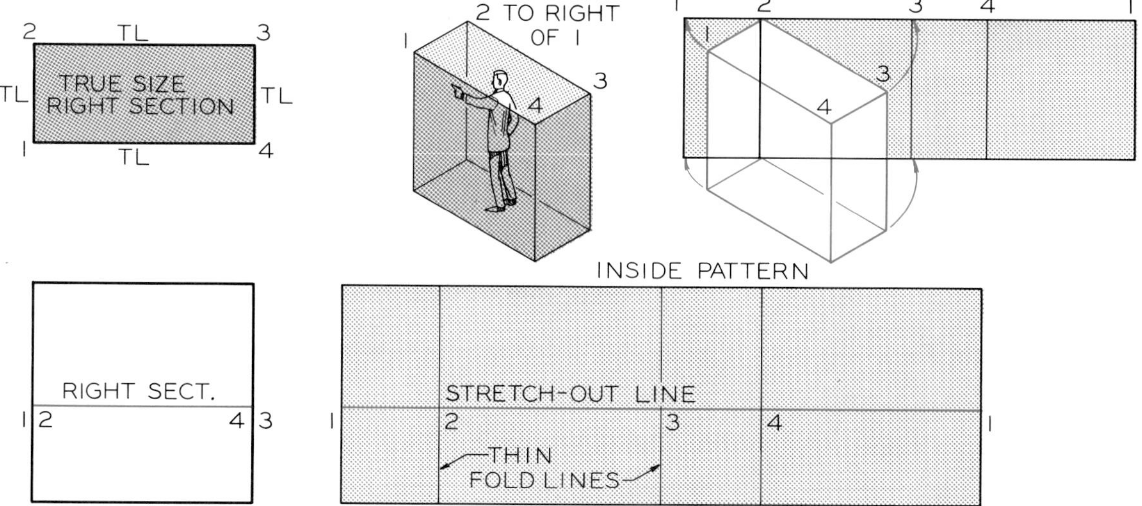

Fig. 9–3. Development of a rectangular prism to give an inside pattern.

ment parallel to the edge view of the right section.

3. Locate the distances between the lateral corner edges by measuring from the true-size views in the right section and transferring these measurements to the stretch-out line. Letter these points.
4. Construct the lateral fold lines perpendicular to the stretch-out line.
5. Establish the lengths of the fold lines by projecting from the view in which the right section appears as an edge.
6. Make sure that the line where the development will be spliced is the shortest line, so that the least amount of welding or joining effort will be required.
7. Connect all points in the proper sequence to give the complete pattern.
8. Verify that the point where the development ends is the same point as the beginning point on the right section.
9. Indicate by a note whether the development is an inside or an outside pattern.

These rules have been applied to the problem in Fig. 9–3. Note that the lateral fold lines of the prism are true length in the front view and that the right section appears as an edge in this view also. The stretch-out line is drawn parallel to the edge view of the right section, beginning with point 1. If an inside pattern is desired, it is necessary to select the point that lies to the right of point 1, since the development will be laid out in this direction. The observer assumes that he is inside the prism, as illustrated pictorially in Fig. 9–3, and that he is looking at the inside view of fold line 1. Note that the top view of the prism can be used for this analysis. Point 2 is seen to lie to the right of line 1, whereas line 4 is to the left; consequently, distance 1–2 is transferred from the top view to the stretch-out line, with point 2 to the right of point 1. Note that all lines on the surface of the right section are true length in the top view. Lines 2–3, 3–4, and 4–1 are then laid out in sequence along the stretch-out line. The length of each fold line is found by projecting its true length from the front view. The ends of the fold lines are connected to form the limits of the developed surface. Fold lines are drawn as thin lines on the development.

The body of the toaster shown in Fig. 9–4 is an example of the application of the principle of developments in the design of a household appliance. Construction is more economical when it consists of forming one continuous piece of material that is bent into shape than it

Fig. 9–4. The surface of this toaster is an application of the development of a rectangular shape. (Courtesy of General Electric Company.)

Fig. 9–6. The all-aluminum body of this coal hauler is an example of an industrial development application. (Courtesy of ALCOA.)

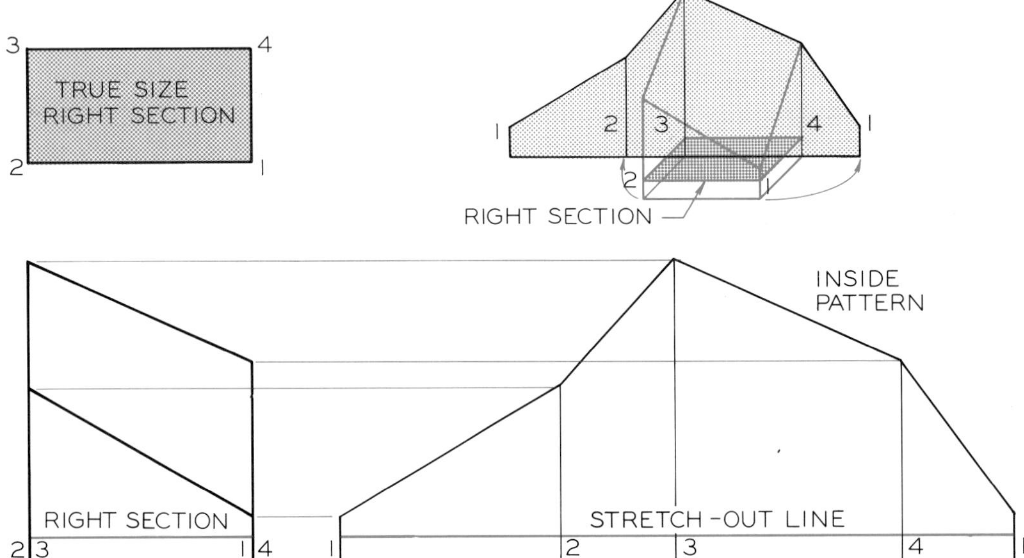

Fig. 9–5. Development of a rectangular prism with a beveled end to give an inside pattern.

is when the operation demands joining a series of sections together. The forms that are used for the pouring of concrete are applications of developments of a different type.

A prism with a beveled end can be developed in the same manner as was the example in Fig. 9–3, except that the lengths of the fold lines will have to be determined by projecting from the front view. In this case, the lines will be unequal in length, resulting in a pattern such as that shown in Fig. 9–5. The use of a lettering system will assist in identifying the points of projection, as shown in this example. Note that the right section is used for the direction of the stretch-out line and as a source of measurements for determining the space between fold lines.

The mammoth coal hauler in Fig. 9–6 was developed as a flat pattern and joined to form this finished shape. Regardless of the size of the problem, the principles of solution are identical. The material used in this example was sheet aluminum, which can support ten times its weight.

9–3 DEVELOPMENT OF AN OBLIQUE PRISM

The standard views of an oblique prism may be presented in a preliminary sketch which must be analyzed for further refinement. For example, an inclined prism might not show the right section as an edge nor a surface area as true size in the standard views. If this is the case, an auxiliary view must be used to provide the additional information necessary to complete the development.

A prism is shown inclined to the horizontal plane, but parallel to the frontal plane, in Fig. 9–7. The lateral corner edges of the prism appear true length in the front view, in which they are frontal lines. The right section can be drawn as an edge in the front view perpendicular to the fold lines. An auxiliary view of the edge view of the section will show the true size of the right section (step 1). The pattern is laid out in the conventional manner, which was covered in Article 9–2, and the stretch-out line is drawn parallel to the edge view of the right section in the front view. The measurements between the fold lines are transferred from the true size of the right section to the development (step 2). The developments of the end pieces can be found by a secondary auxiliary view, which is projected perpendicular to the edge view of the end of the prism. These projections are drawn as part of the total pattern in the developed view.

A prism which does not project true length in either view, but which is oblique to the principal planes, can be developed as illustrated in Fig. 9–8. From the front view we find the true length of the lateral corners by projecting them to an auxiliary view in a direction perpendicular to that of the lateral corners in the front view. The right section will appear as an edge in the primary auxiliary view. The stretch-out line is drawn parallel to this edge view. The true size of the right section is found in an auxiliary view projected perpendicularly from the edge view of the right section. The fold lines are located on the stretch-out line by measuring around the right section in the secondary auxiliary view. The lengths of the fold lines are then projected to the development from the primary auxiliary view.

FIGURE 9–7. DEVELOPMENT OF AN OBLIQUE PRISM

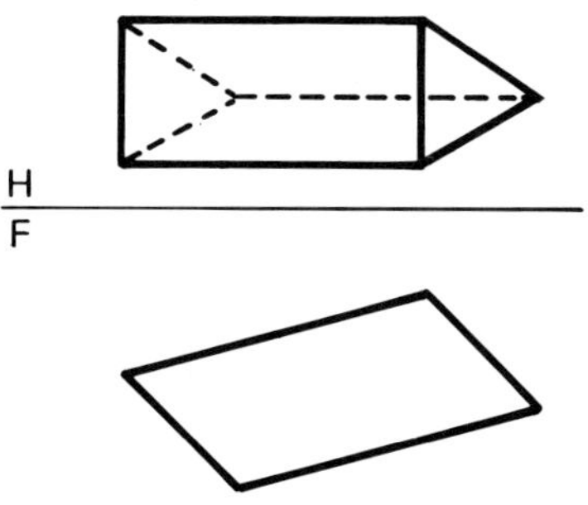

Given: The top and front views of an oblique prism.

Required: The inside pattern of the developed surface of the prism and the end sections.

Reference: Article 9–3.

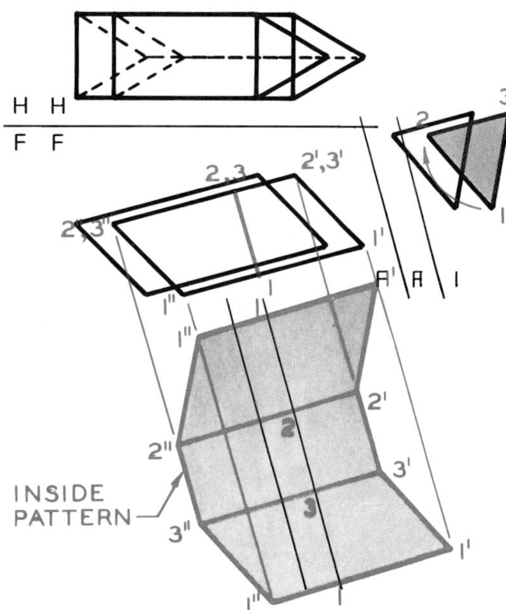

Step 2: Since the pattern is developed toward the right, beginning with line 1′–1″, the next point is found to be line 2′–2″ by referring to the auxiliary view. Transfer true-length lines 1–2, 2–3, and 3–1 from the right section to the stretch-out line to locate the elements. Determine the lengths of the bend lines by projection.

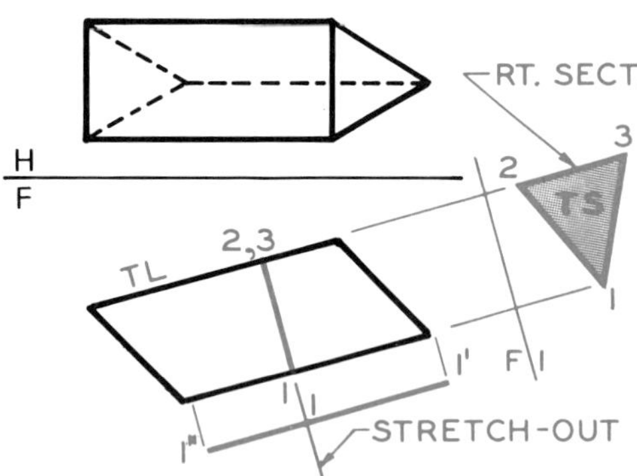

Step 1: The edge view of the right section will appear as perpendicular to the true-length axis of the prism in the front view. Determine the true-size view of the right section by constructing an auxiliary view. Draw the stretch-out line parallel to the edge view of the right section. Project bend line 1′–1″ as the first line of the development.

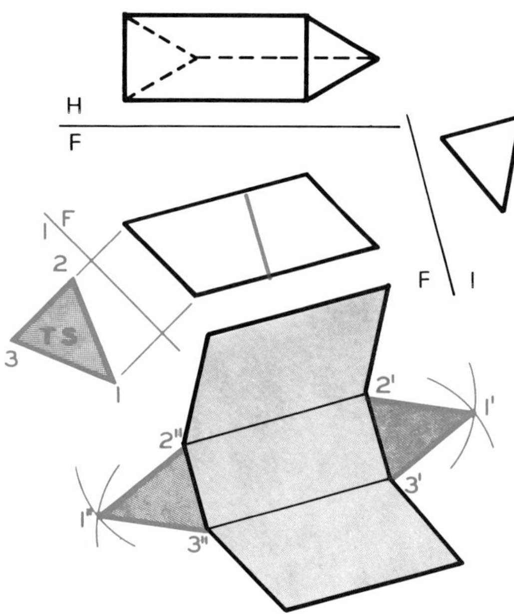

Step 3: Find the true-size views of the end pieces by projecting auxiliary views from the front view. Connect these surfaces to the development of the lateral sides to form the completed pattern. Fold lines are drawn with thin lines, while outside lines are drawn as regular object lines.

9–4 DEVELOPMENT OF A CYLINDER

Cylinders are basic shapes that are used extensively in practically all areas of technology. The large storage tanks shown in Fig. 9–9 are typical of the cylindrical developments that can be found in the petroleum industry. These tanks are efficient and economical vessels for storing petroleum products. Note that these cylinders have no fold lines, since their surfaces are smooth and curving; however, they were developed by using a series of lines on their surface as though they were fold lines.

The example in Fig. 9–10 illustrates the manner in which an inside pattern of a cylinder is developed. The axis of the cylinder appears true length in the front view, which allows the right section to be seen as an edge, since it is perpendicular to the axis. The stretch-out line is drawn parallel to the edge view of the section, and point 1 is chosen as the beginning point, since it is on the shortest possible line on the surface. Since an inside pattern is desired, the observer must assume that he is standing inside the cylinder in the top view, as illustrated pictorially in Fig. 9–10. The pattern will be laid out to the right, so the observer is interested in determining which lines are to the right of point 1. The first point to the right of point 1 is point

Fig. 9–9. Storage tanks are designed through the use of cylindrical developments. (Courtesy of Shell Oil Company.)

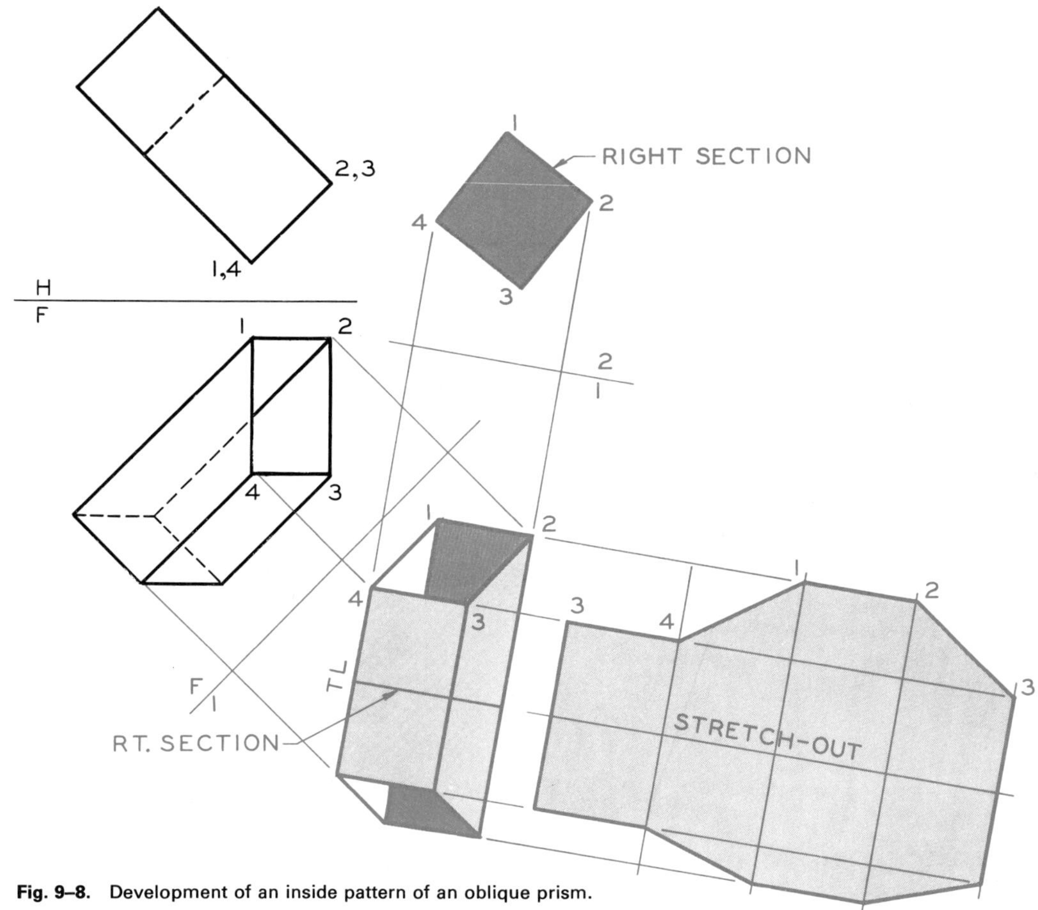

Fig. 9–8. Development of an inside pattern of an oblique prism.

2, which establishes the sequence of points to be followed in laying out the distances between the lines in the development along the stretch-out line. These distances are transferred from the true-size right section in the top view as chordal distances to approximate the circumference around the cylinder. The closer the intervals between the lines on the right section, the closer the graphical solution will be to the theoretical circumference. If accuracy is a critical factor, the circumference can be determined mathematically, laid out true length along the stretch-out line, and divided into the number of divisions desired. The ends of the lines on the surface are projected from the top to the front view. The ends of these lines on the beveled end of the cylinder are projected to their respective lines in the developed view, where they are then connected with a smooth curve.

A practical application of this principle is shown in Fig. 9–11, in which an air-conditioning vent duct is shown as used on an automobile. This vent was developed through the use of the same principles as those covered in this article. The Hydra 5 test vehicle (Fig. 9–12) is composed of a number of cylindrical developments that required graphical solutions.

Gun ranges that are used to simulate meteoroid impact on spacecraft utilize cylindrical forms, as shown in Fig. 9–13. Other applications of cylindrical developments can be seen in the background of this laboratory.

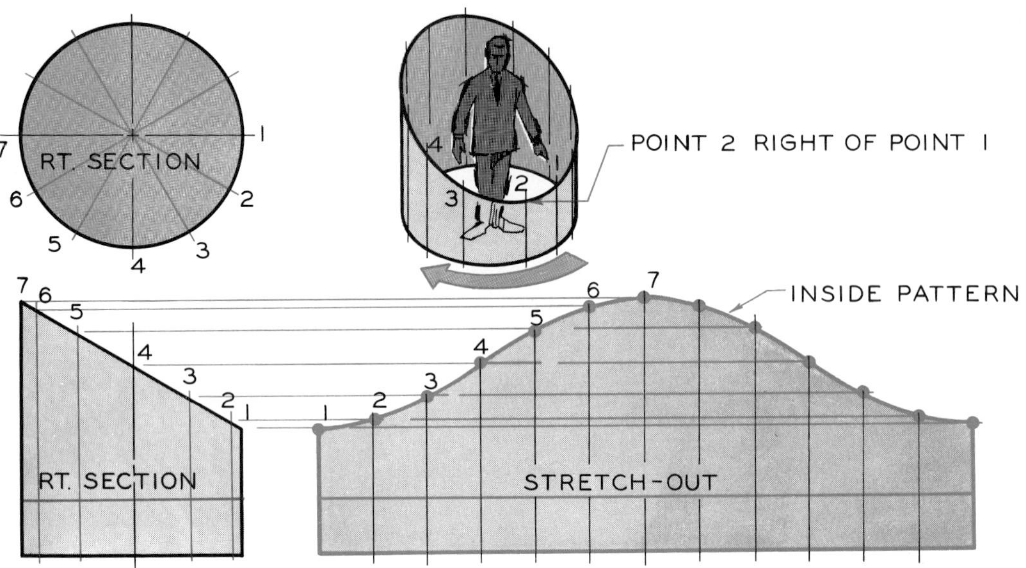

Fig. 9–10. Development of a cylinder.

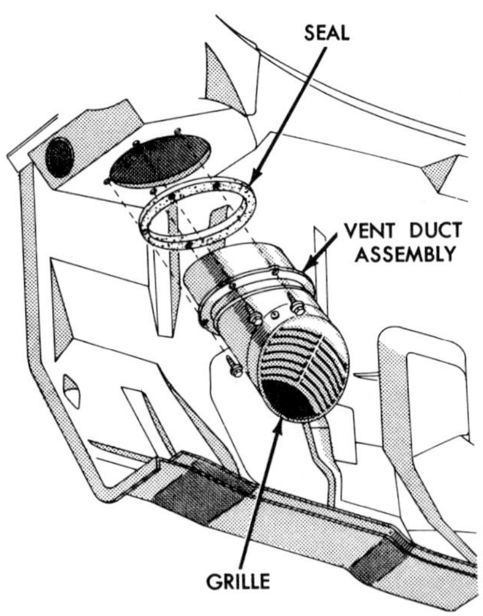

Fig. 9–11. This ventilator air duct was designed through the use of development principles. (Courtesy of Ford Motor Company.)

Fig. 9–12. Cylindrical developments were necessary in the design of the Hydra 5 launch vehicle. (Courtesy of the U.S. Navy.)

Fig. 9–13. Gun ranges, which are used to simulate meteoroid impact on aircraft, are examples of cylindrical developments. (Courtesy of Arnold Engineering Development Center.)

9–5 DEVELOPMENT OF AN OBLIQUE CYLINDER

The oblique cylinder in Fig. 9–14 appears true length in the front view, in which its right section projects as an edge that is perpendicular to its center line. The stretch-out line for the development is drawn parallel to the edge view of the right section (step 1). The true-size view of the right section is found in an auxiliary view. Lines lying on the surface of the cylinder are projected from the right section to the front view. These elements are spaced the same distance apart on the stretch-out line in the development view as they were in the right section in the auxiliary view. All element lengths are projected to the development from the front view, where they are true length. The ends of the elements are connected with a smooth curve (step 2). The true size of one elliptical end of the cylinder is found by auxiliary view, as shown in step 3. This shape is drawn attached to the development. The development of the opposite end can be found by auxiliary view in the same manner.

Figure 9–15 is an example of a cylinder that is oblique to the principal planes in both views. The edge view of the right section is found in an auxiliary view, where the elements on the surface of the cylinder project true length. The stretch-out line is drawn parallel to this edge view. The elements are separated on the stretch-out line by the distance between the point views of the elements in the secondary auxiliary view, where the right section appears true size. The lengths of the elements in the developments are found by projecting from the true-length view of the elements in the primary auxiliary view. These points are connected by a smooth curve. The elliptical development of the beveled end can be found by a secondary auxiliary view which is projected from the primary auxiliary view in a manner similar to that shown in step 3 of Fig. 9–14.

Applications of cylindrical developments vary in size from small to large. The 16-foot wind tunnels shown in Fig. 9–16 are examples of the application of development principles. Regardless of their size, however, all applications of cylindrical developments are solved in the same manner.

9–6 DEVELOPMENT OF A PYRAMID

Corners of pyramids and elements on cones can be found true length by revolution, as previously covered. This technique is reviewed in Fig. 9–17, where fold line 0–5 and element 0–6 are found true length. Note that all elements of a cone will lie on the outside element of the cone in the triangular view.

The development of a pyramid is given in Fig. 9–18. Since all fold lines will have point 0 as a common point, the stretch-out line will not be used on this type of problem; instead, a series of adjacent triangles will be drawn in the development. Recall that *all lines* in a development must be true length.

Lines 1–0 and 2–0 are revolved into the frontal plane in the top view so that their true length will be seen in the front view, as shown. All bend lines are equal in length, since the pyramid is a right pyramid; consequently, line 1–0 is used as a radius in the development for constructing an arc that will contain all corner points lying on the base of the pyramid. The

FIGURE 9–14. DEVELOPMENT OF AN OBLIQUE CYLINDER

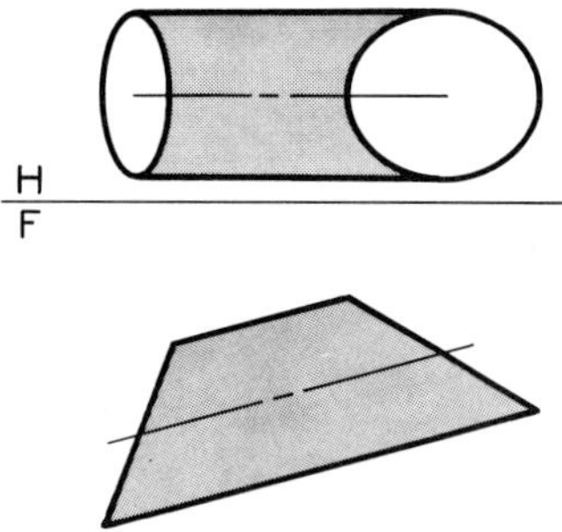

Given: The top and front views of an oblique cylinder.
Required: Find an inside development of the cylinder and its end pieces.
Reference: Article 9–5.

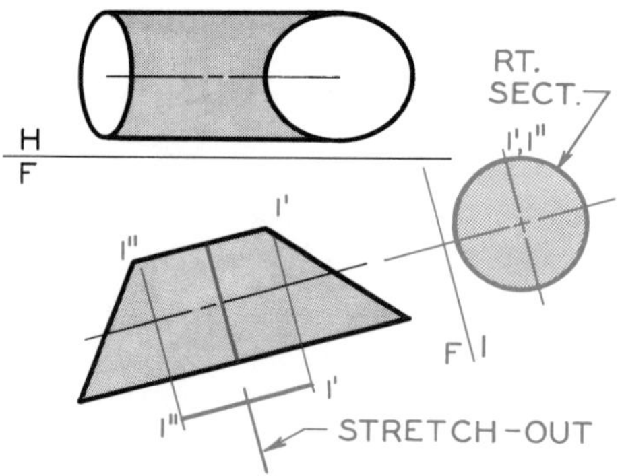

Step 1: The right section appears as an edge in the front view, in which it is perpendicular to the true-length axis. Construct an auxiliary view to determine the true size of the right section. Draw a stretch-out line parallel to the edge view of the right section. Locate element 1′–1″.

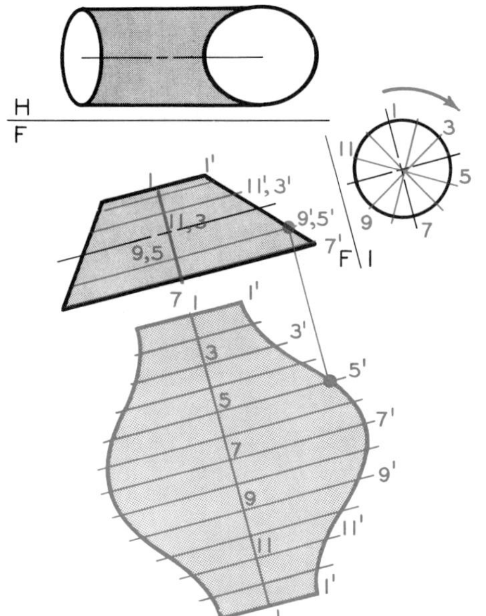

Step 2: Divide the true-size right section into equal points which represent the point views of elements on the cylinder's surface. Project these elements to the front view. Transfer measurements between the lines from the auxiliary view to the stretch-out line to locate the elements in the development. Determine the lengths of the elements by projection to complete the development.

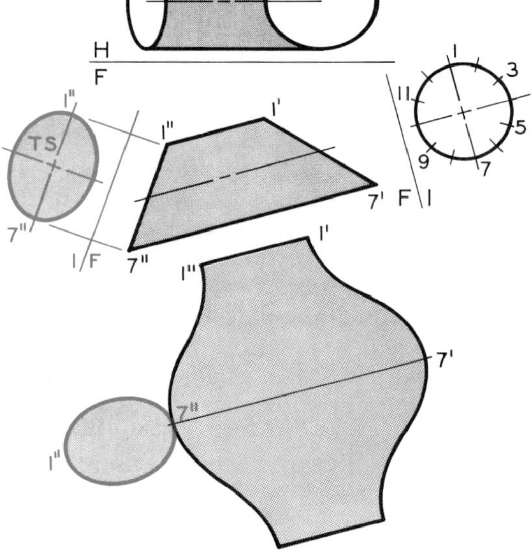

Step 3: The development of the end pieces will require auxiliary views that project these surfaces as ellipses, as shown for the left end. Attach this true-size ellipse to the pattern at a point on the pattern. Note that the line of departure for the pattern was made along line 1″–1′, the shortest element, for economy.

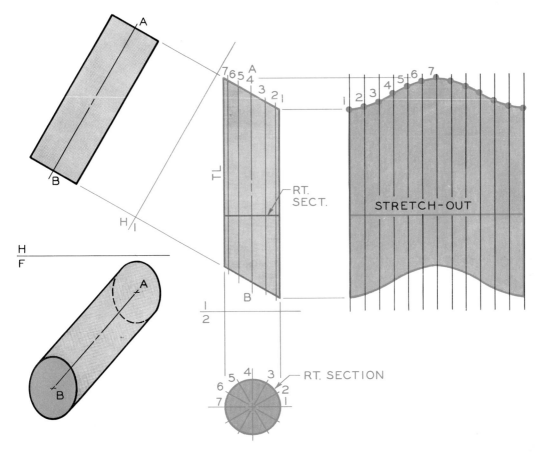

Fig. 9–15. Development of an inside pattern of an oblique cylinder.

Fig. 9–16. Wind tunnels are designed through the extensive application of cylindrical developments. (Courtesy of Arnold Engineering Development Center.)

lines of the base appear true length in the top view since the base is a horizontal plane. Distance 1–2 is measured in the top view and transferred to the development, where it is a chord on the arc from point 1 to point 2. Lines 2–3, 3–4, and 4–1 are found in the same manner. The bend lines are drawn with thin lines from the base to the apex, point 0.

A variation of this problem is given in Fig. 9–19, in which the pyramid has been truncated or cut at an angle to its axis. The development of the inside pattern is found in the same manner as covered previously; however, an additional step is required to establish the upper lines of the development. The development is first laid out as though it were a continuous

FIGURE 9–17. TRUE LENGTH OF ELEMENTS BY REVOLUTION

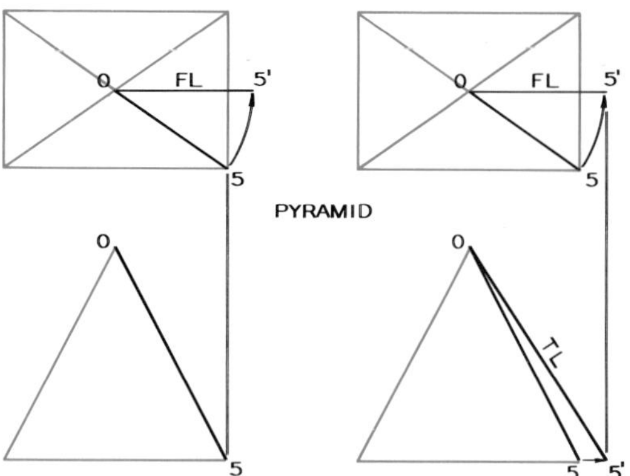

Step 1: Fold line 0–5 is revolved until it is a front line in the top view.

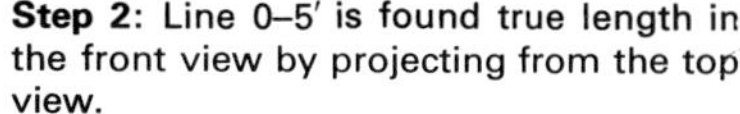

Step 2: Line 0–5′ is found true length in the front view by projecting from the top view.

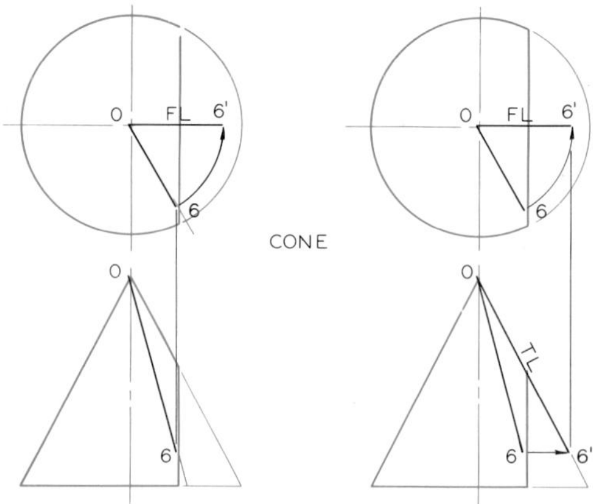

Step 1: Element 0–6 is revolved until it is a frontal line in the top view.

Step 2: Element 0–6′ is found true length in the front view by projecting from the top view.

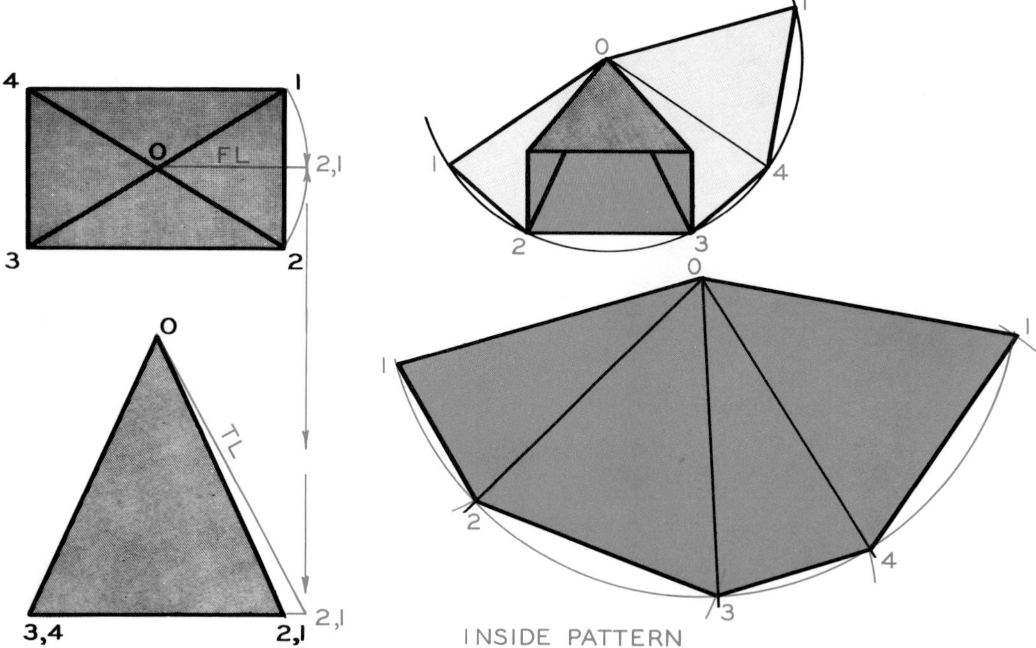

Fig. 9–18. Development of a right pyramid.

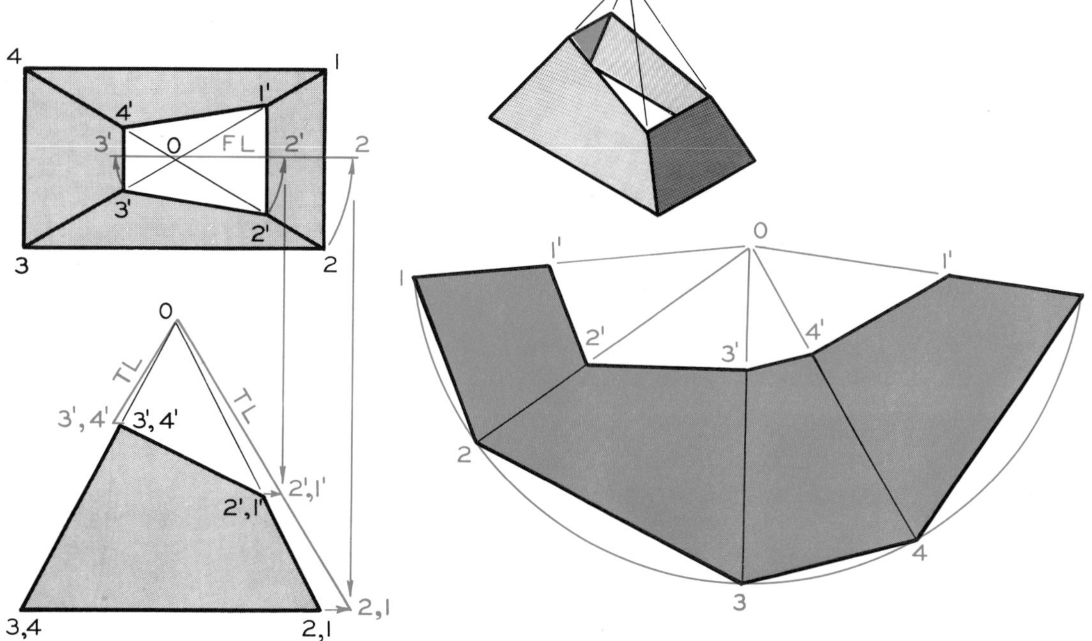

Fig. 9–19. Development of an inside pattern of a truncated pyramid.

pyramid from the apex 0 to the base. The true-length lines from the apex to points 1′, 2′, 3′, and 4′ are found by revolution, as shown. These true-length distances are measured along their respective lines from point 0 to locate the upper limits of the development. These points are then connected to complete the inside development of the truncated pyramid.

The mounting pads in Fig. 9–20 are sections of pyramids that intersect an engine body. This is an example of a design problem involving both intersections and developments.

An oblique pyramid is developed in sequential steps in Fig. 9–21 to illustrate the procedure for constructing the development. The true lengths of all bend lines are determined in step 1 by revolving the lines to the frontal plane and projecting them to the front view. These lines are found to vary in length, since the pyramid is not a right pyramid. The planes of each triangular surface of the pyramid are shown true size in the development by triangulation, in which the revolved lengths and

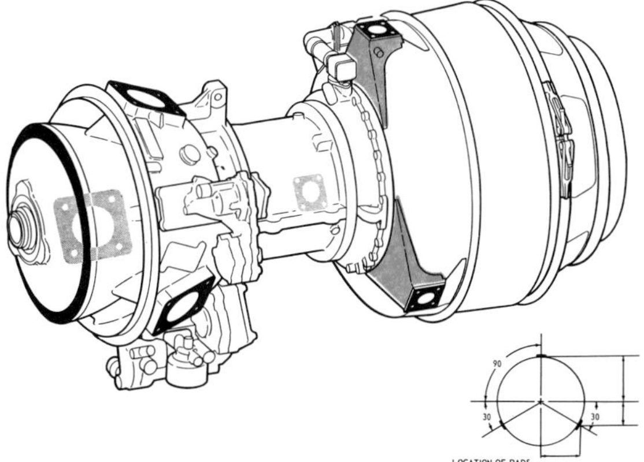

Fig. 9–20. Examples of pyramid shapes in the design of mounting pads for an engine. (Courtesy of Avco Lycoming.)

the true-length base lines taken from the top view are used. The triangles are drawn adjacent to each other, with point 0 common to each (step 2). To determine the upper limits of the developed surface (step 3), we find the true-

FIGURE 9–21. DEVELOPMENT OF AN OBLIQUE PYRAMID

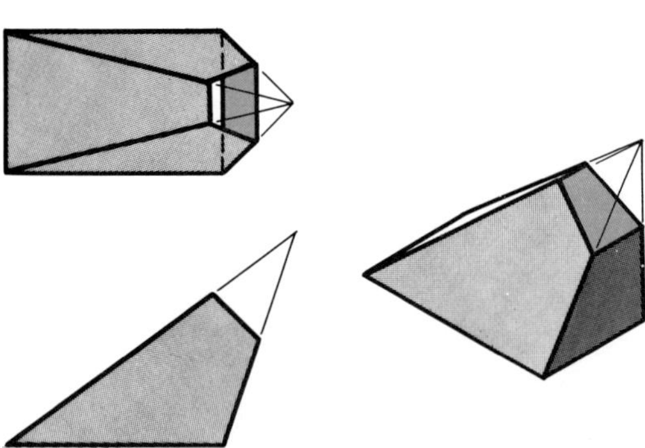

Given: The top and front views of an oblique, truncated pyramid.

Required: Find the inside development of the pyramid's surface.

Reference: Article 9–6.

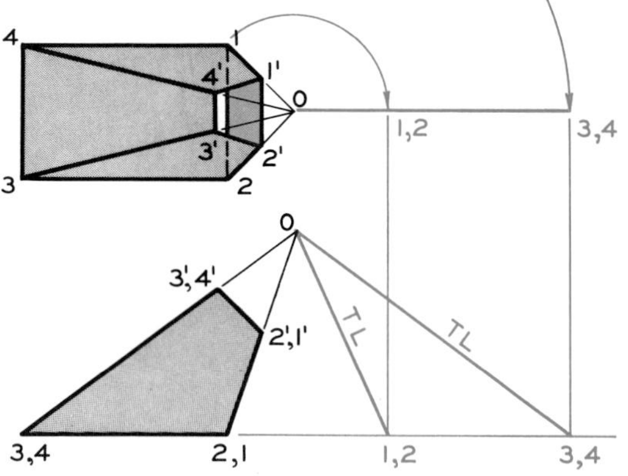

Step 1: Revolve each of the bend lines in the top view until they are parallel to the frontal plane. Project to the front view where the true-length views of the revolved lines can be found. Let point 0 remain stationary but project points 1, 2, 3, and 4 horizontally in the front view to the projectors from the top view.

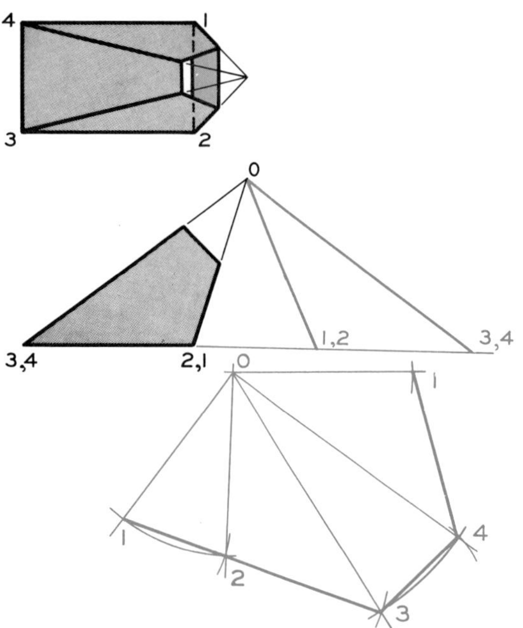

Step 2: The base lines appear true length in the top view. Using these true-length lines from the top view and the revolved lines in the front view, draw the development triangles. All triangles have one side and point 0 in common. This gives a development of the surface, excluding the truncated section.

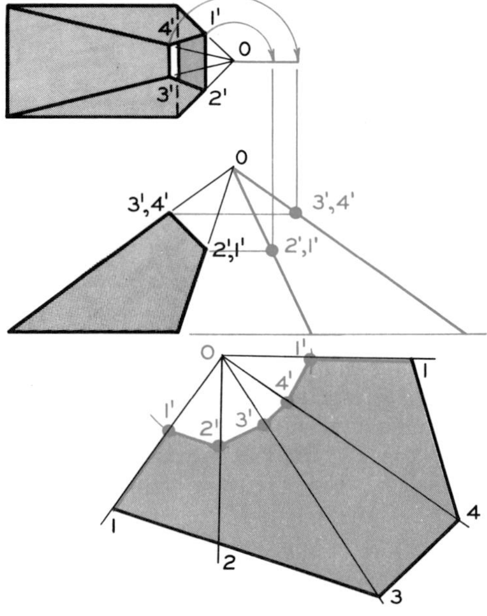

Step 3: The true lengths of the lines from point 0 to the points 1′, 2′, 3′, and 4′ are found by revolving these lines. These distances are laid off from point 0 along their respective lines to establish points along the upper edge of the developed pattern. The points are then sequentially connected by straight lines to complete the development.

length distances from point 0 to points 1′, 2′, 3′, and 4′ by revolution and transfer these lengths to the respective bend lines in the developed view. The limits of the development are connected with straight lines, and fold lines are indicated with thin lines.

9–7 DEVELOPMENT OF A CONE

The Apollo command module, shown in Fig. 9–22, is an example of a conical development. Many other examples of cones and other irregular shapes can be seen in the Charger aircraft and its missiles, shown in Fig. 9–23. Development principles were used to fabricate these irregular shapes from flat materials.

Cones are developed by a procedure similar to that used to develop pyramids. A series of triangles is constructed on the surface through the use of the elements of the cone and a chordal connection between points on the base of the cone. Figure 9–24 illustrates the division of the surface of the cone into triangular sections in the top and front views. The element 0–10 appears true length in the front view, since it is a frontal line in the top view. All elements on a right cone are equal; therefore, line 0–10 will be used to construct the arc upon which the developed base will lie. The inside pattern of the cone is drawn beginning with point 1 and moving toward the right. The point to the right of point 1 is point 2, which is found by inspection of the top view and the pictorial view of the cone. Point 2 is found in the top view and in the development by measuring the true-length chordal distance from point 1 along the arc. Successive triangles are found in this manner until point 1 is again reached at the extreme edge of the development. The base of the development is drawn as an arc rather than a series of chords along the arc that were connected by triangulation.

A more accurate approximation of the distance between the base points on the arc can be determined by finding the circumference of the base by mathematics and laying off this distance in equal increments along the arc formed by radius 0–1. The graphical approximation is sufficient in most cases.

Fig. 9–22. The Apollo command module is an example of a conical development. (Courtesy of NASA.)

Conical developments are used as an integral part of the wind tunnel design shown in Fig. 9–25. Cylindrical and spherical sections were also developed during the design of this facility.

A cone that has been truncated as shown in Fig. 9–26 can be developed by applying the principles illustrated in Fig. 9–24. It is advisable to construct the total development as though it were a complete cone that had not been modified. This portion of the development is identical to that shown in Fig. 9–24. A conical section has been removed from the upper portion of the cone. This part of the pattern can be removed by constructing an arc in the development, using as the radius the true-length line 0–7′, which is found in the front view. The true-length measurements from point 0 to the limits of the development, on the hyperbolic surface that is formed by the modification of the cone in the front view, are found by revolution. Lines 0–2′ and 0–3′ are projected horizontally to the extreme element, 0–1 in the front view, where they will appear true length. These distances are measured along their respective lines in the development to establish points through which the smooth curve will be drawn to outline the development.

Fig. 9–23. The body of this aircraft and the irregularly shaped missiles were fabricated through the application of development principles. (Courtesy of General Dynamics Corporation.)

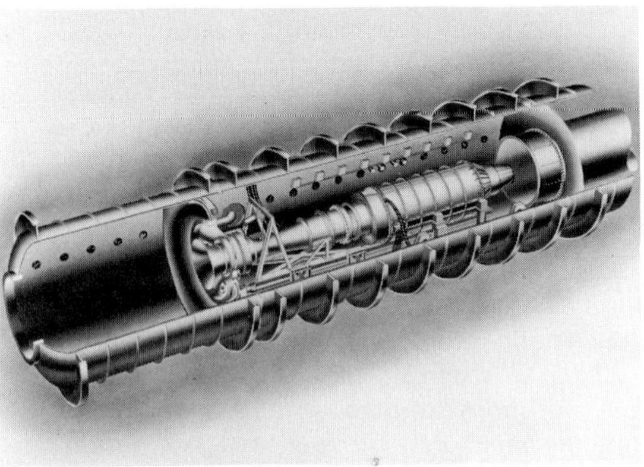

Fig. 9–25. Conical developments were used as an integral part of this wind tunnel design. (Courtesy of Arnold Engineering Development Center.)

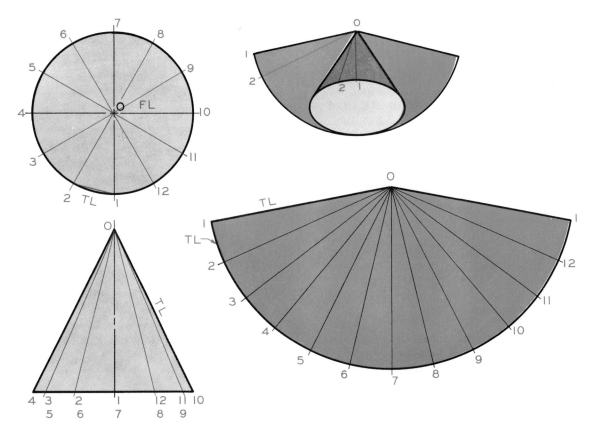

Fig. 9–24. Development of an inside pattern of a right cone.

INSIDE PATTERN

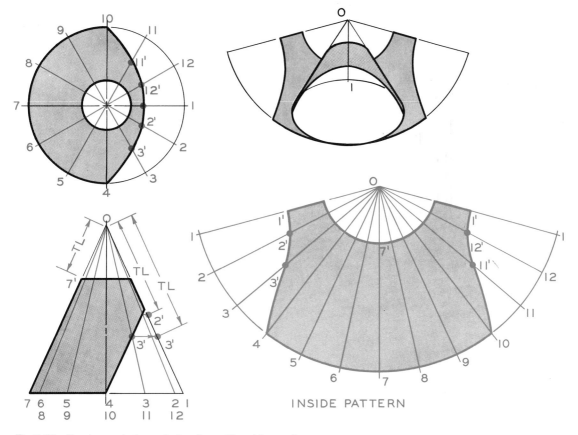

Fig. 9–26. Development of a conical surface with a side opening.

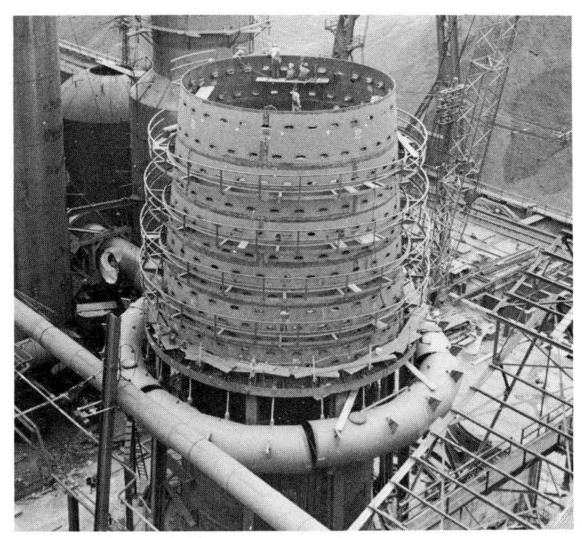

◀ **Fig. 9–27.** Huge conical developments are used in the construction of a blast furnace. (Courtesy of Jones & Laughlin Steel Corporation.)

Huge conical developments are necessary in the construction of a blast furnace (Fig. 9–27). Cylindrical developments are also frequently used in structures of this type. The developments must be carefully constructed to enable on-the-site assembly with considerable accuracy.

The development of an oblique cone is shown in Fig. 9–28. Elements on this cone are of varying lengths, but the resulting development will be symmetrical, since the top view is symmetrical. Elements in the given views are revolved into the frontal plane, as shown, so that their true lengths can be determined in the

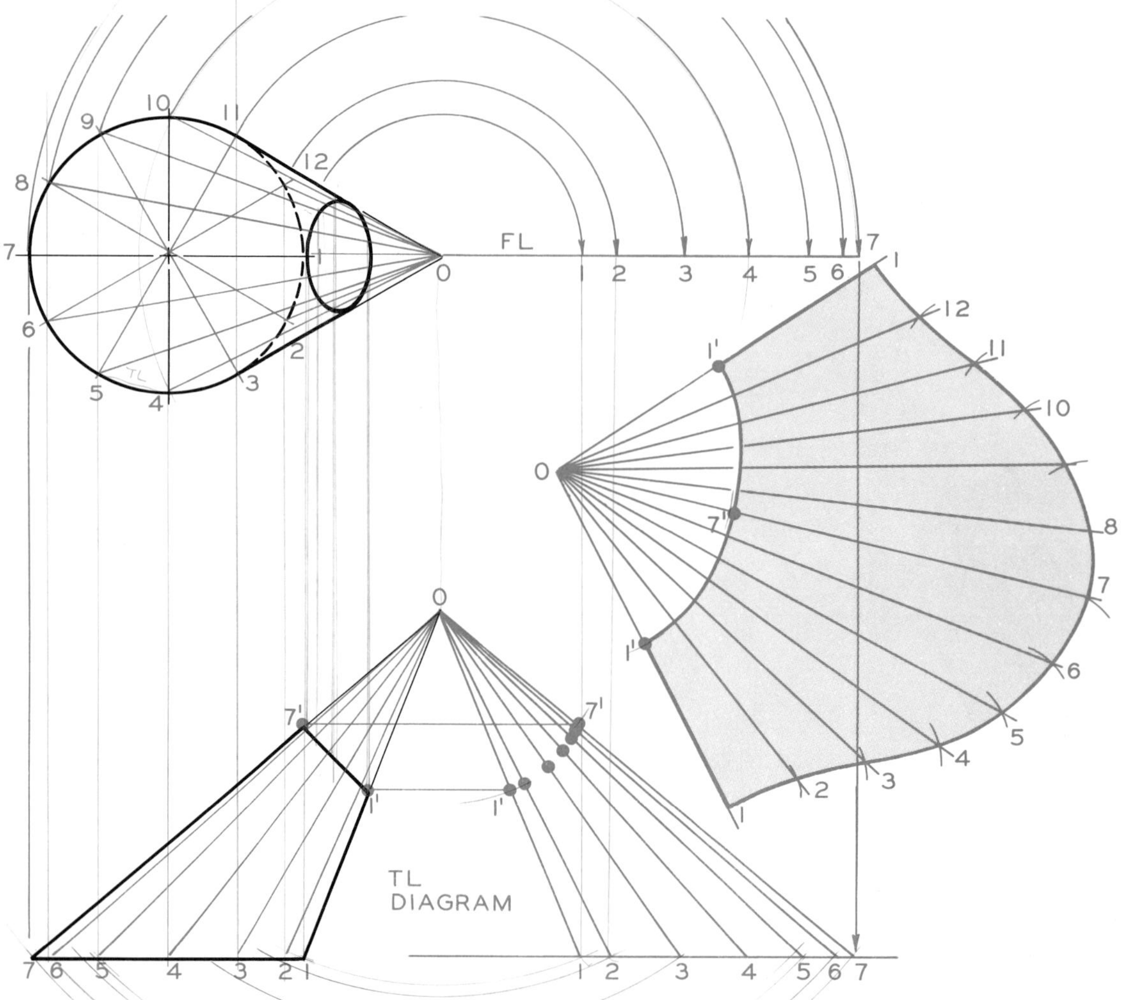

Fig. 9–28. Development of an oblique cone.

front view. The development is begun by constructing a series of triangles which are composed of elements and the chordal distances found on the base. The line of separation for the cone is chosen to be 0–1, since this is the shortest line on the cone's surface. The base is connected with a smooth curve. The true-length lines from apex 0 to the upper surface of the approximate cone are found by projecting from the front view to the true-length diagram. Points 1′ and 7′ are projected from the front view to their true-length lines in the true-length diagram found by revolution. Lines 0–1′ and 0–7′ are shown in the development, where they are used to locate points along the upper edges of the developed surfaces. These points are connected with a smooth line, but this line will not be an arc, since the geometric shape is not a true right cone and the edge view of the plane through points 7′ and 1′ in the front view is not perpendicular to the axis of the cone.

9–8 DEVELOPMENT OF A WARPED SURFACE

The geometric shape shown in Fig. 9–29 is an approximate cone with a warped surface and is

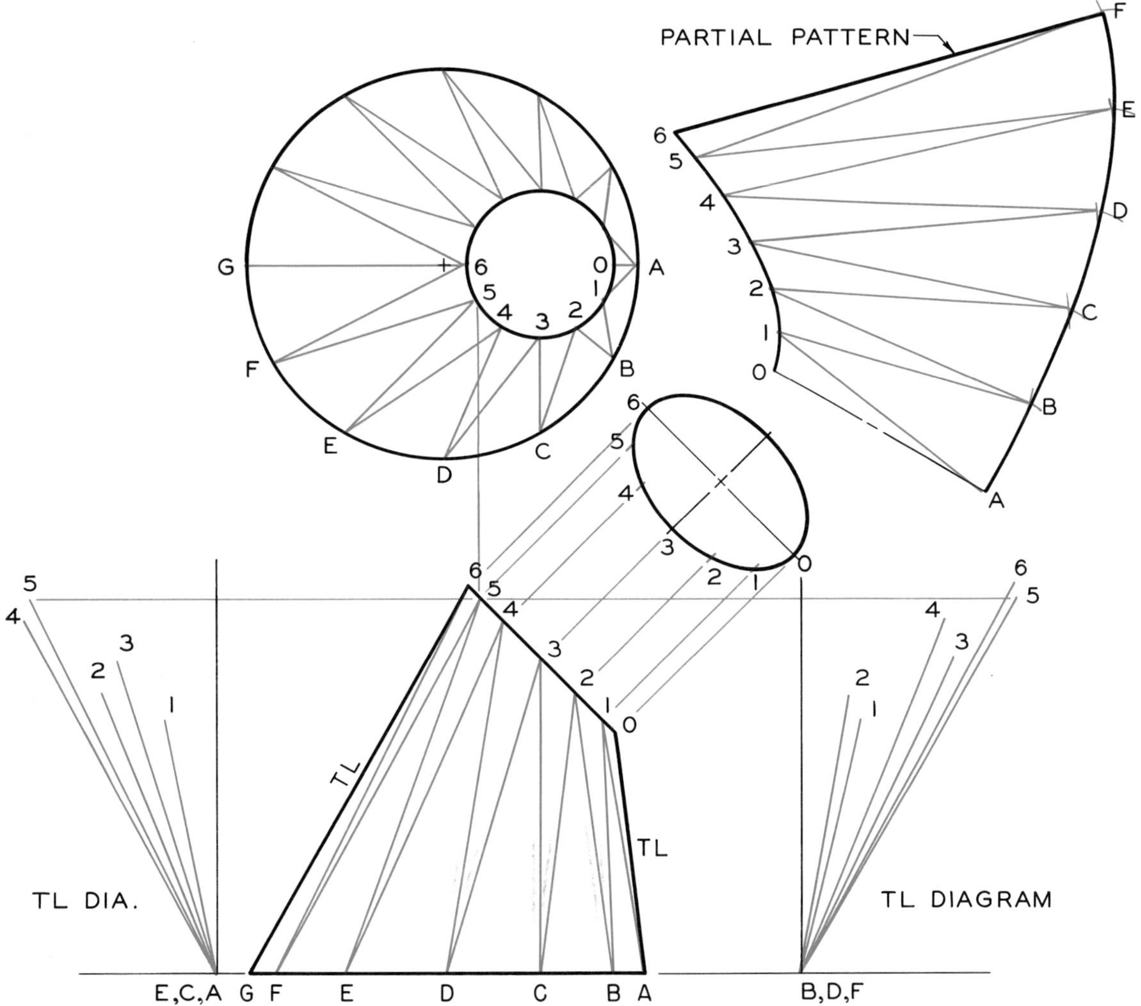

Fig. 9–29. Development of a partial inside pattern of a warped surface.

similar to the oblique cone shown in Fig. 9–28. The development of this surface will be merely an approximation, since a truly warped surface cannot be laid out on a flat surface. The surface is divided into a series of triangles in the top and front views by dividing the upper and lower views as shown. The true lengths of all lines are found in the true-length diagrams, which are drawn on each side of the front view, by projecting horizontally from the front view the vertical distances between the lines. To complete the true-length views of the lines, the horizontal distance between the ends of the lines is measured along the actual projection of the top view of the lines. A true-length line found in this manner is equivalent to a line that has been revolved, such as those illustrated in Fig. 9–28.

The chordal distance between the points on the base appears true-length in the top view since the base is horizontal. The chordal distance between the points on the upper edge of the lateral surface will appear true-length in a view that shows a true-size plane of this end. The developed surface is found by triangulation using true-length lines from (1) the true-length diagram, (2) the horizontal base in the top view, and (3) the primary auxiliary view. Each point should be carefully lettered to facilitate construction in all views.

9–9 DEVELOPMENT OF A TRANSITION PIECE

A transition piece is a figure that transforms its section at one end to a different shape at the other. This change is made gradually and uniformly. A duct with a rectangular cross section is connected to a cylinder with a transition piece in Fig. 9–30. Another example of a transition piece is the interior of the supersonic circuit of the wind tunnel shown in Fig. 9–31. Note that the cross section of the tunnel is changed from a rectangle to a circle at this point with a transition piece. Many other examples of these shapes can be seen in concrete structures.

The problem in Fig. 9–32 is solved by steps. The circular view of the transition piece is divided into equal units from which radial lines are drawn to each corner of the base. The true lengths of these lines are found by revolution (step 1). The chordal lines between the points on the circular section appear true length and the lines on the rectangular base appear true length, since these planes are horizontal. The line of separation for the development is line 1–A, the shortest line.

A portion of the development is laid out by triangulation in step 2, utilizing the true lengths of the lines. The remaining planes of the surface are found in step 3 to complete half of the symmetrical development. The upper points are connected with a smooth curve and the points on the base are connected with straight lines. Thin fold lines are given to indicate the curving surface at the corners.

Fig. 9–30. Transition-piece developments are used to join a circular shape with a rectangular section. (Courtesy of Western Precipitation Group, Joy Manufacturing Company.)

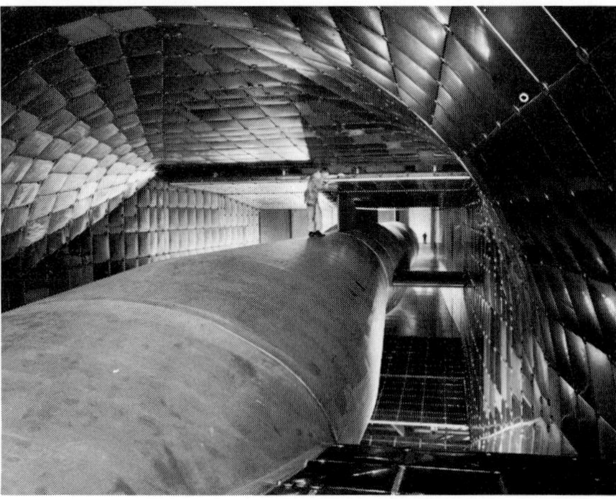

Fig. 9–31. An example of a transition application is the interior of this supersonic circuit of a wind tunnel. (Courtesy of Arnold Engineering Development Center.)

9–10 DEVELOPMENT OF A SPHERE—ZONE METHOD

The zone method is a conventional method of developing a sphere on a flat surface. A series of parallels, called latitudes in mapping, are drawn in the front view of Fig. 9–33. The parallels are spaced so that they establish equal arcs, D, on the surface of the sphere in the front view. Note that until D was determined mathematically and set off on the sphere. This

FIGURE 9–32. DEVELOPMENT OF A TRANSITION PIECE

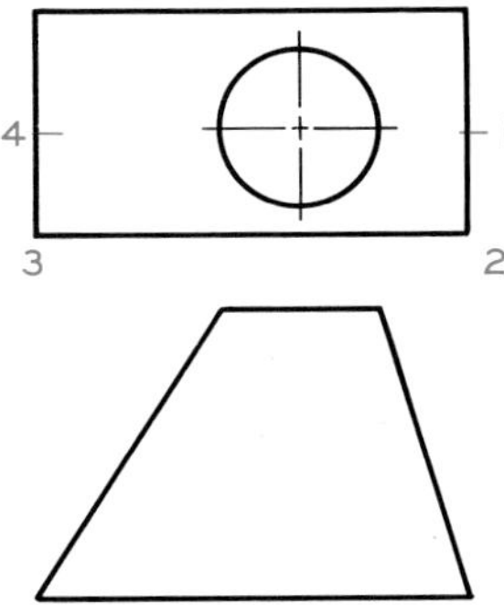

Given: The top and front views of a transition piece.
Required: Find an inside development of the surface from point 1 to point 4.
Reference: Article 9–9.

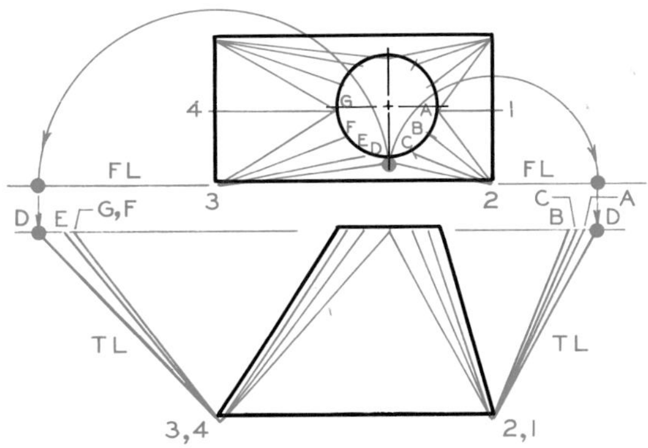

Step 1: Divide the circular edge of the surface into equal parts in the top view. Connect these points with bend lines to the corner points, 2 and 3. Find the true length of these lines by revolving them into a frontal plane and projecting them to the front view. These lines represent elements on the surface of an oblique cone.

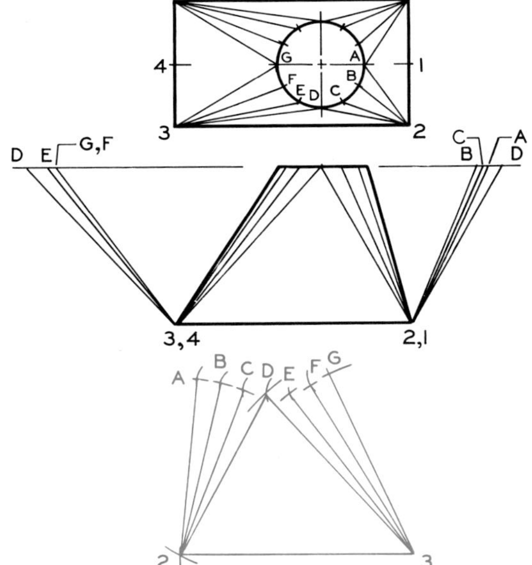

Step 2: Using the true-length lines found in the TL diagram and the lines on the circular edge in the top view, draw a series of triangles, which are joined together at common sides, to form the development. *Example:* Arcs 2*D* and 2*C* are drawn from point 2. Point *C* is found by drawing arc *DC* from point *D* to find point *C*. Line *DC* is true length in the top view.

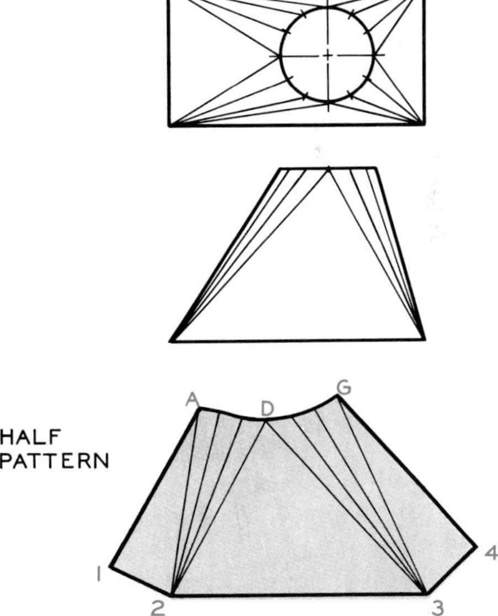

Step 3: Construct the remaining planes, *A*–1–2 and *G*–3–4, by triangulation to complete the inside half-pattern of the transition piece. Draw the fold lines as thin lines at the places where the surface is to be bent slightly. The line of departure for the pattern is chosen along *A*–1, the shortest possible line, for economy.

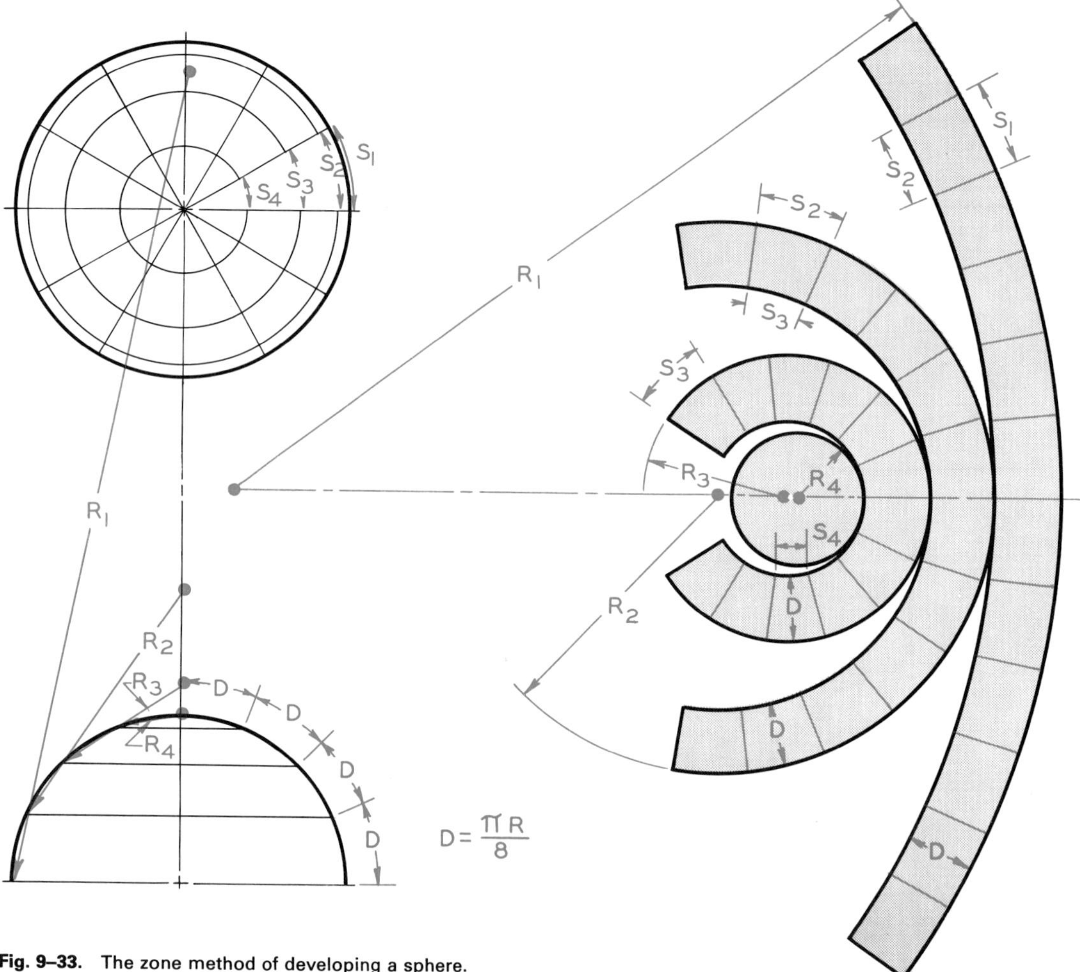

Fig. 9–33. The zone method of developing a sphere.

was done to establish uniformity in the development so that each of the developed zones would be equal in breadth. Cones are passed through the sphere's surface so that they form truncated cones in which one parallel serves as a base of a cone, and the other as the truncated top. The largest cone, which has an element equal to R_1, is found by extending line R_1 through the points where the equator and the next parallel intersect the sphere's surface in the front view, until R_1 intersects the extended center line of the sphere. Spherical elements R_2, R_3, and R_4 are found by repeating this process. The development is begun by laying out the largest zone, using R_1 as the radius of an arc which represents the base of an imaginary cone. The breadth of the zone is found by transferring distance D from the front view to the development and drawing the upper portion of the zone with a radius equal to R_1–D, using the same center. No regard is given to finding the arc lengths at this point. The next zone is drawn using the radius R_2 with its center located on a line through the center of arc R_1. The center of

Fig. 9–34. The giant dome of the United States Exhibit at Expo 67 is an example of a geodesic dome formed by straight structural members. (Courtesy of Rohm and Haas Company.)

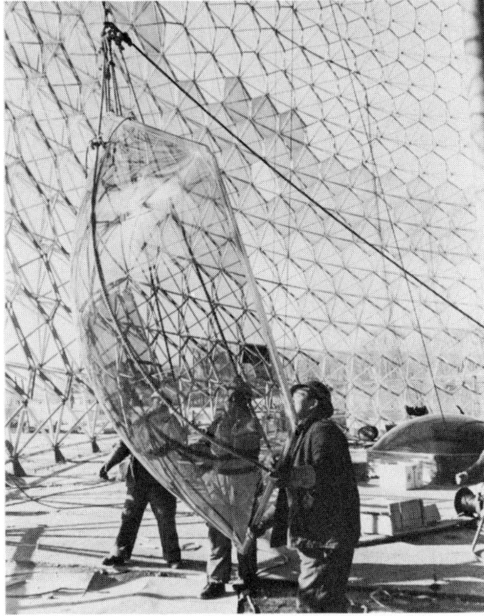

Fig. 9–35. Individual panels of Plexiglas® are installed in the giant dome. (Courtesy of Rohm and Haas Company.)

R_2 is positioned along this line such that the arc to be drawn will be tangent to the preceding arc, which was drawn with radius R_1–D. The upper arc of this second zone is drawn with a radius R_2–D. The remaining zones are constructed successively in this manner. The last cone will appear as a circle with R_4 as its radius.

The lengths of the arcs can be established by dividing the top view with vertical cutting planes that radiate through the poles. These lines, which lie on the surface of the sphere, are called longitudes in cartography. Arc distances S_1, S_2, S_3, and S_4 are found on each parallel in the top view. These distances are measured off on the constructed arcs in the development. In this case, there are twelve divisions, but smaller divisions would provide a more accurate measurement. A series of zones found in this manner can be joined to give an approximate sphere.

The giant dome of the United States Exhibit at the International Exhibition in Montreal is an example of a geodesic dome formed by straight structural members. This dome, shown in Fig. 9–34, is 250 feet in diameter and 187 feet high. Individual panels of Plexiglas® are shown being installed in Fig. 9–35. Most panels measured 10 feet by 12 feet. This dome is another example of a unique application of the sphere. Domes of this type have been considered as possible enclosures for entire cities to control weather conditions and environment.

9–11 DEVELOPMENT OF A SPHERE—GORE METHOD

Figure 9–36 is an alternative method of developing a flat pattern for a sphere. This method uses a series of spherical elements called gores. Equally spaced vertical cutting planes are passed through the poles in the top view. Parallels are located in the front view by dividing the surface into equal zones of dimension D. A front view of one of the gores is projected to the front view. A true-size view of one of the gores is developed by projecting from the top, which represents an approximation of the surface between two of the vertical cutting planes. Dimensions can be checked mathematically at all points.

The Unisphere® was designed by determining chordal lengths of longitudes and latitudes on the surface of the sphere, as shown in Fig. 9–37. The chordal lengths made it possible to fabricate the structure with straight

πR

0 1 2 3 4

D D D D

PARTIAL PATTERN

$D = \frac{\pi R}{8}$

D D D D

0 1 2 3 4

TYPICAL GORE

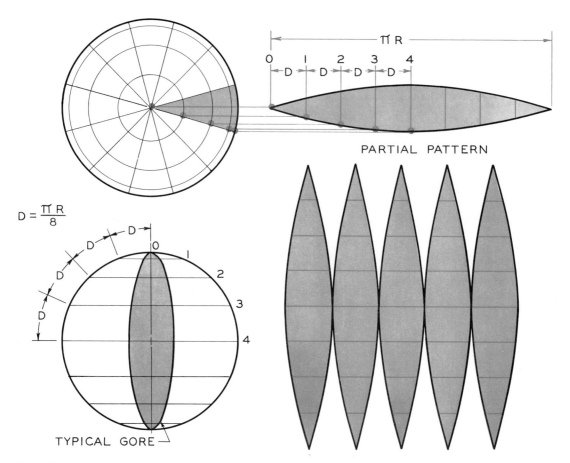

Fig. 9–36. The gore method of developing a sphere.

Fig. 9–37. The Unisphere® was designed by determining chordal lengths of longitudes and latitudes on the surface of the sphere. This method is similar to the gore method of development. (Courtesy of U.S. Steel Corporation.)

Fig. 9–38. Surface areas are being attached to the structural frame of the Unisphere®. (Courtesy of U.S. Steel Corporation.)

FIGURE 9–39. STRAP DEVELOPMENT

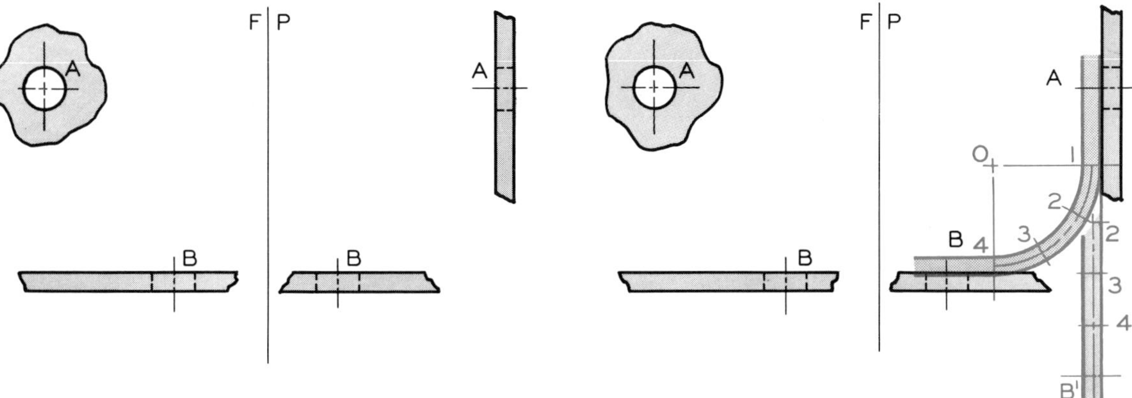

Given: The front and side views of two planes that are to be connected at points *A* and *B* with a metal strap.

Required: Find the true development of the strap and show it in both views.

Reference: Article 9–12.

Step 1: Construct the edge view of the strap in the side view using the specified radius of bend. Locate points 1, 2, 3, and 4 on the neutral axis at the bend. Revolve this portion of the strap into the vertical plane and measure the distances along this view of the neutral axis. Check the arc distances by mathematics. The hole is located at *B'* in this view.

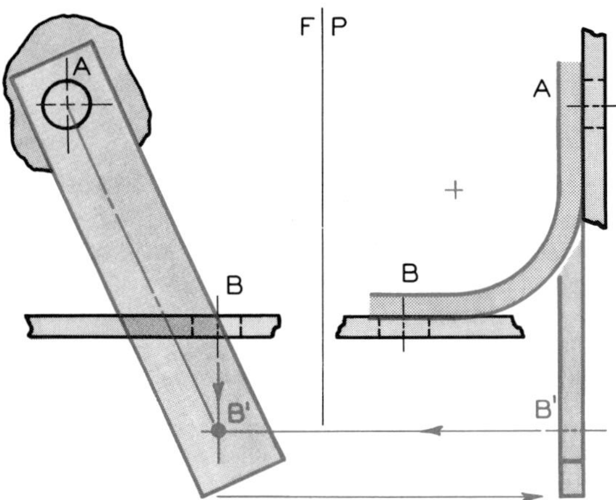

Step 2: Construct the front view of *B'* by revolving point *B* parallel to the profile plane until it intersects the projector from *B'* in the side view. Draw the center line of the true-size strap from *A* to *B'* in the front view. Add the outline of the strap around this center line and around the holes at each end, allowing enough material to provide sufficient strength.

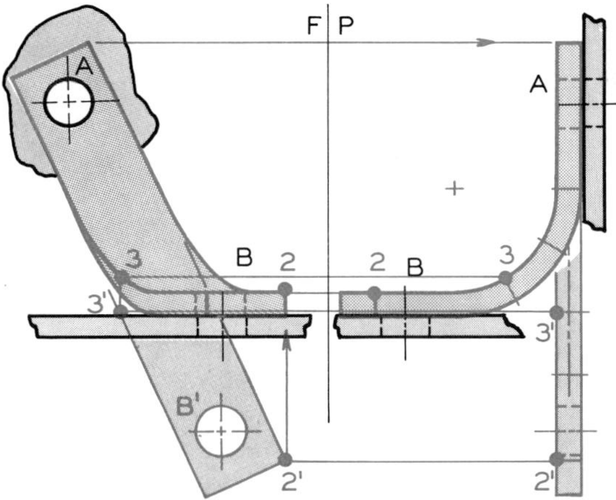

Step 3: Determine the projection of the strap in the front view by projecting points from the given views. Points 3 and 2 are shown in the views to illustrate the system of projection used. The ends of the strap are drawn in each view to form true projections.

members. The land areas attached to the sphere were developed by a method similar to the gore method. Segments of these surfaces can be seen in Fig. 9–38, where they are being attached to the framework for later assembly on the site.

9–12 DEVELOPMENT OF A SUPPORT STRAP

Strap metal is universally used in mass-produced products such as brackets, connectors, and supports. It is more economical to form these shapes to the desired configuration by stamping and bending than by any other fabrication method. Almost all designs contain a variety of oblique surfaces and structures which must be connected by brackets that have been stamped. Figure 9–39 illustrates the steps necessary to the design of a developed view of a support bracket. The bracket is to connect two surfaces in different planes whose points of connection are oblique to each other. The strap is drawn in the view in which the planes appear as edges (step 1), using the specified radius of bend. The arc of the bend is divided into smaller arcs and developed as a straight strap without a bend in the side view. Point B' is found in this view to indicate the location of the hole. The hole in the front view, shown at B, is projected to its position on the developed strap, as shown in step 2. This permits the true-size development of the strap to be drawn, and allowance to be made for the appropriate amount of metal on each side of the hole for strength. The projected front view of the strap in its bent position is constructed in step 3 to indicate its final configuration. An accurate design that reduces surplus material would result in considerable savings when mass produced.

Observe that principles of revolution have been applied to this development, as well as the techniques of three-view projection. Most industrial problems tend toward a combination of graphical principles rather than the application of a single concept. Mathematics could also be used to verify the arc measurements in step 2. The designer should develop versatility in applying every tool at his disposal to the solving and checking of problems.

Figure 9–40 illustrates many examples of stamped metal components used in the body of an automobile. Each of these components was developed during the design process to obtain flat patterns from which the finished shapes could be fabricated.

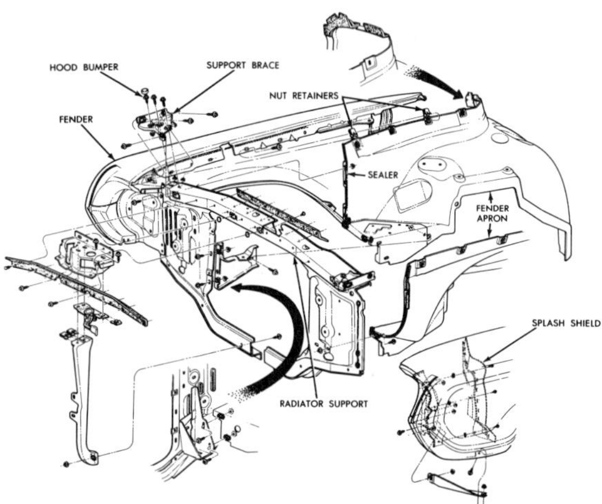

Fig. 9–40. Many examples of stamped metal developments can be seen in this exploded assembly drawing of a portion of an automobile body. (Courtesy of Ford Motor Company.)

9–13 SUMMARY

The examples and applications covered in this chapter should serve to illustrate the many uses of the principles of developments. Essentially all engineering and technological problems are concerned with a wide assortment of geometric shapes that must be constructed from flat materials. Developments are made possible through the application of basic graphical and descriptive geometry principles in conjunction with mathematics.

Principles of intersections are closely related to developments, since different shapes must be joined together in many instances. Any development problem will be easier to solve if it is first resolved into its basic geometric elements; this process will facilitate the application of the principles covered in this chapter. A lettering system should be utilized in laying out a pattern, to avoid confusion with the projections and constructions.

PROBLEMS

Problems for this chapter should be constructed with instruments from the given sketches on $8\frac{1}{2}'' \times 11''$ sheets, as illustrated in the accompanying figures. Each grid represents $\frac{1}{4}''$. All reference planes and figure points should be labeled, using $\frac{1}{8}''$ letters with guidelines. Solutions should be sufficiently noted and labeled to explain all construction.

1. (A through C) Using Fig. 9–41, lay out an inside pattern for the prisms in parts A, B, and C of the figure. Number representative points.

2. (A and B) Using Fig. 9–42, lay out an inside pattern for each of the prisms in parts A and B of the figure. Show all construction and number the points.

3. (A through C) Lay out an inside half pattern for each cylinder in parts A, B, and C of Fig. 9–43.

4. (A and B) Lay out an inside pattern for each cylinder in parts A and B of Fig. 9–44. Show all construction and number the points.

5. (A and B) Lay out an inside half pattern for each pyramid in parts A and B of Fig. 9–45.

6. (A and B) In Fig. 9–46 lay out inside half patterns for each of the cones in parts A and B.

7. Lay out an inside pattern for the warped surface in Fig. 9–47. Show a half development.

8. (A and B) Lay out inside half patterns for the transition pieces in parts A and B of Fig. 9–48.

9. (A) Lay out an inside developed pattern of the sphere in Fig. 9–49 using the gore method. (B) Using a separate sheet of paper, lay out an inside developed pattern of the sphere in Fig. 9–49 by the zone method.

10. Complete the front and side views of the $1\frac{1}{4}''$ strap which is shown in Fig. 9–50 bent into position at holes *A* and *B*. Give the complete development of the strap, including squared-off ends that extend $\frac{3}{4}''$ beyond the center line of the holes.

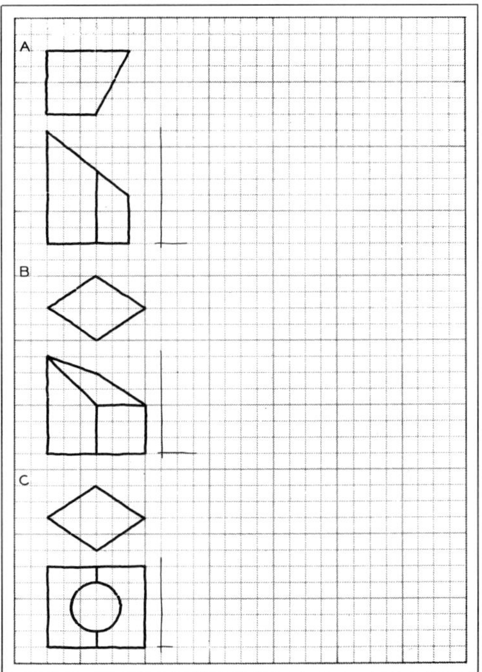

Fig. 9–41. Development of prisms.

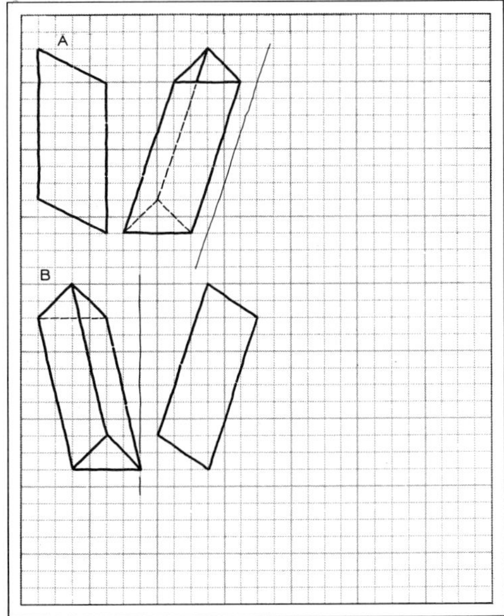

Fig. 9–42. Development of prisms.

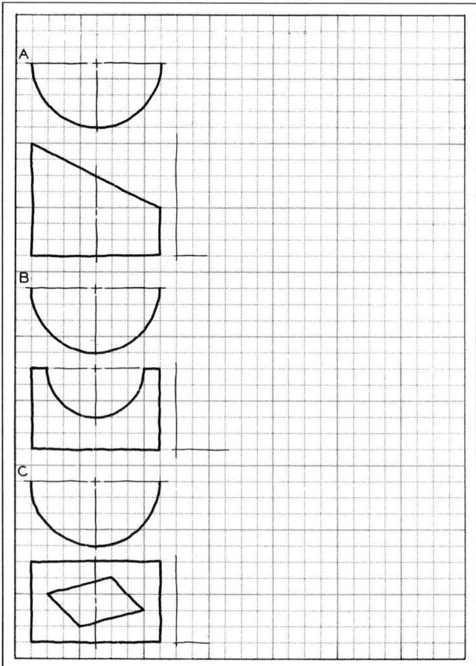

Fig. 9–43. Development of cylinders.

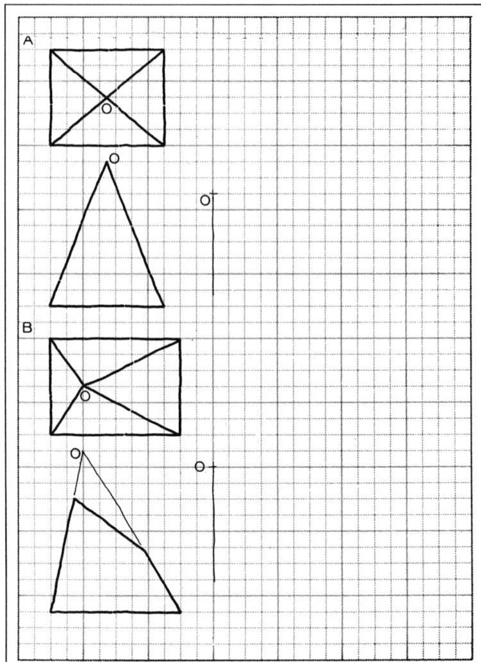

Fig. 9–45. Development of pyramids.

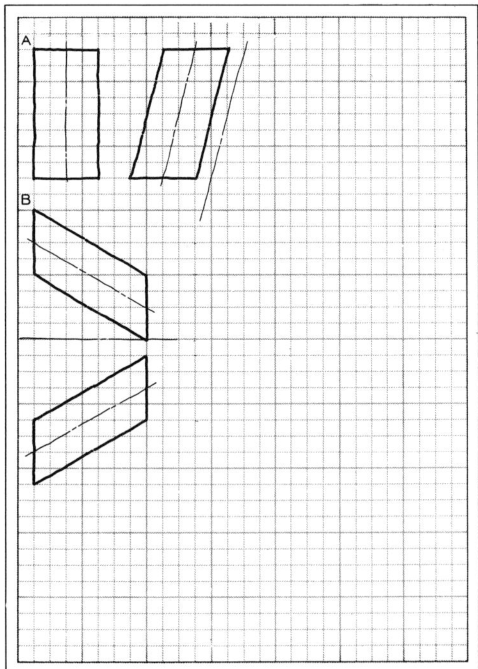

Fig. 9–44. Development of cylinders.

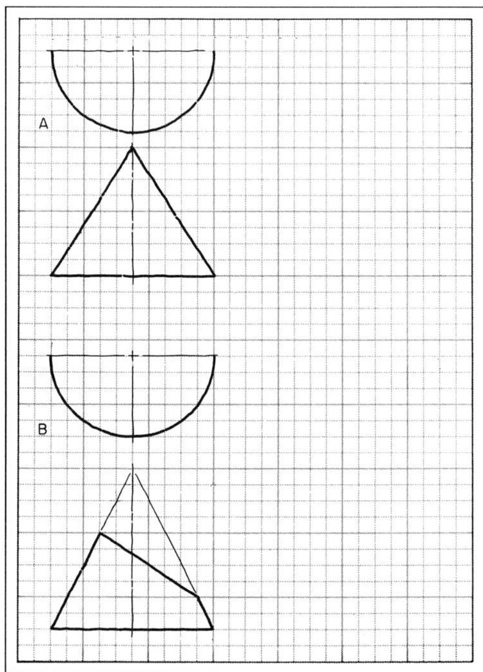

Fig. 9–46. Development of a cone.

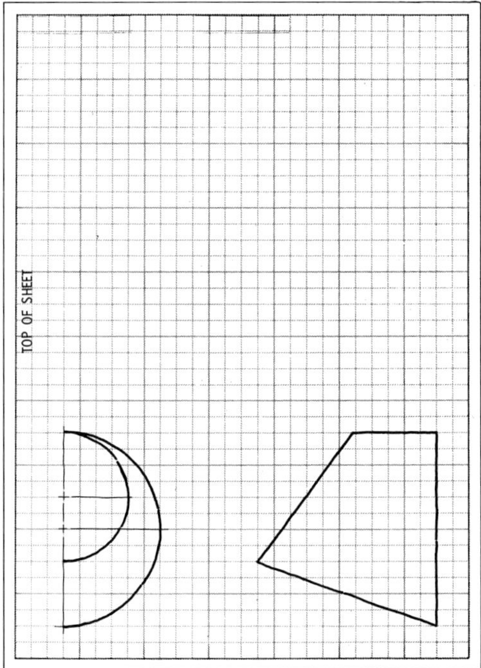

Fig. 9–47. Development of a warped surface.

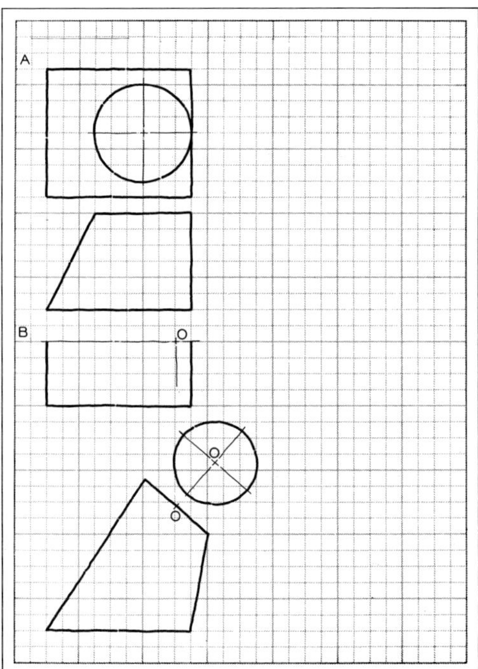

Fig. 9–48. Development of transition pieces.

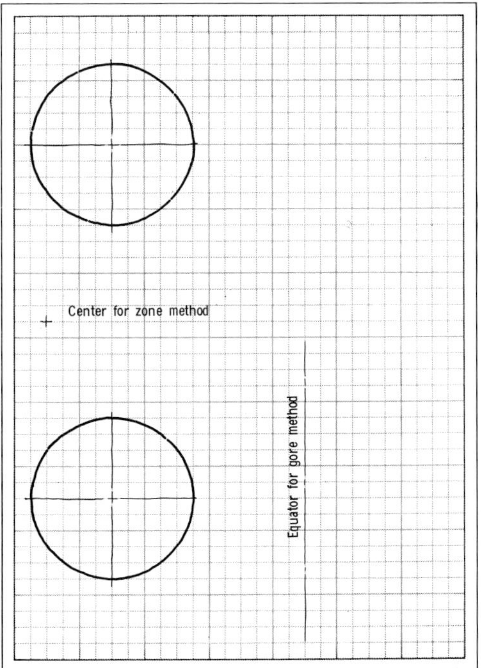

Fig. 9–49. Development of a sphere.

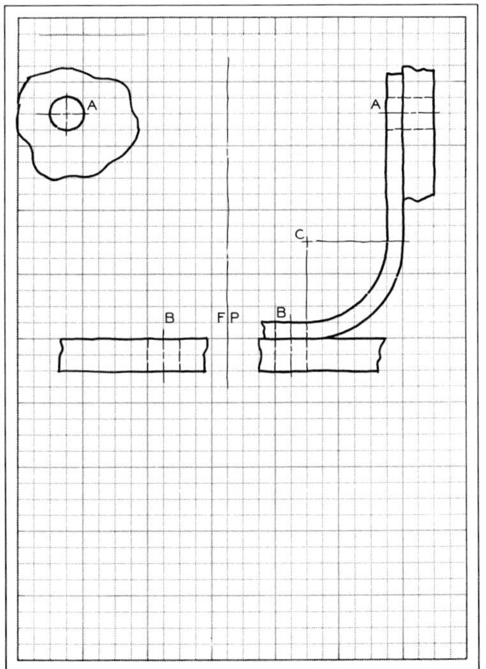

Fig. 9–50. Development of a strap.

10

Vector Analysis

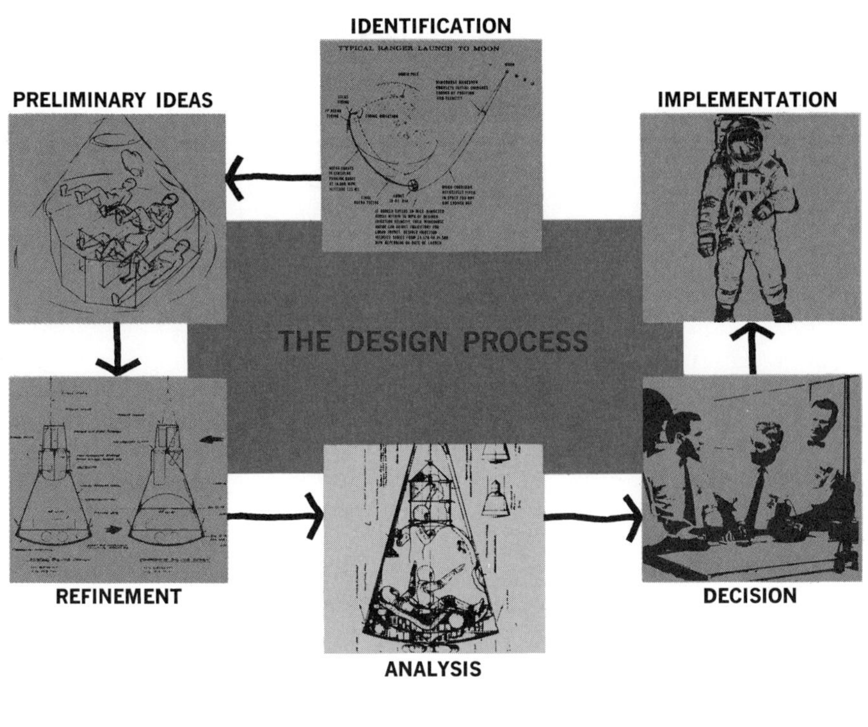

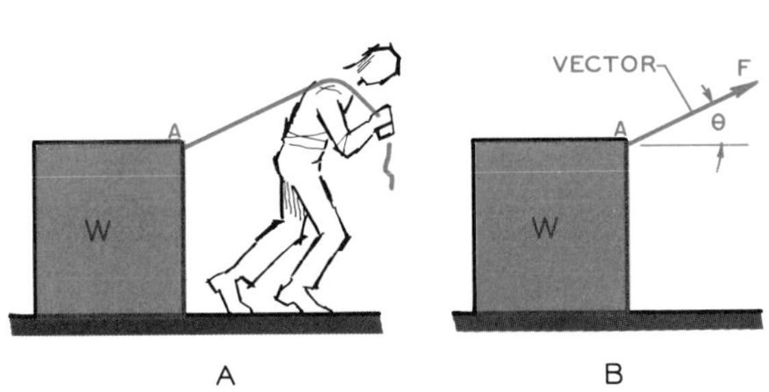

Fig. 10–1. Representation of a force by a vector.

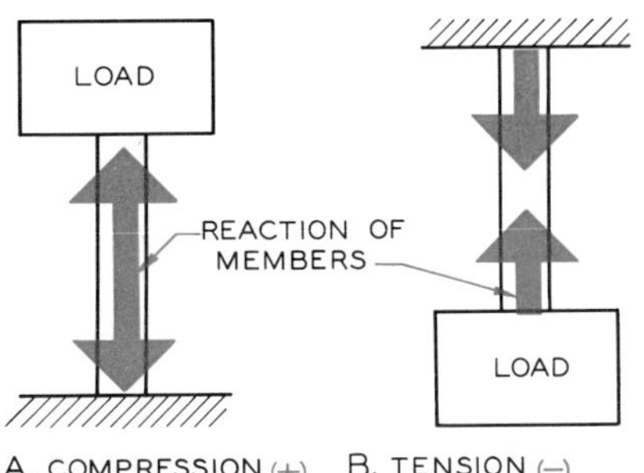

Fig. 10–2. Comparison of tension and compression in a member.

10–1 INTRODUCTION

After a design of a structural system has been refined and drawn to scale, and angular and linear dimensions have been determined, it is necessary to analyze the system for strength and stresses. When the forces are known, members of an appropriate size may be selected to withstand the forces within the system.

In analyzing a system for strength it is necessary to consider the forces of tension and compression within the system. These forces are represented by vectors. Vectors may also be used to represent other quantities such as distance, velocity, and electrical properties.

Graphical methods are useful in the solution of vector problems, which are often very complicated to solve by conventional trigonometric and algebraic methods. Each method can serve as an effective check on the solutions determined by other methods.

10–2 BASIC DEFINITIONS

A knowledge of the terminology of graphical vectors is prerequisite to an understanding of the techniques of problem solving with vectors. The following definitions will be used throughout this chapter.

Force. A push or a pull that tends to produce motion. All forces have (1) magnitude, (2) direction, (3) a point of application, and (4) sense. A force is represented by the rope being pulled in Fig. 10–1A.

Vector. A graphical representation of a quantity of force which is drawn to scale to indicate magnitude, direction, sense, and point of application. The vector shown in Fig. 10–1B represents the force of the rope pulling the weight, *W*.

Magnitude. The amount of push or pull. In drawings, this is represented by the length of the vector line. Magnitude is usually measured in pounds of force.

Direction. The inclination of a force (with respect to a reference coordinate system).

Point of Application. The point through which the force is applied on the object or member. This is point *A* in Fig. 10–1A.

Sense. Either of the two opposite ways in which a force may be directed, i.e., toward or away from the point of application. The sense is shown by an arrowhead attached to one end of the vector line. In Fig. 10–1A, the sense of the force is away from point A. It is shown in part B of the figure by the arrowhead at *F*.

Compression. The state created in a member by subjecting it to opposite pushing forces. A member tends to be shortened by compression (Fig. 10–2A). Compression is represented by a plus sign (+) or the letter C.

Tension. The state created in a member by subjecting it to opposite pulling forces. A member tends to be stretched by tension, as shown in Fig. 10–2B. Tension is represented by a minus sign (−) or the letter T.

FIGURE 10–3. RESULTANT BY THE PARALLELOGRAM METHOD

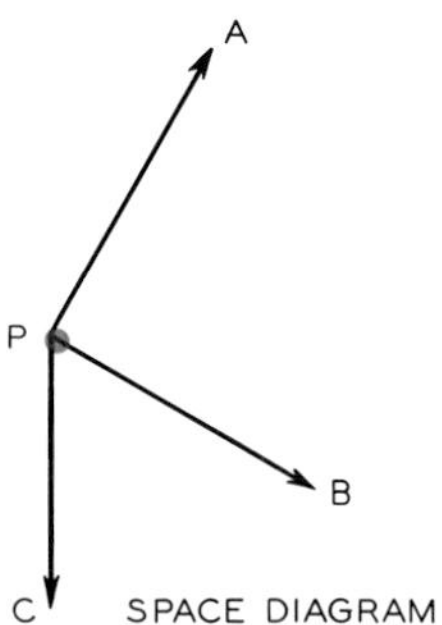

Required: Find the resultant of this coplanar, concurrent force system by the parallelogram method.

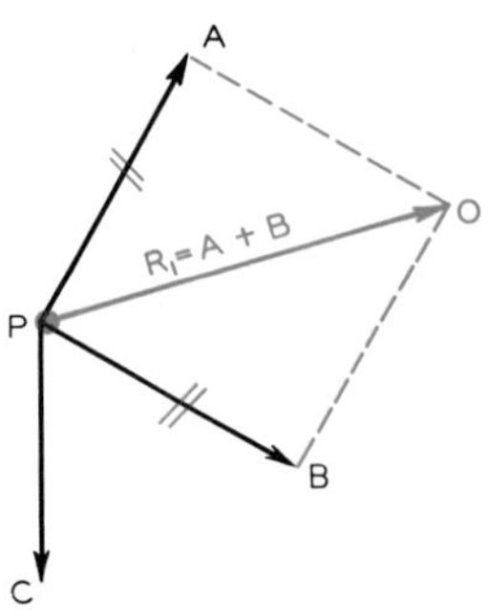

Step 1: Draw a parallelogram with its sides parallel to vectors A and B. The diagonal, R_1, drawn from point P to point O is the resultant of forces A and B.

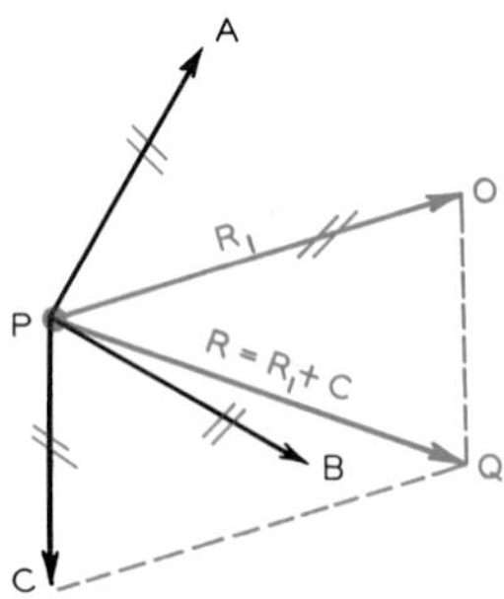

Step 2: Draw a parallelogram using vectors R_1 and C to find diagonal R from P to Q. This is the resultant that can replace forces A, B, and C.

Force System. The combination of all forces acting on a given object. Figure 10–3 shows a force system.

Resultant. A single force that can replace all the forces of a force system and have the same effect as the combined forces.

Equilibrant. The opposite of a resultant; it is the single force that can be used to counterbalance all forces of a force system.

Components. Any individual forces which, if combined, would result in a given single force. For example, Forces A and B are components of resultant R_1 in step 1 of Fig. 10–3.

Space Diagram. A diagram depicting the physical relationship between structural members. The force system in Fig. 10–3 is given as a space diagram.

Vector Diagram. A diagram composed of vectors which are scaled to their appropriate lengths to represent the forces within a given system. The vector diagram is used to solve for unknowns that are required in the solution of the problem. A vector diagram may be a polygon or a parallelogram.

Statics. The study of forces and force systems that are in equilibrium.

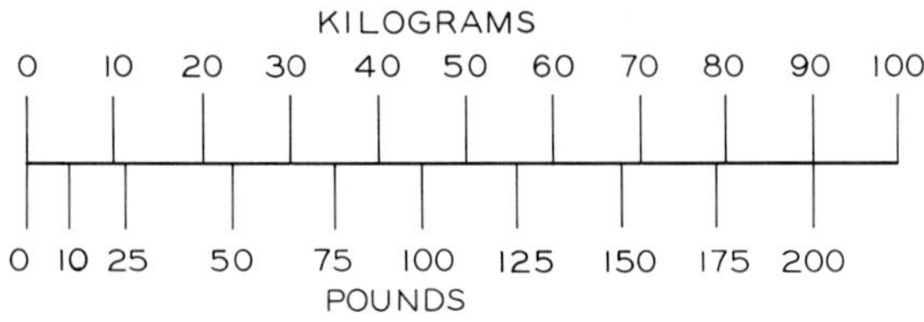

Fig. 10–4. The kilogram (Kg) is the standard metric unit for representing forces, which are represented by pounds in the English system.

Metric Units. The kilogram (kg) is the standard unit for indicating mass (loads). A comparison of kilograms with pounds is shown in Fig. 10–4. The metric ton is 1000 kilograms.

10–3 COPLANAR, CONCURRENT FORCE SYSTEMS

When several forces, represented by vectors, act through a common point of application, the system is said to be *concurrent.* Vectors A, B, and C act through a single point in Fig. 10–3; therefore this is a concurrent system. When only one view is necessary to show the true length of all vectors, as in Fig. 10–3, the system is *coplanar.*

Engineering designs are analyzed to determine the total effect of the forces applied in a system. Such an analysis requires that the known forces be resolved into a single force—the *resultant*—that will represent the composite effect of all forces on the point of application. The resultant is found graphically by two methods: (1) the parallelogram method and (2) the polygon method. In either case, the selection of a proper scale is important to the final solution. A larger drawing will result in a higher degree of accuracy.

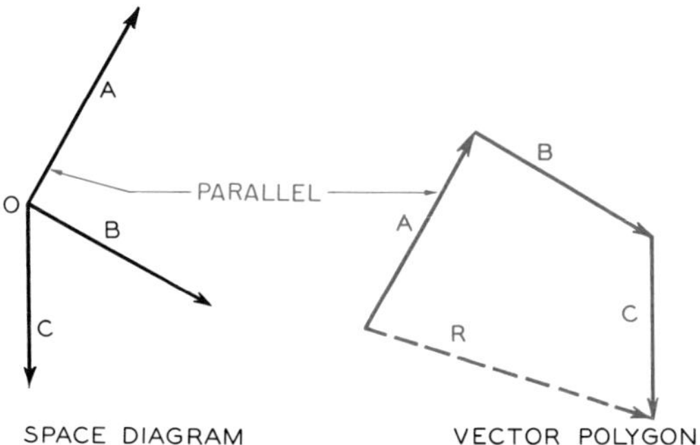

Fig. 10–5. Resultant of a coplanar, concurrent system as determined by the polygon method.

10–4 RESULTANT OF A COPLANAR, CONCURRENT SYSTEM—PARALLELOGRAM METHOD

In the system of vectors shown in Fig. 10–3, all the vectors lie in the same plane and act through a common point. The vectors are scaled to a known magnitude.

The vectors for a force system must be known and drawn to scale in order to apply the parallelogram method to determine the resultant. Two vectors are used to find a parallelogram; the diagonal of the parallelogram is the resultant of these two vectors and has its point of origin at point P (Fig. 10–3). Resultant R_1 can be called the *vector sum* of vectors A and B.

Since vectors A and B have been replaced by R_1, they can be disregarded in the next step of the solution. Again, resultant R_1 and vector C are resolved by completing a parallelogram, i.e., by drawing a line parallel to each vector. The diagonal of this parallelogram, PQ, is the resultant of the entire system and is the vector sum of R_1 and C. This resultant, R, can be analyzed as though it were the only force acting on the point; therefore the analysis of a particular point-of-force application is simplified by finding the resultant.

10–5 RESULTANT OF A COPLANAR, CONCURRENT SYSTEM—POLYGON METHOD

The system of forces shown in Fig. 10–3 is shown again in Fig. 10–5, but in this case the resultant is found by the polygon method. The forces are drawn to scale and in their true directions, with each force being drawn head-to-tail to form the polygon. In this example, the vectors are drawn in a counterclockwise sequence, beginning with vector A. Vector B is drawn with its tail at the arrowhead end of vector A and vector C is similarly attached to B. Note that the polygon does not close; this means that the system is not in *equilibrium*. In other words, it would tend to be in motion, since the forces are not balanced in all directions. The resultant R is drawn from the tail of vector A to the head of vector C to close the polygon. It can be seen by inspection that the resultant is equal in length, direction, and sense to the resultant found by the parallelogram method of the previous article.

10–6 RESULTANT OF A COPLANAR, CONCURRENT SYSTEM—ANALYTICAL METHOD

Vectors can be solved analytically by application of algebra and trigonometry. The graphical method is generally much faster, and presents less chance of error due to an arithmetical mistake. The designer should, however, be well versed in all methods, since each will have advantages over the others in certain situations. The example in Fig. 10–6 is given to afford a comparison between the graphical and analytical methods.

FIGURE 10–6. RESULTANT BY THE ANALYTICAL METHOD

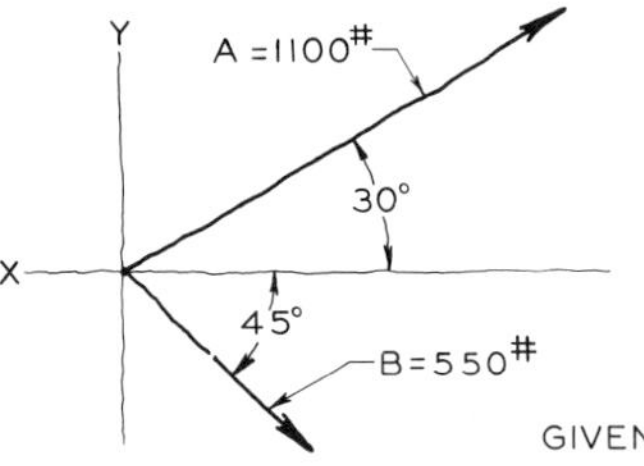

Required: An unscaled, freehand sketch of two forces is given. Find the resultant using the analytical method.

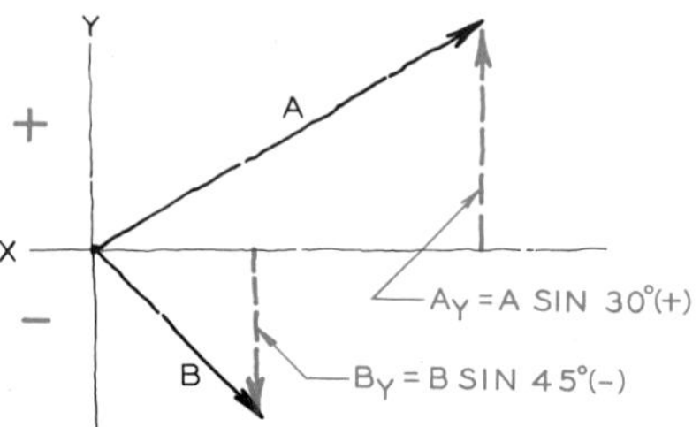

Step 1: The *Y*-components (vertical components) are found to be the sine functions of the angles the vectors make with the *X*-axis. The *Y*-component of *A* is a positive and the *Y*-component of *B* is negative.

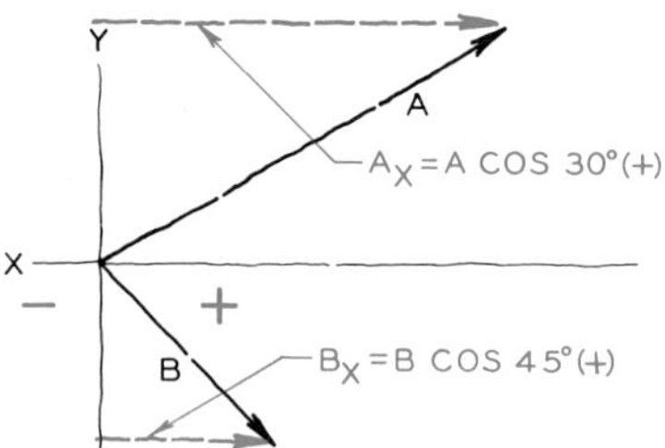

Step 2: The *X*-components (horizontal components) are the cosine functions of 30° and 45° in this case, both in the positive direction.

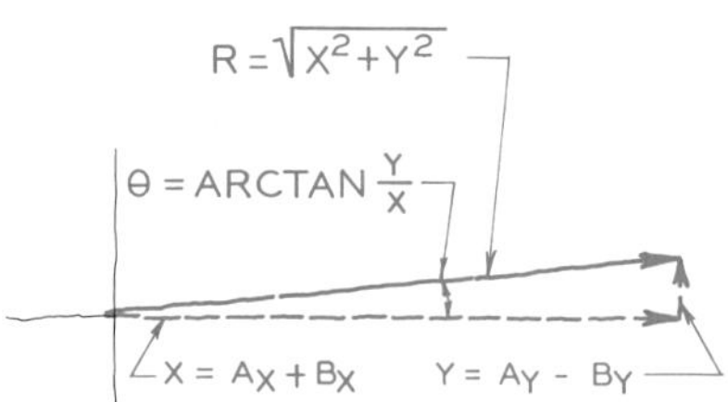

Step 3: The *Y*-components and *X*-components are summed to find the components of the resultant, *X* and *Y*. The Pythagorean theorem is applied to find the magnitude of the resultant. Its angle with the *X*-axis is the arctangent of *Y*/*X*.

Since the analytical approach will be used, it is unnecessary for the vectors and angles to be drawn accurately to scale. A freehand sketch is sufficient.

In step 1, the vertical components, which are parallel to the *Y*-axis, are drawn from the ends of both vectors to form right triangles. The lengths of these components are found through the use of the trigonometric functions of the angles the vectors make with the *X*-axis.

The horizontal component of each vector is drawn parallel to the *X*-axis through the end of each vector. The lengths of these components are found to be the cosine functions of the given vectors in step 2.

The *Y*-components of each vector, A_y and B_y, can be added, since each lies in the same direction (step 3). The resulting value is $Y = A_y - B_y$, since the components have opposite senses. The horizontal component is $X = A_x + B_x$, since both components have equal directions and senses. A right triangle is sketched using the *X*- and *Y*-distances that were found trigonometrically. The vertical and horizontal components are laid off head-to-tail and the head of the horizontal component is connected to the tail of the vertical to form a three-sided polygon of forces. The magnitude of the resultant is found by the Pythagorean theorem,

$$R = \sqrt{X^2 + Y^2}$$

The direction of the resultant is

$$\text{angle } \theta = \arctan Y/X$$

and it is measured from the horizontal *X*-axis. The sense is in the direction of the hypotenuse from the point of application to the head of the vertical component.

Law of Sines. The law of sines is illustrated in Fig. 10–7A. When any three values are known, the remaining unknowns of a triangle can be computed. An example is given (Fig. 10–7B) where two sides of a triangle are vectors of known magnitude and direction. This enables you to find the resultant mathematically, as shown.

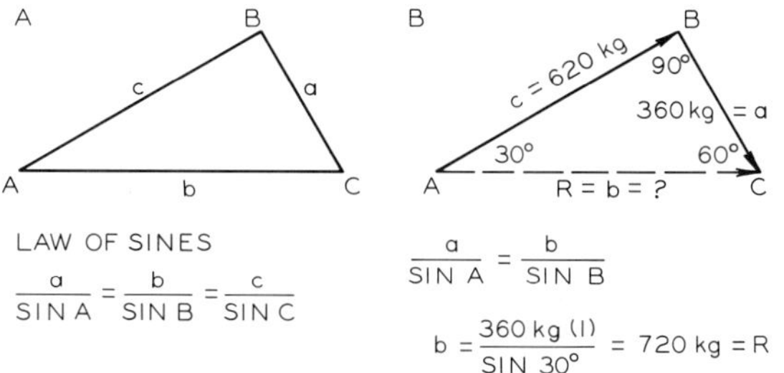

Fig. 10–7. The law of sines is illustrated in part A. This principle is used in part B to solve for resultant *R*(*b*) when two angles and vectors are known.

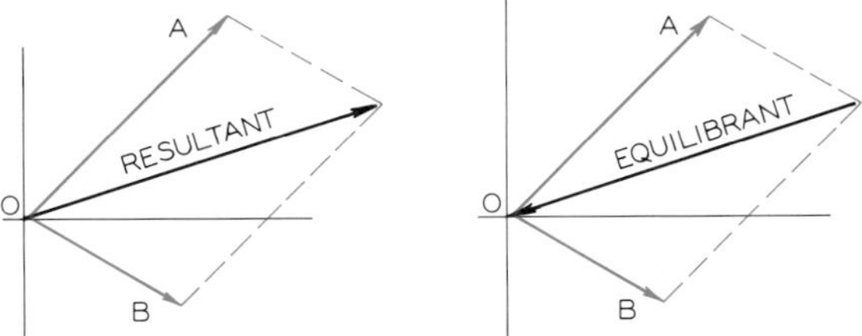

Fig. 10–8. The resultant and equilibrant are equal in all respects except in sense.

An *equilibrant* has the same magnitude, direction, and point of application as the *resultant* in any system of forces. The difference is the sense. Note that the resultant of the system of forces shown in Fig. 10–8 is solved for through the parallelogram method. The sense of the resultant is toward point *C* along the direction *OC*, the diagonal of the parallelogram. The equilibrant is drawn so that its arrowhead is at the opposite end, toward point *O*. The equilibrant can be applied at point *O* to balance the forces *A* and *B* and thereby cause the system to be in a state of equilibrium.

10–7 RESULTANT OF NONCOPLANAR, CONCURRENT FORCES—PARALLELOGRAM METHOD

When vectors lie in more than one plane of projection, they are said to be *noncoplanar*; therefore more than one view is necessary to analyze their spatial relationships. The resultant of a system of noncoplanar forces can be found, regardless of their number, if their true projections are given in two adjacent orthographic views. The solution of an example of this type is shown through sequential steps of the parallelogram method in Fig. 10–9.

Vectors 1 and 2 are used to construct the top and front views of a parallelogram. The diagonal of the parallelogram, R_1, is found in both views. As a check, the front view of R_1 must be an orthographic projection of its top view; if it is not, there is an error in construction. Since R_1 is used to replace vectors 1 and 2, they may be omitted in further construction.

In step 2, resultant R_1 and vector 3 are resolved to form resultant R_2 by the parallelogram method in both views. The top and front views of R_2 must project orthographically if there is no error in construction. Resultant R_2 can be used to replace vectors 1, 2, and 3. Since R_2 is an oblique line that is not true length, its true length can be found by auxiliary view, as shown in Fig. 10–8 or by revolution, as previously covered.

10–8 RESULTANT OF NONCOPLANAR. CONCURRENT FORCES—POLYGON METHOD

The same system of forces that was given in Fig. 10–9 is given in Fig. 10–10. In this instance, we are required to solve for the resultant of the system by the polygon method.

In step 1, the given orthographic views of the vectors are transferred to a vector polygon, in which each vector is laid head-to-tail in a clockwise direction, beginning with vector 1. The vectors are drawn in each view to be orthographic projections at all times (step 2). Since the vector polygon did not close, the system is not in equilibrium. The resultant *R* is constructed from the tail of vector 1 to the head of vector 3 in both views.

Resultant *R* is an oblique line and so requires an auxiliary view to find its true length. The magnitude of the resultant can be measured in the true-length auxiliary view by using the same scale as was used to draw the original views.

FIGURE 10–9. RESULTANT BY THE PARALLELOGRAM METHOD

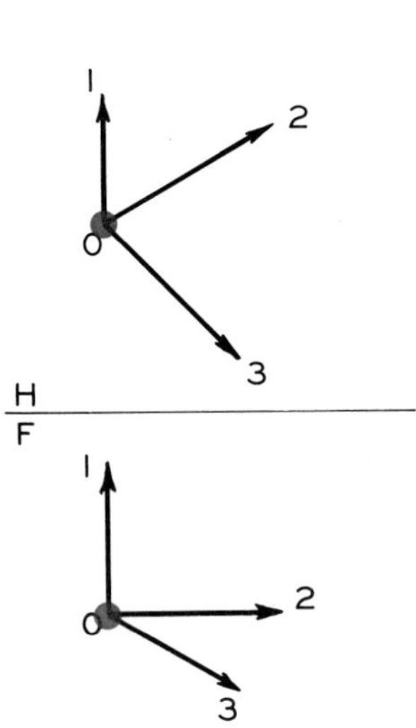

Required: Find the resultant of this non-coplanar, concurrent system of forces by the parallelogram method.

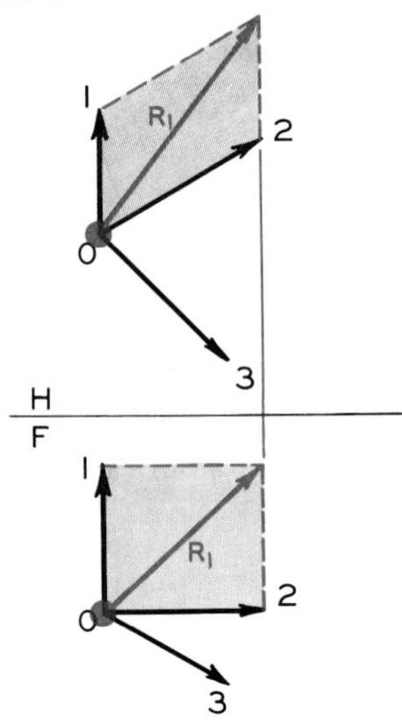

Step 1: Vectors 1 and 2 are used to construct a parallelogram in the top and front views. The diagonal, R_1, is the resultant of these two vectors.

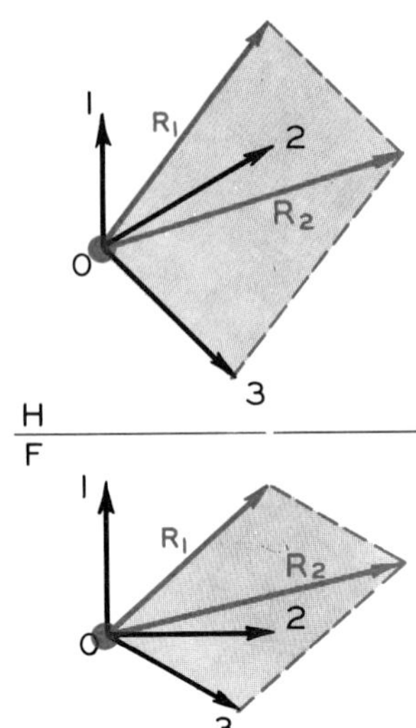

Step 2: Vectors 3 and R_1 are used to construct a second parallelogram to find the views of the overall resultant, R_2.

FIGURE 10–10. RESULTANT BY THE POLYGON METHOD

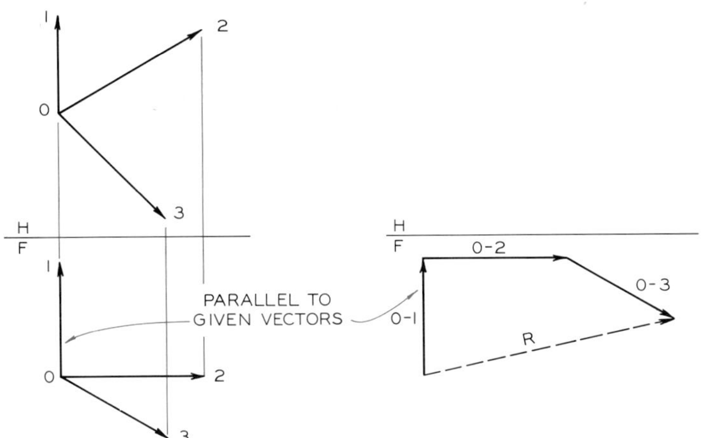

Required: Find the resultant of this system of concurrent, noncoplanar forces by the polygon method.

Step 1: Each vector is laid head-to-tail in the front view. The front view of the resultant is the vector drawn.

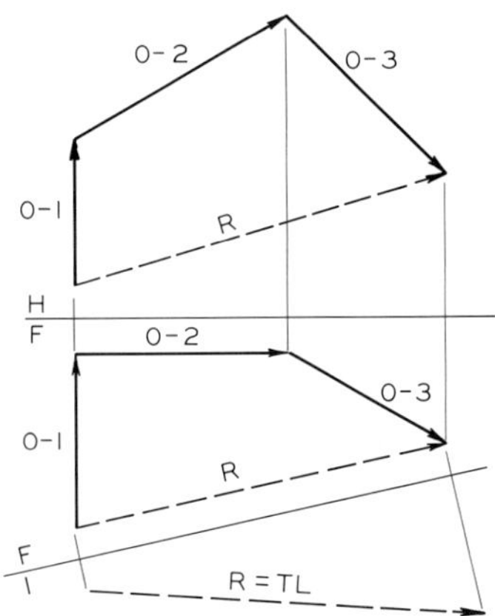

Step 2: The same vectors are laid head-to-tail in the top view to complete the three-dimensional polygon. The resultant is found true length by an auxiliary view.

FIGURE 10–11. RESULTANT BY THE ANALYTICAL METHOD

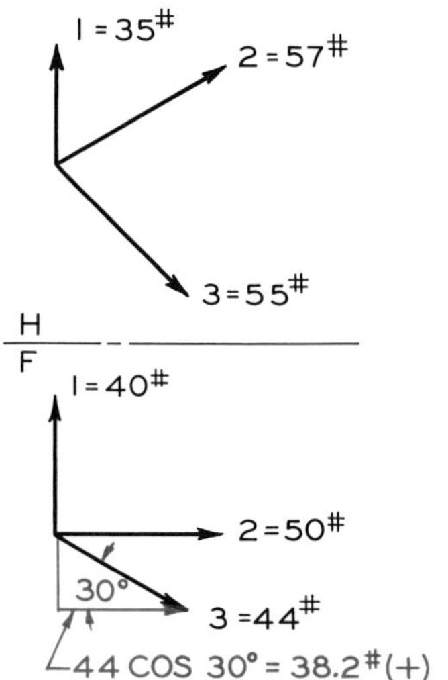

Step 1: The *X*-component is found in the front or top view. The *X*-components are found to be: force 1, 0 lb; force 2, 50 lb; force 3, 44 lb cos 30°. These values are positive.

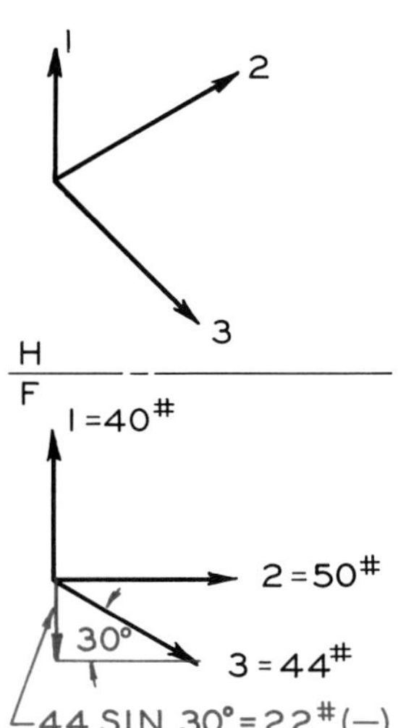

Step 2: The *Y*-component must be found in the front view. The *Y*-components are found to be: force 1, 40 lb; force 2, 0 lb; force 3, 44 lb sin 30°.

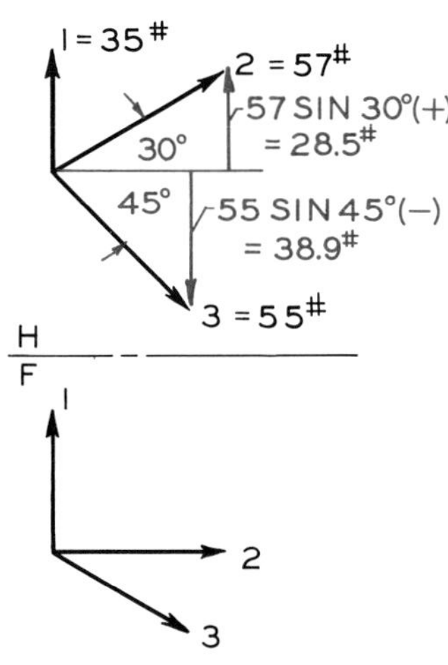

Step 3: The *Z*-component must be found in the top view. The *Z*-components are found to be: force 1, 35 lb; force 2, 57 lb sin 30°; force 3, 55 lb sin 45°. The resultant is found in Fig. 10–12.

10–9 RESULTANT OF NONCOPLANAR, CONCURRENT FORCES—ANALYTICAL METHOD

The same system of forces that was given in Fig. 10–9 is given in Fig. 10–11. We are required to solve for the resultant of the system by the analytical method, using trigonometry and algebraic equations. The projected lengths of the vectors are known in both views, as indicated in Fig. 10–11.

In step 1, the summation of the forces in the X-direction is found in the front view. Since this left and right direction can be seen in either the top view or front view, either view can be used for finding the X-component of the system. The summation in the X-direction is expressed in the following equation:

$$\sum F_x = (2) + (3)\cos 30^\circ$$
$$= 50 + 44\cos 30^\circ = 88.2\text{ lb }(+)$$

The X-component is found to be 88.2 lb in the positive direction, which is toward the right. Vector 1 is vertical and consequently has no component in the X-direction.

The summation of forces in the Y-direction is found in the front view. This summation is expressed in the following equation:

$$\sum F_y = (1) - (3)\sin 30^\circ$$
$$= 40 - 44\sin 30^\circ = 18\text{ lb }(+)$$

Vector 2 is horizontal and has no vertical component.

The summation of forces in the Z-direction is found in the top view. Positive direction is considered to be backward, and negative to be forward. This summation is expressed in the following equation:

$$\sum F_z = (1) + (2)\sin 30^\circ - (3)\sin 45^\circ$$
$$= 35 + 57\sin 30^\circ - 55\sin 45^\circ$$
$$= 24.16\text{ lb }(+)$$

The resultant that can be used to replace vectors 1, 2, and 3 can be found from these three components by the following equation:

$$R = \sqrt{X^2 + Y^2 + Z^2}$$

By substitution of the X-, Y-, and Z-components found in the three previous summations, the equation can be solved as follows:

$$R = \sqrt{88.2^2 + 18^2 + 24.6^2} = 93.3 \text{ lb}$$

The resultant force of 93.3 lb is of no value unless its direction and sense are known. To find this information, we must refer to the two orthographic views of the force system, as shown in Fig. 10–12. The X- and Z-components, 88.2 lb and 24.6 lb, are drawn to form a right triangle in the top view. The hypotenuse of this triangle depicts the direction and sense of the resultant in the top view. The angular direction of the top view of the resultant is found in the following equation:

$$\tan\theta = \frac{24.6}{88.2} = 0.279; \qquad \theta = 15.6°$$

The angular direction of the resultant is found in the front view by constructing a triangle with the X- and Y-components, 88.2 lb and 18 lb. The hypotenuse of this right triangle is the direction of the resultant. The direction of the resultant in the front view is expressed in the following equation:

$$\tan\phi = \frac{18}{88.2} = 0.204; \qquad \phi = 11.5°$$

These two angles, found in the top and front views, establish the direction of the resultant vector, whose sense can be described as upward, to the right, and back. The force system and the various steps of solution need not be drawn to scale for analytical solution, since no attempt is made to measure lines or angles.

The advantages of the graphical solution of this problem should be apparent after this example is completed in its entirety. Errors are more likely in the analytical solution due to the number of components involved.

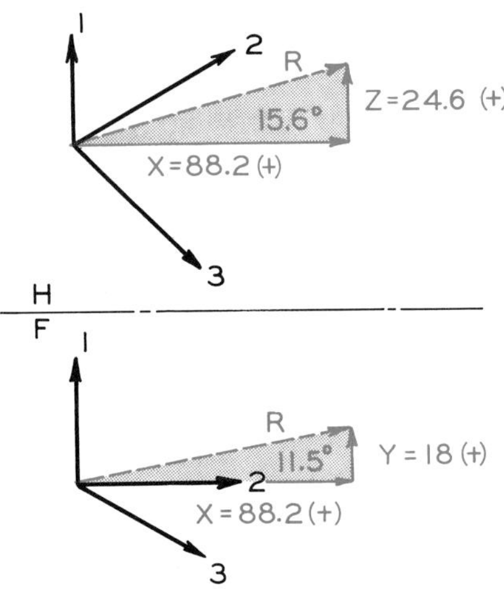

Fig. 10–12. The three components, X, Y, and Z, found in Fig. 10–11 are used to find the resultant, $R = \sqrt{X^2+Y^2+Z^2}$.

10–10 STRUCTURES IN EQUILIBRIUM

In the previous examples, the vectors were drawn from given or known magnitudes and directions. The same principles can be applied to structural systems in which the magnitudes and senses are not given. All engineering structures are analyzed for their loads as the first step in designing and selecting members to adequately support the loads for which the structure is designed. An example of coplanar structures in equilibrium can be seen in the loading cranes in Fig. 10–13, which are used for the handling of cargo on board ship. These are coplanar, concurrent force systems.

The coplanar, concurrent structure given in Fig. 10–14. is designed to support a load of $W = 1000$ kg. The maximum loading in each is used to determine the type and size of structural members used in the structural design.

In step 1, the only known force, $W = 1000$ kg, is laid off parallel to the given direction. Unknown forces A and B are drawn as vectors to close the force polygon. Each vector must be drawn head-to-tail. The vectors can be scaled in this view.

Fig. 10–13. The cargo cranes on the cruise ship *Santa Rosa* are examples of coplanar, concurrent force systems that are designed to remain in equilibrium. (Courtesy of Exxon Corporation.)

In step 2, vectors *A* and *B* can be analyzed to determine whether they are in tension or compression. Vector *B* points upward to the left, which is toward point *O* when transferred to the structural diagram. A vector which acts toward a point of application is in compression. Vector *A* points away from point *A* when transferred to the structural diagram and is, therefore, in tension.

The length versus the cross section of a member will be considered when selecting a member, but the determination of force in the member is found in the same manner in the vector polygon regardless of member length.

A similar example of a force system involving a pulley is solved in Fig. 10–15 to determine the loads in the structural members caused by the weight of 100 lb. The only difference between this solution and the previous one is the construction of two equal vectors to represent the loads in the cable on both sides of the pulley.

FIGURE 10–14. COPLANAR FORCES IN EQUILIBRIUM

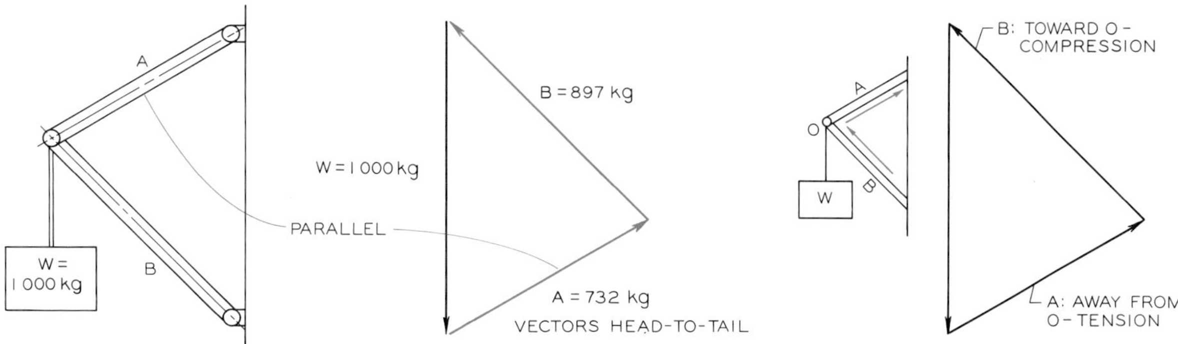

Required: Find the forces in the two structural members caused by the load of 1000 kg.

Step 1: Draw the known load of 1000 kg as a vector. Draw the vectors *A* and *B* to the same scale and parallel to their directions. Arrowheads are drawn head-to-tail.

Step 2: Vector *A* points away from point *O* when transferred to the structural diagram. Therefore, vector *A* is in tension. Vector *B* points toward point *O* and is in compression.

FIGURE 10–15. DETERMINATION OF FORCES IN EQUILIBRIUM

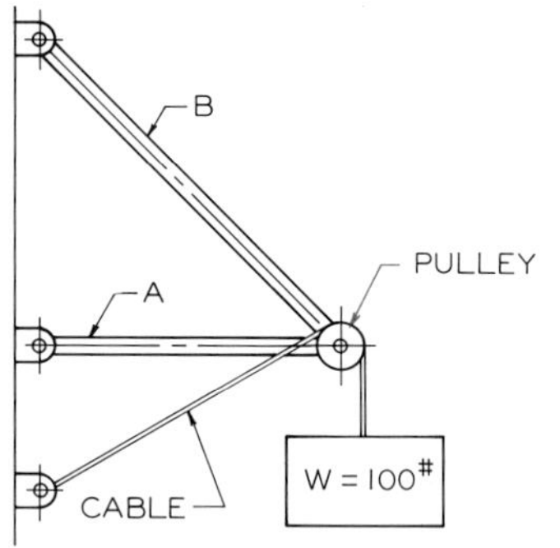

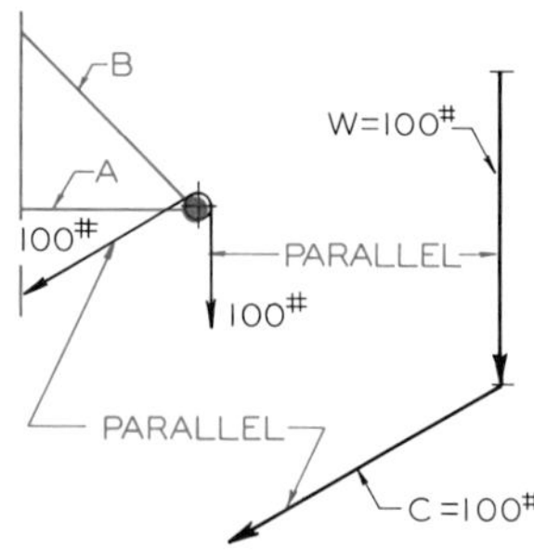

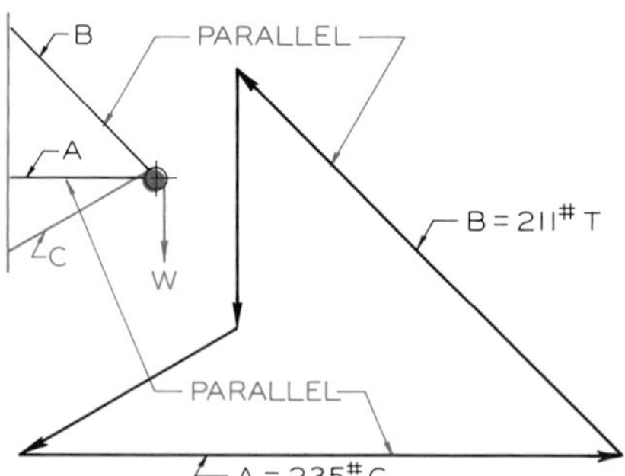

Required: Find the forces in the members caused by the load of 100 lb supported by the pulley.

Step 1: The force in the cable is equal to 100 lb on both sides of the pulley. These two forces are drawn as vectors head-to-tail parallel to their directions in the space diagram.

Step 2: The remaining unknowns, *A* and *B*, are drawn to close the polygon, and arrowheads are placed to give a head-to-tail arrangement. The sense of *A* is toward the point of application, and *A* is in compression; *B* is away from the point and is in tension.

10–11 FORCES IN EQUILIBRIUM—ANALYTICAL SOLUTION

The force system shown in Fig. 10–14 has been sketched in Fig. 10–16 for solution by the analytical method. The load *W* must be known along with the angular directions of members *A* and *B*. Forces *A* and *B* are unknown; however, they can be used in the equations where the *X*- and *Y*-forces will be summed. Since the system is in equilibrium, the summation of forces in any direction will equal zero, which indicates balance or equilibrium. The summation of forces in the *Y*-direction and *X*-direction can be expressed in the following equations:

$$\sum F_y = A \sin 30° + B \sin 45° - 1000 = 0 \quad (1)$$

$$F_x = A \cos 30° - B \cos 45° = 0 \quad (2)$$

We can solve Eq. (2) for *A* by rearranging the equation to the following form:

$$A = \frac{B \cos 45°}{\cos 30°} = 0.816\,(B) \quad (3)$$

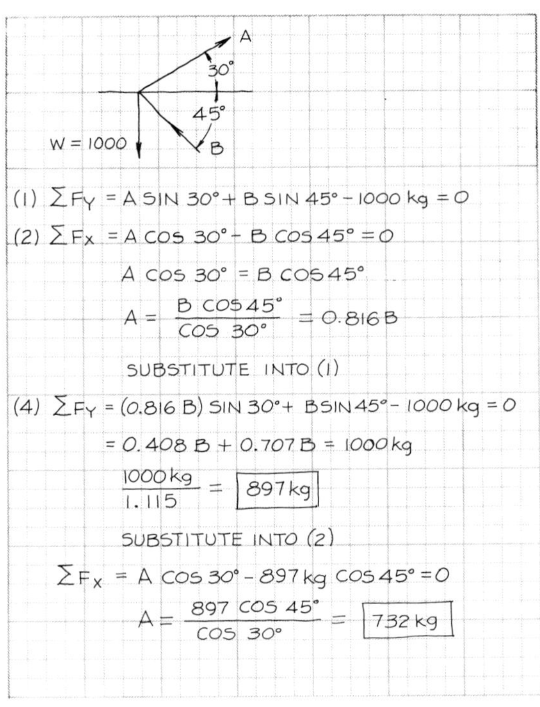

Fig. 10–16. Analytical determination of the forces in a concurrent, coplanar system in equilibrium.

When Eq. (3) is substituted into Eq. (1), the equation can be rewritten as:

$$\sum F_y = 0.816(B) \sin 30° + B \sin 45° - 1000$$
$$= 0 = 0.408\ B + 0.707\ B - 1000;$$
$$B = \frac{1000}{1.115} = 897 \text{ kg} \qquad (4)$$

The value of $B = 897$ lb is substituted into Eq. (2) so that we may solve for A in the following manner:

$$\sum F_x = A \cos 30° - 897 \cos 45° = 0;$$
$$A = \frac{897 \cos 45°}{\cos 30°} = 732 \text{ kg} \qquad (5)$$

In comparing the graphical and analytical methods, we should note that both methods are limited to two unknowns.

10–12 TRUSS ANALYSIS

Vector polygons can be used to analyze a structural truss to determine the loads in each member by two graphical methods: (1) joint-by-joint analysis, and (2) Maxwell diagrams.

Joint-by-Joint Analysis. The truss shown in Fig. 10–17 is called a Fink truss, and is loaded with forces of 3000 lb that are concentrated at joints of the structural members. A special method of designating forces, called *Bow's notation*, is used. The exterior forces applied to the truss are labeled with letters placed between the forces. Numerals are placed between the interior members.

Each vector used to represent the load in each member is referred to by the number on each of its sides by reading in a clockwise direction. For example, the first vertical load at the left is called *AB*, with *A* at the tail and *B* at the head of the vector.

FIGURE 10–17. JOINT ANALYSIS OF A TRUSS

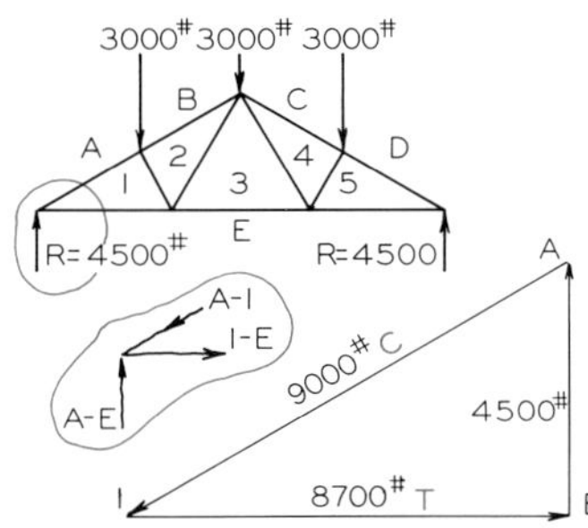

Step 1: The truss is labeled using Bow's notation, with letters between the exterior loads and numbers between interior members. The lower left joint can be analyzed, since it has only two unknowns, *A*–1 and 1–*E*. These vectors are found by drawing them parallel to the unknown vectors from both ends of the known reaction of 4500 lb. The vectors are laid off in a head-to-tail order.

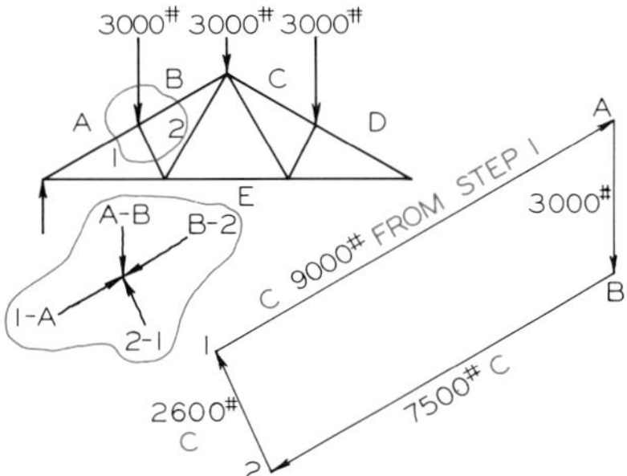

Step 2: Using the vector of 1–*A* found in step 1 and load *AB*, the two unknowns *B*–2 and 2–1 can be found. The known vectors are laid out beginning with vector 1–*A* and moving clockwise about the joint. Vectors *B*–2 and 2–1 close the polygon. If the sense of a vector is toward the point of application, it is in compression; if away from the point, it is in tension.

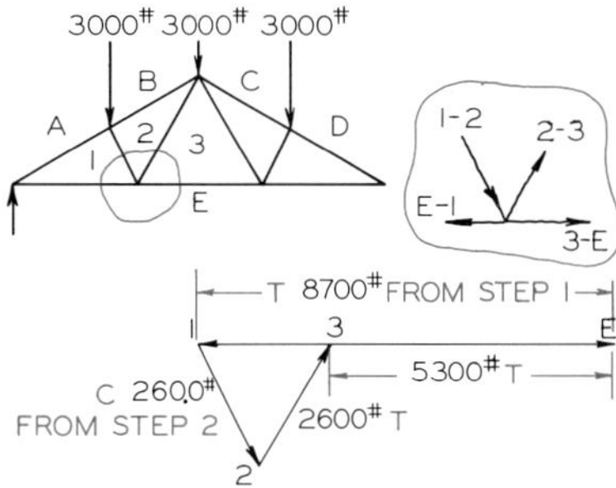

Step 3: The third joint can be analyzed by laying out the vectors *E*–1 and 1–2 from the previous steps. Vectors 2–3 and 3–*E* close the polygon and are parallel to their directions in the space diagram. The senses of 2–3 and 3–*E* are away from the point of application; these vectors are in tension.

The first joint that is analyzed is the one at the left, where the reaction of 4500 pounds (denoted by #) is known. This force, reading in a clockwise direction about the joint, is called *EA* with an upward sense. The tail is labeled *E* and the head *A*. Continuing in a clockwise direction, the next force is *A*–1 and the next 1–*E*, which closes the polygon and ends with the beginning letter, *E*. The arrows are placed, beginning with the known vector *EA*. The arrows are placed, beginning with the known vector *EA*, in a head-to-tail arrangement. Tension and compression can be determined by relating the sense of each vector to the original joint. For example, *A*–1 has a sense toward the joint and is in compression, while 1–*E* is away and in tension.

Since the truss is symmetrical and equally loaded, the loads in the members on the right will be equal to those on the left.

The other joints are analyzed in the same manner in steps 2 and 3; the procedure is to begin with known vectors found in the previous polygons and then solve for the unknowns. Note that the sense of the vectors is opposite at each end. Vector *A*–1 has a sense toward the left in step 1, and toward the right in step 2.

Maxwell Diagrams. The Maxwell diagram is exactly the same as the joint-by-joint analysis except that the polygons are positioned to overlap, with some vectors common to more than one polygon; separate polygons are not used for each joint. Again, Bow's notation is used to good advantage.

The first step (Fig. 10–18) is to lay out the exterior loads beginning clockwise about the truss—*AB*, *BC*, *CD*, *DE*, and *EA*—head-to-tail. A letter is placed at each end of the vectors. Since they are parallel, this polygon will be a straight line.

The structural analysis begins at the joint through which reaction *EA* acts. A free-body diagram is drawn to isolate this joint for easier analysis. The two unknowns are members *A*–1 and 1–*E*. These vectors are drawn parallel to their direction in the truss in step 1 of Fig. 10–18, with *A*–1 beginning at point *A* and 1–*E* beginning at point *E*. These directions are extended to a point of intersection, which locates point 1. Since this joint is in equilibrium, as are all joints of a system in equilibrium, the vectors must be drawn head-to-tail. Because resultant *EA* has an upward sense, vector *A*–1 must have its tail at *A*, giving it a sense toward point 1. By relating this sense to the free-body diagram, we can see that the sense is toward the point of application, which means that *A*–1 is a compression member. Vector 1–*E* has a sense away from the joint, which means that it is a tension member. The vectors are coplanar and can be scaled to determine their loads as tabulated.

In step 2 we select the next adjacent joint to take advantage of the load found in vector *A*–1. Vectors 1–*A* and *AB* are known, while vectors *B*–2 and 2–1 are unknown. Since there are only two unknowns it is possible to solve for them. A free-body diagram showing the joint to be analyzed is sketched. Vector *B*–2 is drawn parallel to the structural member through point *B* in the Maxwell diagram and the line of vector 2–1 is extended from point 1 until it intersects with *B*–2, where point 2 is located. The sense of each vector is found by laying off each vector head-to-tail. Both vectors *B*–2 and 2–1 have a sense toward the joint in the free-body diagram; therefore, they produce compression. Their magnitudes are scaled and tabulated.

The next joint is analyzed in sequence to find the stresses in 2–3 and 3–*E* (steps 3). The truss will have equal forces on each side, since it is symmetrical and is loaded symmetrically. The total Maxwell diagram is drawn to illustrate the completed work in step 3. If all the polygons in the series do not close at every point with perfect symmetry, there is an error in construction. If the error of closure is very slight, it can be disregarded, since safety factors are generally applied in derivation of working stresses of structural systems to assure safe construction. Arrowheads are usually omitted on Maxwell diagrams, since each vector will have opposite senses when applied to different joints.

FIGURE 10–18. TRUSS ANALYSIS

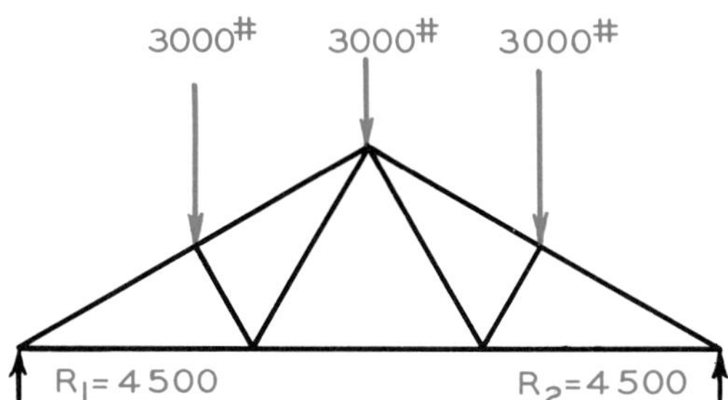

Given: A Fink truss loaded as shown.
Required: Find the stresses in each member of the truss and indicate whether each member is in compression or tension.
Reference: Article 10–12.

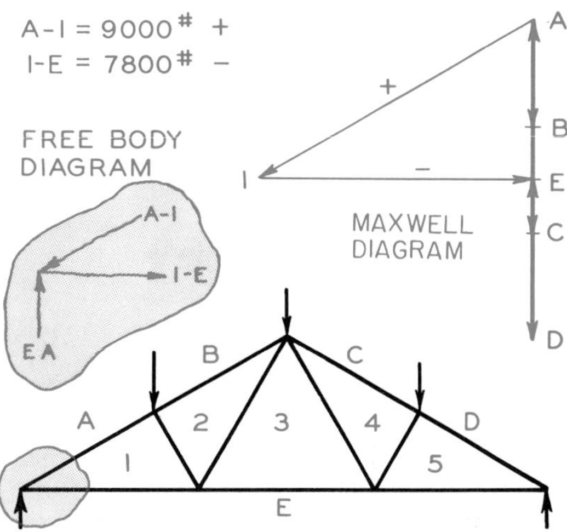

Step 1: Label the portions of supports between the outer forces of the truss with letters and the internal portions with numbers, using Bow's notation. Add the given load vectors graphically in a Maxwell diagram, and sketch a free-body diagram of the first joint to be analyzed. Using vectors *EA*, *A*–1, and 1–*E* drawn head-to-tail, draw a vector diagram to find their magnitudes. Vector *A*–1 is in compression (+) because its sense is toward the joint, and 1–*E* is in tension (−) because its sense is away from the joint.

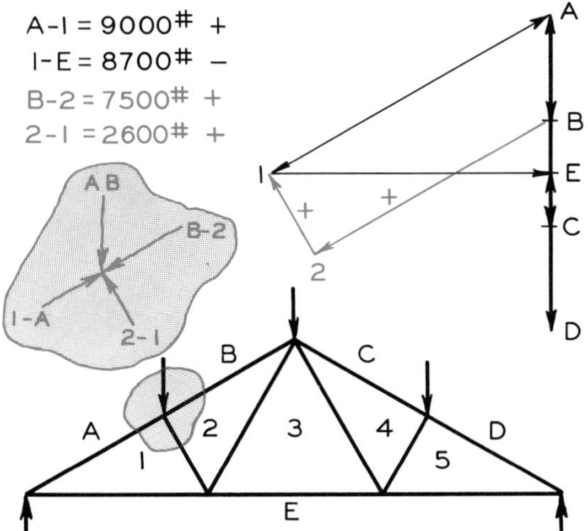

Step 2: Draw a sketch of the next joint to be analyzed. Since *AB* and *A*–1 are known, we have to determine only two unknowns, 2–1 and *B*–2. Draw these parallel to their direction, head-to-tail, in the Maxwell diagram using the existing vectors found in step 1. Vectors *B*–2 and 2–1 are in compression since each has a sense toward the joint. Note that vector *A*–1 becomes 1–*A* when read in a clockwise direction.

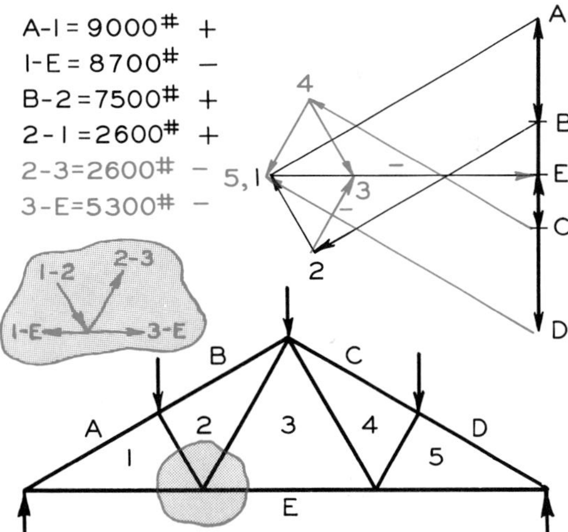

Step 3: Sketch a free-body diagram of the next joint to be analyzed. The unknowns in this case are 2–3 and 3–*E*. Determine the true length of these members in the Maxwell diagram by drawing vectors parallel to given members to find point 3. Vectors 2–3 and 3–*E* are in tension because they act away from the joint. This same process is repeated to find the loads of the members on the opposite side.

The analytical solution to this problem could be found by applying algebraic and trigonometric methods, as discussed in Article 10–11. The graphical method gives a visible indication of error in projection when the polygons do not close.

10–13 NONCOPLANAR STRUCTURAL ANALYSIS—SPECIAL CASE

Structural systems that are three-dimensional require the use of descriptive geometry, since it is necessary to analyze the system in more than one plane. The manned flying system (MFS) in Fig. 10–19 can be analyzed to determine the forces in the support members (Fig. 10–20). Weight on the moon can be found by multiplying earth weight by a factor of 0.165. A tripod that must support 182 lb on earth has to support only 30 lb on the moon. This is a special case that will serve as an introduction to the general noncoplanar problem. Since members *B* and *C* lie in the same plane and appear as an edge in the front view, we need to determine only two unknowns; that is what makes this a special case.

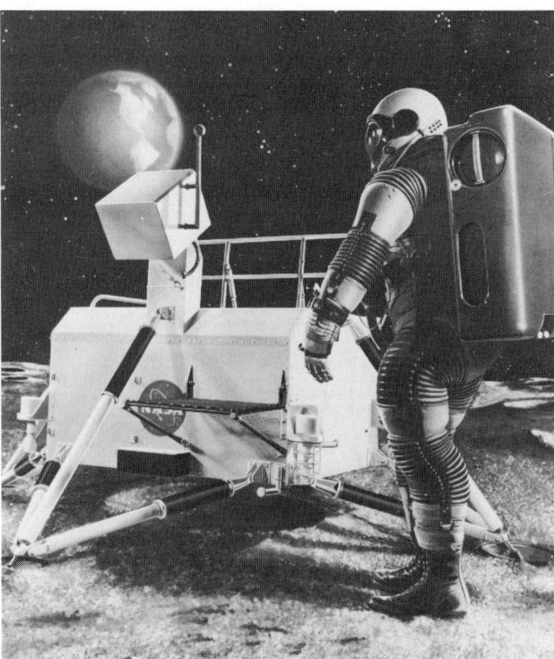

Fig. 10–19. The structural members of this tripod support for a moon vehicle can be analyzed graphically to determine design loads. (Courtesy of NASA.)

A vector polygon is constructed in the front view in step 1 of Fig. 10–20 by drawing force *F* as a vector and using the other vectors as the other sides of the polygon. One of these vectors is actually a summation of vectors *B* and *C*. The top view is drawn using the vectors *B* and *C* to close the polygon from each end of vector *A*. In step 2, the front view of vectors *B* and *C* are found.

The true lengths of the vectors are found in a true-length diagram in step 3. (Refer to Article 5–6 to review construction of true-length diagrams.) The vectors are measured to determine their loads. Vector *A* is found to be in compression because its sense is toward the point of concurrency. Vectors *B* and *C* are in tension.

10–14 NONCOPLANAR STRUCTURAL ANALYSIS—GENERAL CASE

The structural frame shown in Fig. 10–21 is attached to a vertical wall to support a load of $W = 600$ lb. The loads in each member must be determined prior to the selection of the structural shapes. Since there are three unknowns in each of the views, we are required to construct an auxiliary view that will give the edge view of a plane containing two of the vectors, thereby reducing the number of unknowns to two (step 1). We no longer need to refer to the front view. A vector polygon is drawn by constructing vectors parallel to the members in the auxiliary view (step 1). An adjacent orthographic view of the vector polygon is also drawn by constructing its vectors parallel to the members in the top view (step 2). A true-length diagram is used in step 3 to find the true length of the vectors so they can be scaled to determine their magnitudes.

A three-dimensional system is the side-boom tractors used for lowering pipe into a ditch during pipeline construction (Fig. 10–22).

FIGURE 10–20. NONCOPLANAR STRUCTURAL ANALYSIS—SPECIAL CASE

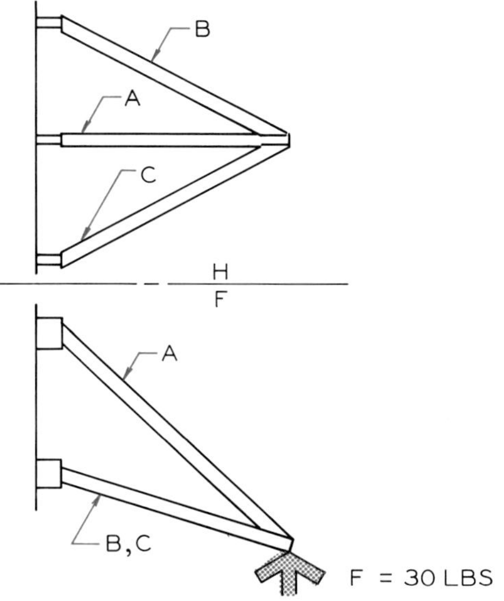

Given: The top and front views of a structural landing gear to support a portion of a craft on the moon. This footing must support 30 lb of force.
Required: Find the forces in each of the structural members, *A*, *B*, and *C*.
Reference: Article 10–13.

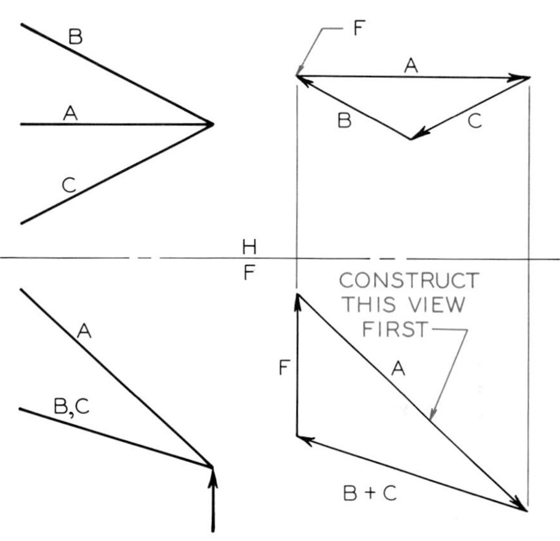

Step 1: Two forces, *B* and *C*, coincide in the front view, resulting in only two unknowns in this view. Vector *F* (30 lb) is drawn, and the other two unknowns are drawn parallel to their front view to complete the front view of the vector polygon. The top view of *A* can be found by projection, from which vectors *B* and *C* can be found.

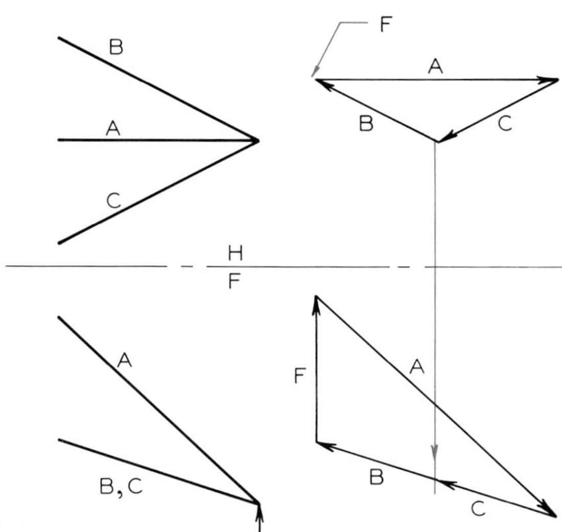

Step 2: The point of intersection of vectors *B* and *C* in the top view is projected to the front view to separate these vectors. All vectors are drawn head-to-tail. Note that the sense of a vector in the polygon is related to the point of application in the space diagram. Vectors *B* and *C* are in tension because their vectors are acting away from the point, while *A* is in compression.

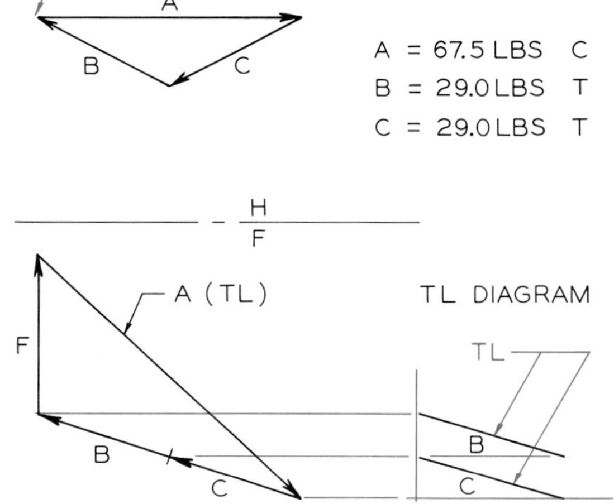

Step 3: The completed top and front views found in step 2 do not give the true lengths of vectors *B* and *C*, since they are oblique. The true lengths of these lines are determined by a true-length diagram. These lines are scaled to find the forces in each member. Refer to Article 5–6 for a review of TL diagrams.

FIGIRE 10–21. NONCOPLANAR STRUCTURAL ANALYSIS—GENERAL CASE

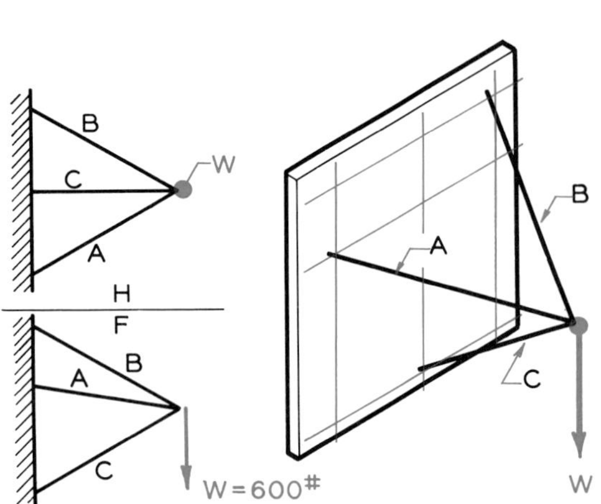

Given: The top and front views of a three-member frame which is attached to a vertical wall in such a way that it can support a maximum weight of 600 lb.
Required: Find the loads in the structural members.
References: Articles 10–12 and 10–14.

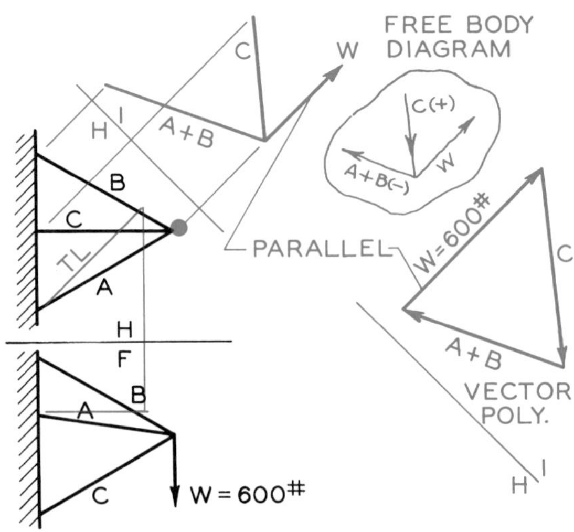

Step 1: To limit the unknowns to two, construct an auxiliary view to find two vectors lying in the edge view of a plane. Use the auxiliary view and top view in the remainder of the problem. Draw a vector polygon parallel to the members in the auxiliary view in which $W = 600$ lb is the only known vector. Sketch a free-body diagram for preliminary analysis.

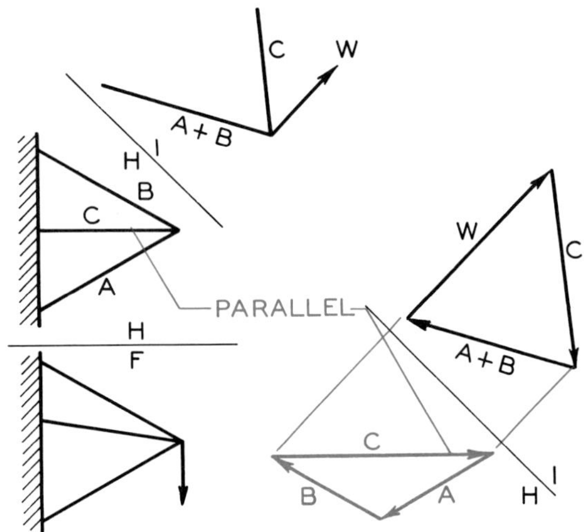

Step 2: Construct an orthographic projection of the view of the vector polygon found in step 1 so that its vectors are parallel to the members in the top view. The reference plane between the two views is parallel to the *H*–1 plane. This portion of the problem is closely related to the problem in Fig. 10–20.

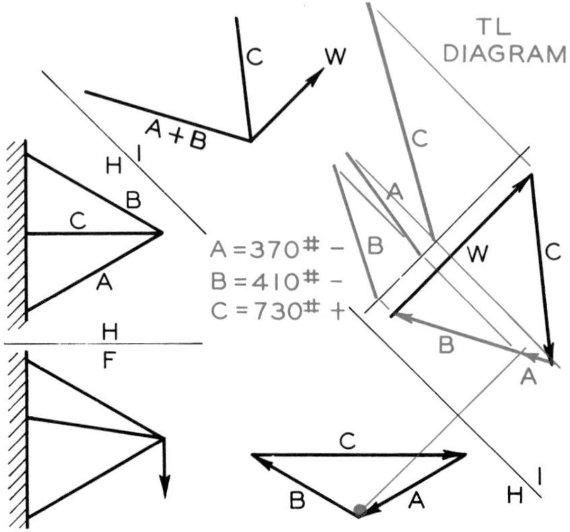

Step 3: Project the intersection of vectors *A* and *B* in the horizontal view of the vector polygon to the auxiliary view polygon to establish the lengths of vectors *A* and *B*. Determine the true lengths of all vectors in a true-length diagram and measure them to determine their magnitudes. Analyze for tension or compression, as covered in Article 10–12.

Fig. 10–22. Tractor sidebooms represent noncoplanar, concurrent systems of forces that can be solved graphically. (Courtesy of Trunkline Gas Company.)

10–15 NONCONCURRENT, COPLANAR VECTORS

Forces *may* be applied in such a manner that they are not concurrent, as illustrated in Fig. 10–23. Bow's notation can be used to locate the resultant of this type of nonconcurrent system.

In step 1, the vectors are laid off to form a vector diagram in which the closing vector is the resultant, $R = 68$ lb. Each vector is resolved into two components by randomly locating point O on the interior or exterior of the polygon and connecting point O with the end of each vector. The components, or strings, from point O are equal and opposite components of adjacent vectors. For example, component o–b is common to vectors AB and BC. Since the strings from point O are equal and opposite, the system has not changed statically.

In step 2, each string is transferred to the space diagram of the vectors where it is drawn between the respective vectors to which it applies. (The figure thus produced is called a *funicular diagram.*) For instance, string o–b is drawn in the area between vectors AB and BC. String o–c is drawn in the C-area to connect at the intersection of o–b and vector BC. The point of intersection of the last two strings, o–a and o–d, locates a point through which the resultant R will pass. The resultant has now been determined with respect to magnitude, sense, direction, and point of application, thus completing the solution of the problem.

FIGURE 10–23. RESULTANT OF NONCONCURRENT FORCES

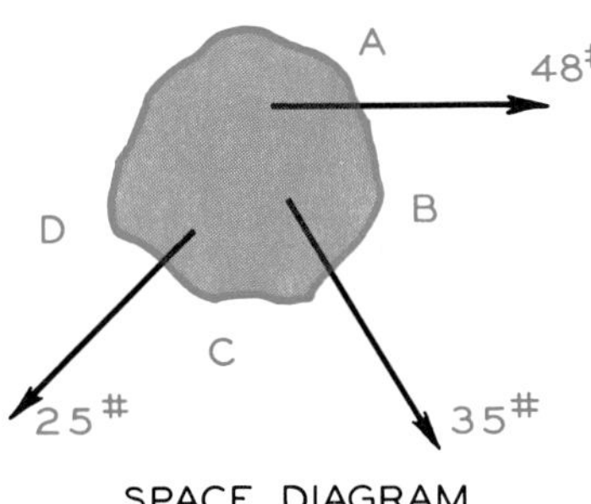

Required: Find the resultant of the known forces applied to the above object. The forces are nonconcurrent.

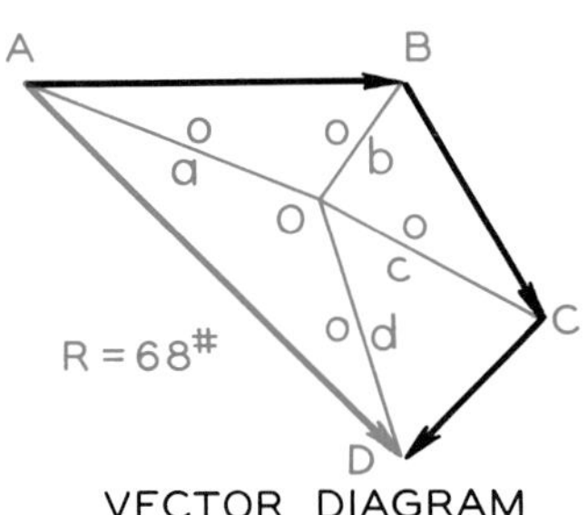

Step 1: The vectors are drawn head-to-tail to find resultant *R*. Point *O* is conveniently located for the construction of strings to the ends of each vector.

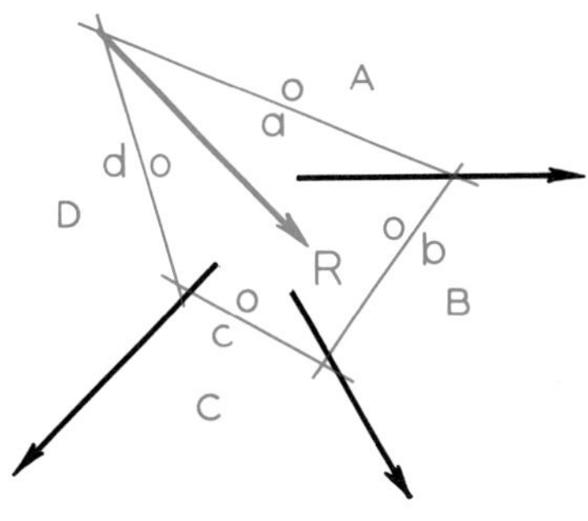

Step 2: Each string is drawn between the two vectors to which it applies in the original space diagram. *Example*: *o–c* between *BC* and *CD*. These strings are connected in sequence until the strings *o–a* and *o–d* establish the position of *R*, which was found in step 1.

10–16 NONCONCURRENT SYSTEMS RESULTING IN COUPLES

A *couple* is the descriptive name given to two parallel, equal, and opposite forces which are separated by a distance and are applied to a member in such a manner that they cause the member to rotate. The handwheels in Fig. 10–24 are examples of mechanical systems which take advantage of this method of force application.

An important quantity associated with a couple is its *moment.* The moment of any force is a measure of its rotational effect. An example is shown in Fig. 10–25, in which two equal and opposite forces are applied to a wheel. The forces are separated by the distance D. The moment of the couple is found by multiplying one of the forces by the perpendicular distance between it and a point on the line of action of the other: $F \times D$. If the force is 20 lb and the distance is 3 ft, the moment of the couple would be given as 60 ft-lb.

A series of parallel forces is applied to a beam in Fig. 10–26. The spaces between the vectors are labeled with letters which follow Bow's notation. We are required to determine the resultant.

Fig. 10–24. Handwheels are designed for operation by the application of forces in the form of couples. (Courtesy of Standard Oil Company.)

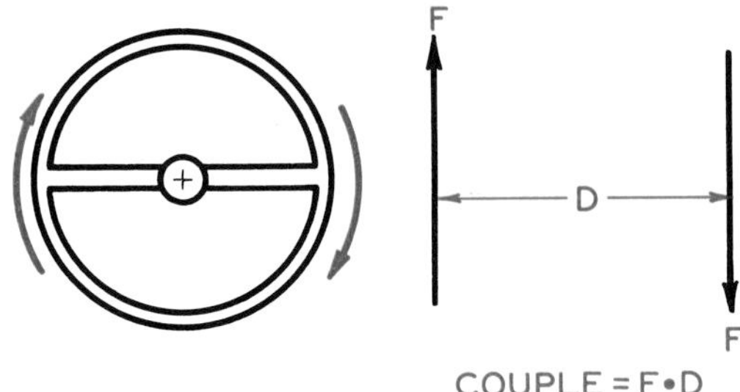

Fig. 10–25. Representation of a couple or moment.

FIGURE 10–26. COUPLE RESULTANTS

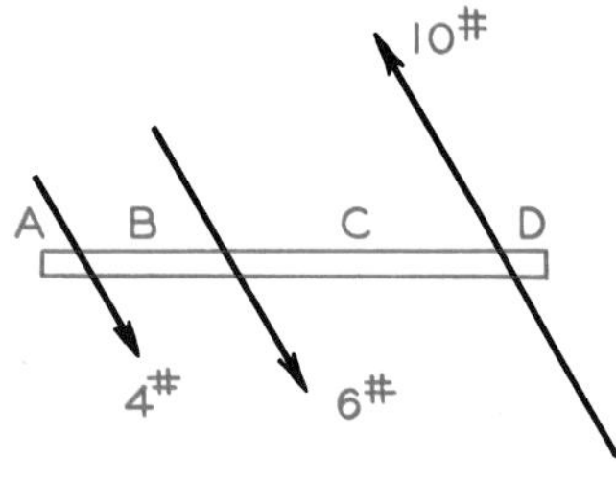

SPACE DIAGRAM

Required: Find the resultant of these nonconcurrent forces applied to this beam.

Step 1: The spaces between each force are labeled in Bow's notation.

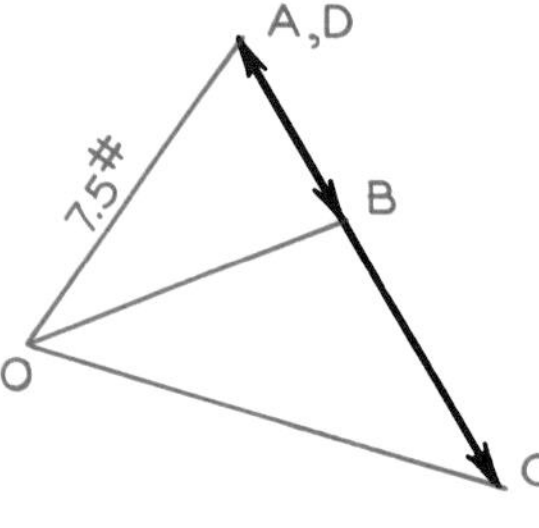

VECTOR DIAGRAM

Step 2: The vectors are laid out head-to-tail; they will lie in a straight line since they are parallel. Pole point O is located in a convenient location and the ends of each vector are connected with point O.

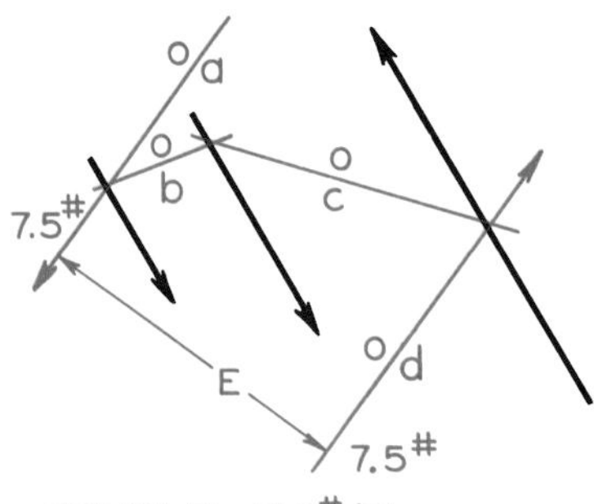

FUNICULAR DIAGRAM

Step 3: Strings *o–a*, *o–c*, and *o–d* are successively drawn between the vectors to which they apply. Since strings *o–a* and *o–d* are parallel, the resultant will be a couple equal to 7.5 lb × E where E is the distance between *o–a* and *o–d*.
o–e.

After constructing a vector diagram, we have a straight line which is parallel to the direction of the forces and which closes at point A. We then locate pole point O and draw the strings of a funicular diagram.

The strings are transferred to the space diagram, where they are drawn in their respective spaces. For example, o–c is drawn in the C-space between vectors BC and CD. The last two strings, o–d and o–a do not close at a common point, but are found to be parallel; the result is therefore a couple. The distance between the forces of the couple is the perpendicular distance, E, between strings o–a and o–d in the space diagram, using the scale of the space diagram. The magnitude of the force is the scaled distance from point O and A and D in the vector diagram, using the scale of the vector diagram. The moment of the couple is equal to 7.5 lb $\times$ E in a counterclockwise direction.

10–17 RESULTANT OF PARALLEL, NONCONCURRENT FORCES

Forces applied to beams, such as those shown in Fig. 10–27, are parallel and nonconcurrent in many instances, and they may have the effect of a couple, tending to cause a rotational motion. When the loads exerted on the beams are known, the magnitude and location of the total resultant or equilibrant can be found. This will provide the designer with a better understanding of where supports should be placed.

The beam in Fig. 10–28 is on a rotational crane that is used to move building materials in a limited area. The magnitude of the weight W is unknown, but the counterbalance weight is known to be 2000 lb; column R supports the beam as shown. Assuming that the support cables have been omitted, we desire to find the weight W that would balance the beam.

This problem can be solved by the application of the law of moments, i.e., the force is multiplied by the perpendicular distance to its line of action from a given point, or $F \times A$. If the beam is to be in balance, the total effect of the moments must be equal to zero, or $F \times A = W \times B$.

Fig. 10–27. The boom of this crane can be analyzed for its resultant as a parallel, nonconcurrent system of forces when the cables have been disregarded.

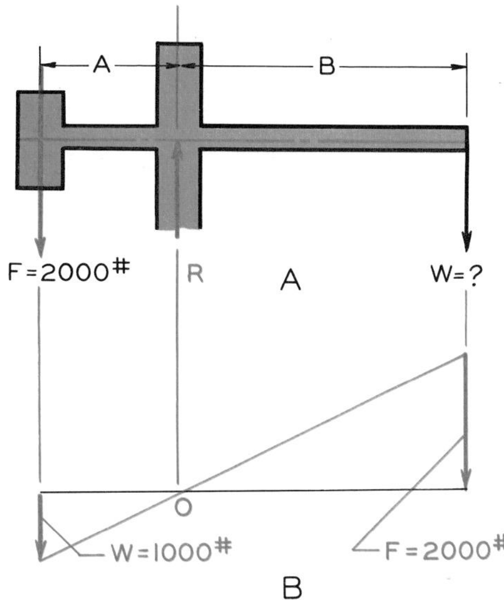

Fig. 10–28. Determining the resultant of parallel, nonconcurrent forces.

The graphical solution (Fig. 10–28B) is found by constructing a line to represent the total distance between the forces F and W. Point O is projected from the space diagram to this line. Point O is the point of balance where the summation of the moments will be equal to

FIGURE 10–29. BEAM ANALYSIS WITH PARALLEL LOADS

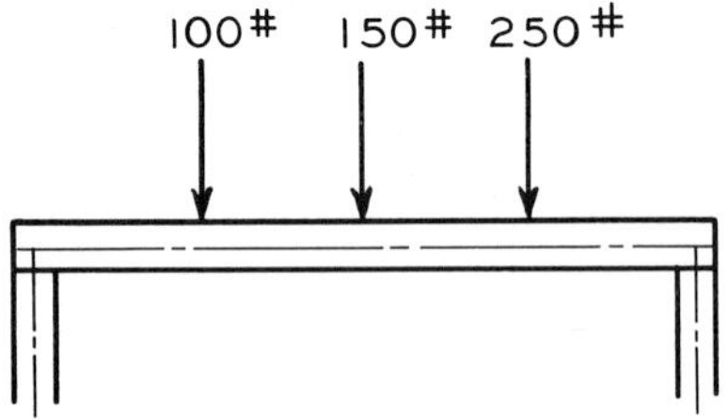

Given: A beam that is loaded with three parallel, unequal loads.
Required: Find the reactions R_1 and R_2 and the total resultant that will replace the parallel loads.
Reference: Article 10–18.

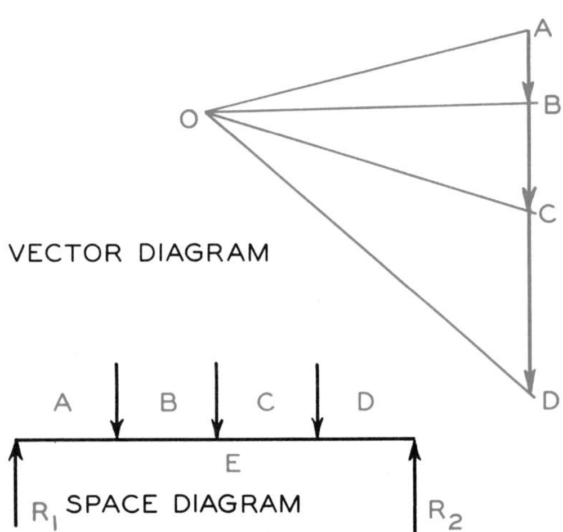

Step 1: Letter the spaces between the loads with Bow's notation. Find the graphical summation of the vectors by drawing them head-to-tail in a vector diagram at a convenient scale. Locate pole point *O* at a convenient location and draw strings from point *O* to each end of the vectors.

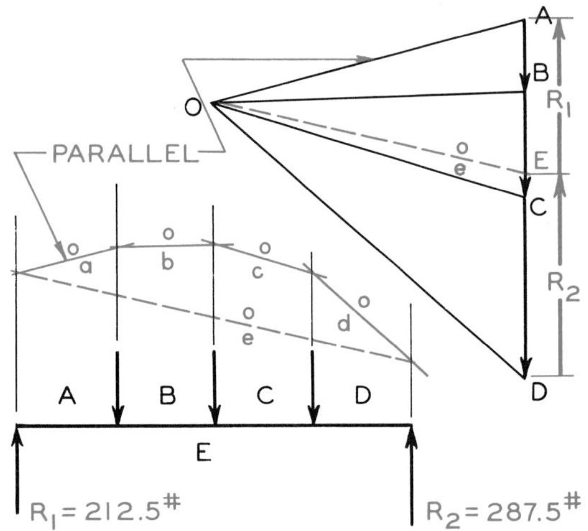

Step 2: Extend the lines of force in the space diagram, and draw a funicular diagram with string *o–a* in the *A*-space, *o–b* in the *B*-space, *o–c* in the *C*-space, etc. The last string, which is drawn to close the diagram, is *o–e*. Transfer this string to the vector polygon and use it to locate point *E*, thus establishing the lengths of R_1 and R_2, which are *EA* and *DE*, respectively.

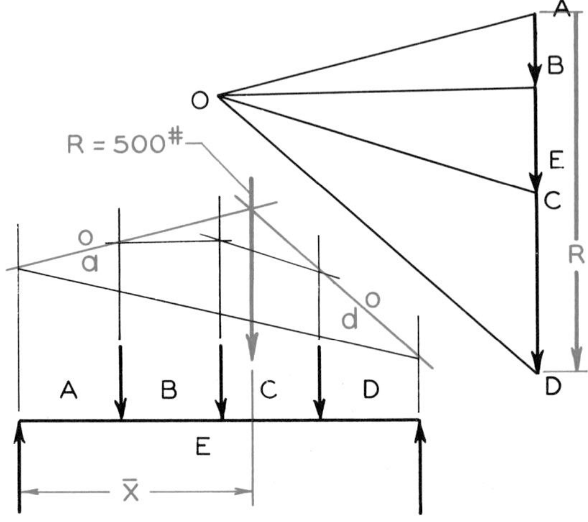

Step 3: The resultant of the three downward forces will be equal to their graphical summation, line *AD*. Locate the resultant by extending strings *o–a* and *o–d* in the funicular diagram to a point of intersection. The resultant $R = 500$ lb will act through this point in a downward direction. $\bar{X}$ is a locating dimension.

zero. Vectors F and W are drawn to scale at each end of the line by transposing them to the opposite ends of the beam. A line is drawn from the end of vector F through point O and extended to intersect the direction of vector W. This point represents the end of vector W, which can be scaled, resulting in a magnitude of 1000 lb.

10–18 RESULTANT OF PARALLEL, NON-CONCURRENT FORCES ON A BEAM

When two or more forces are applied to a beam that is supported at more than one point, a somewhat different approach is taken to locate the resultant of the system.

The beam given in Fig. 10–29 is supported at each end and must in turn support three given loads. We are required to determine the magnitude of each support, R_1 and R_2, along with the resultant of the loads and its location. The spaces between all vectors are labeled in a clockwise direction with Bow's notation in the space diagram in step 1.

In step 2 the lines of force in the space diagram are extended and the strings from the vector diagram are drawn in their respective spaces, parallel to their original direction. *Example*: String *oa* is drawn parallel to string *OA* in space *A* between forces *EA* and *AB*, and string *ob* is drawn in space *B* beginning at the intersection of *oa* with vector *AB*. The last string, *oe*, is drawn to close the funicular diagram. The direction of string *oe* is transferred to the force diagram, where it is laid off through point *O* to intersect the load line at point *E*. Vector *DE* represents support R_2 (refer to Bow's notation as it was applied in step 1). Vector *EA* represents support R_1.

The magnitude of the resultant of the loads (step 3) is the summation of the vertical downward forces, or the distance from *A* to *D*, or 500 lb. The location of the resultant is found by extending the extreme outside strings in the funicular diagram, *oa* and *od*, to their point of intersection. The resultant is discovered to have a magnitude of 500 lb, a vertical direction, a downward sense, and a point of application established by $\bar{X}$. This location would be important to an engineer if he intended to locate a third support under the beam.

Fig. 10–30. Structural beams are examples of parallel, non-concurrently loaded beams. (Courtesy of Jones & Laughlin Steel Corporation.)

Figure 10–30 shows a number of beams that had to be analyzed to determine their resultants and reactions. In a structure of this type, most of the forces applied are in a vertical direction, and the support members are also vertical.

10–19 SUMMARY

The product design example which has been used to illustrate the application of principles covered in each chapter is used here to illustrate the application of graphical principles to its analysis. The problem is restated below.

Hunting Seat—Problem. Many hunters, especially deer hunters, hunt from trees to obtain a better vantage point. Sitting in a tree for several hours can be uncomfortable and hazardous to the hunter, thus indicating a need for a hunting seat that could be used to improve this situation. Design a seat that would provide the hunter with comfort and safety while he is hunting from a tree and that would meet the general requirements of economy and limitations of hunting.

Hunting Seat—Analysis. The refinement of the design was discussed in Chapter 6 to illustrate how graphical methods apply to that stage

PROBLEMS

Problems should be presented in instrument drawings on $8\frac{1}{2}'' \times 11''$ paper, grid or plain, using the format introduced in Article 3–6. Each grid square represents $\frac{1}{4}''$. All notes, sketches, drawings, and graphical work should be neatly prepared in keeping with good practices. Written matter should be legibly lettered using $\frac{1}{8}''$ guidelines.

1. In Fig. 10–32A, determine the resultant of the force system by the parallelogram method at the left of the sheet. Solve the same system using the vector polygon method at the right of the sheet. Scale: $1'' = 100$ lb (note that each grid square equals $\frac{1}{4}''$). (B) In part B of the figure, determine the resultant of the concurrent, coplanar force system shown at the left of the sheet by the parallelogram method. Solve the same system using the polygon method at the right of the sheet. Scale $1'' = 100$lb.

2. (A and B) In Fig. 10–33, solve for the resultant of each of the concurrent, noncoplanar force systems by the parallelogram method at the left of the sheet. Solve for the resultant of the same systems by the vector polygon method at the right of the sheet. Find the true length of the resultant in both problems. Letter all construction. Scale: $1'' = 600$ lb.

3. (A and B) In Fig. 10–34, the concurrent, coplanar force systems are in equilibrium. Find the loads in each structural member. Use a scale of $1'' = 300$ lb in part A and a scale of $1'' = 200$ lb in part B. Show and label all construction.

4. In Fig. 10–35, solve for the loads in the structural members of the truss. Vector polygon scale: $1'' = 2000$ lb. Label all construction.

5. In Fig. 10–36, solve for the loads in the structural members of the concurrent, noncoplanar force system. Find the true length of all vectors. Scale: $1'' = 300$ lb.

6. In Fig. 10–37, solve for the loads in the structural members of the concurrent, noncoplanar force system. Find the true length of all vectors. Scale: $1'' = 400$ lb.

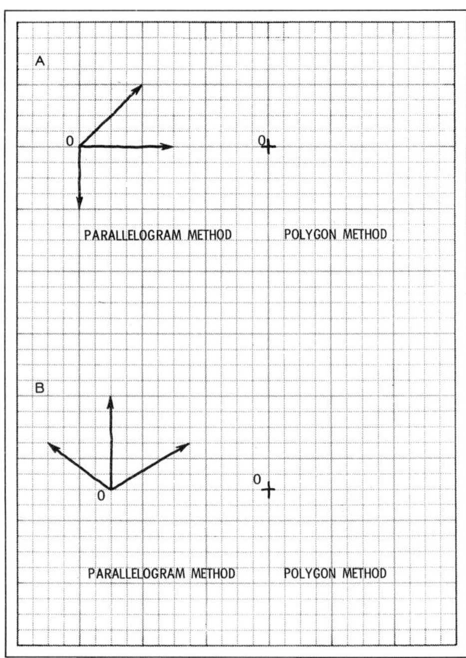

Fig. 10–32. Resultant of concurrent, coplanar vectors.

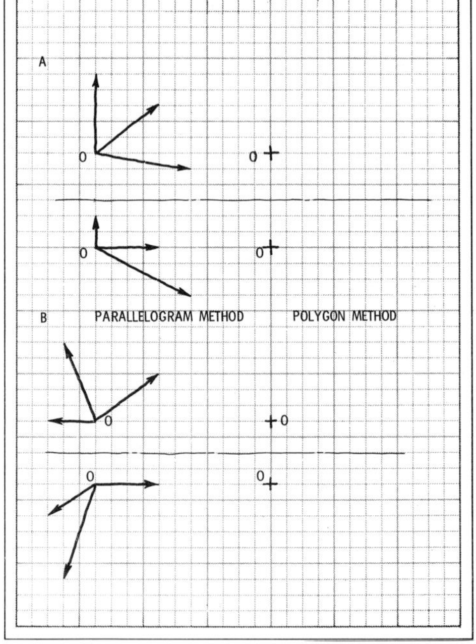

Fig. 10–33. Resultant of concurrent, noncoplanar vectors.

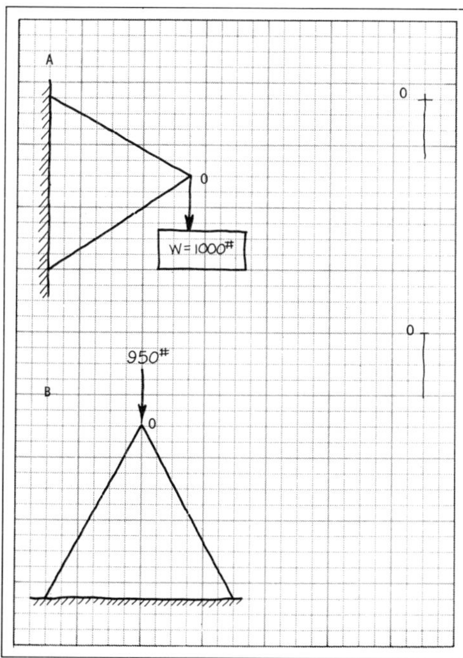

Fig. 10–34. Coplanar, concurrent forces in equilibrium.

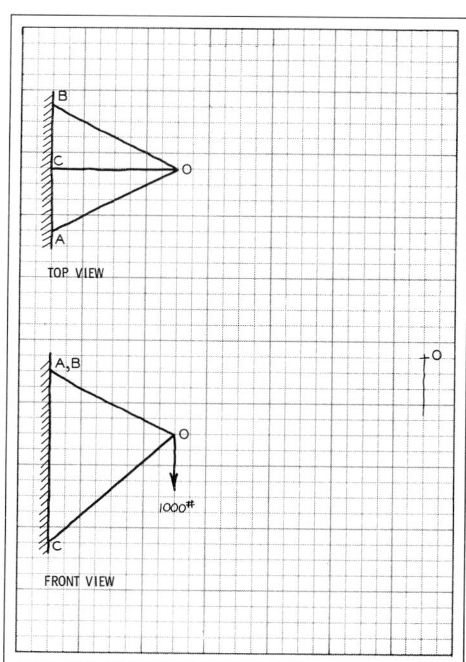

Fig. 10–36. Noncoplanar, concurrent forces in equilibrium.

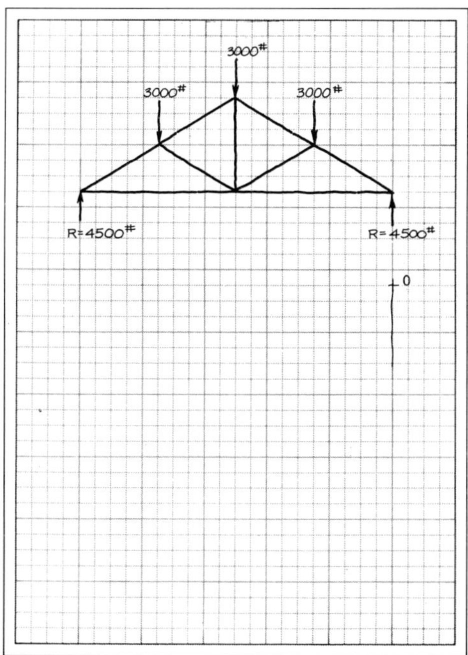

Fig. 10–35. Truss analysis.

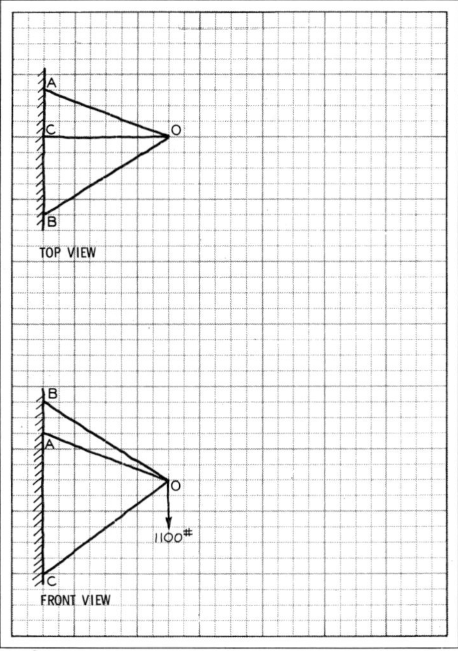

Fig. 10–37. Noncoplanar, concurrent forces in equilibrium.

7. (A) In Fig. 10–38, find the resultant of the coplanar, nonconcurrent force system. The vectors are drawn to a scale of 1″ = 100 lb. (B) In part B of the figure, solve for the resultant of the coplanar, nonconcurrent force system. The vectors are given in their true positions and at the true distances from each other. The space diagram is drawn to scale of 1″ = 1.0′. Draw the vectors to a scale of 1″ = 30 lb. Show all construction.

8. (A) In Fig. 10–39, determine the force that must be applied at *A* to balance the horizontal member supported at *B*. Scale 1″ = 100 lb. (B) In part B of the figure, find the resultants at each end of the horizontal beam. Find the resultant of the downward loads and determine where it would be positioned. Scale: 1″ = 600 lb.

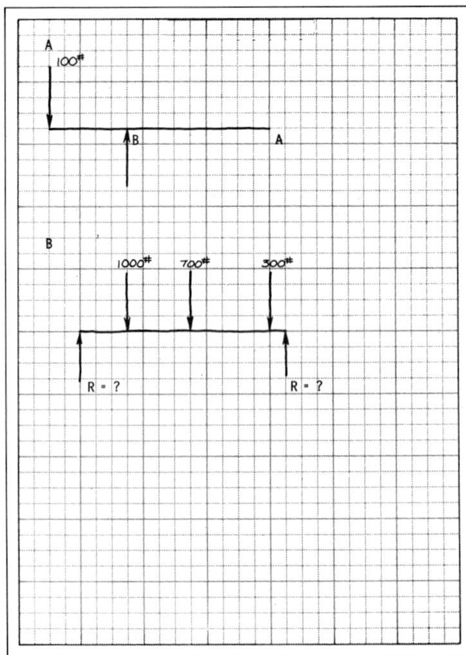

Fig. 10–39. Beam analysis.

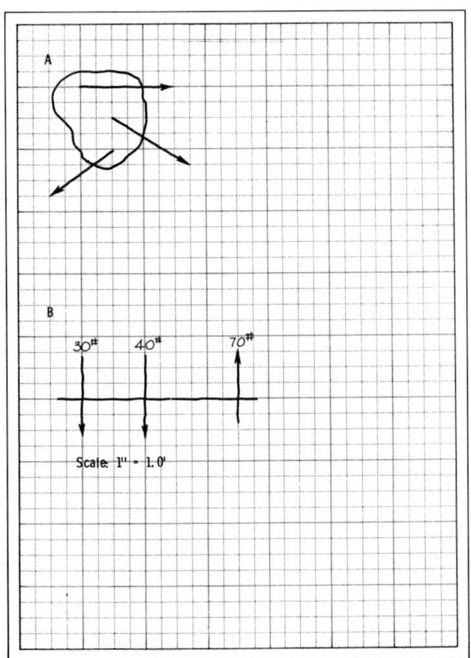

Fig. 10–38. Coplanar, nonconcurrent forces.

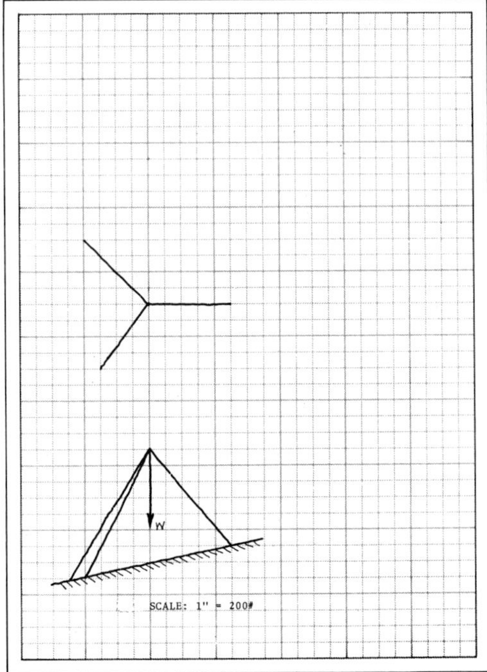

Fig. 10–40. Beam analysis.

9. Determine the forces in the three members of the tripod in Fig. 10–40. The tripod supports a load of $W = 250$ lb. Find the true lengths of all vectors.

10. The vectors in Fig. 10–41 each make an angle of 60° with the structural member on which they are applied. Find the resultant of this force system. Refer to Article 14–16.

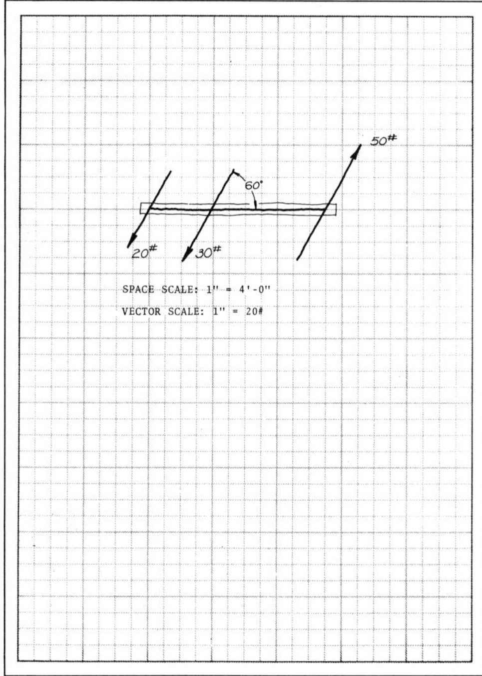

Fig. 10–41. Noncoplanar, concurrent forces in equilibrium.

11

Analysis of Design Data

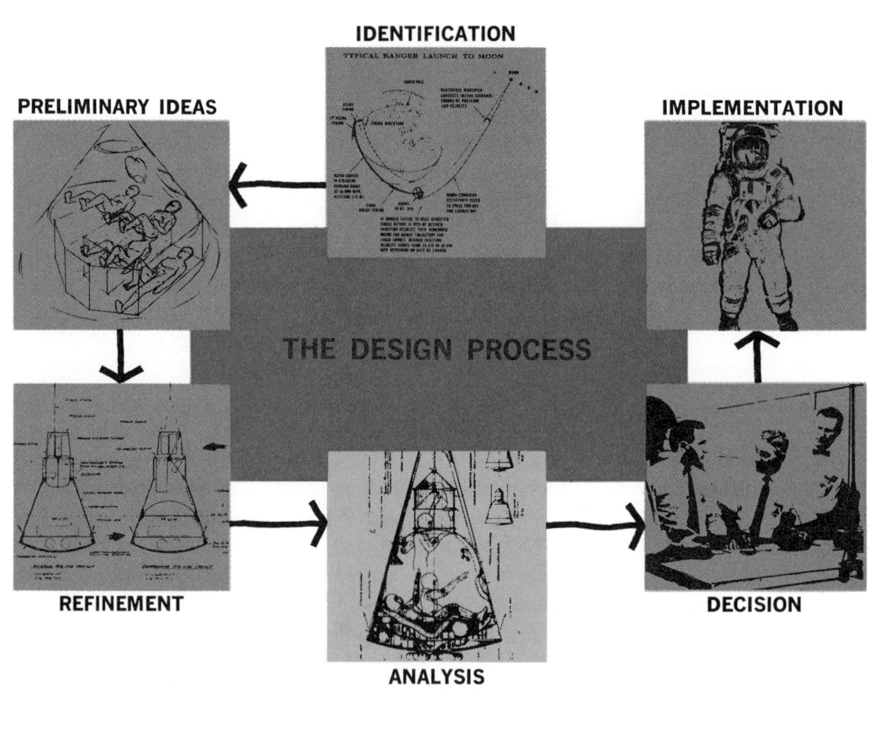

Fig. 11–1. Engineering aspects of a toy engine are discussed by an industrial team. All aspects of the design and its analysis must be considered in detail before the design can be accepted for mass production. (Courtesy of Mattel, Inc.)

Fig. 11–2. Most products are tested extensively as a means of gathering data for analysis of the design; this testing is also important to effective quality control. (Courtesy of Mattel, Inc.)

11–1 GENERAL

Before a proposed design is accepted, it must be subjected to careful analysis. During this process, data provided in many forms must be evaluated and interpreted. Most frequently, data are submitted in numerical form whose interpretation is often a lengthy and difficult procedure. Thus, to ensure that each member of the design team understands all aspects of the project (Fig. 11–1), it is customary to convert numerical data to a more convenient form.

Design information may be analyzed in a variety of forms. The more fundamental ones are (1) graphs, (2) empirical equations, (3) mechanisms, (4) graphical calculus, and (5) nomograms. We shall discuss these areas of analysis and the graphical techniques used to improve their analysis.

Many data that are obtained from laboratory experiments or physical relationships can be expressed in terms of mathematical equations. This approach is helpful in that it establishes mathematical relationships that might not be apparent in the initial data. Graphical techniques can be used to advantage in determining the equation form of empirical data when such an equation exists. Using mathematical and analytical procedures (Fig. 11–2), mechanisms can be analyzed graphically for motion, function, clearance, and interference. Calculus problems can be solved graphically within the limits of reasonable accuracy. Thus the designer has at his disposal a variety of graphical procedures to supplement his analytical approach to studying a design.

11–2 INTRODUCTION TO GRAPHS

Any design can be evaluated to a considerable degree by reviewing the data that pertain to it. These data may fall into one of the following different categories: (1) field data, (2) market data, (3) design-performance data, and (4) comparative data.

Field data may affect a design directly or indirectly. A traffic engineer must gather information about traffic flow, driving habits, peak periods of volume, and traffic speed before he can prepare a new design for a traffic system at a given location. On-the-site observations and counts are made during representative periods and tabulated. Often, field data can be obtained from existing agencies, such as records of average temperatures, rainfall, and other weather data usually maintained by local weather departments.

Market data are evaluated to determine the probable acceptance of an engineering project whether it is a supersonic aircraft or a household appliance. It is necessary to obtain information about the characteristics of the prospective users of the design, such as numbers, needs, average incomes, and so on. Data about existing competition in the field are also of considerable value. Data concerning populations, incomes, areas of population density, and statistical information are available from the Department of Labor and the U. S. Census Bureau, as well as state and local agencies.

Design performance must be studied to determine the effectiveness of a finished design. Frequently, a prototype is constructed specifically for testing the operation of the design prior to all-out production. Extreme conditions that a design is likely to be exposed to must be simulated prior to actual exposure, since it may not be possible to test the product under actual conditions without considerable danger and expense. The organization of these data into graphical form permits efficient analysis and evaluation.

Comparative data are used to establish relationships between two or more variables to improve the chances for making a correct decision. For example, to choose between two machines that will be used to produce the same product, the engineer must compare the operational expenses required by each and their relative outputs, as well as the predicted life of each machine and its estimated maintenance expense. For specific applications, the engineer may compare the advantages of one material versus those of another or the effectiveness of one fuel versus that of another.

11–3 TYPES OF GRAPHS

The types of graphs emphasized in this chapter are primarily those used to analyze data that will aid in the final decision on a design. Although graphs do not make decisions or solve problems, they give the designer a picture of the background information and thus help him to familiarize himself with all aspects of the problem.

The basic types of graphs are (1) linear (including rectangular, logarithmic, and semilogarithmic grids), (2) bar graphs, (3) pie or circular graphs, (4) polar graphs, (5) schematics and graphical diagrams, and (6) computation graphs and nomograms.

11–4 LINEAR GRAPHS—RECTANGULAR GRIDS

The linear graph is the one most commonly used to present information either to the general public or to a group of technically oriented people. Graphs of this kind may be drawn in their entirety, including the grid, or the data may be plotted on commercially prepared graph paper. A typical graph is shown in Fig. 11–3; the more important parts are properly labeled. A graph should be prepared with the same care and precision that would be exercised in any other portion of the design.

11–5 DRAWING A LINEAR GRAPH

A graph similar to the example in Fig. 11–3 is drawn in Fig. 11–4 to illustrate the steps of construction and the important elements of a rectilinear graph. The data for this graph are given in tabular form in the first part of the figure.

Selection of a Grid. A graph can be drawn on commercially printed graph paper, or the grid may be drawn to fit the specific needs of a particular graph. If the graph is to be reproduced in a formal report or a report that is to be published in large quantities, the entire graph should be drawn in ink.

To permit easy analysis and a minimum of clutter, only important divisions on the graph should be shown. Judgment must be used in selecting the scales to be assigned along the horizontal axis (*abscissa* or *X-axis*) and the vertical axis (*ordinate* or *Y-axis*). Variation in either of these scales can exaggerate or minimize fluctuations.

Plotting the Data. The data should be plotted on the graph with symbols such as circles, triangles, rectangles, or crosses to indicate the

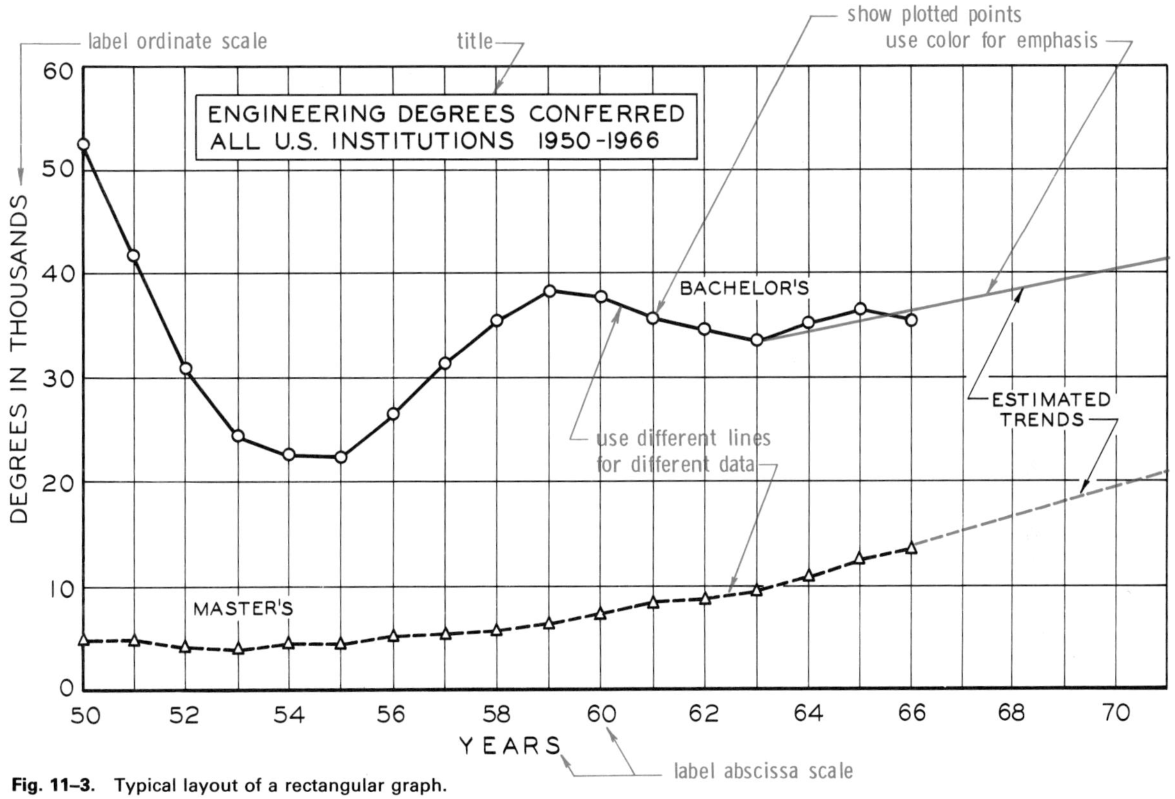

ENGINEERING DEGREES CONFERRED ALL U.S. INSTITUTIONS 1950-1966

Fig. 11–3. Typical layout of a rectangular graph.

FIGURE 11–4. CONSTRUCTION OF A BROKEN-LINE GRAPH

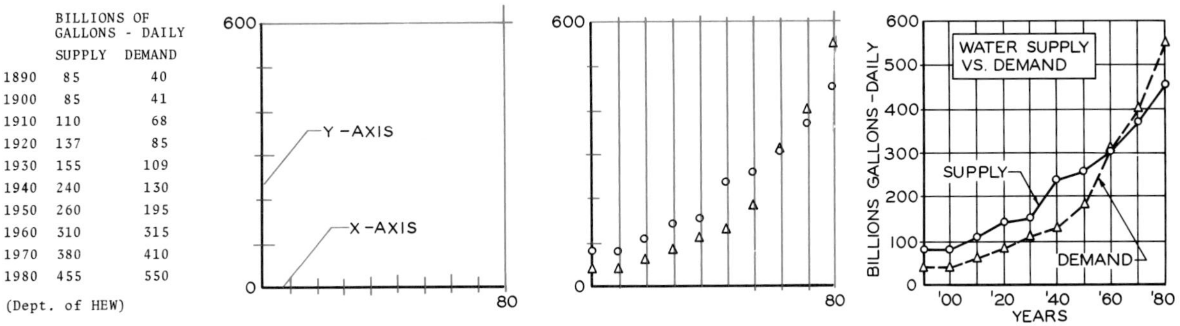

	BILLIONS OF GALLONS - DAILY	
	SUPPLY	DEMAND
1890	85	40
1900	85	41
1910	110	68
1920	137	85
1930	155	109
1940	240	130
1950	260	195
1960	310	315
1970	380	410
1980	455	550

(Dept. of HEW)

Given: A record of water supply and water demand since 1890 has been obtained to determine the future relationships that may occur. These data are to be plotted as a line graph.

Step 1: The vertical and horizontal axes are laid off to provide adequate space for the years and the largest values.

Step 2: The points are plotted directly over the respective years. Different symbols are used for each curve.

Step 3: The data points are connected with straight lines, the axes are labeled, the graph is titled, and the lines are strengthened.

actual data values used. These symbols should be drawn with a template to ensure uniformity.

Drawing the Curve. The data presented on a linear graph will be in either of two forms, *discrete* or *continuous.* Discrete data are connected with straight lines to give a broken-line appearance. The data in Fig. 11–4 are discrete because there is no continuous change in the given data from year to year between the plotted points; that is, the supply and demand did not change at a smooth, continuous rate. Consequently the points are connected with straight lines. The discrete data points of Fig. Fig. 11–3 were also connected with straight lines, since no uniform rate of change between points on the graph could be assumed. For example, when you say that 52,000 students graduated in 1950 and 42,000 in 1951 you cannot assume that there was a uniform, continuous reduction of students during this one-year period.

Continuous data are connected with a smooth curve from point to point, rather than a broken-line curve. The curves in Fig. 11–5 represent continuous data, since it is understood that there is an infinite number of speeds between 30 mph and 31 mph, for example, 31.01 mph, 31.02 mph, 31.03 mph, and so on.

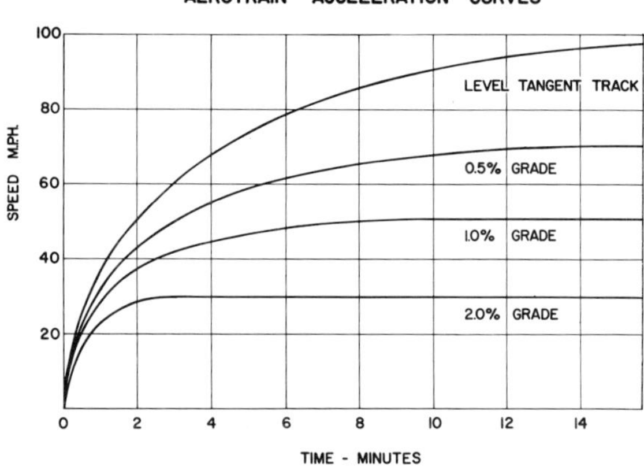

Fig. 11–5. A rectangular graph for presenting continuous data is connected with a smooth curve. (General Motors Corporation, *Engineering Journal* **3,** No. 4 (1956), p. 15.)

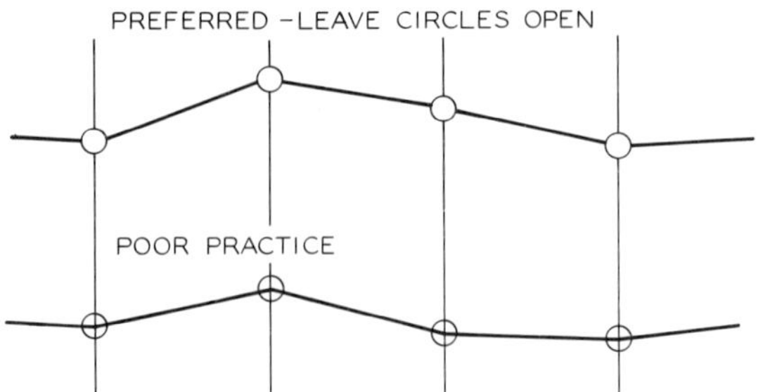

Fig. 11–6. When a curve is drawn through points on a graph, the curve should not overlap the symbols used to plot the points.

To attain a speed of 30 mph, one must pass through every speed from 0 to 30 mph in a continuous order.

When a curve is drawn through points on a graph, the line of the curve should not overlap the symbols so that they cannot be seen (Fig. 11–6).

Labeling the Axes. The independent variable is customarily plotted along the horizontal axis and the dependent variable along the vertical axis. The dependent variable is best evaluated if its initial point, where the two axes intersect, is zero.

Only the important divisions on a graph should be labeled. The values to be labeled should be chosen so that interpolation of values between those that are labeled will be easy. Familiar multiples of numbers should be used, such as 2,4, 6, etc., or 0, 5, 10, 15, etc. In addition, each axis should be labeled in general terms to define clearly what the units represent. In Fig. 11–4, these axes are labeled "Billions Gallons—Daily" and "Years."

Title. All graphs should have a title (or caption) that will clearly identify the graph and its contents. The title can usually be placed within the grid area of the graph to conserve space and to give a pleasing appearance. When placed on the grid, the title should be surrounded by a box (Fig. 11–4). Where space does not permit inclusion within the grid area, the title should be located prominently at the top or bottom of the graph.

11-6 APPLICATIONS OF LINEAR GRAPHS WITH RECTANGULAR GRIDS

General Graphs. The graph in Fig. 11–7 compares the variation in cost of premium gasoline with that of methanol. Notice that each broken-line curve is labeled, the vertical and horizontal scales are labeled, and a title is given to identify the data being shown.

The graph in Fig. 11–8 is an example of continuous data plotting. The percent of compressive strength is plotted against the number of days of curing time for portland cement. This is a gradual, continuing process; therefore, the data are continuous and the points are connected with a smooth curve.

The design of an automobile's power system is easily analyzed by referring to Fig. 11–9. Four types of data are compared to indicate the usable horsepower available at various velocities. The horsepower *available* at the rear wheels versus the horsepower *required* at the rear wheels is the critical information that must receive first study, since this factor will determine the performance of the automobile. The optimum speed is between 50 and 65 mph, which is the average highway driving speed.

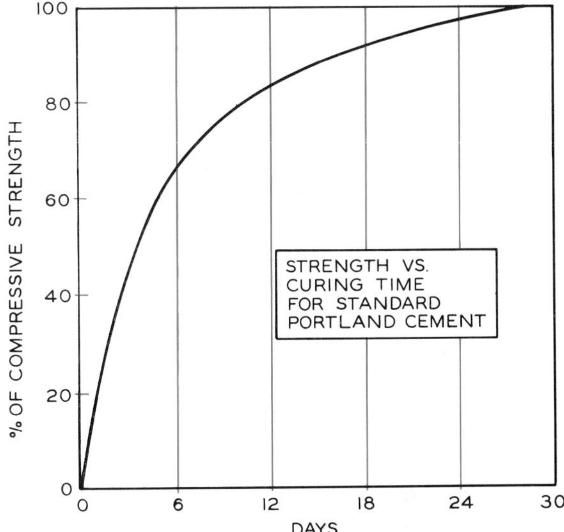

Fig. 11–8. When the process that is graphed involves gradual, continuous changes of relationships, the curve should be drawn as a smooth line.

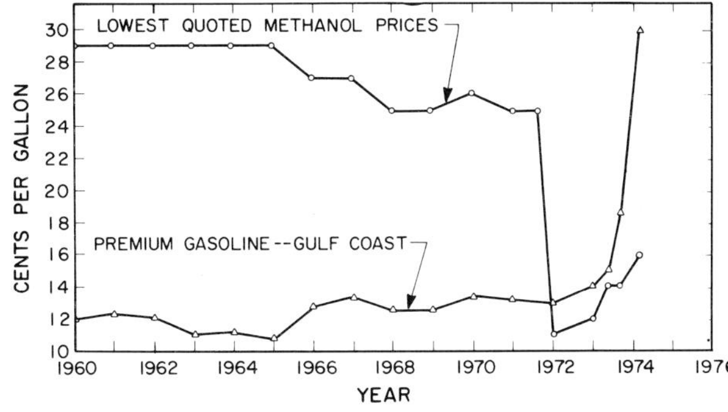

Fig. 11–7. A typical rectangular graph with a broken-line curve. (Courtesy of *The Oil and Gas Journal.*)

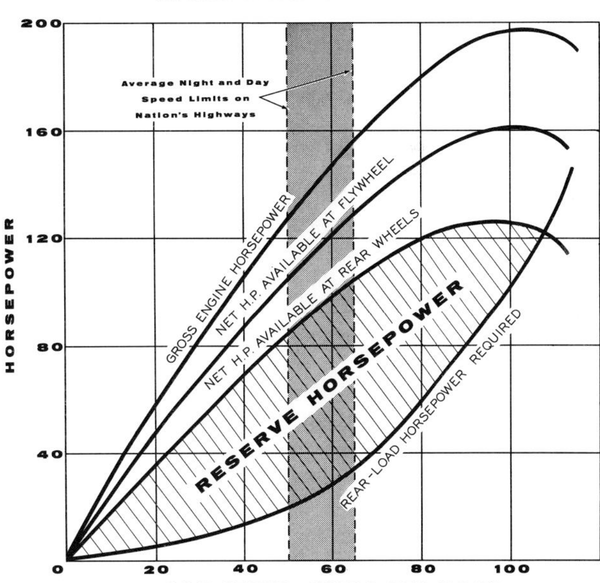

Fig. 11–9. A rectangular graph used to analyze data affecting the design of an automobile's power system. (Courtesy of General Motors Corporation.)

The Best Curve. Some data points may have built-in errors due to the measuring instruments used or to slightly faulty methods of collecting the data. When it is known that the data should produce a smooth, continuous relationship, the curve is drawn as the *best curve.* The best curve cannot pass through each point that has been plotted; instead, it represents an approximation of the data as if there were no errors.

The data points plotted in Fig. 11–10 are experimental data obtained from field tests of two engines. To obtain a smooth curve, it was necessary to draw the curve through some of the points and near others. The compressive strength of structural clay tile is related to the absorption characteristics of this material in Fig. 11–11. The curve of this graph does not pass through the points, but represents the average trend shown by the somewhat scattered data.

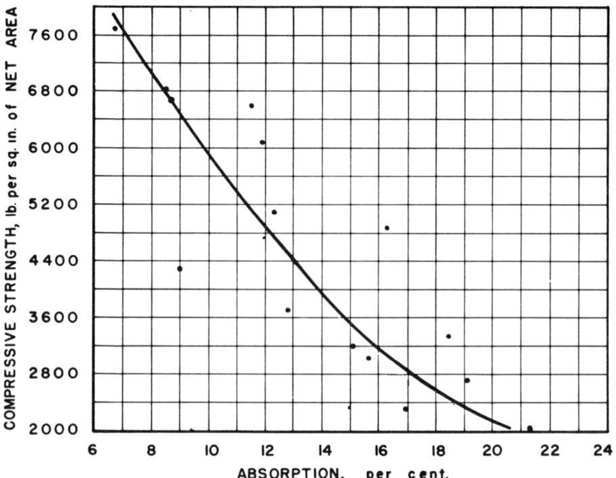

Fig. 11–11. An example of an approximate curve that represents scattered data points. (Courtesy of the Structural Clay Products Institute.)

Break-Even Graphs. Rectilinear graphs are useful in analyzing the marketing and manufacturing costs that will be involved in the development of a product. A type of break-even graph is constructed in steps in Fig. 11–12. The Y-axis represents thousands of dollars; the X-axis, shows units in thousands to be manufactured. Once the cost of development, design, and planning has been figured ($20,000 in this case), the curve can be plotted. The manufacturer estimates that he can produce the product at $1.50 per unit if 10,000 are produced. Thus the amount of $15,000 is added to the $20,000 cost at the 10,000-unit division in step 1. If the manufacturer wishes to break even after 10,000 are sold, the selling price must be $3.50 each. The break-even point is plotted, and a straight line is drawn connecting this point to the origin. This line is extended past the break-even point to the edge of the grid and is labeled "Gross Income." If zero units are produced, the cost will be the $20,000 spent on development. At the break-even point, by definition, the cost will equal the gross income (step 3). Consequently, a line connecting the $20,000 point on the Y-axis with the break-even point will represent the manufacturer's costs as the number of units produced is increased from 0 to 10,000. This cost line should be extended to the opposite edge of the grid. The distances between the income and

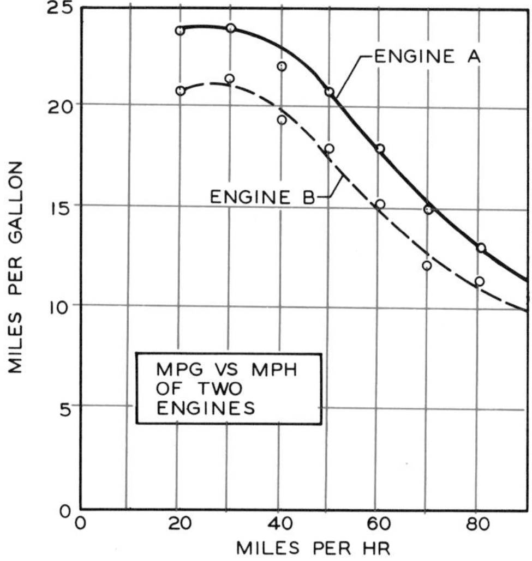

Fig. 11–10. These curves are "best curves," which approximate the data without necessarily passing through each data point.

FIGURE 11–12. BREAK-EVEN GRAPH

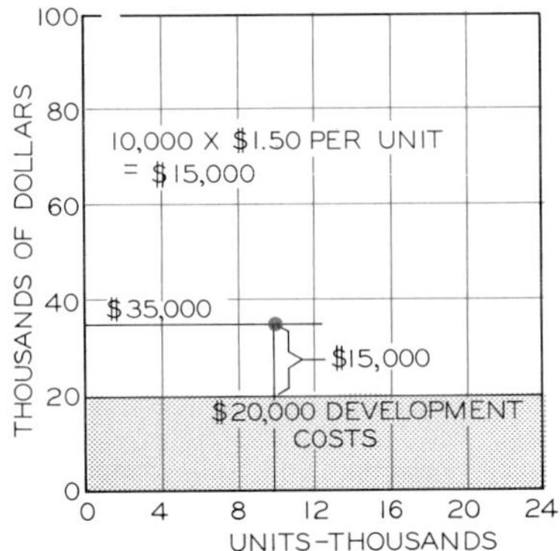

Step 1: The graph is drawn to show the cost ($20,000 in this case) of developing the product. It is determined that each unit would cost $1.50 to manufacture if the total quantity were 10,000. This is a total investment of $35,000 for 10,000 units.

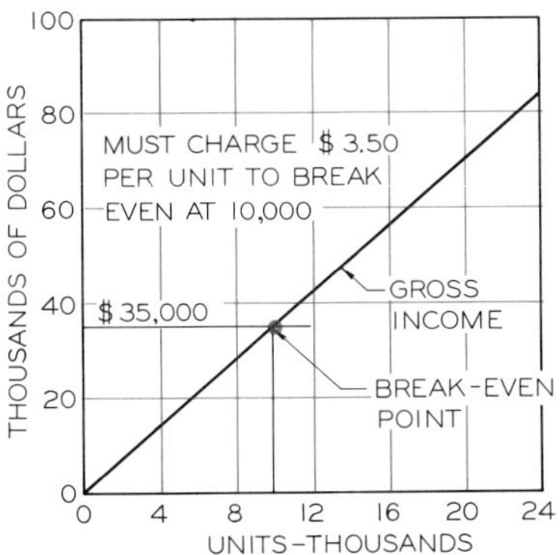

Step 2: In order for the manufacturer to break even at 10,000, the units must be sold for $3.50 each. Draw a line from zero through the break-even point for $35,000.

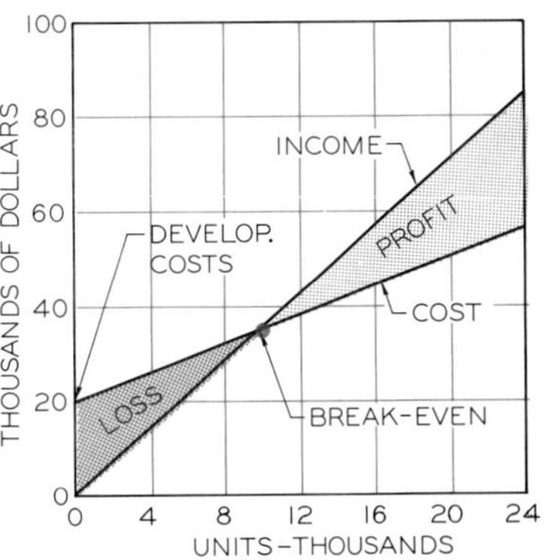

Step 3: The manufacturer's loss is $20,000 at zero units and becomes progressively less until the break-even point is reached. The profit is the difference between the cost and income to the right of the break-even point.

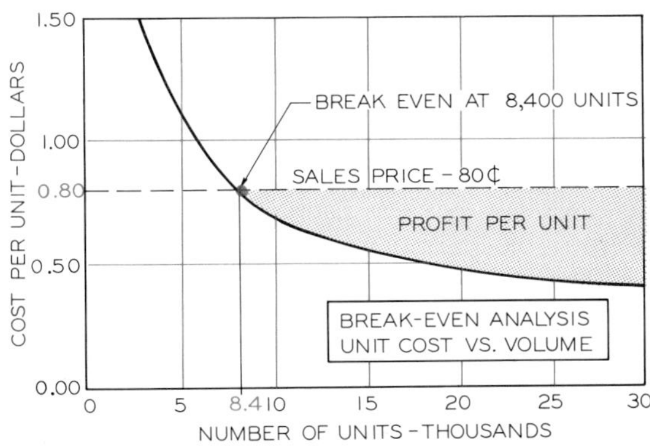

Fig. 11–13. The break-even point can be found on a graph that shows the relationship between the cost per unit, which includes the development cost, and the number of units produced. The sales price is a fixed price. The break-even point is reached when 8400 units have been sold at 80¢ each.

cost lines represent the manufacturer's losses and profits.

A second type of break-even graph (Fig. 11–13) uses the cost of manufacture per unit of a product versus the number of units produced. In this example, the development costs must be incorporated into the unit costs. The manufacturer can determine how many units must be sold to break even at a given price, or the price per unit if a given number is selected. In this example, a sales price of 80¢ requires that 8400 units be sold to break even.

11–7 LOGARITHMIC GRAPHS

The logarithmic graph is a type of rectangular graph in which the scales are graduated with logarithmic divisions along the ordinate and the abscissa. Commercially prepared logarithmic graphs are available in many forms and cycles to fulfill most needs.

Data that vary from small to very large numbers can be presented on logarithmic graphs in less space than would be required by a conventional rectangular grid. There is no zero point on this type of graph, just as there is no zero on a slide rule scale. Each cycle is raised by a factor of 10. For example, in Fig. 11–14, the ordinate begins at 10 and ends at 100 for the first cycle. The second cycle is from 100 to 1000, but only 200 is shown at the top of the graph. The abscissa is a two-cycle grid.

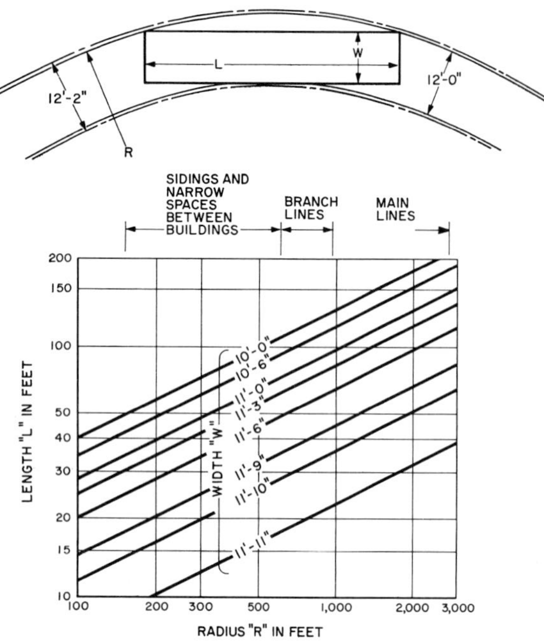

Fig. 11–14. This logarithmic graph shows the maximum load projection of 12 feet in relation to the length of a railroad car and the radius of the curve. (Courtesy of *Plant Engineering.*)

The curves plotted in this graph were derived from a number of calculations involving the geometry of standard railroad tracks and the sizes of cars used on them. The graph can be used to determine the relationship of various curves in a railroad to the lengths of railroad cars if a maximum projection width of 12 feet is allowed.

11–8 SEMILOGARITHMIC GRAPHS

The semilogarithmic graph is referred to as a ratio graph or a rate-of-change graph because one scale, usually the vertical scale, is logarithmic, while the other, usually the horizontal scale, is arithmetic (divided into equal divisions). Whereas the arithmetic graph gives a picture of absolute amounts of change, the semilogarithmic graph shows the relative rate of change. These two types of graphs are shown in Fig. 11–15, where the same data are plotted on each type of grid. The rate of change on the

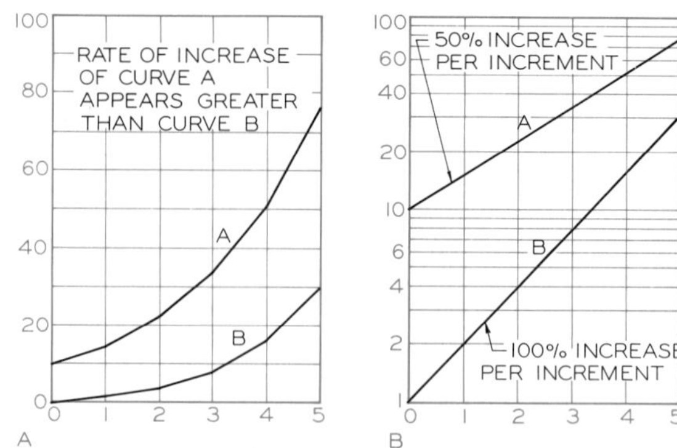

Fig. 11–15. When plotted on a standard grid, curve *A* appears to be increasing at a greater rate than curve *B*. However, the true rate of increase can be seen when the same data are plotted on a semilogarithmic graph in part B.

arithmetic graph can be computed from a common base point, but not from point to point as in the case of the semilogarithmic graph. Note that curve *A* in Fig. 11–15A appears to be increasing at a greater rate than curve *B*; however, the true comparison between the rates of changes of the two curves is shown in part B of the figure. Curve *A* increases 50 percent each increment marked on the *X*-axis and 125 percent each double period. Curve *B* increases 100 percent each period, 300 percent each double period.

The relationship between the arithmetic scale used on the conventional rectilinear graph and the logarithmic scale used on the semilogarithmic graph can be seen in Fig. 11–16. Note that the equal divisions along the arithmetic scale have unequal ratios, and that the unequal divisions along the logarithmic scale have equal ratios. Examples of three-cycle, two-cycle, and one-cycle grids are shown in Fig. 11–17. Each cycle increases in magnitude by a factor of 10. When logarithmic scales are needed on a graph, commercially printed grid paper can be used in transferring dimensions to a scale of any length, as shown in Fig. 11–17C.

It can be seen in Fig. 11–18A that the numbers along a logarithmic scale are separated by the difference of their logarithms. The length of the scale is multiplied by the log of each number to find its location above the number

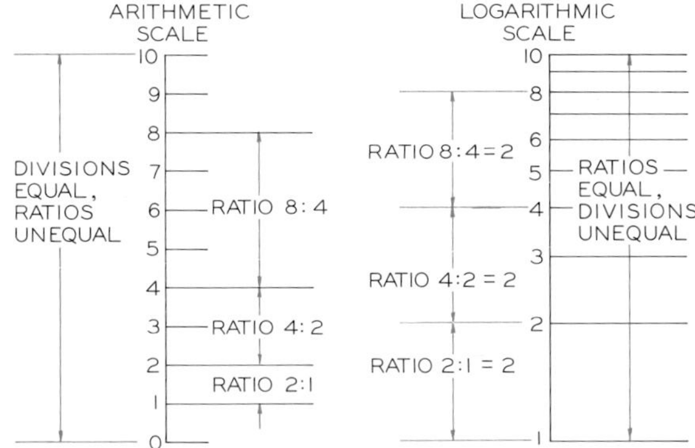

Fig. 11–16. The spacings on an arithmetic scale are equal, with unequal ratios between points. The spacings on logarithmic scales are unequal, but equal spaces represent equal ratios.

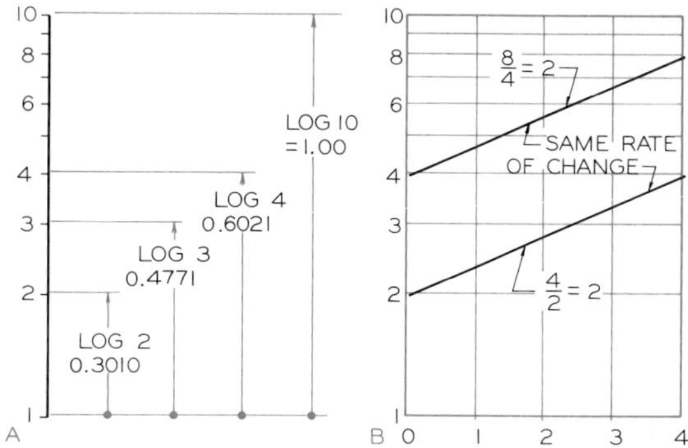

Fig. 11–18. A number's logarithm is used to locate its position on a log scale (A). This makes it possible to see the true rate of change at any location on a semilogarithmic graph (B).

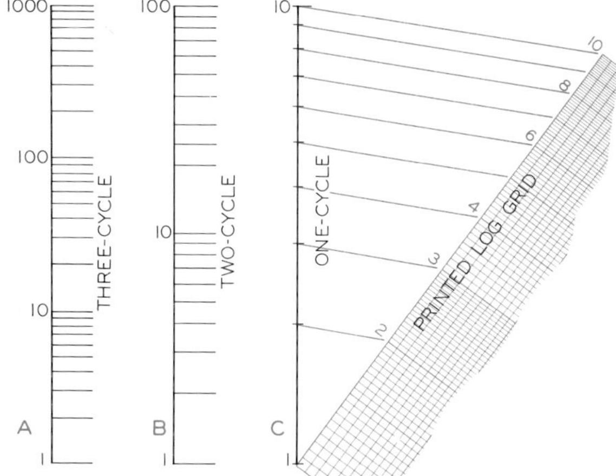

Fig. 11–17. Logarithmic paper can be purchased or drawn using several cycles. Three-, two-, and one-cycle scales are shown here. Calibrations can be drawn on a scale of any length by projecting from a printed scale as shown in part C.

one. This makes it possible to determine the angle of the line which represents the rate of change of the data. It can be seen in Fig. 11–18B that parallel slopes represent equal rates of change; this was not the case with the arithmetic graph in Fig. 11–15A.

The semilogarithmic graph has certain fundamental advantages and disadvantages that must be considered before the type of grid best suited to the purpose is chosen. The advantages are:*

1. The semilogarithmic graph presents a picture that cannot be shown on an arithmetic scale chart.
2. It converts absolute data into a relative comparison, without computing.
3. It shows the relative change from any point to any succeeding point in a series.
4. It retains the actual units of measurement of the absolute data.
5. It reveals whether or not the data follow a consistent relative-change pattern.

The disadvantages must also be considered:*

1. The semilogarithmic graph presents a picture that many people misunderstand and mistakenly read as an arithmetic graph.
2. It cannot be used for data that include a zero or a negative value.
3. It does not provide a scale from which percentage changes can be read directly.
4. It requires a comparison of angles of change, which are difficult to compare by eye.
5. It gives a percentage decrease at a different angle of change than the same percentage increase.

---

* Extracted from ANSI Time-Series Charts (ANSI; Y15.2–1960).

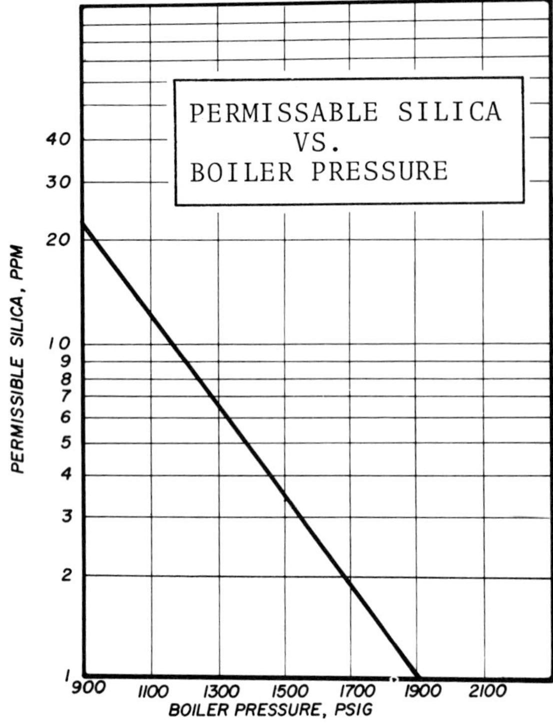

Fig. 11–19. A semilogarithmic graph is used to compare the permissible silica (parts per million) in relation to the boiler pressure. (Courtesy of *Power Engineering.*)

The same general methods that are used to construct a rectilinear arithmetic graph are applied to a semilogarithmic graph (Article 11–5). Plotted points should be indicated on the graph by circles or other geometric symbols to provide a visual impression of the actual data.

An example of a semilogarithmic graph is given in Fig. 11–19.

Percentage Charts. Special applications of semilogarithmic grids can take advantage of the logarithmic plots of the data. The percent that one value is of another value on the graph can be determined with a pair of dividers with no calculations. Likewise, the percent of increase of one value with respect to another value can be found with similar ease.

Data are shown plotted in step 1 of Fig. 11–20. To find the percent that 30 is of 60, measure the vertical distance between these two points. This distance is the difference between the logarithms of these numbers; it is transferred with dividers and is subtracted from

FIGURE 11–20. PERCENTAGE GRAPHS

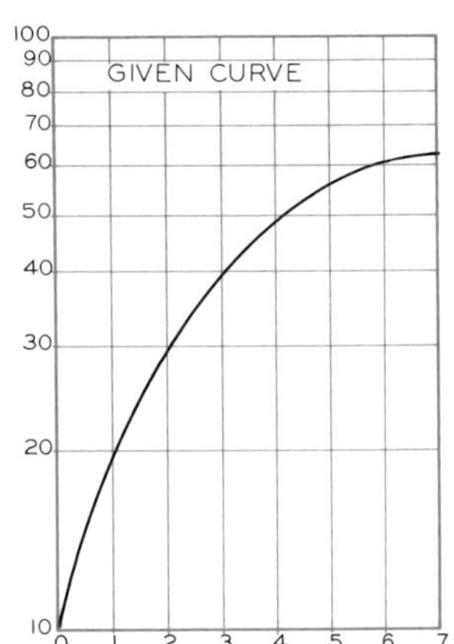

Given: The data are plotted on a semilogarithmic graph to enable you to determine percentages and ratios in much the same manner that you use a slide rule.

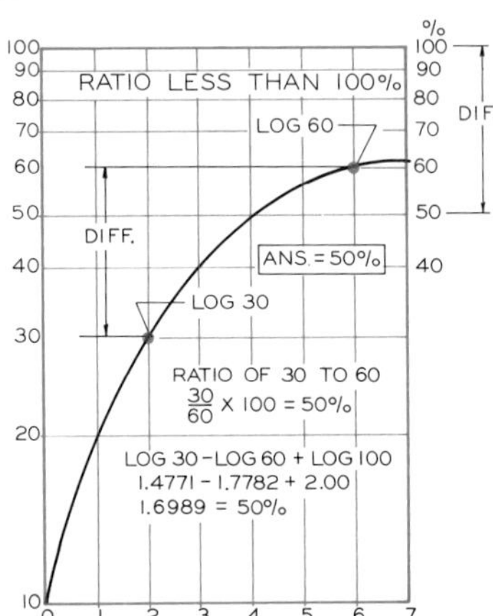

Step 1: In finding the percent that a smaller number is of a larger number, you know that the percent will be less than 100%. The log of 30 is subtracted from the log of 60 with dividers and this dimension is transferred to the percent scale at the right, where 30 is found to be 50% of 60.

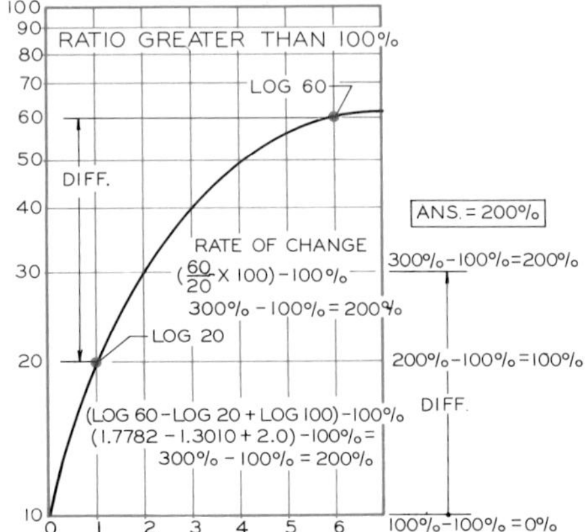

Step 2: To find the rate of increase, a smaller number is divided into a larger number to give a value greater than 100%. The difference between the logs of 60 and 20 is found with dividers, and this distance is measured upward from 100% at the right, to find that the rate of increase is 200%.

the log of 100 at the right of the graph. This gives a value of 50%, which is obviously correct. Relate this procedure to the operation of the log scales of your slide rule, which gives the same answer.

In step 2, the rate of increase is found in much the same manner, but the percent increase is measured upward from the other end of the log scale, since the increase will be greater than 100 percent. The origin is considered to be 100 percent; consequently 100 percent is subtracted from the percent of increase that is found at the right of the scale.

You can see that for a working graph used to determine percentage relationships, this ease of calculation has many advantages.

11–9 BAR GRAPHS

Bar graphs are commonly used to compare a wide variety of variables, since they are readily understood by the general public. The bars may be vertical or horizontal (Fig. 11–21). Bar graphs are easier to interpret if the bars are arranged either in descending or ascending order according to their lengths or in chronological order. Often the amounts represented by the bars are also given numerically to provide exact information. The steps of constructing a bar graph are illustrated in Fig. 11–22.

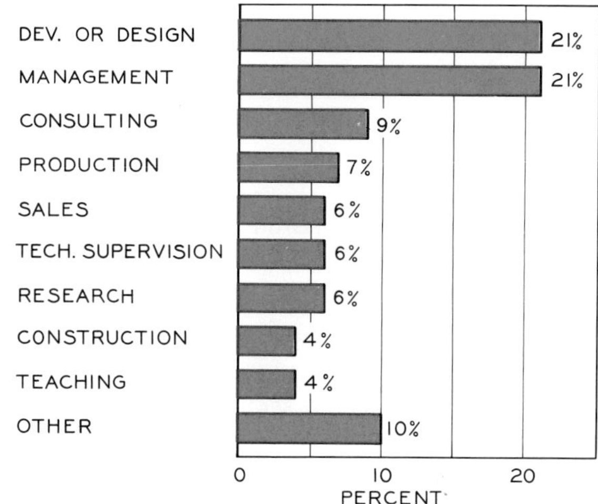

Fig. 11–21. The employment functions of engineers are shown in a bar graph. (Courtesy of the U.S. Department of Labor.)

Figure 11–23 is an application of a bar graph to performance levels of various transportation systems. The space between the bars and the widths of the bars should *not* be the same so that the bars can easily be distinguished.

FIGURE 11–22. CONSTRUCTION OF A BAR GRAPH

DIVIDENDS PAID BY THE AJAX COMPANY

YEAR	AMOUNT
A	$.44
B	.63
C	1.03

Given: These data are to be plotted as a bar graph.

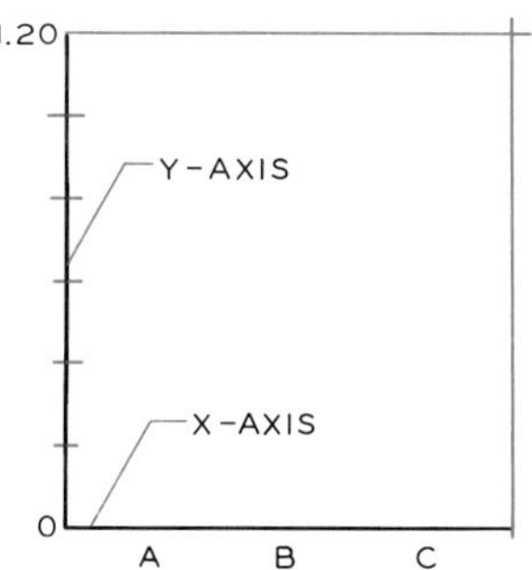

Step 1: Lay off the vertical and horizontal axes so that the data will fit on the grid. Make the bars begin at zero.

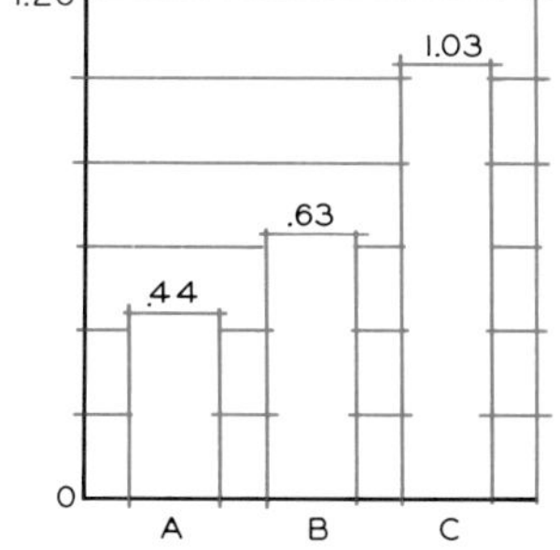

Step 2: Construct and label the bars. The width of the bars should be different from the space between the bars. Horizontal grid lines should not pass through the bars.

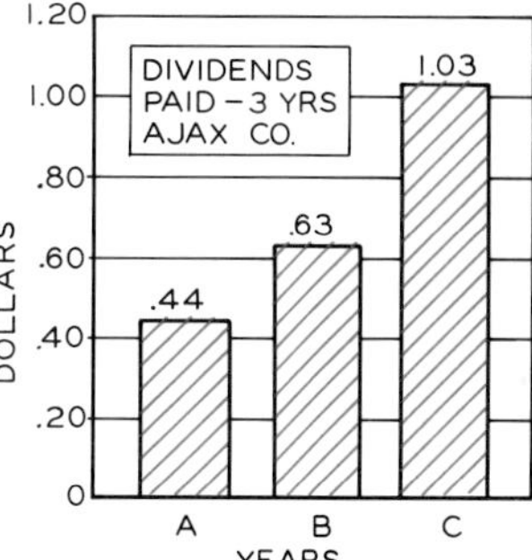

Step 3: Strengthen lines, title the graph, label the axes, and crosshatch the bars.

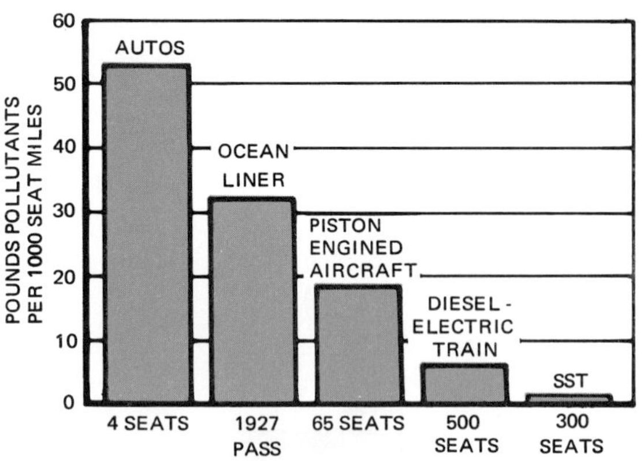

Fig. 11–23. A bar graph is used here to compare the pollution caused by various types of transportation systems. (Courtesy of Boeing Company.)

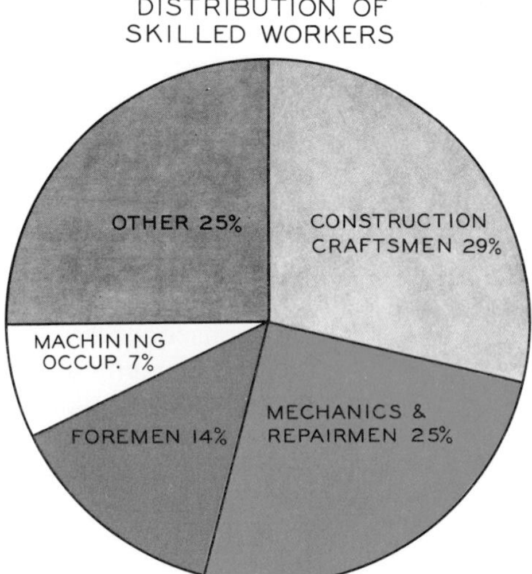

Fig. 11–24. The distribution of skilled workers presented in a pie graph. (Courtesy of the U.S. Department of Labor.)

11–10 PIE GRAPHS

Pie graphs are used to compare the relationship of parts to a whole when there are not very many parts. Figure 11–24 shows the distribution of skilled workers employed in industry. The sectors are found by determining the percentage each part is of the whole and multiplying by 360°. For example, 25 percent of 360° is 90°, which is the size of the sector for mechanics and repairmen. To facilitate lettering within narrow spaces, the narrow sectors should be placed in a horizontal position. Where there is not enough space, labels may be placed outside the sector. The actual percentages should be given in all cases, and, depending on the use of the graph, it may be desirable to give the actual numbers involved. Pie graphs are often used to present the expenditure of budgeted funds and other information to the general public.

11–11 POLAR GRAPHS

Polar graphs are composed of a series of concentric circles with the origin at the center. Lines are drawn from the center toward the perimeter of the graph, where data can be plotted through 360° by measuring quantities from the origin. The illumination of a lamp is shown in Fig. 11–25. The maximum lighting of the lamp is 550 lumens at 35° from vertical. This form of graph is commonly used to plot the areas of illumination of lighting fixtures. Polar graph paper is available commercially.

11–12 SCHEMATICS

Designs and complicated systems may be more easily analyzed if schematics are used to separate major components. Figure 11–26 is a block diagram schematic that is useful in describing steps of a project. The diagram in Fig. 11–27 illustrates the design of a pressure gauge. Note that neither of these diagrams is drawn to scale or with any great degree of detail. Instead they are kept simple and symbolic in order to emphasize the relationship of the components of the system. Similar schematics can be used to present more detailed components within each major section of the schematic. Diagrams of this type are used to illustrate various steps in production, or personnel organization, or any related sequence of components or activities.

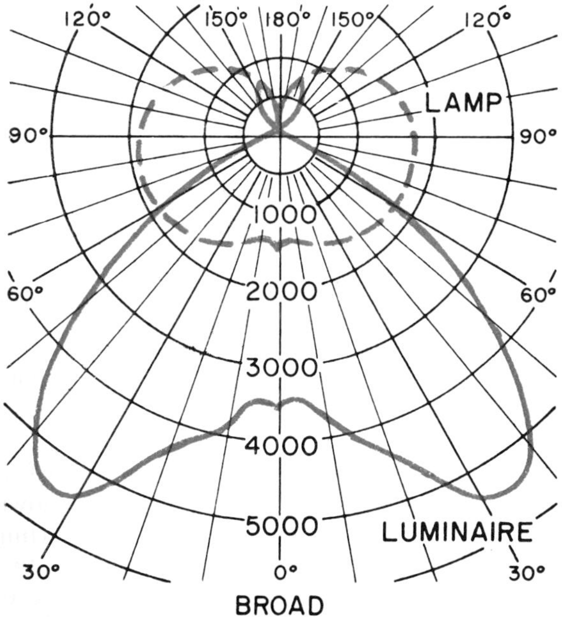

Fig. 11–25. A polar graph is used to show the illumination characteristics of luminaires.

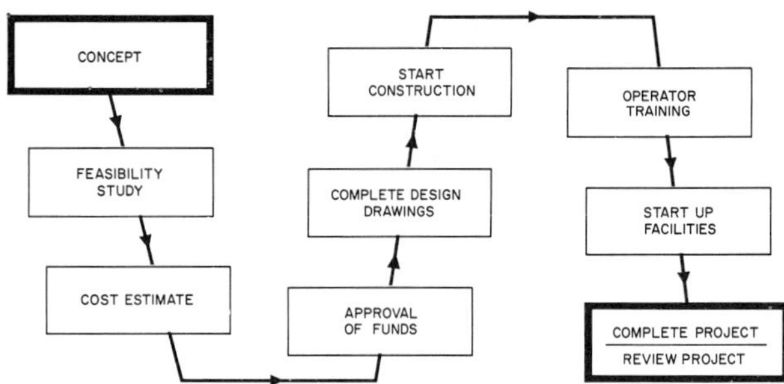

Fig. 11–26. This schematic shows a block diagram of the steps required to complete a project. (Courtesy of *Plant Engineering.*)

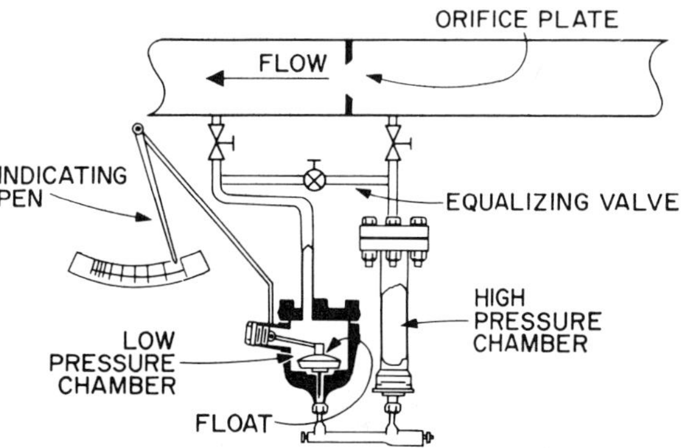

Fig. 11–27. A schematic showing the components of a gauge that measures the flow in a pipeline. (Courtesy of *Plant Engineering.*)

11–13 EMPIRICAL DATA

Data gathered from laboratory experiments and tests of prototypes or from actual field tests are called empirical data. Often empirical data can be transformed to equation form by means of one of three types of equations to be covered here.

The analysis of empirical data begins with the plotting of the data on rectangular grids, logarithmic grids, and semilogarithmic grids. Curves are then sketched through each point to determine which of the grids renders a straight-line relationship (Fig. 11–28). When the data plots as a straight line, its equation may be determined. Note that in the figure, three sets of empirical data are plotted and that curves are sketched to connect them. Each curve appears as a straight line in one of the graphs. We use this straight-line curve to write an equation for the data.

11–14 SELECTION OF POINTS ON A CURVE

Two methods of finding the equation of a curve are (1) the selected-points method and (2) the slope-intercept method. These are compared on a semilogarithmic graph in Fig. 11–29.

Selected-Points Method. Two widely separated points, such as (2, 30) and (4, 50), can be selected on the curve. These points are substituted in the equation below:

$$\frac{\log Y - \log 30}{X - 2} = \frac{\log 50 - \log 30}{4 - 2}$$

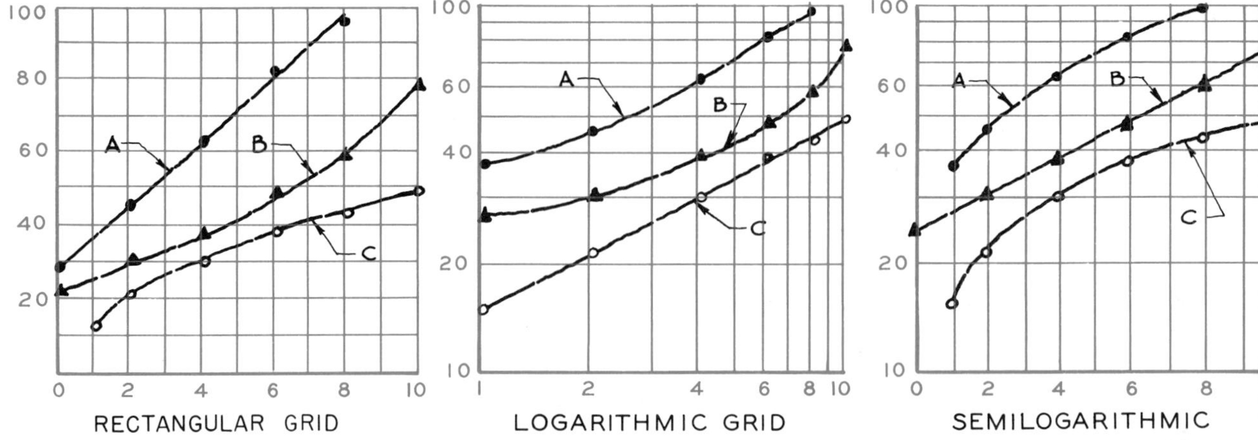

Fig. 11–28. Empirical data are plotted on each of these types of grids to determine which will render a straight-line plot. If the data can be plotted as a straight line on one of these grids, their equation can be found.

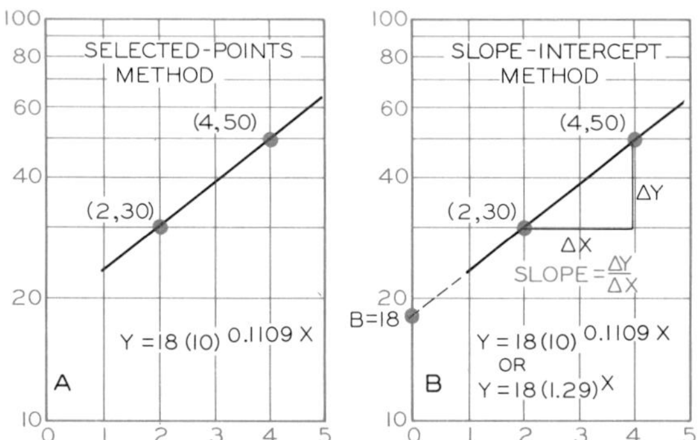

Fig. 11–29. The equation of a straight line on a grid can be determined by selecting any two points on the line (A). The slope-intercept method requires that the intercept be found where $X = 0$ on a semilog grid (B). This requires the extension of the curve to the Y-axis.

Note that the values in the Y-direction are logarithms and must be handled as such. The resulting equation for the data is

$$Y = 18(10)^{0.1109X}$$

Slope-Intercept Method. To apply the slope-intercept method, the intercept on the Y-axis where $X = 0$ must be known. If the X-axis is logarithmic, then the log of $X = 1$ is 0 and the intercept must be found above the value of $X = 1$.

In Fig. 15–29B, the data do not intercept the Y-axis; therefore, the curve must be extended to find the intercept $B = 18$. The slope of the curve is found ($\Delta Y/\Delta X$) and substituted into the slope-intercept form to give the equation as

$$Y = 18(10)^{0.1109X} \quad \text{or} \quad Y = 18(1.29)^X$$

The base-e logarithms can be used just as effectively as the base-10 logarithms, as shown in these examples. Other methods of converting data to equations are used, but the two methods illustrated here make the best use of the graphical process and are the most direct methods of introducing these concepts.

11–15 THE LINEAR EQUATION: $Y = MX + B$

The curve fitting the experimental data plotted in Fig. 15–30 is a straight line; therefore, we

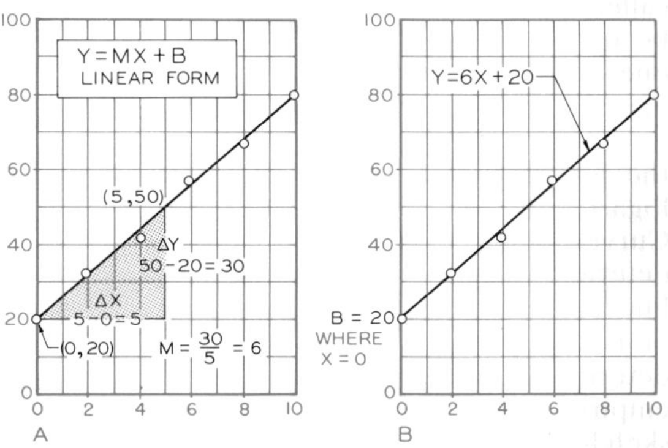

Fig. 11–30. (A) A straight line on an arithmetic grid will have an equation in the form $Y = MX + B$. The slope, M, is found to be 6. (B) The intercept, B, is found to be 20. The equation is written as $Y = 6X + 20$.

may assume that these data are linear, meaning that each measurement along the Y-axis is directly proportional to X-axis units. We may use the slope-intercept form or the selected-points method to illustrate one method of writing the equation for the data.

In the slope-intercept method, two known points are selected along the curve. The vertical and horizontal differences between the coordinates of each of these points are determined to establish the right triangle shown in part A of the figure. In the slope-intercept equation, $Y = MX + B$, M is the tangent of the angle between the curve and the horizontal, B is the intercept of the curve with the Y-axis where where $X = 0$, and X and Y are variables. In this example $M = \frac{30}{5} = 6$ and the intercept is 20. If the curve has sloped downward to the right, the slope would have been negative. By substituting this information into the slope-intercept equation, we obtain $Y = 6X + 20$, from which we can determine values of Y by substituting any value of X into the equation.

The selected-points method could also have been used to arrive at the same equation if the intercept were not known. By selecting two widely separated points such as (2, 32) and (10, 80), one can write the equation in this form:

$$\frac{Y - 32}{X - 2} = \frac{80 - 32}{10 - 2}, \qquad \therefore Y = 6X + 20$$

which results in the same equation as was found by the slope-intercept method ($Y = MX + B$).

11–16 THE POWER EQUATION: $Y = BX^M$

Since the data shown plotted on a rectangular grid in Fig. 11–31 do not form a straight line, they cannot be expressed in the form of a linear equation. However, when the data are plotted on a logarithmic grid, they are found to form a straight line (step 1). Therefore, we express the data in the form of a power equation in which Y is a function of X raised to a given power or $Y = BX^M$. The equation of the data is obtained in much the same manner as was the linear equation, using the point where the curve intersects the Y-axis where $X = 0$, and letting M equal the slope of the curve. Two known points are selected on the curve. Any linear scale in decimal units, such as the 20-scale on

FIGURE 11–31. THE POWER EQUATION, $Y = BX^M$

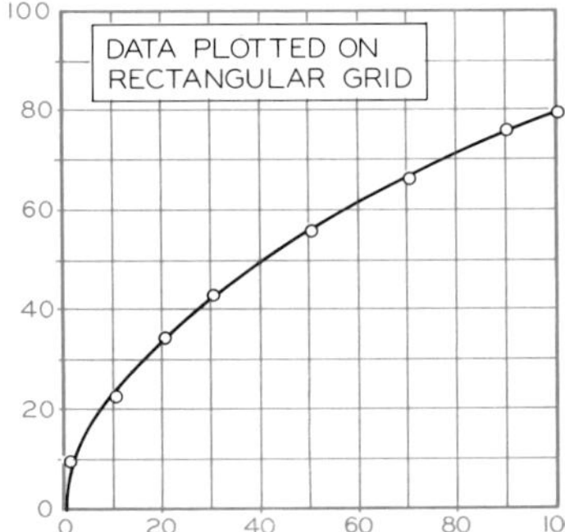

Given: The data plotted on the rectangular grid give an approximation of a parabola. Since the data does not form a straight line on the rectangular grid, the equation will not be linear.

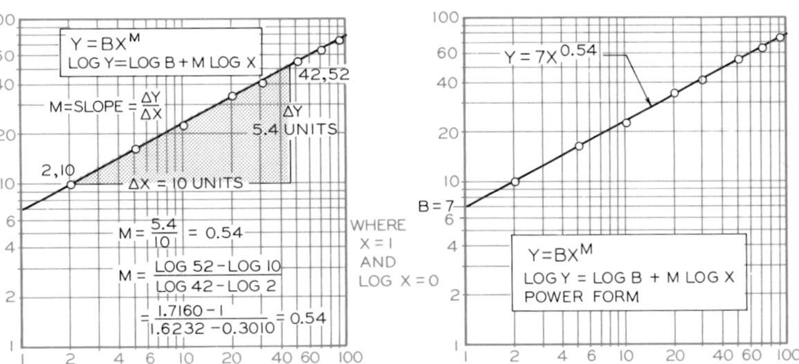

Step 1: The curve of the data forms a straight line on a logarithmic grid, making it possible to find the equation of the data. The slope, M, can be found graphically with an engineer's scale, setting dX at 10 units and measuring the slope (dY) using the same scale.

Step 2: The intercept $B = 7$ is found where $X = 1$. The slope and intercept are substituted into the equation, which then becomes $Y = 7X^{0.54}$.

the engineers' scale, can be used, when the cycles along the X- and Y-axes are equal, to measure the vertical and horizontal differences between the coordinates of the two points.

If the horizontal distance of the right triangle is drawn to be 1 or 10 or a multiple of 10, the vertical distance can be read off directly. In step 2, the slope M (tangent of the triangle) is found to be 0.54. The intercept B is 7; thus the equation is $Y = 7X^{0.54}$, which can be evaluated for each value of Y by converting this power equation into the logarithmic form of $\log Y$:

$$\log Y = \log B + M \log X,$$
$$\log Y = \log 7 + 0.54 \log X$$

Note that when the slope-intercept method is used, the intercept can be found on the Y-axis where $X = 1$. In Fig. 11–32, the Y-axis at the left of the graph has an X-value of 0.1; consequently, the intercept is located midway across the graph where $X = 1$. This is analogous to the linear form of the equation, since the log of 1 is 0. The curve slopes downward to the right; thus the slope, M, is negative. The selected-points method can be applied to find the equation of the data as discussed in the previous article.

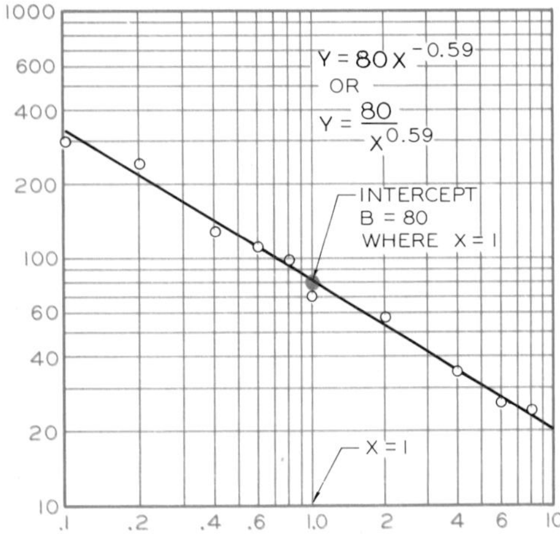

Fig. 11–32. When the slope-intercept equation is used, the intercept can be found only where $X = 1$. Therefore, in this example the intercept is found at the middle of the graph.

Base-10 logarithms are used in these examples, but natural logs could be used with e (2.718) as the base.

11–17 THE EXPONENTIAL EQUATION: $Y = BM^X$

The experimental data plotted in Fig. 11–33 form a curve, indicating that they are not linear. When the data are plotted on a semilogarithmic grid, as has been done in step 1 of the figure, they approximate a straight line for which we can write the equation $Y = BM^X$, where B is the Y-intercept of the curve and M is the slope of the curve. The procedure for deriving the equation is shown in step 2, in which two points are selected along the curve so that a right triangle can be drawn to represent the differences between the coordinates of the points selected. The slope of the curve is found to be

$$\log M = \frac{\log 40 - \log 6}{8 - 3} = 0.1648$$

or

$$M = (10)^{0.1648} = 1.46$$

The value of M can be substituted in the equation in the following manner:

$$Y = BM^X \quad \text{or} \quad Y = 2(1.46)^X,$$
$$Y = B(10)^{MX} \quad \text{or} \quad Y = 2(10)^{0.1648X}$$

where X is a variable that can be substituted into the equation to give an infinite number of values for Y. We can write this equation in its logarithmic form, which enables us to solve it readily for the unknown value of Y for any given value of X. The equation can be written as

$$\log Y = \log B + X \log M$$

or

$$= \log 2 + X \log 1.46$$

The same methods are used to find the slope of a curve with a negative slope. The curve of the data in Fig. 11–34 slopes downward to the right; therefore, the slope is nega-

FIGURE 11–33. THE EXPONENTIAL EQUATION: $Y = BM^X$

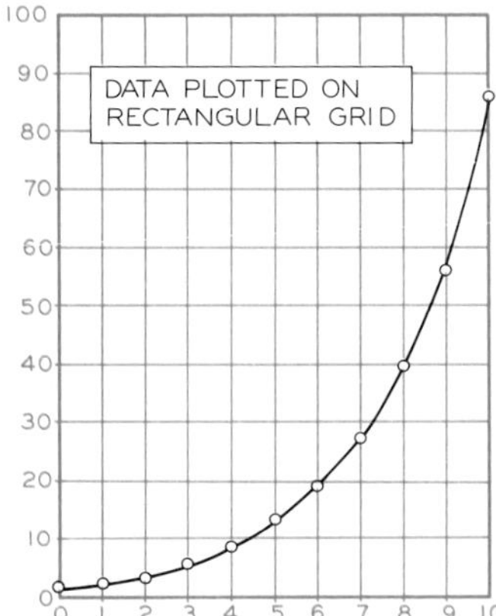

Y = B 10^MX
OR
Y = BM^X
8,40
M = SLOPE = ΔY/ΔX
ΔY
LOG 40 − LOG 6
3,6
ΔX = 8 − 3
LOG M = (LOG 40 − LOG 6)/(8 − 3)
M = 10^0.1648 X OR M = 1.46

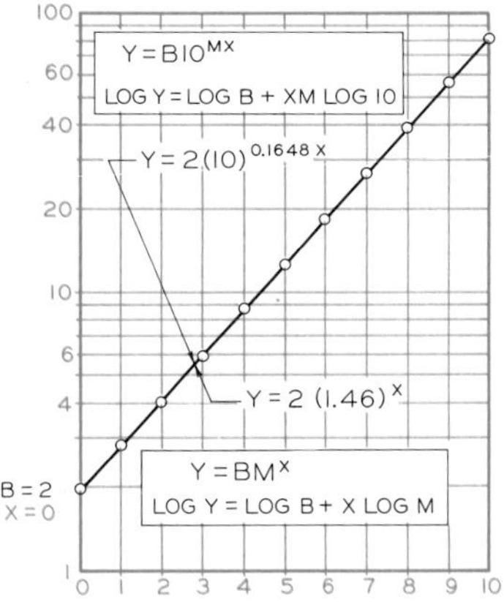

Given: These data do not give a straight line on either a rectangular grid or a logarithmic grid. However, when plotted on a semilogarithmic grid, they do give a straight line.

Step 1: The slope must be found by mathematical calculations; it cannot be found graphically. The slope may be written in either of the forms shown here.

Step 2: The intercept $B = 2$ is found where $X = 0$. The slope, M, and the intercept, B, are substituted into the equation to give $Y = 2(10)^{0.1648X}$ or $Y = 2(1.46)^X$.

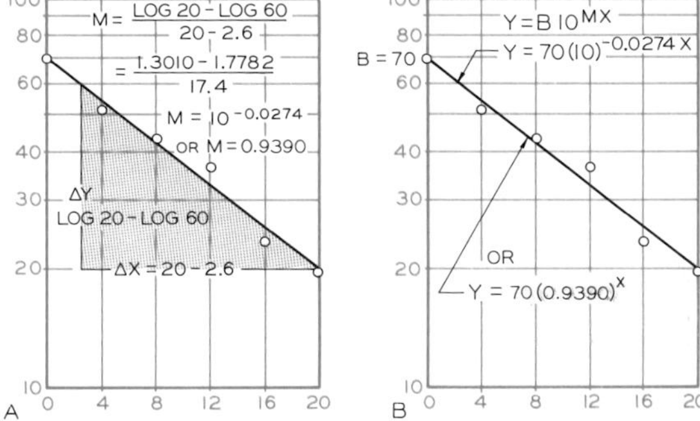

Fig. 11–34. When a curve slopes downward to the right, its slope is negative as calculated in part A. Two forms of the final equation are shown in part B by substitution.

tive. Two points are selected in order to find order according to their lengths or in which is the antilog of −0.0274. The intercept, 70, can be combined with the slope, M, to find the final equations as illustrated in step 2 of Fig. 11–34.

11–18 APPLICATIONS OF EMPIRICAL GRAPHS

Figure 11–35 is an example of how empirical data can be plotted to compare the transverse

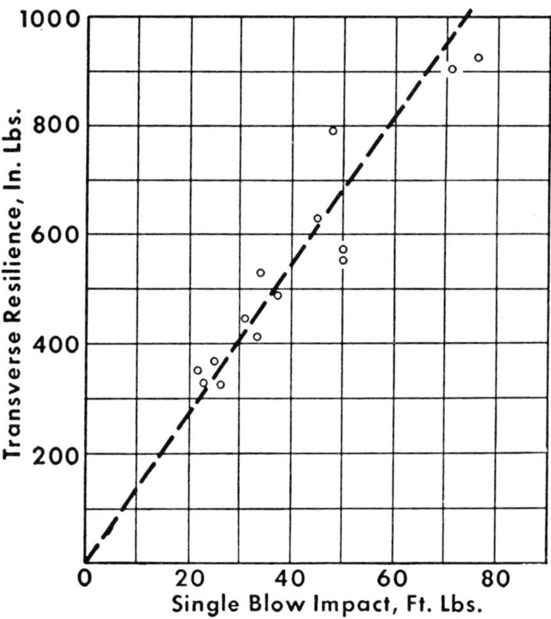

Fig. 11–35. The relationship between the transverse strength of gray iron and impact resistance results in a straight line with an equation of the form $Y = MX + B$.

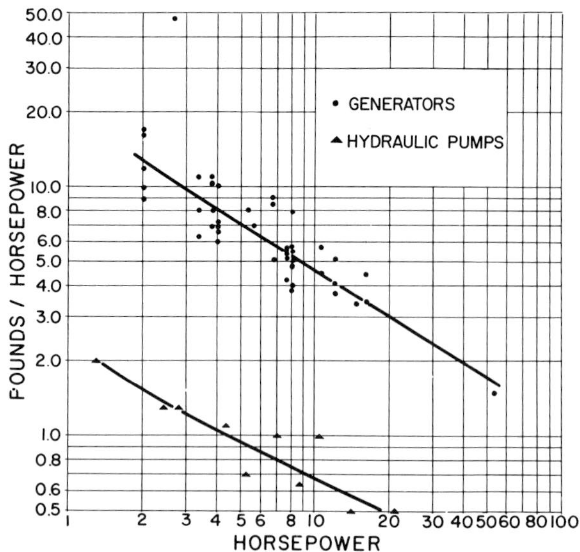

Fig. 11–36. Empirical data plotted on a logarithmic grid, showing the specific weight versus the horsepower of electric generators and hydraulic pumps. The curve is the average of joints plotted. (Courtesy of General Motors Corporation, *Engineering Journal.*)

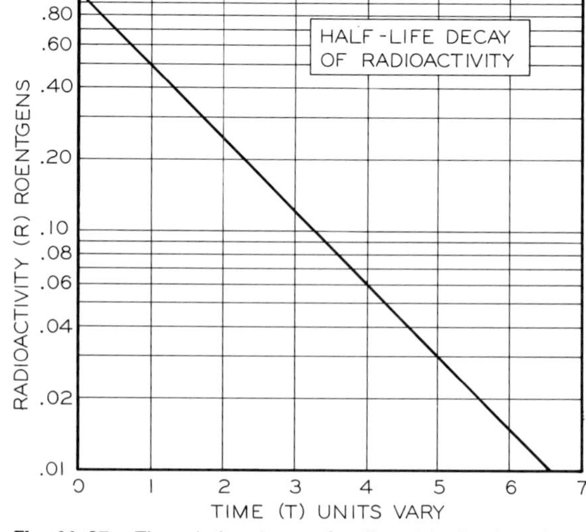

Fig. 11–37. The relative decay of radioactivity is plotted as a straight line on this semilog graph, making it possible for its equation to be found in the form $Y = BM^X$.

strength and impact resistance of gray iron. Note that the data are somewhat scattered, but the best curve is drawn. Since the curve is a straight line on a linear graph, the equation of these data can be found by the equation

$$Y = MX + B$$

Figure 11–36 is an example of how empirical data can be plotted to compare the specific weight (pounds per horsepower) of generators and hydraulic pumps versus horsepower. Note that the weight of these units decreases linearly as the horsepower increases. Therefore, these data can be written in the form of the power equation

$$Y = BX^M$$

We obtain the equation of these data by applying the procedures covered in Article 11–15 and thus mathematically analyze these relationships.

The half-life decay of radioactivity is plotted in Fig. 11–37 to show the relationship of decay to time. Since the half-life of different isotopes varies, different units would have to be assigned to time along the X-axis; however, the curve would be a straight line for all isotopes. The exponential form of the equation discussed in Article 11–17 can be applied to find the equation for these data in the form of

$$Y = BM^X$$

11–19 INTRODUCTION TO MECHANISMS AND LINKAGES

Mechanisms are used to produce force or motion through a series of interrelated components. A mechanism is a combination of components based on rotation, leverage, or the inclined plane. A linkage is a type of mechanism relying primarily on leverage. An alternative system could be operated electronically with a minimum of mechanical links. Linkages and mechanisms are universally used in machinery (Fig. 11–38), jigs and fixtures, and, to some extent, in practically all designs. The analysis of mechanisms is often referred to as *mechanics*

Fig. 11–38. This metal-forming machine illustrates the many linkages and mechanisms used in standard equipment. (Courtesy of A. H. Nilson Company.)

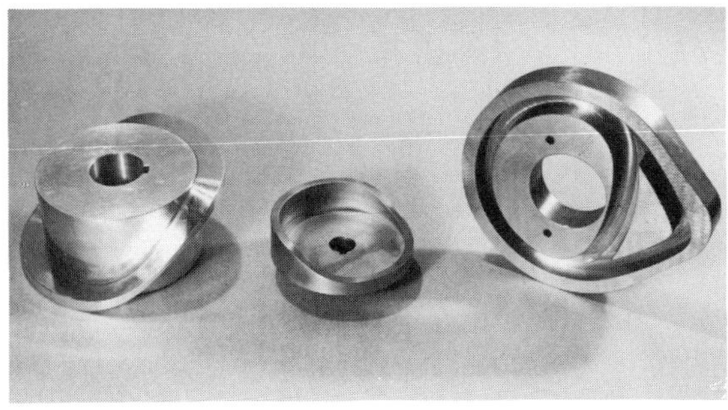

Fig. 11–39. Examples of machined cams. (Courtesy of Ferguson Machine Company.)

and is undertaken to determine the effects of forces upon the mechanisms.

The following definitions are fundamental to the study of linkages and mechanisms involved in a design.

Statics. The study of the effect of forces upon bodies or parts which are at rest or moving at uniform velocity.

Dynamics. The study of the effect of forces that cause a change in the motion of machine components or material bodies. Includes *kinematics* and *kinetics.*

Kinetics. The study of the effect of forces that cause a change in the motion of machine components or other bodies.

Kinematics. The study of motion without regard to forces.

These areas of analysis are very critical to the final analysis of a design. However, as a supplement to the usual analytical procedures, graphical methods can be used as a basis for initial steps toward analyzing a design.

The mechanisms that are graphically analyzed in this chapter are cams and linkages. Several examples are given that are closely related to variations of rotary motion coupled with linkages. This motion is plotted on rectangular graphs to permit analysis.

11–20 CAMS

Cams (grooved cams, plate cams, or cylindrical cams) are components that produce motion in a single plane, usually up and down (Fig. 11–39). The cam revolving about an eccentric center produces a rise and fall in the follower during rotation. The configuration of the cam is analyzed graphically prior to the preparation of the specifications for its manufacture. Only plate cams are covered in the brief review of this mechanism. Cams utilize the principle of the inclined wedge, with the surface of the cam causing a change in the slope of the plane, thereby producing the desired motion.

11–21 CAM MOTION

Cams are designed primarily to produce (1) uniform or linear motion, (2) harmonic motion, (3) gravity motion, or (4) combinations of these. Some cams are designed to fit special needs that do not fit these patterns, but are instead based on particular design requirements. Displacement diagrams are used to represent the travel of the follower relative to the rotation of the cam.

Uniform motion is shown in the displacement diagram in Fig. 11–40A. Displacement diagrams represent the motion of the cam

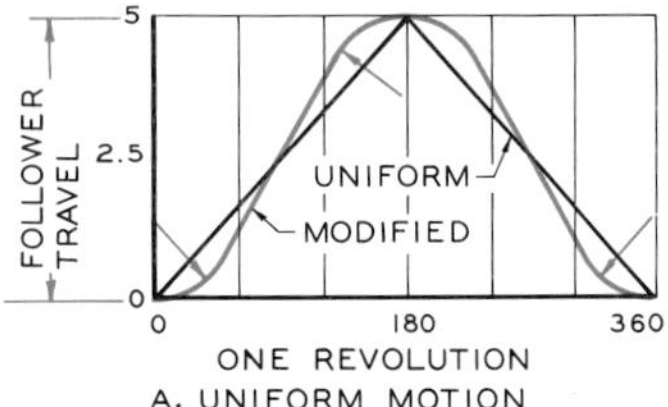

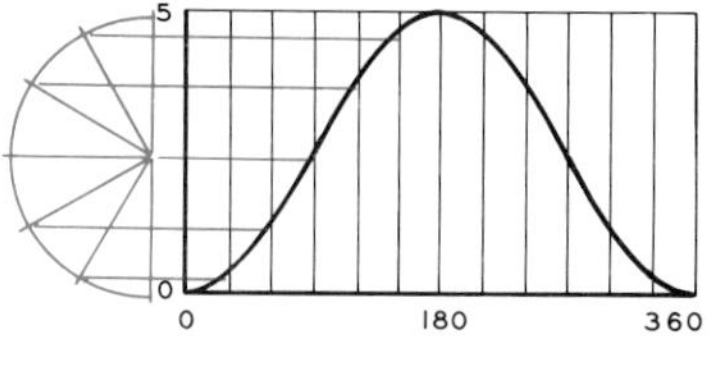

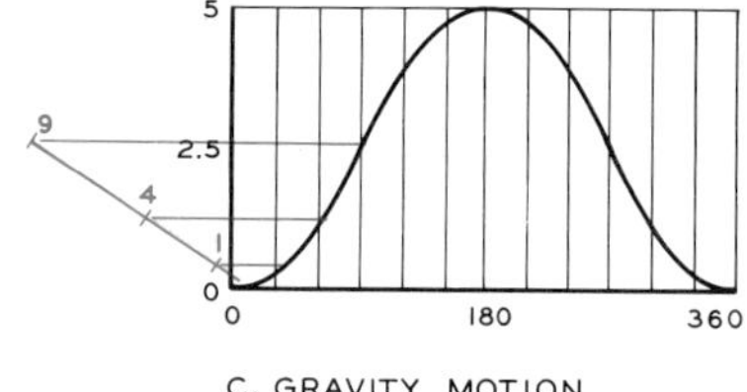

Fig. 11–40. The methods of plotting the basic motions of cams—uniform, harmonic, and gravity.

follower as the cam rotates through 360°. It can be seen that the uniform-motion curve has sharp corners, indicating abrupt changes of velocity at two points. This is impractical and inefficient, since it causes the follower to bounce. Hence this motion is usually modified with arcs that tend to smooth this change and thus the operation. The radius of the modifying arc can vary up to a radius of one-half the total displacement of the follower, depending upon the speed of operation. Usually a radius of about one-third to one-fourth total displacement is best.

Harmonic motion, plotted in part B of the figure, is a smooth continuous motion based on the change of position of the points on the circumference of a circle. At moderate speeds, this displacement results in a smooth operation. Note the method of drawing a semicircle to establish points on the displacement diagram.

Gravity motion (uniform acceleration), plotted in part C, is commonly used for high speed operation. The variation of displacement is analogous to the force of gravity exerted on a falling body, with the difference in displacement being 1;3;5;5;3;1, based on the square of the number. For instance, $1^2 = 1$; $2^2 = 4$; $3^2 = 9$. This same motion is repeated in reverse order for the remaining half of the movement of the follower. Intermediate points can be found by squaring fractional increments, such as $(2.5)^2$. The gravity fall of the follower is designed to conform to the shape of the cam so that its contact with the surface will provide smooth operation.

Cam Followers. Three basic types of cam followers are (A) the flat surface, (B) the roller, and (C) the knife edge, as shown in Fig. 11–41. The flat-surface and knife-edge followers are limited to use with slow-moving cams where the minimum of force will be exerted by the friction that is caused during rotation. The roller is the most often used form of follower, since it can withstand higher speeds and transmit greater forces.

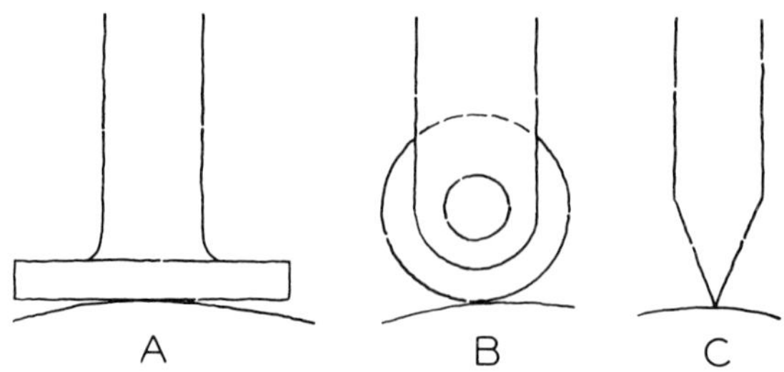

Fig. 11–41. Three basic types of cam followers—the flat surface, the roller, and the knife edge.

11–22 CONSTRUCTION OF A CAM

Plate Cam—Harmonic Motion. The steps oɪ constructing a plate cam with harmonic motion are shown in Fig. 11–42. The draftsman must known certain information before he can design

FIGURE 11–42. CONSTRUCTION OF A PLATE CAM WITH HARMONIC MOTION

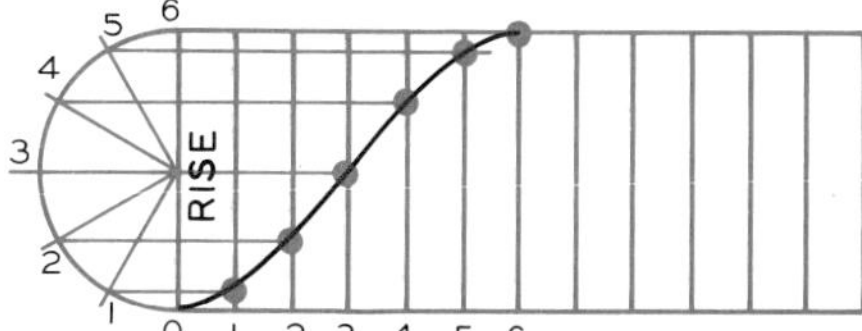

Step 1: Construct a semicircle whose diameter is equal to the rise of the follower. Divide the semicircle into the same number of divisions as there are between 0° and 180° on the horizontal axis of the displacement diagram. Plot half of the displacement curve in the displacement diagram.

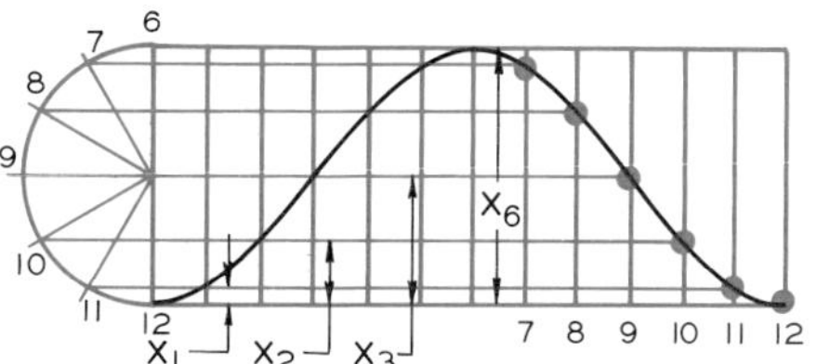

Step 2: Continue the process of plotting points by projecting from the semicircle, starting from the top of the semicircle and proceeding to the bottom. Complete the curve symmetrical to the left half.

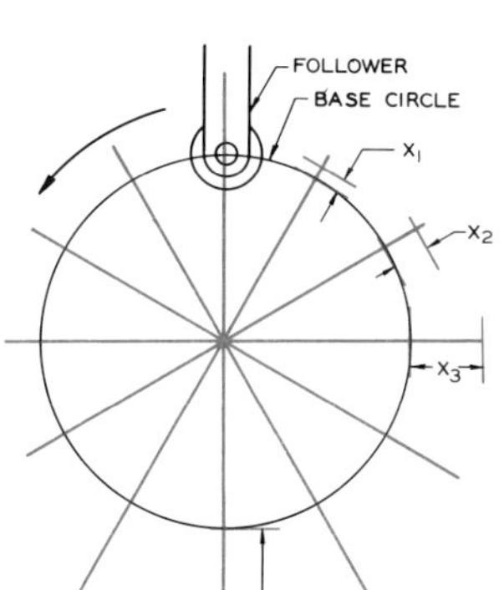

Step 3: Construct the base circle and draw the follower. Divide the base circle into the same number of sectors as there are divisions on the displacement diagram. Transfer distances from the displacement diagram to the respective radial lines of the base circle, measuring outward from the base circle.

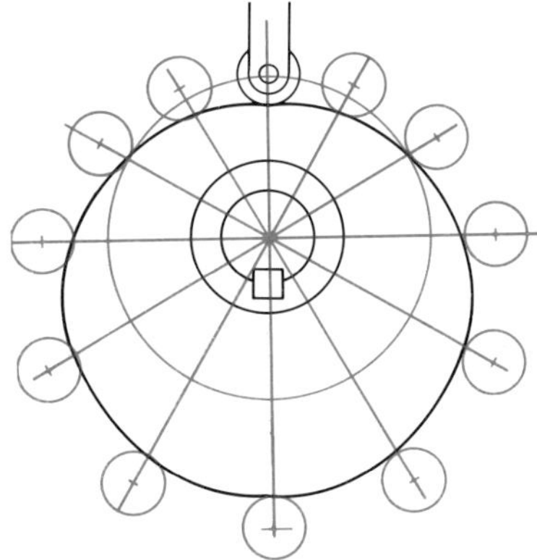

Step 4: Draw circles to represent the positions of the roller as the cam revolves in a counterclockwise direction. Draw the cam profile tangent to all the rollers to complete the drawing.

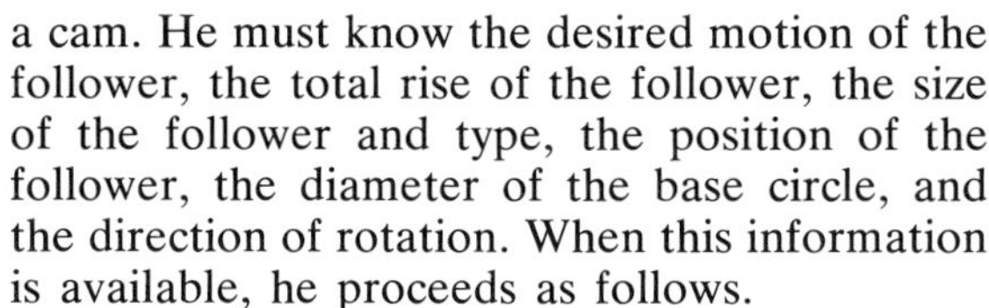

a cam. He must know the desired motion of the follower, the total rise of the follower, the size of the follower and type, the position of the follower, the diameter of the base circle, and the direction of rotation. When this information is available, he proceeds as follows.

Step 1. The displacement diagram is laid out. The vertical axis represents the rise of the follower from its lowest point. The horizontal axis is divided into equal divisions representing degrees of rotation of the cam (each division usually represents 15° or 30°). A semicircle with a diameter equal to the rise of the follower is constructed and divided into the same number of equal units as are drawn between 0° and 180° on the horizontal axis of the displacement diagram. The points on the semicircle are projected to their respective lines drawn vertically through the divisions of the horizontal axis. These points are connected with a smooth curve.

Step 2. The same semicircle is used to find points on the curve from 180° to 360° starting from the top of the semicircle (point 6) and proceeding downward to point 12. The points are projected to their respective lines and are connected with a smooth irregular curve. The right side and left side of the displacement diagram are symmetrical.

Step 3. The base circle is drawn from given specifications and the follower is drawn with its center on the base circle. The circle is divided into the same number of sectors as shown on the displacement diagram. There are twelve in this example, since the circle is divided into 30° sectors. The displacement of the cam follower is taken from the displacement diagram, and since the motion of the cam is counterclockwise, the displacement is plotted to the right of the follower. For example, the distances X_1, X_2, and X_3 are measured from the base circle outward. Points are located in this manner all the way around the base circle.

Step 4. Construction circles representing the roller follower are drawn with the points plotted in step 3 as centers. The profile of the cam is drawn with an irregular curve to be tangent to each of the roller constructions. Additional intervals can be found to construct a more accurate profile. The cam hub and keyway are drawn to given specifications.

Plate Cam—Uniform Acceleration. This construction is the same as the previous example except for the displacement diagram and the knife-edge follower. The steps involved in constructing a plate cam with uniform acceleration are shown in Fig. 11–43.

Step 1. The displacement diagram is drawn with each division on the horizontal axis representing 30° and with the vertical axis equal to the rise of the follower. The rate of travel of the follower changes constantly, producing acceleration and deceleration. The changes in rise are based on the square of each division. Note that these divisions are laid off on the construction line and are projected back to the vertical axis. The follower accelerates from 0° to 180° and decelerates from 180° to 360°. One half of the curve is plotted as an irregular curve.

Step 2. The same construction is used to find the displacement curve from 180° to 360° to complete the symmetrical curve.

Step 3. The base circle is drawn to represent the lowest position of the knife-edge follower. The circle is divided into 30° sectors—the same number of divisions as in the displacement diagram. Since the rotation of the cam is counterclockwise, the displacement is plotted outward from the base circle to the right of the follower in a clockwise direction.

Step 4. The profile of the cam is drawn with a smooth irregular curve through the plotted points. The cam hub and keyway are added to complete the drawing.

Plate Cam—Combination Motion. In Fig. 11–44 a knife-edge follower is used with a plate cam to produce harmonic motion from 0° to 180°, uniform acceleration from 180° to 300°, and dwell (no follower motion) from 300° to 360°. We are to draw a cam that will give this motion from the given base circle.

Step 1. The harmonic portion of the displacement diagram is constructed by drawing a semicircle whose circumference is divided into the same number of equal parts as there are horizontal divisions on the displacement diagram between 0° and 180°—six in this case. The uniform acceleration (four divisions in Fig. 11–44) of the follower is found by dividing the number of horizontal divisions by 2; that is, $4 \div 2 = 2$. Then 1^2 would give a travel of 1 during the first 30°, and 2^2 would give a travel of 4 from the peak, or a fall of 3 units between 210° and 240°. From 300° to 360°, where a dwell condition exists, the follower does not move; consequently, this portion of the curve is a horizontal line.

Step 2. Radial lines are drawn from the center of the base circle to correspond to the intervals used on the horizontal scale of the displacement diagram. The displacement is measured outward along the radial lines from the base circle with dividers. Distance X is shown as an example.

FIGURE 11–43. CONSTRUCTION OF A PLATE CAM WITH UNIFORM ACCELERATION

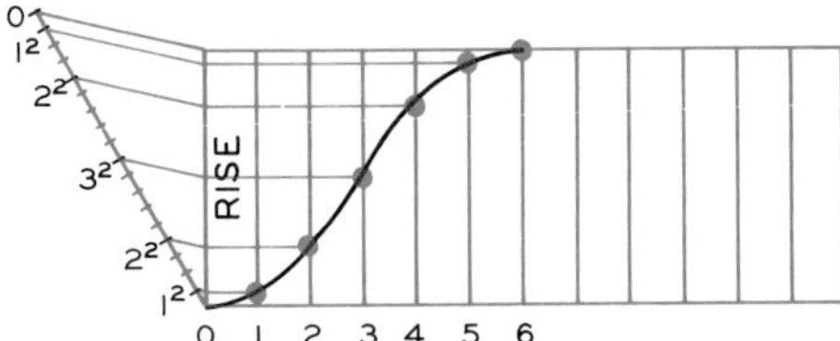

Step 1: Construct a displacement diagram to represent the rise of the follower. Divide the horizontal axis into angular increments of 30°. Draw a construction line through point 0; locate the 1^2, 2^2, and 3^2 divisions and project them to the vertical axis to represent half of the rise. The other half of the rise is found by laying off distances along the construction line with descending values.

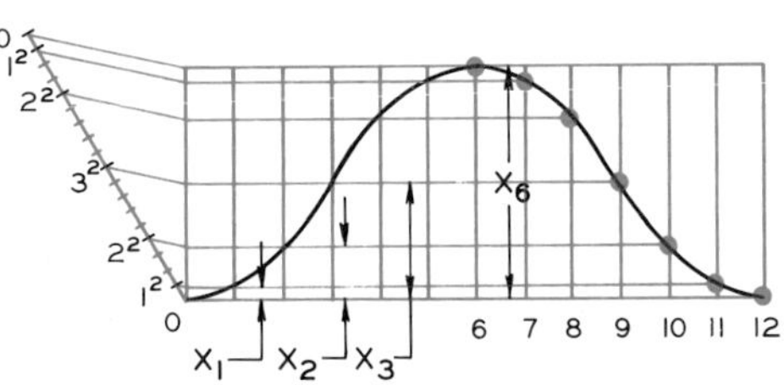

Step 2: Use the same construction to find the right half of the symmetrical curve.

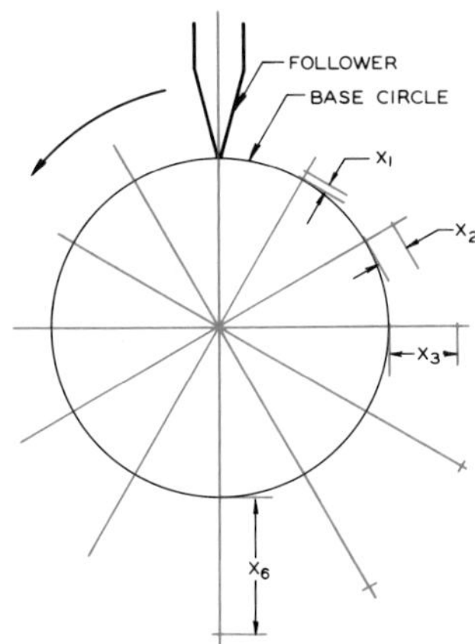

Step 3: Construct the base circle and draw the knife-edge follower. Divide the circle into the same number of sectors as there are divisions in the displacement diagram. Transfer distances from the displacement diagram to the respective radial lines of the base circle, measuring outward from the base circle.

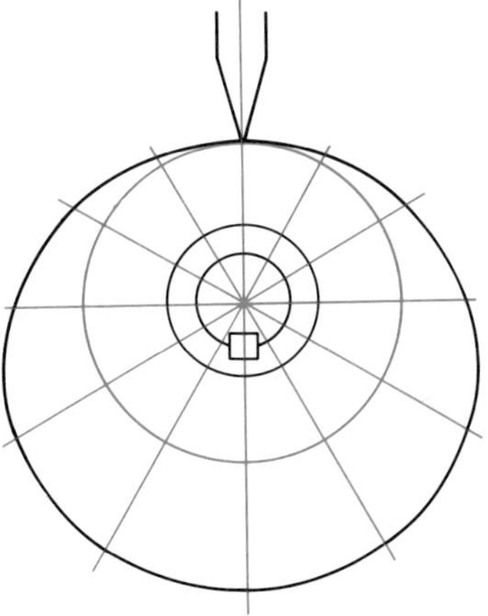

Step 4: Connect the points found in step 3 with a smooth curve to complete the cam profile. Show also the cam hub and keyway.

Step 3. The points on the radial lines are connected with a smooth, irregular curve to form the cam profile that will produce the specified motion. The hub and keyway are drawn to complete the construction.

11–23 CONSTRUCTION OF A CAM WITH AN OFFSET ROLLER FOLLOWER

The cam in Fig. 11–45 is required to produce harmonic motion through 360°. This motion

FIGURE 11–44. CONSTRUCTION OF A PLATE CAM WITH COMBINATION MOTIONS

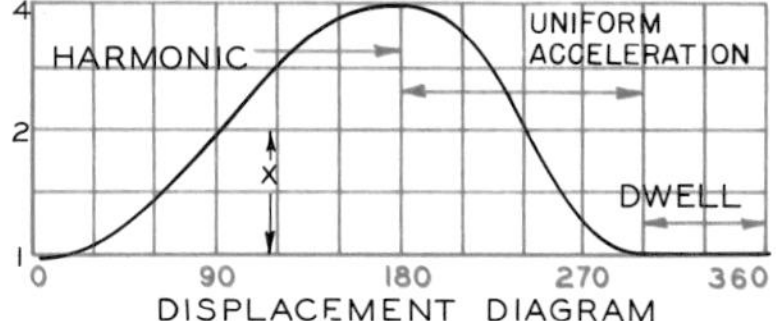

Step 1: The cam is to rise 4″ in 180° with harmonic motion, fall 4″ in 120° with uniform acceleration, and dwell for 60°. These motions are plotted on the displacement diagram.

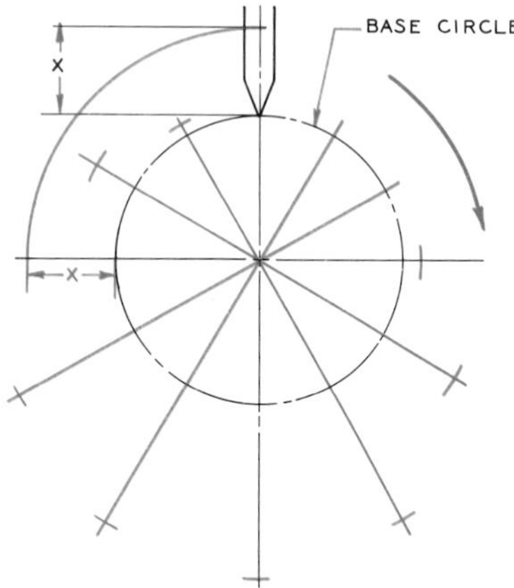

Step 2: Construct the base circle and draw the knife-edge follower. Transfer distances from the displacement diagram to the respective radial lines of the base circle, measuring outward from the base circle.

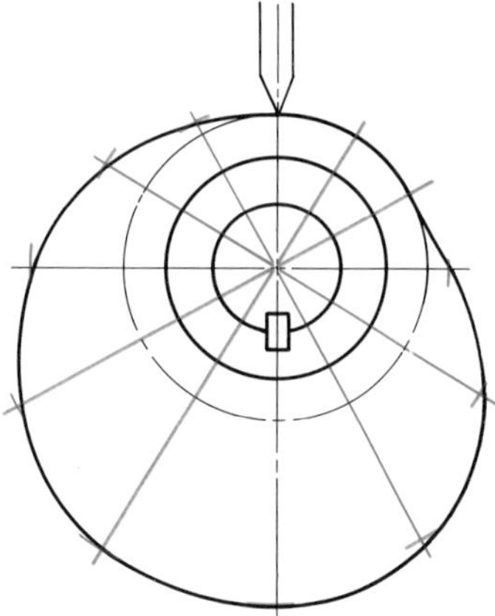

Step 3: Draw a smooth curve through the points found in step 2 to complete the profile of the cam. Show also the cam hub and keyway.

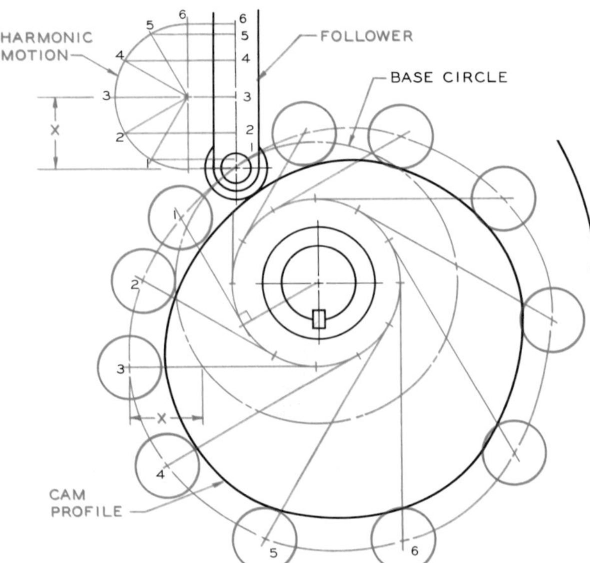

Fig. 11–45. Construction of a plate cam with an offset roller follower.

can be plotted directly from the follower rather than on a displacement diagram, since there are no combinations of motion involved. A semicircle is drawn with its diameter equal to the total motion desired in the follower. In this case, the base circle is the center of the roller of the follower. The center line of the follower is extended down, and a circle is drawn with its center at the center of the base circle so that it is tangent to the extension of the follower center line. This circle is divided into 30° intervals to establish points of tangency for all positions of the follower as it revolves through 360°. These tangent lines can be accurately constructed by drawing them perpendicular to the 30° interval lines extended from the center of the circle to the points on the circumference. The distances are transferred from the harmonic motion diagram to each subsequent tangent line. Distance X is located as an example of this procedure. The circular roller is drawn in all views and the profile of the cam is constructed to be tangent to the rollers at all positions as shown.

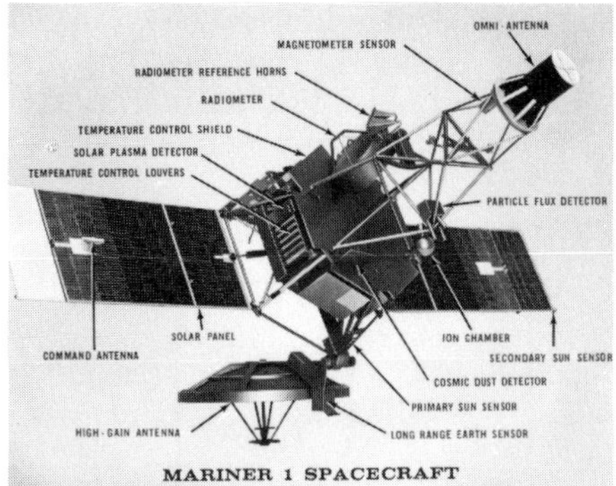

Fig. 11–46. The Mariner I spacecraft uses a linkage system that allows the craft to be collapsed into a minimum of space during its flight through the earth's atmosphere. (Courtesy of NASA.)

Fig. 11–47. This pusher, which is used to force pipes under roadways, is a linkage system that uses hydraulic power. (Courtesy of Arnold Engineering Development Center.)

11–24 LINKAGES

The spacecraft shown in Fig. 11–46 employs a unique linkage system which is used to position the long-range earth sensor and the solar panels after the vehicle is in space. These linkages must be designed so that the components can be withdrawn into a position that will require minimum space during launching. The hydraulic pusher in Fig. 11–47, which was designed to force pipe under roadways, is based on a linkage system utilizing hydraulic power. Linkages are universally used in mechanisms of all sizes. The initial analysis of a linkage system can be approached graphically prior to the final mathematical analysis.

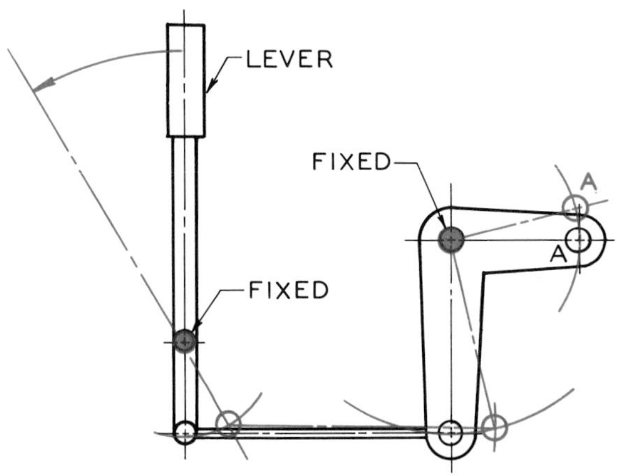

Fig. 11–48. A basic three-bar linkage.

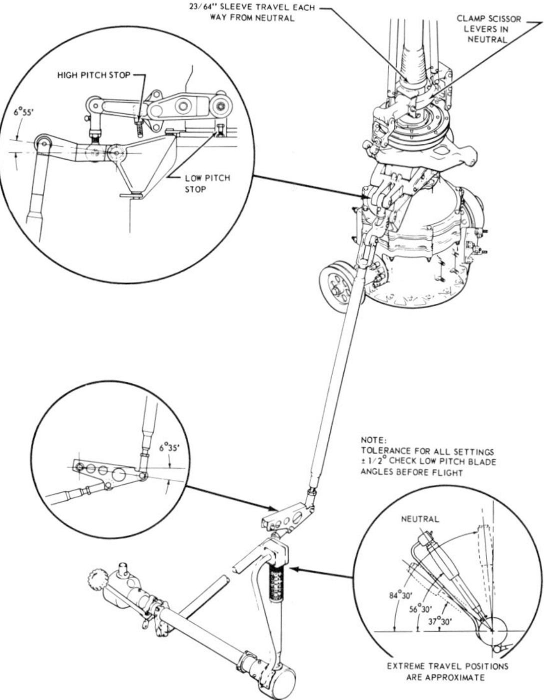

Fig. 11–49. Example of a linkage system used to control the blade pitch on helicopters. (Courtesy of Bell Helicopter Corporation.)

Fig. 11–50. This gate lock utilizes a combination of linkages and cams. (Courtesy of General American Transportation Corporation.)

11–25 A BASIC LINKAGE SYSTEM

The simple linkage system shown in Fig. 11–48 illustrates the graphical analysis of a linkage, in which fixed points are given from which the components must pivot. The motion in this case is transmitted through levers to produce motion of point *A*. To locate the linkage components, all measurements are made by swinging arcs from the fixed pivot points. This is a three-bar linkage composed of three moving parts. An application of this type of system is shown in Fig. 11–49, which illustrates the linkage control system for a helicopter. Another application of a linkage system of this type is shown in Fig. 11–50.

11–26 LINKAGE APPLICATIONS

The revolution of a crankshaft is shown in Fig. 11–51, in which a connecting rod and a piston are analyzed for motion in a displacement diagram. This displacement can be found graphically with a high degree of accuracy if a sufficiently large scale is used. The graphical analysis assists the engineer in applying principles of mechanics to the analytical solution. A typical steering linkage system for trailers used in industrial materials handling is shown in Fig. 11–52. Steering systems can be analyzed graphically to determine the motion of the components during the turning process.

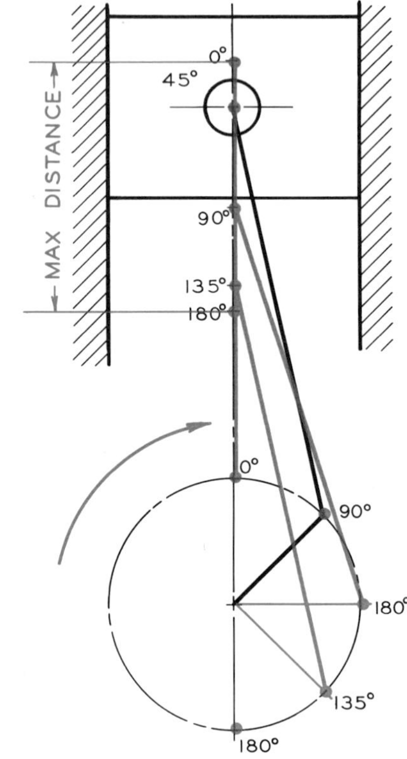

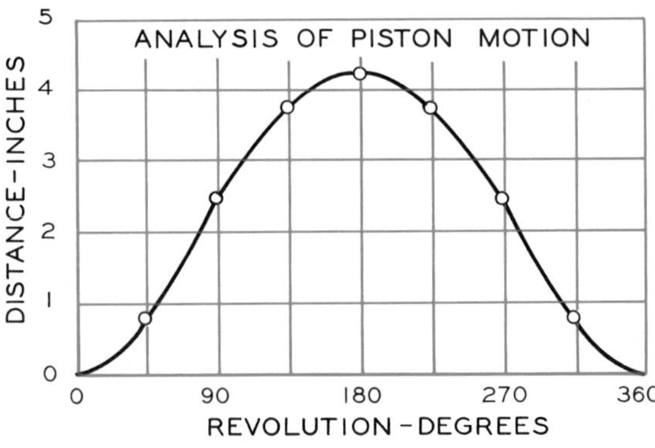

Fig. 11–51. The displacement of the travel of a piston caused by a crankshaft linkage. This is called a *slider-crank* mechanism.

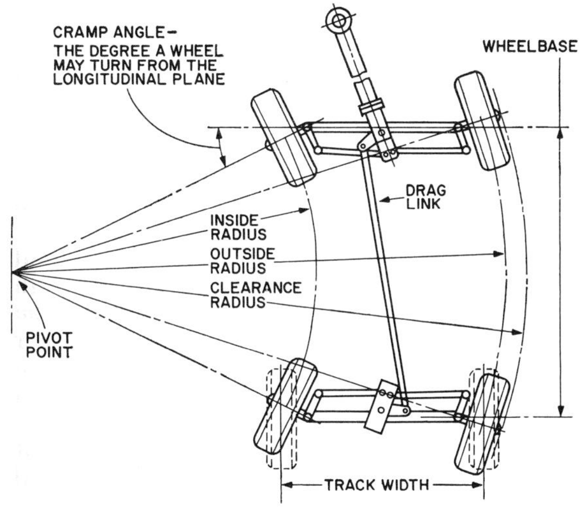

Fig. 11–52. This linkage ensures that an in-plant cargo trailer will tow properly within the available traffic lanes. (Courtesy of *Plant Engineering.*)

11–27 ANALYSIS OF A CLAMPING DEVICE

Many linkage systems are used in the design of clamping mechanisms which, as components of jigs and fixtures, are necessary for holding parts being machined. An example of such a device is shown in Fig. 11–53. The handle is shown in its closed position, position A, and in its revolved position, position B. A partial view of the handle is shown in position C, where the handle makes contact with the working surface. The graphical analysis of the linkage indicates that the plunger will be in its most withdrawn position at position C, where the handle is revolved θ degrees from the horizontal. If utilization of the maximum travel of the plunger were critical, the handle would have to be modified to permit full operation and hand clearance.

The advantage of the graphical method of analysis of a linkage system of this type is rather obvious. With a minimum of effort, the

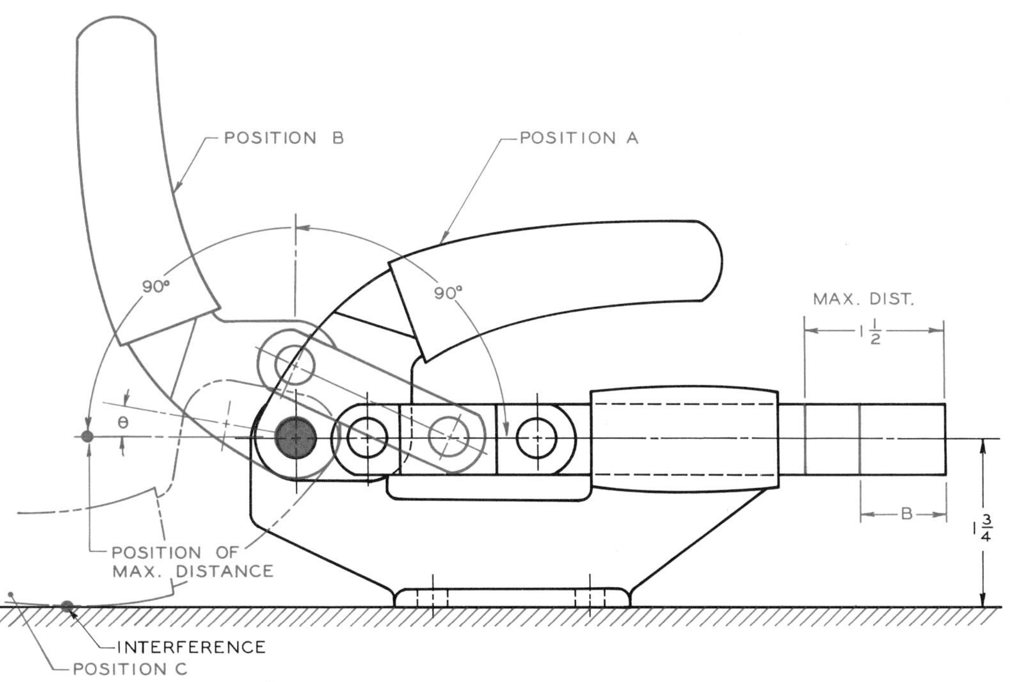

Fig. 11–53. Graphical analysis of a hand-operated clamping device that incorporates linkage principles. (Courtesy of Universal Engineering Corporation.)

designer can determine the effects of the dimensions of each link with respect to the motion produced an the available clearance. The design of the handle must consider the position of the operator and the comfort of operation provided by its configuration.

11–28 ANALYSIS FOR CLEARANCE

A counterweight for an engine is shown in Fig. 11–54. For optimum operation, the counterweight should be as large as possible. The maximum radius of the counterweight is determined primarily by the path of the nose of exhaust cam 1. It is permissible to have a close clearance between the nose of the cam and the counterweight, since the centers of each are held closely and the surfaces are finished to an acceptable tolerance. However, the clearance between the piston skirt and the counterweight should be questioned.

The determination of the clearance between these parts can be solved graphically. A series of sections, 1, 2, 3, and 4, are constructed from the center of the counterweight. The points are projected to the top and then to the side view, where the radial sections of the piston skirt can be seen. This view indicates that the application of a chamfer to the counterweight will provide the necessary clearance without interfering with effectiveness.

The path described by the connecting rod in Fig. 11–55 must be analyzed, since this path will influence the location of the camshaft, the crankcase walls, the width of oil pans, and other limiting factors. The clearance between the camshaft and the connecting rod bolt head can be close because these are finished surfaces whose centers are closely held. On the other hand, the clearance between the connecting rod and the walls and the oil pan must be somewhat greater to allow for imperfections in the rough forgings and rough crankcase walls. The designer can position the camshaft and establish the size of the crankcase by plotting the extreme path of the connecting rod. The graphical method lends itself to this form of analysis more than any other technique.

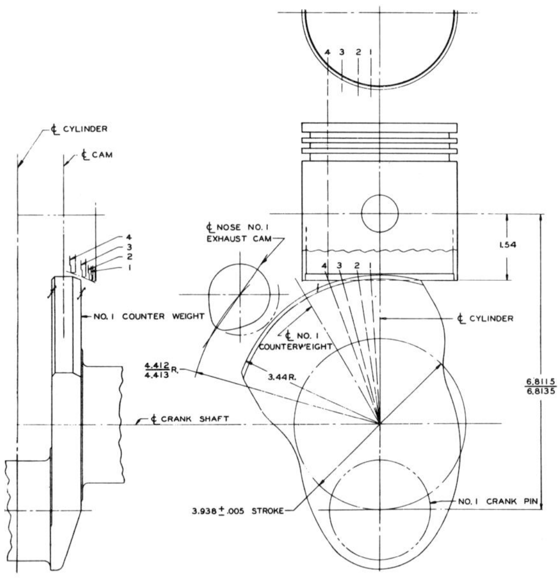

Fig. 11–54. Graphical determination of the clearance between a counterweight and a piston skirt. (Courtesy of Chrysler Corporation.)

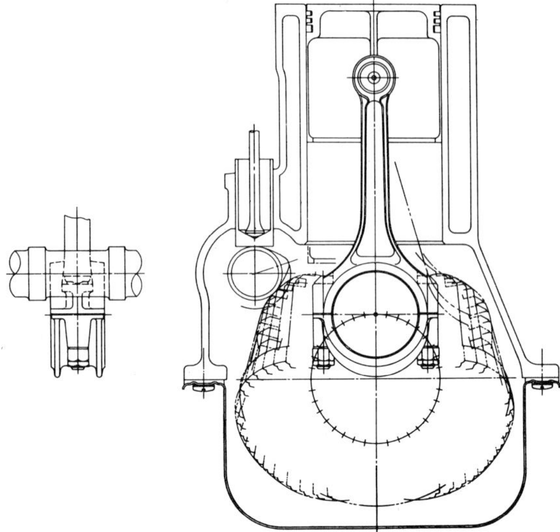

Fig. 11–55. Graphical clearances between the path of a connecting rod and the interior of a crankcase. (Courtesy of Chrysler Corporation.)

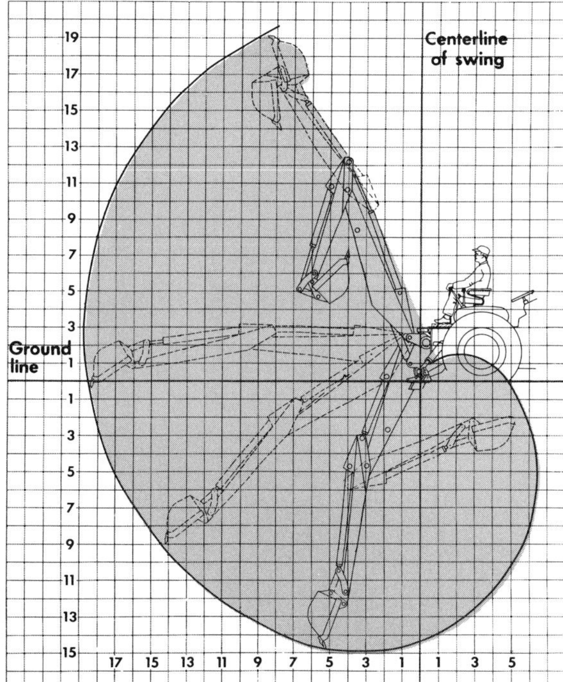

Fig. 11–56. Graphical analysis of the limits of operation of a 3141 Backhoe plotted on graph paper. (Courtesy of International Harvester Company.)

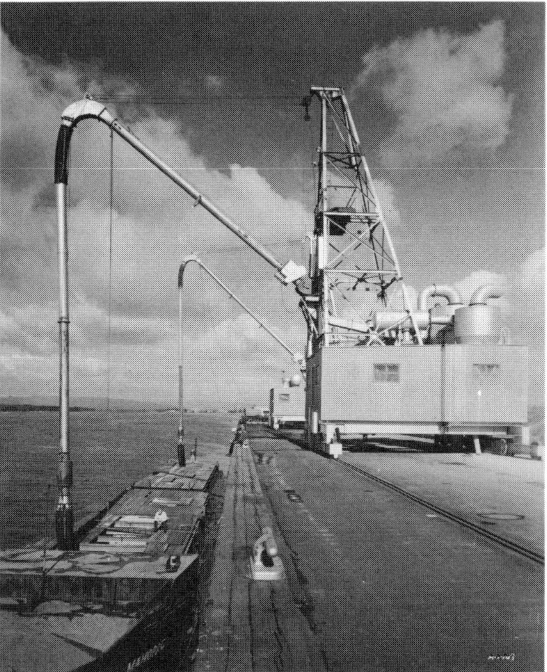

Fig. 11–57. These vacuum systems, which are used for unloading barges, can be analyzed graphically to determine loads and clearances. (Courtesy of General American Transportation Company.)

In Fig. 11–56, the operating positions for a backhoe are plotted in graphical form to permit visual analysis of its linkage system. The graphical presentation of these positions can be used to communicate its operating characteristics and its limits of efficiency.

11–29 EMPIRICAL ANALYSIS OF A LINKAGE

Linkage systems are often designed to be movable so that the equipment may assume a variety of positions. An example of such a design can be seen in the vacuum booms for unloading barges shown in Fig. 11–57. When the booms are raised or lowered in different positions, the loads in them and in the support cables will change. The maximum loads that will be supported by the boom, the strength of the cable, and the boom position will affect the limiting positions of the boom within which it can be safely operated. The boom is analyzed graphically in Fig. 11–58 to establish the load in the cable when the boom is in a number of positions. The data thus obtained can be used to determine the safe operating area.

The boom is shown in seven positions in the space diagram loaded with the maximum load of 1000 lb, which will be a vertical force. Force diagrams are drawn for each of the seven positions. For example, when the boom is at 15°, the load in the cable is 360 lb. The stress in the cable versus the position of the boom is plotted in a graph. These points are connected with a smooth, irregular curve since the change in load is continuous. Given that the cable is designed for an 1800-lb load, the critical point is found to be at 110°. If the boom is lowered

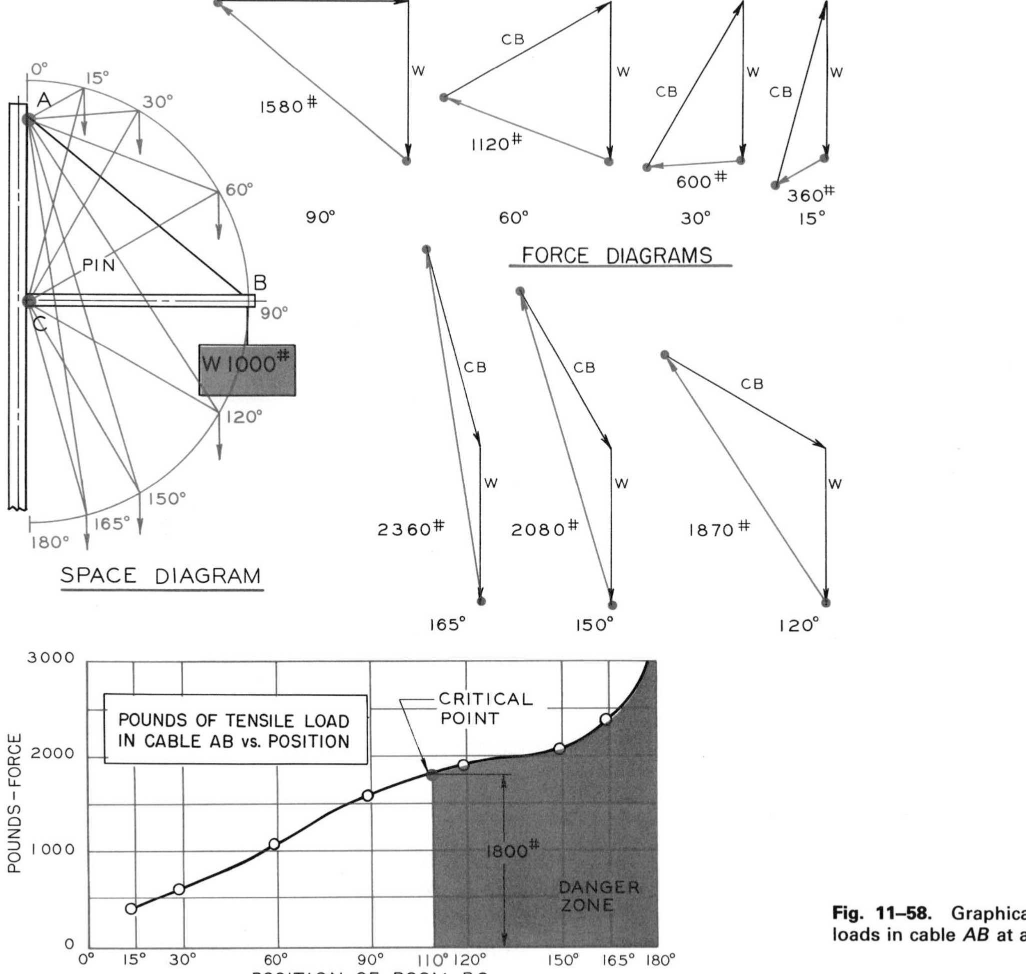

Fig. 11–58. Graphical analysis of the loads in cable *AB* at all positions.

beyond this point, the cable is subject to failure. Inspection of this graph will help the designer select a cable strong enough to withstand the stress imposed it the boom must be designed for lowering beneath the 110° zone. These data are empirical because, only selected points were used.

11–30 INTRODUCTION TO GRAPHICAL CALCULUS

The engineer, designer, and technician must often deal with relationships between variables that must be solved using the principles of calculus. If the equation of the curve is known, traditional methods of calculus will solve the problem. However, many engineering data cannot be converted to standard equations. In these cases, it is desirable to use the graphical method of calculus which provides relatively accurate solutions to irregular problems.

The two basic forms of calculus are (1) differential calculus, and (2) integral calculus. Differential calculus is used to determine the rate of change of one variable with respect to another. For example, the curve plotted in Fig.

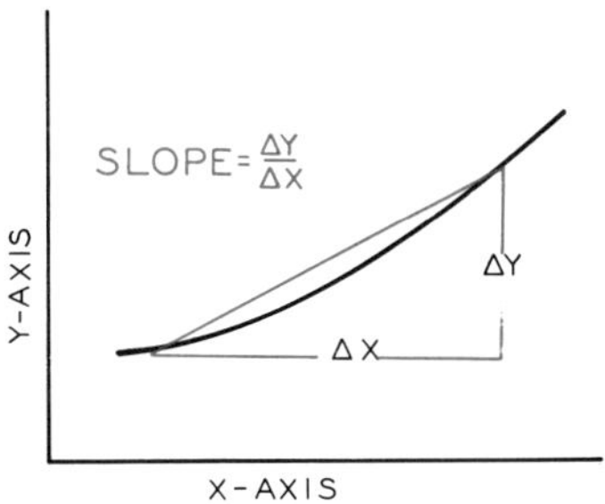

Fig. 11–59. The derivative of a curve is its rate of change at any point, which is the slope of curve $\Delta Y/\Delta X$.

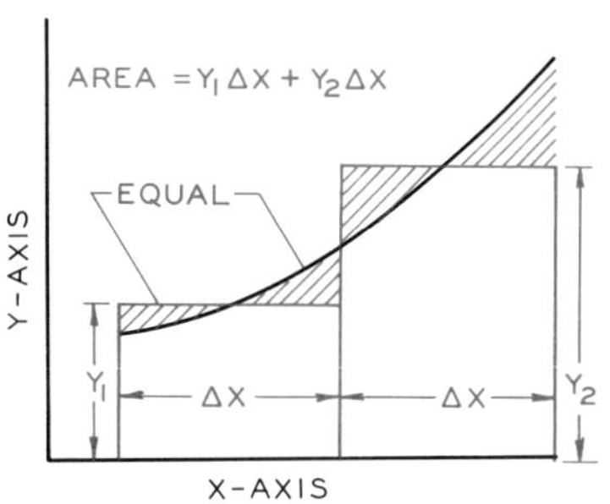

Fig. 11–60. The integral of a curve is the cumulative area enclosed by the curve, which is the product of the two variables.

11–59 represents the relationship between two variables. Note that the Y-variable increases as the X-variable increases. The rate of change at any instant along the curve is the slope of a line that is tangent to the curve at that particular point. This exact slope is often difficult to determine graphically; consequently, it can be approximated by constructing a chord at a given interval, as shown in Fig. 11–59. The slope of this chord can be measured by finding the tangent of $\Delta Y/\Delta X$. These measurements give a general estimate of the slope of the line through this interval if the intervals are selected to be sufficiently small to minimize error. This slope can represent miles per hour, weight versus length, or a number of other meaningful rates that are important to the analysis of data.

Integral calculus is the reverse of differential calculus. Integration is the process of finding the area under a given curve, which can be thought of generally as the product of the two variables plotted on the X- and Y-axes. If one of the variables is area and the other is linear distance, the resulting integral is volume. The area under a curve is approximated by dividing one of the variables into a number of very small intervals, which become small rectangular areas at a particular zone under the curve, as shown in Fig. 11–60. The bars are extended so that as much of the square end of the bar is under the curve as above the curve and the average height of the bar is, therefore, near its midpoint.

11–31 GRAPHICAL DIFFERENTIATION

Graphical differentiation is defined as the determination of the rate of change of two variables with respect to each other at any given point. Figure 11–61 illustrates the preliminary construction of the derivative scale that would be used to plot a continuous derivative curve from the given data.

Step 1. The original data are plotted graphically and the axes are labeled with the proper units of measurement. The maximum scale required for the ordinate on the derivative grid will be equal to the maximum slope of the original data. A chord can be constructed to estimate the maximum slope by inspection. In the given curve, the maximum slope is estimated to be 2.3. A vertical scale is constructed in excess of this to provide for the plotting of slopes that may exceed the estimate. This ordinate scale is drawn to a convenient scale to facilitate measurement.

Step 2. A known slope is plotted on the given data grid. This slope need not be related to the curve in any way. In this case, the slope can be read directly as 1.

Step 3. The pole can be found by drawing a line from the ordinate of 1 (the known slope) on the derivative scale parallel to the slope line. These similar triangles are used to obtain the pole, which will be used in determining the derivative curve.

FIGURE 11–61. SCALES FOR GRAPHICAL DIFFERENTIATION

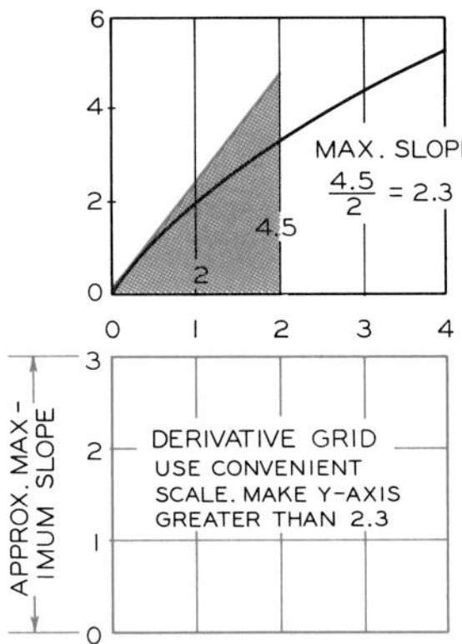

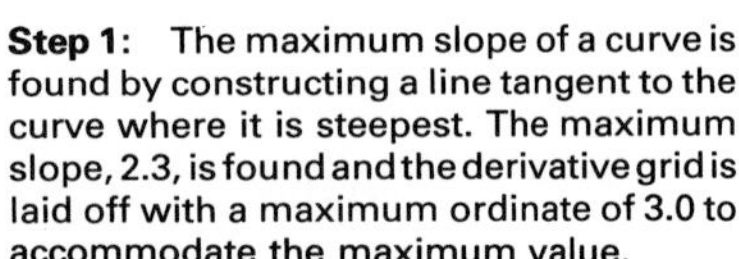

Step 1: The maximum slope of a curve is found by constructing a line tangent to the curve where it is steepest. The maximum slope, 2.3, is found and the derivative grid is laid off with a maximum ordinate of 3.0 to accommodate the maximum value.

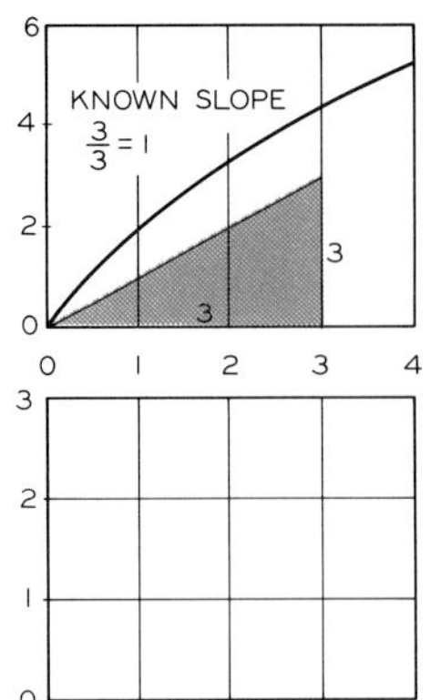

Step 2: A known slope is found on the given grid; this value is 1 in this example. The known slope has no relationship to the curve at this point.

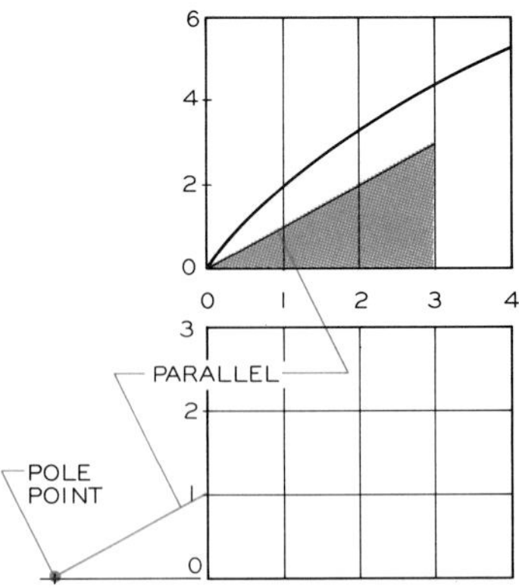

Step 3: Construct a line from 1 on the Y-axis of the derivative grid that is parallel to the slope of the triangle constructed in the given grid. This line locates the pole point where it crosses the extension of the X-axis.

The steps in completing the graphical differentiation are given in Fig. 11–62. Note that the same horizontal intervals used in the given curve are projected directly beneath on the derivative scale. The maximum slope of the data curve is estimated to be slightly less than 12. A scale is selected that will provide an ordinate that will accommodate the maximum slope. A line is drawn from point 14 on the ordinate axis of the derivative grid that is parallel to the known slope on the given curve grid. The point of intersection of this line and the extension of the X-axis is the pole point.

A series of chords are constructed on the given curve. These can be varied in length or interval to best approximate the curve. Lines are constructed parallel to these chords through point P and extended to the Y-axis of the derivative grid to locate bars at each interval. Notice that the interval between 0 and 1 was divided in half to provide a more accurate plot. The curve is the sharpest in this interval. A smooth curve is constructed through the top of these bars in such a manner that the area above the horizontal top of the bar is the same as that below it. This curve represents the derivative of the given data. The rate of change, $\Delta Y/\Delta X$, can be found at any interval of the variable X by reading directly from the graph at the value of X in question.

This system of graphical differentiation can be used to find the derivative curve of irregular and empirical data that cannot be expressed by a standard equation. Graphical differentiation is important, since many engineering data do not fit algebraic forms.

11–32 APPLICATIONS OF GRAPHICAL DIFFERENTIATION

The mechanical handling shuttle shown in Fig. 11–63 is used to convert rotational motion into controlled linear motion. A scale drawing of the linkage components is given so that graphical analysis can be applied to determine the motion resulting from this system.

The linkage is drawn to show the end positions of point P, which will be used as the zero point for plotting the travel versus the degrees of revolution. Since rotation is constant at one revolution per three seconds, the degrees of revolution can be converted to time, as shown in the data curve given at the top of Fig. 11–64. The drive crank, R_1, is revolved at 30° intervals, and the distance that point P travels from its end position is plotted on the graph, as shown in the given data. This gives the distance-versus-time relationship.

We determine the ordinate scale of the derivative grid by estimating the maximum slope of the given data curve, which is found to be a little less than 100 in/sec. A convenient

FIGURE 11–62. GRAPHICAL DIFFERENTIATION

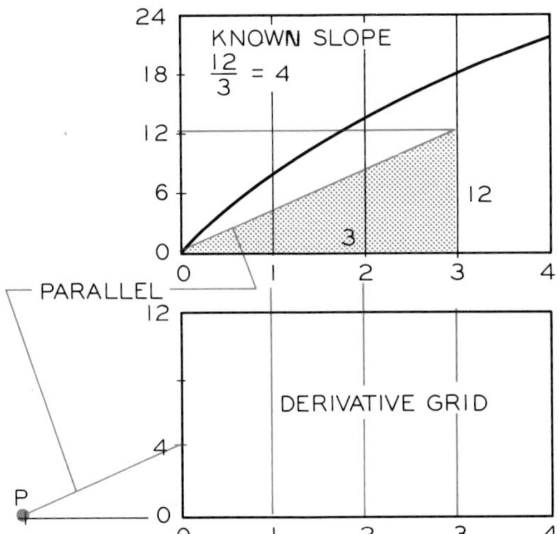

Required: Find the derivative curve of the given data.

Step 1: Find the derivative grid and the pole point using the construction illustrated in Fig. 11–61.

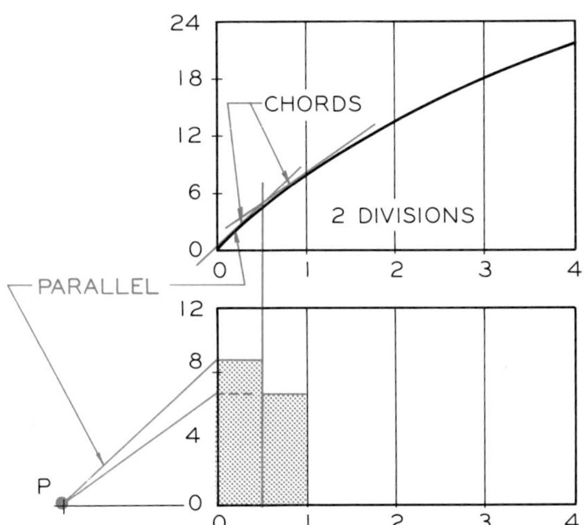

Step 2: Construct a series of chords between selected intervals on the given curve and draw lines parallel to these chords through point *P* on the derivative grid. These lines locate the heights of bars in their respective intervals. Notice that the first interval is divided into two intervals, since the curve is changing sharply in this interval.

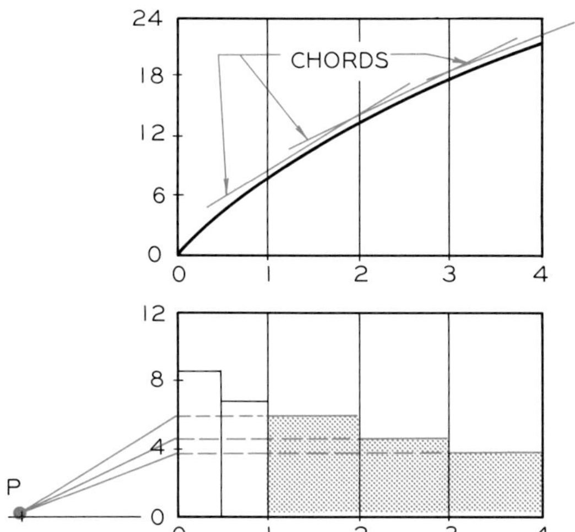

Step 3: Additional chords are drawn in the last two intervals. Lines parallel to these chords are drawn from the pole point to the *Y*-axis to find additional bars in their respective intervals.

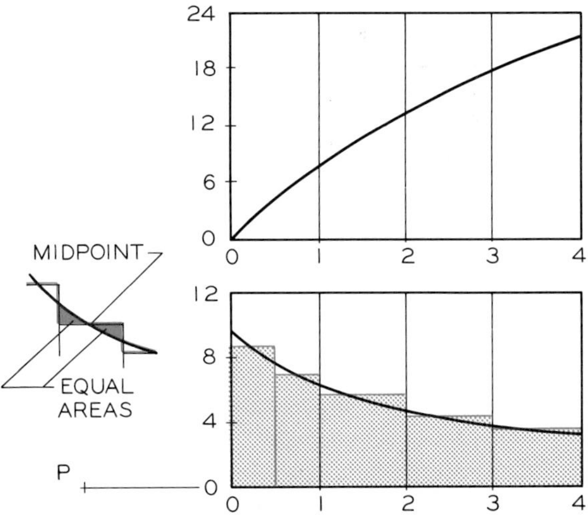

Step 4: The vertical bars represent the different slopes of the given curve at different intervals. The derivative curve is drawn through the midpoints of the bars so that the areas under and above the bars are approximately equal.

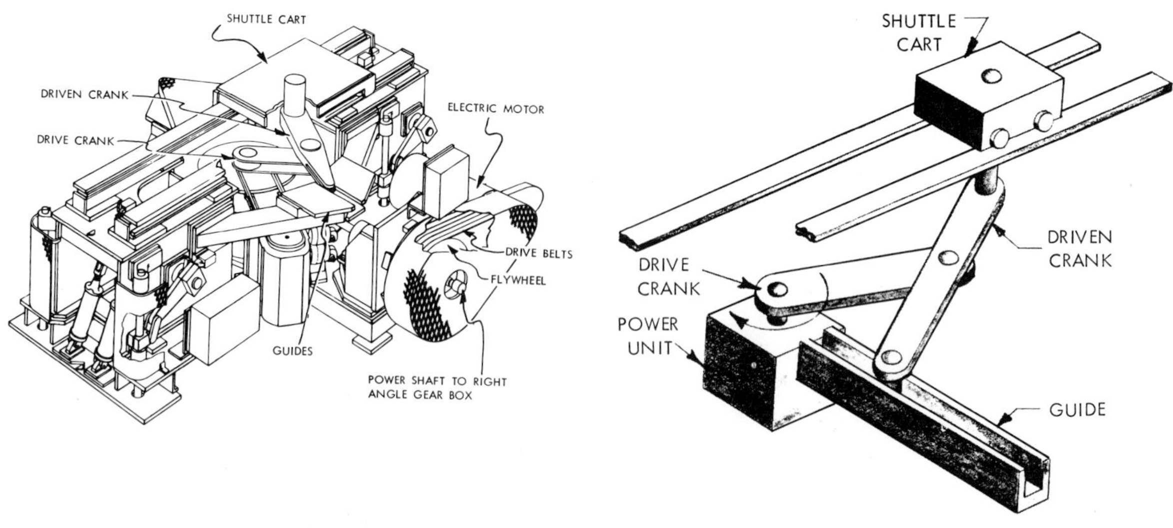

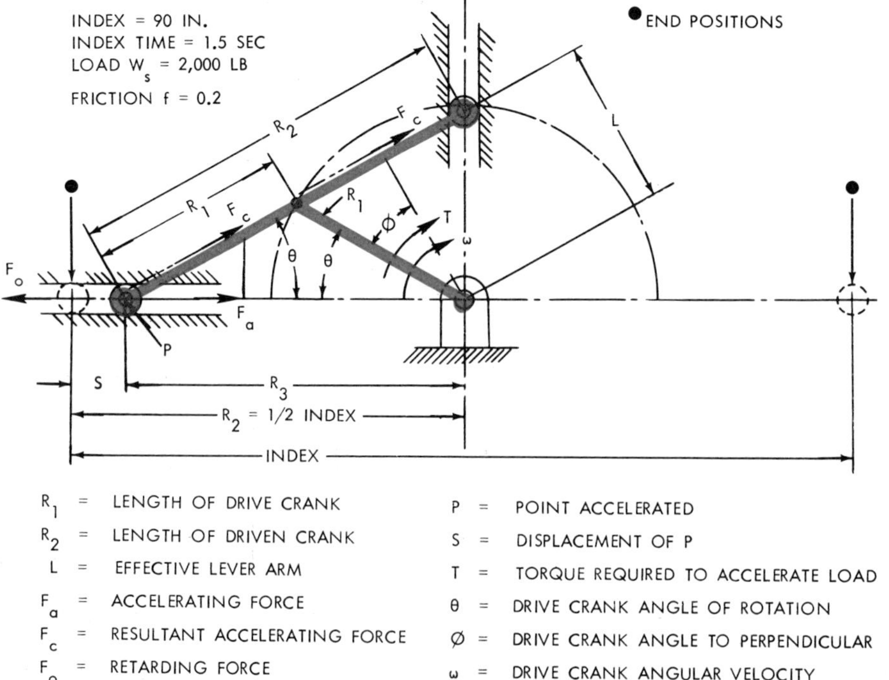

R_1	=	LENGTH OF DRIVE CRANK	P	=	POINT ACCELERATED
R_2	=	LENGTH OF DRIVEN CRANK	S	=	DISPLACEMENT OF P
L	=	EFFECTIVE LEVER ARM	T	=	TORQUE REQUIRED TO ACCELERATE LOAD
F_a	=	ACCELERATING FORCE	θ	=	DRIVE CRANK ANGLE OF ROTATION
F_c	=	RESULTANT ACCELERATING FORCE	ϕ	=	DRIVE CRANK ANGLE TO PERPENDICULAR
F_o	=	RETARDING FORCE	ω	=	DRIVE CRANK ANGULAR VELOCITY

Fig. 11–63. A pictorial and scale drawing of an electrically powered mechanical handling shuttle used to move automobile parts on an assembly line. (Courtesy of General Motors Corporation.)

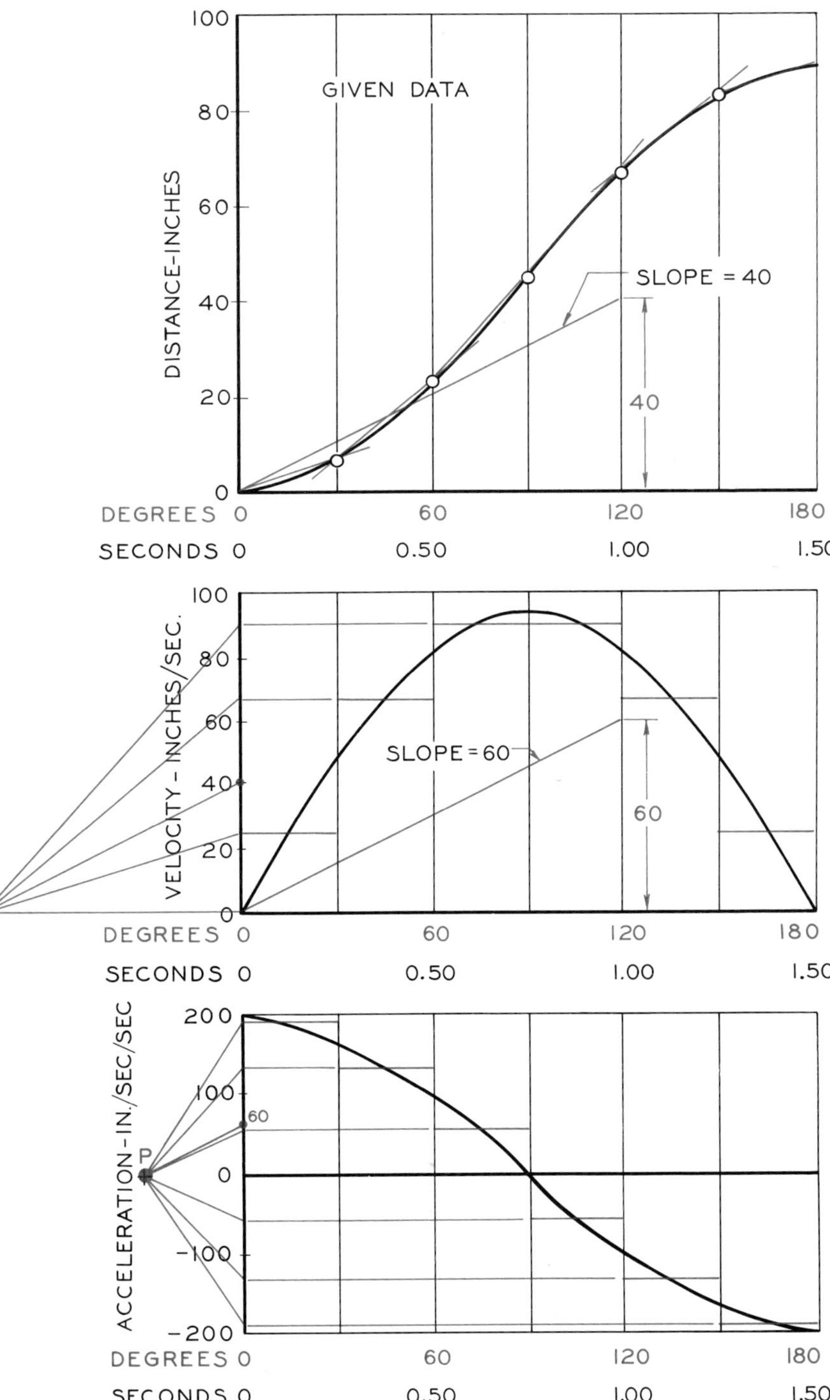

Fig. 11–64. Graphical determination of velocity and acceleration of the mechanical handling shuttle by differential calculus.

scale is chosen that will be used for the derivative curve; the maximum limit is 100 units. A slope of 40 is drawn on the given data curve; this will be used in determining the location of pole *P* in the derivative grid. From point 40 on the derivative ordinate scale, we draw a line parallel to the known slope, which is found on the given grid. Point *P* is the point where this line intersects the extension of the *X*-axis.

A series of chords are drawn on the given curve to approximate the slope at various points. Lines are constructed through point *P* of the derivative scale parallel to the chord lines of the given curve and extended to the ordinate scale. The points thus obtained are then projected across to their respective intervals to form vertical bars. A smooth curve is drawn through the top of each of the bars to give an average of the bars. This curve can be used to find the velocity of the shuttle in inches per second at any time interval.

The construction of the second derivative curve, the acceleration, is very similar to that of the first derivative. By inspecting the first derivative, we estimate the maximum slope to be 200 in./sec/sec. An easily measured scale is established for the ordinate scale of the second derivative curve. Point *P* is found in the same manner as the first derivative.

Chords are drawn at intervals on the first derivative curve. Lines are drawn parallel to these chords from point *P* in the second derivative curve to the *Y*-axis, where they are projected horizontally to their respective intervals to form a series of bars. A smooth curve is drawn through the tops of the bars to give a close approximation of the average areas of the bars. Note that a minus scale is given for the acceleration curve to indicate deceleration.

The maximum acceleration is found to be at the extreme endpoints and the minimum acceleration is at 90°, where the velocity is the maximum. It can be seen from the velocity and acceleration plots that the parts being handled by the shuttle are accelerated at a rapid rate until the maximum velocity is attained at 90°, at which time deceleration begins and continues until the parts come to rest.

Cam displacement diagrams can be analyzed to determine the velocity and acceleration of the follower at any instant during the cams' revolution. The velocity and acceleration of the connecting rod in Fig. 11–51 can be found by graphical differentiation.

11–33 GRAPHICAL INTEGRATION

Integration is the process of determining the area (product of two variables) under a given curve. For example, if the *Y*-axis were pounds and the *X*-axis were feet, the integral curve would give the product of the variables, foot-pounds, at any interval of feet along the *X*-axis. Figure 11–65 depicts the method of constructing scales for graphical integration.

Step 1. It is customary to locate the integral curve above the given data curve, since the integral will be an equation raised to a higher power. A line is drawn through the given data curve to approximate the total area under the curve. This line is estimated to go through point 5 on the ordinate to give approximately equal areas above and below the curve. The approximate area is 4 × 5, or 20, square units of area. The ordinate scale is drawn on the integral curve in excess of 20 units to provide a margin for any overage. The horizontal scale intervals are projected from the given curve to the integral grid.

Step 2. The ordinate at any point on the integral scale will have the same numerical value as the area under the curve as measured from the origin to that point on the given data grid. The ordinate at point 2 on the *X*-axis directly above the rectangle must be equal to its area of 8. A slope is drawn from the origin to the ordinate of 8.

Step 3. Point *P* is found by drawing a line from point 4 on the given grid parallel to the slope established in the integral grid. This line intersects the extension of the *X*-axis at point *P*. This point will be used to find the integral curve.

FIGURE 11–65. SCALES FOR GRAPHICAL INTEGRATION

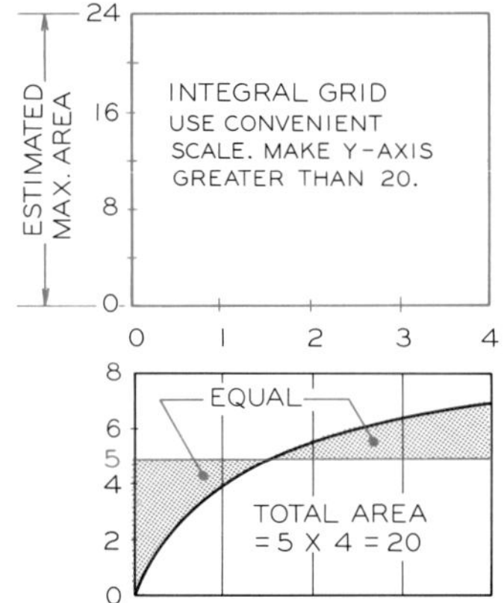

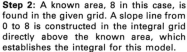

Step 1: To determine the maximum value on the *Y*-axis of the integral curve, a line is drawn to approximate the area under the given grid. This is found to be 20, and the *Y*-axis of the integral grid is constructed with 24 as the maximum value.

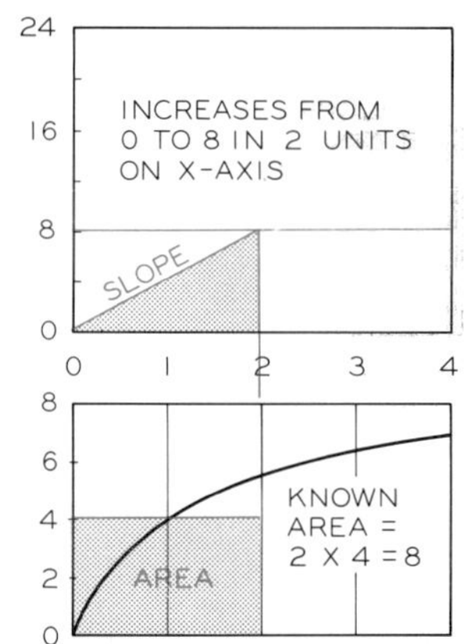

Step 2: A known area, 8 in this case, is found in the given grid. A slope line from 0 to 8 is constructed in the integral grid directly above the known area, which establishes the integral for this model.

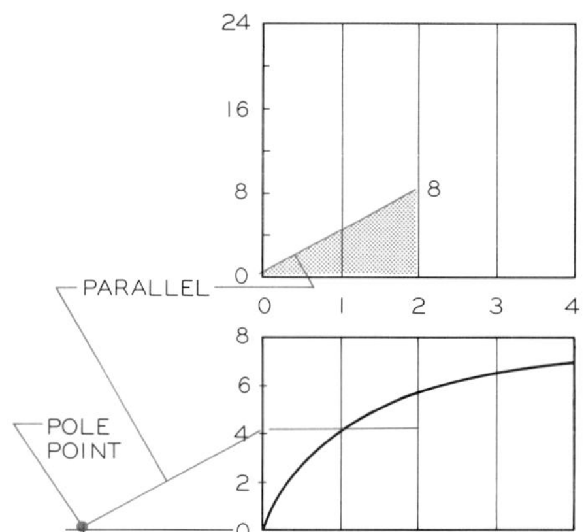

Step 3: A line is drawn from 4 on the *Y*-axis of the given grid parallel to the slope line in the integral grid. This line from 4 crosses the extension of the *X*-axis to locate the pole point.

The technique illustrated in Fig. 11–66 can be applied to most integration problems. The equation of the given curve is $Y = 2X^2$, which can also be integrated mathematically as a check.

From the given grid, the total area under the curve can be estimated to be less than 40 units. This value becomes the maximum height of the Y-axis on the integral curve. A convenient scale is selected, units are assigned to the ordinate, and the pole point, P, is found.

A series of vertical bars are constructed to approximate the areas under the curve at these intervals. The narrower the bars, the more accurate will be the resulting calculations. Notice that the interval between 1 and 2 was divided in half to provide a more accurate plot. The top lines of the bars are extended horizontally to the Y-axis, where the points are then connected by lines to point P. Lines are drawn parallel to AP, BP, CP, DP, and EP in the integral grid to correspond to the respective intervals in the given grid. The intersection points of the chords are connected by a smooth curve—the integral curve. This curve gives the cumulative product of the X- and Y-variables at any value along the X-axis. For example, the area under the curve at $X = 3$ can be read directly as 18.

Mathematical integration gives the following result for the area under the curve from 0 to 3:

$$\text{Area } A = \int_0^3 Y\,dX, \qquad \text{where } Y = 2X^2;$$

$$A = \int_0^3 2X^2\,dX = \tfrac{2}{3}X^3\Big|_0^3 = 18$$

11–34 APPLICATIONS OF GRAPHICAL INTEGRATION

Integration is commonly used in the study of the strength of materials to determine shear, moments, and deflections of beams. An example problem of this type is shown in Fig. 11–67, in which a truck exerts a total force of 36,000 lb on a beam that is used to span a portion of a bridge. The first step is to determine the resultants supporting each of the beam.

A scale drawing of the beam is made with the loads concentrated at their respective positions. A force diagram is drawn, using Bow's

FIGURE 11–66. GRAPHICAL INTEGRATION

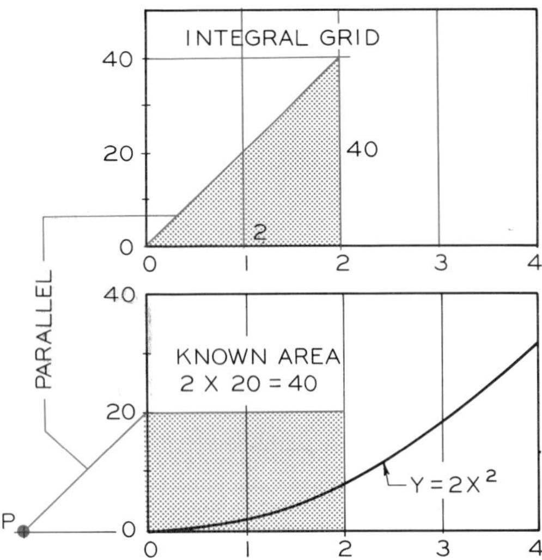

Required: Plot the integral curve of the given data.

Step 1: Find the pole point, *P*, using the technique illustrated in Fig. 11–65.

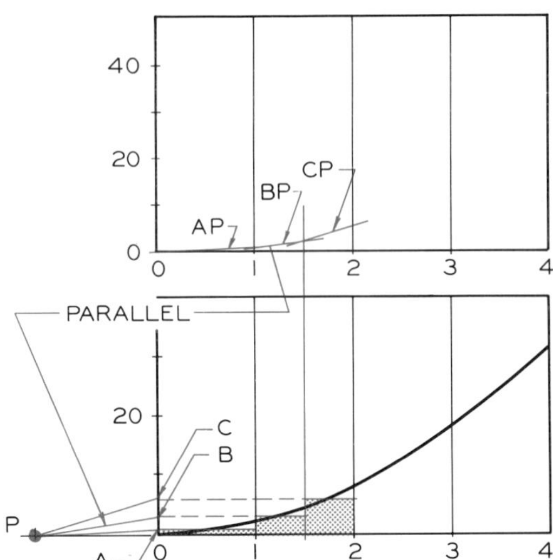

Step 2: Construct bars at intervals to approximate the areas under the curve. Notice that the interval from 1 to 2 was divided in half to improve the accuracy of the approximation. The heights of the bars are projected to the *Y*-axis and lines are drawn to the pole point. Sloping lines *AP*, *BP*, and *CP* are drawn in their respective intervals and parallel to the lines drawn to the pole point.

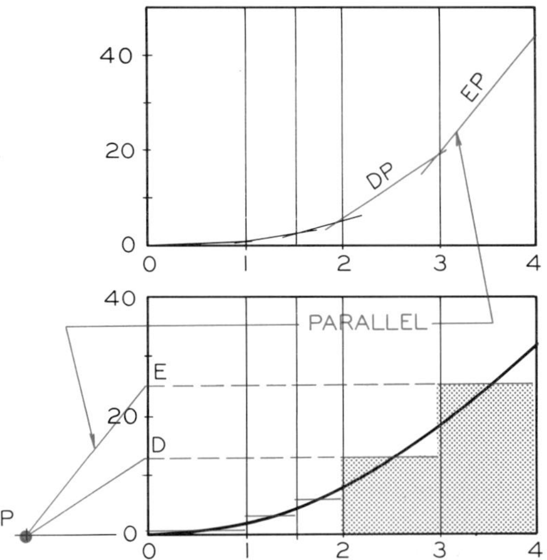

Step 3: Additional bars are drawn from 2 to 4 on the *X*-axis. The heights of the bars are projected to the *Y*-axis and rays are drawn to the pole point, *P*. Lines *DP* and *DE* are drawn in their respective intervals and parallel to their rays in the integral grid.

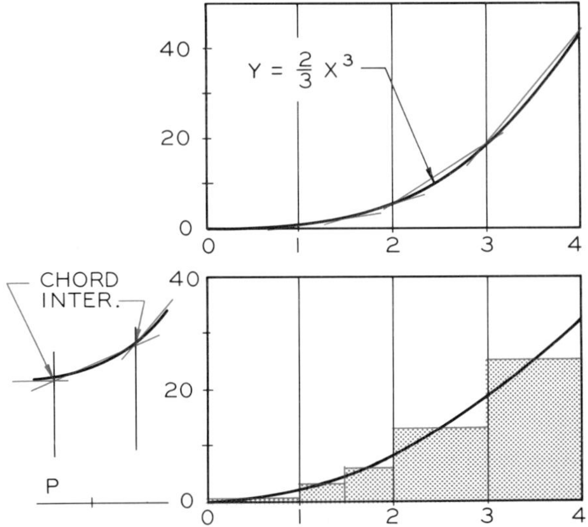

Step 4: The straight lines that were connected in the integral grid represent chords of the integral curve. Construct the integral curve to pass through the points where the chords intersect. Any ordinate value on the integral curve represents the cumulative area under the given curve from zero to that point on the *X*-axis.

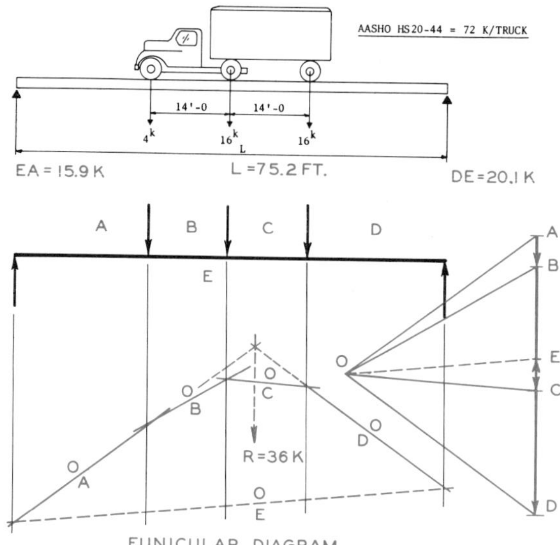

Fig. 11–67. Determination of the forces on a beam of a bridge and its total resultant.

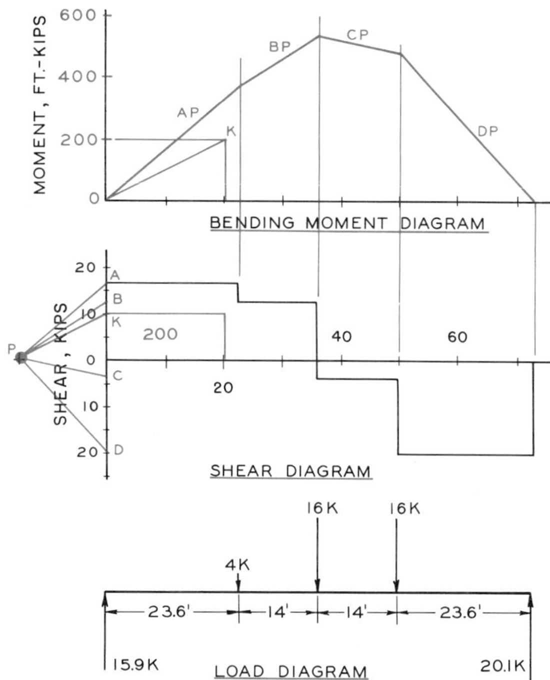

Fig. 11–68. Determination of shear and bending moment by graphical integration.

notational system for laying out the vectors in sequence. Pole point *O* is located, and rays are drawn from the ends of each vector to point *O*. The lines of force in the load diagram at the top of the figure are extended to the funicular diagram. Then lines are drawn parallel to the rays between the corresponding lines of force. For example, ray *OA* is drawn in the *A*-interval in the funicular diagram. The closing ray of the funicular diagram, *OE*, is transferred to the vector diagram by drawing a parallel through point *O* to locate point *E*. Vector *DE* is the right-end resultant of 20.1 kips (one kip equals 1000 lb) and *EA* is the left-hand resultant of 15.9 kips. The origin of the resultant force of 36 kips is found by extending *OA* and *OD* in the funicular diagram to their point of intersection.

From the load diagram shown in Fig. 11–68 we can, by integration, find the shear diagram, which indicates the points in the beam where failure due to crushing is most critical. Since the applied loads are concentrated rather than uniformly applied, the shear diagram will be composed of straight-line segments. In the shear diagram, the left-end resultant of 15.9 kips is drawn to scale from the axis. The first load of 4 kips, acting in a downward direction, is subtracted from this value directly over its point of application, which is projected from the load diagram. The second load of 16 kips also exerts a downward force and so is subtracted from the 11.9 kips (15.9 − 4). The third load of 16 kips is also subtracted, and the right-end resultant will bring the shear diagram back to the *X*-axis. It can be seen that the beam must be designed to withstand maximum shear at each support and minimum shear at the center.

The moment diagram is used to evaluate the bending characteristics of the applied loads in foot-pounds at any interval along the beam. The ordinate of any *X*-value in the moment diagram must represent the cumulative foot-pounds in the shear diagram as measured from either end of the beam.

Pole point *P* is located in the shear diagram by applying the method described in Fig. 11–65. A rectangular area of 200 ft-kips is found in the shear diagram. We estimate the total area to be less than 600 ft-kips; so we select a convenient scale that will allow an ordinate scale of 600 units for the moment diagram. We locate the area of 200 ft-kips in the moment diagram by projecting the 20-foot mark on the X-axis until it intersects with a 200-unit projection from the Y-axis, locating point K. The diagonal, OK, is transferred to the shear diagram, where it is drawn from the ordinate of the given rectangle to point P on the extension of the X-axis. Rays AP, BP, CP, and DP are found in the shear diagram by projecting horizontally from the various values of shear. In the moment diagram, these rays are then drawn in their respective intervals to form a straight-line curve that represents the cumulative area of the shear diagram, which is in units of ft-kips. Maximum bending will occur at the center of the beam, where the shear is zero. The bending is scaled to be about 560 ft-kips. The beam selected for this span must be capable of withstanding a shear of 20.1 kips and a bending moment of 560 ft-kips.

11–35 NOMOGRAPHY*

An additional aid in analyzing data is a graphical computer called a *nomogram* or *nomograph.* Basically, a nomogram or "number chart" is any graphical arrangement of calibrated scales and lines which may be used to facilitate calculations, usually those of a repetitive nature. The most frequently used nomograms are the common line graphs previously discussed in this chapter.

The term "nomogram" is frequently used to denote a specific type of scale arrangement called an alignment chart. Typical examples of alignment charts are shown in Fig. 11–69. Many other types are also used which have curved scales or other scale arrangements for more complex problems. The discussion of nomograms in this chapter will be limited to the simpler conversion, parallel-scale, and N-type charts and their variations.

Using an Alignment Chart. An alignment chart is usually constructed to help solve for one or more unknowns in a formula or empirical relationship between two or more quantities, for example, to convert degrees Celsius to degrees Fahrenheit, to find the size of a structural member to sustain a certain load, and so on. An alignment chart is read by placing a straightedge, or by drawing a line called an *isopleth*, across the scales of the chart and reading corresponding values from the scales on this line. The example in Fig. 11–70 shows readings for the formula $U + V = W$.

* Articles 11–35 through 11–41 were written by Michael P. Guerard, formerly of the Department of Engineering Design Graphics, Texas A&M University.

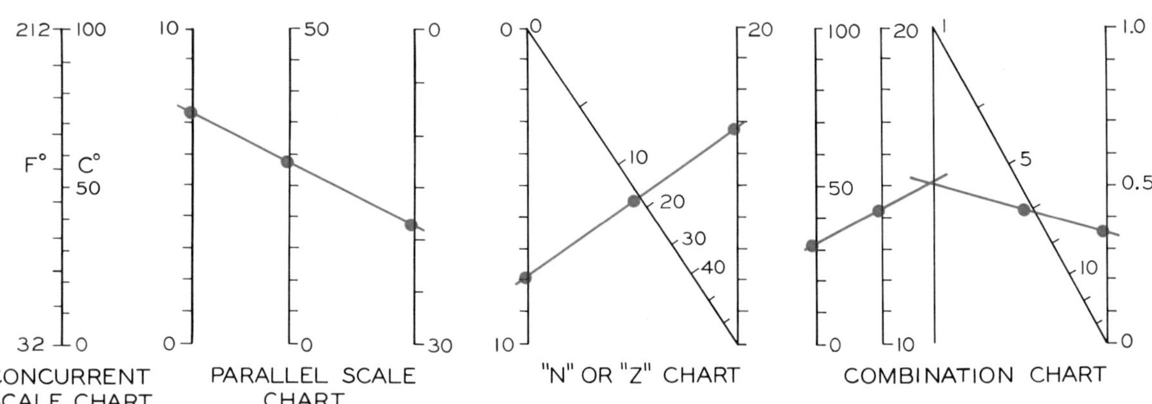

Fig. 11–69. Typical examples of types of alignment charts.

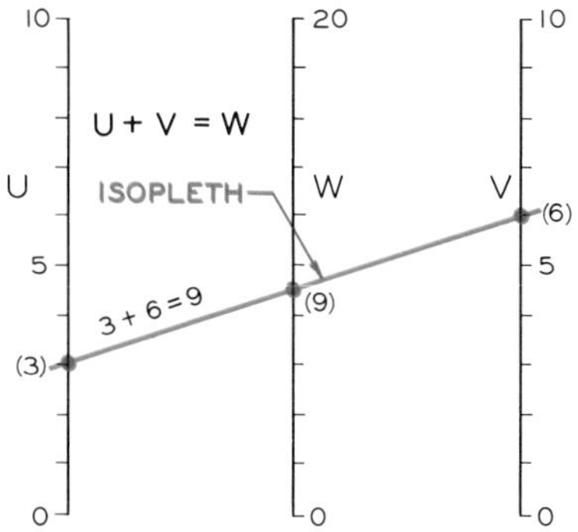

Fig. 11–70. Use of an isopleth to solve graphically for unknowns in the given equation.

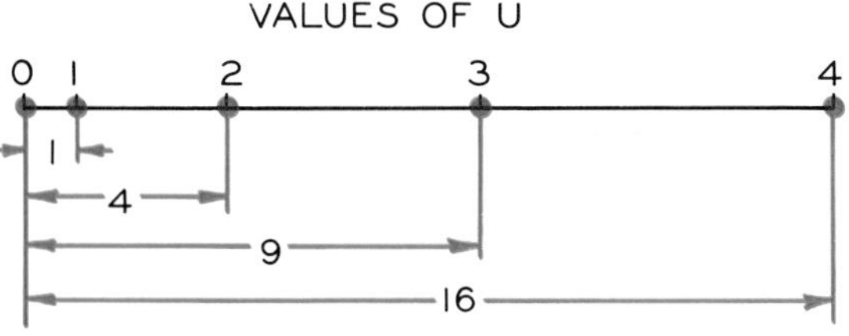

Fig. 11–71. Functional scale for units of measurement that are proportional to $F(U) = U^2$.

11–36 ALIGNMENT CHART SCALES

To construct any alignment chart, we must first determine the graduations of the scales that will be used to give the desired relationships. Alignment-chart scales are called *functional scales.* A functional scale is one that is graduated according to values of some *function* of a variable, but *calibrated* with values of the variable. A functional scale for $F(U) = U^2$ is illustrated in Fig. 11–71. It can be seen in this example that if a value of $U = 2$ was substituted into the equation, the position of U on the functional scale would be 4 units from zero, or $2^2 = 4$. This procedure can be repeated with all values of U by substitution.

The Scale Modulus. Since the graduations on a functional scale are spaced in proportion to values of the function, a proportionality, or scaling factor, is needed. This constant of proportionality is called the *scale modulus* and it is given by the equation

$$m = \frac{L}{F(U_2) - F(U_1)} \tag{1}$$

where

m = scale modulus, in inches per functional unit,

L = desired length of the scale, in inches,

$F(U_2)$ = function value at the end of the scale,

$F(U_1)$ = function value at the start of the scale.

For example, suppose that we are to construct a functional scale for $F(U) = \sin U$, with $0° \leq U \leq 45°$ and a scale 6″ in length. Thus $L = 6''$, $F(U_2) = \sin 45° = 0.707$, $(F(U_1) = \sin 0° = 0$. Therefore Eq. (1) can be written in the following form by substitution:

$$m = \frac{6}{0.707 - 0} = 8.49 \text{ inches per (sine) unit}$$

The Scale Equation. Graduation and calibration of a functional scale are made possible by a *scale equation.* The general form of this equation may be written as a variation of Eq. (1) in the following form:

$$X = m[f(U) - F(U_1)] \tag{2}$$

where

X = distance from the measuring point of the scale to any graduation point,

m = scale modulus,

$F(U)$ = functional value at the graduation point,

$F(U_1)$ = functional value at the measuring point of the scale.

For example, a functional scale is constructed for the previous equation, $F(U) = \sin U$ ($0° \leq U \leq 45°$). It has been determined that $m = 8.49$, $F(U) = \sin U$, and $F(U_1) = \sin 0° = 0$. Thus by substitution the scale equation, (2), becomes

$$X = 8.49\,(\sin U - 0) = 8.49 \sin U$$

Using this equation, we can substitute values of U and construct a table of positions. In this

Table 11–1

U	0°	5°	10°	15°	20°	25°	30°	35°	40°	45°
X	0	0.74	1.47	2.19	2.90	3.58	4.24	4.86	5.45	6.00

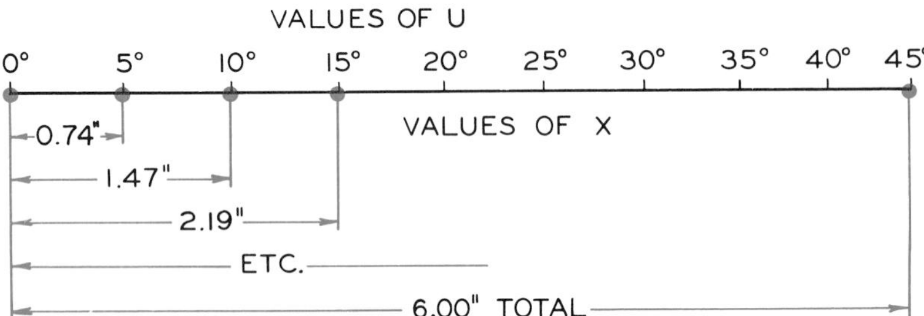

Fig. 11–72. Construction of a functional scale using values from Table 11–1, which were derived from the scale equation.

case, the scale is calibrated at 5° intervals, as reflected in Table 11–1.

The values of X from the table give the positions, in inches, for the corresponding graduations, measured from the start of the scale ($U = 0°$); see Fig. 11–72. It should be noted that the measuring point does *not* need to be at one end of the scale, but it is usually the most convenient point, especially if the functional value is zero at that point.

11–37 CONCURRENT SCALE CHARTS

Concurrent scale charts are useful in the rapid conversion of one value into terms of a second system of measurement. Formulas of the type $F_1 = F_2$, which relate two variables, can be adapted to the concurrent scale format. Typical examples might be the Fahrenheit-Celsius temperature relation,

$$°F = \tfrac{9}{5}°C + 32$$

or the area of a circle,

$$A = \pi r^2$$

Design of a concurrent-scale chart involves the construction of a functional scale for each side of the mathematical formula in such a manner that the *position* and *lengths* of each scale coincide. For example, to design a conversion chart 5″ long that will give the areas of circles whose radii range from 1 to 10, we first write $F_1(A) = A$, $F_2(r) = \pi r^2$, and $r_1 = 1$, $r_2 = 10$. The scale modulus for r is

$$m_r = \frac{L}{F_2(r_2) - F_2(r_1)}$$
$$= \frac{5}{\pi(10)^2 - \pi(1)^2} = 0.0161$$

Thus the scale equation for r becomes

$$X_r = m_r[F_2(r) - F_2(r_1)]$$
$$= 0.0161[\pi r^2 - \pi(1)^2]$$
$$= 0.0161\pi(r^2 - 1)$$
$$= 0.0505(r^2 - 1)$$

A table of values for X_r and r may now be completed as shown in Table 11–2. The r-scale can be drawn from this table, as shown in Fig. 11–73. From the original formula, $A = \pi^2$, the limits of A are found to be $A_1 = \pi = 3.14$ and $A_2 = 100\pi = 314$. The scale modulus for concurrent scales is always the same for equal-length scales; therefore $m_A = m_r = 0.0161$, and the scale equation for A *becomes*

$$X_A = m_A[F_1(A) - F_1(A_1)]$$
$$= 0.0161(A - 3.14)$$

Table 11–2

r	1	2	3	4	5	6	7	8	9	10
X_r	0	0.15	0.40	0.76	1.21	1.77	2.42	3.18	4.04	5.00

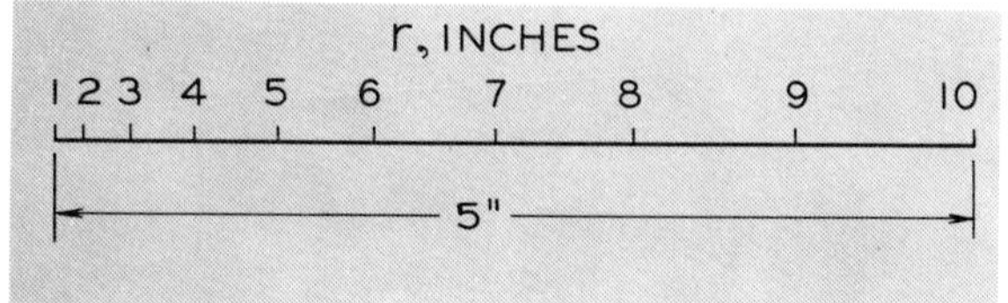

Fig. 11–73. Calibration of one scale of a concurrent scale chart using values from Table 11–2.

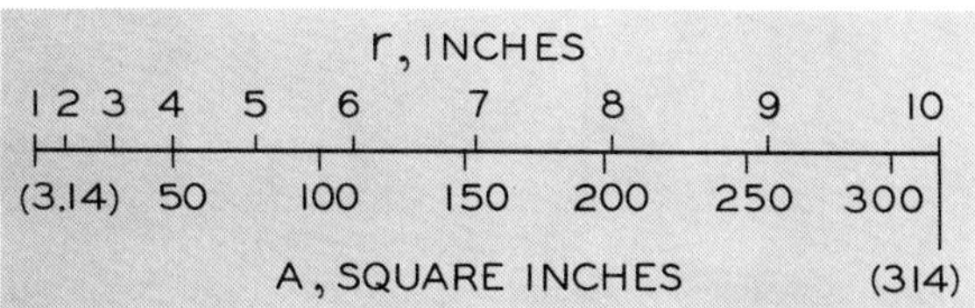

Fig. 11–74. The completed concurrent scale chart for the formula $A = \pi r^2$. Values for the A-scale are taken from Table 11–3.

Table 11–3

A	(3.14)	50	100	150	200	250	300	(314)
X_A	0	0.76	1.56	2.36	3.16	3.96	4.76	5.00

The corresponding table of values is then computed for selected values of A, as shown in Table 11–3.

The A-scale is now superimposed on the r-scale; its calibrations have been placed on the other side of the line to facilitate reading (Fig. 11–74). It may be desired to expand or contract one of the scales, in which case an alternative arrangement may be used, as shown in Fig. 11–75. The two scales are drawn parallel at any convenient distance, and calibrated in *opposite* directions. A different scale modulus and corresponding scale equation must be calculated for each scale if they are *not* the same length.

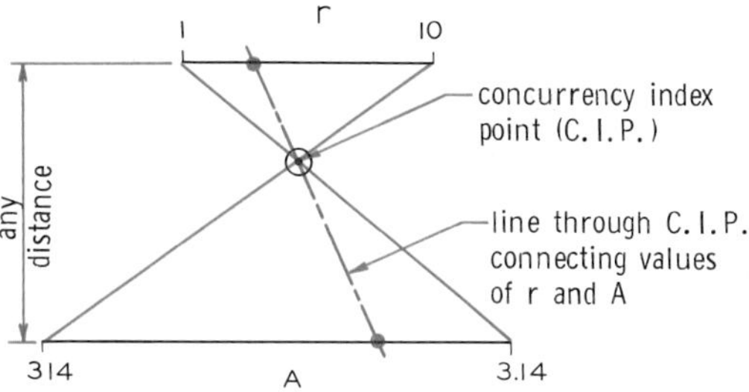

Fig. 11–75. Concurrent scale chart with unequal scales.

11–38 CONSTRUCTION OF ALIGNMENT CHARTS WITH THREE VARIABLES

For a formula of three functions (of one variable each), the general approach is to select the lengths and positions of *two* scales according to the range of variables and size of the chart desired. These are then calibrated by means of the scale equations, as shown in the preceding article. The position and calibration of the third scale will then depend upon these initial constructions. Although definite mathematical relationships exist which may be used to locate the third scale, graphical constructions are simpler and usually less subject to error. Examples of the various forms are presented in the following articles.

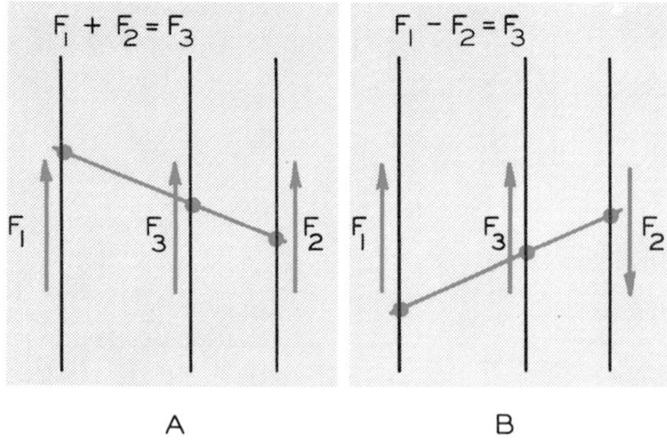

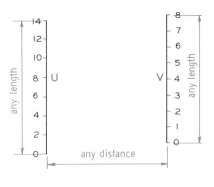

Fig. 11–76. Two common forms of parallel-scale alignment charts.

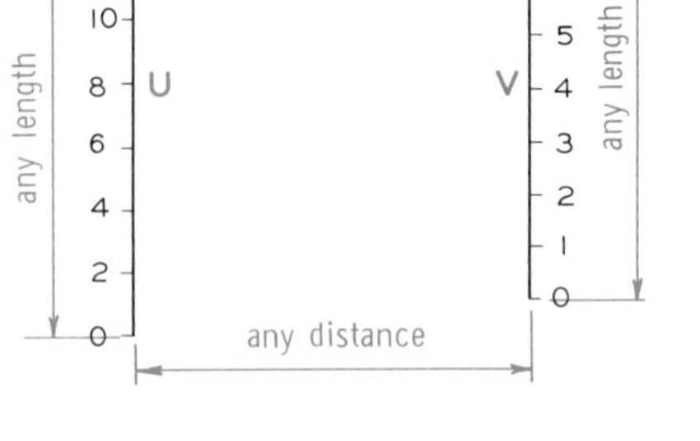

Fig. 11–77. Calibration of the outer scales for the formula $U+2V=3W$, where $0 \leq U \leq 14$ and $0 \leq V \leq 8$.

11–39 PARALLEL-SCALE CHARTS

Many engineering relationships involve three variables that can be computed graphically on a repetitive basis. Any formula of the type $F_1 + F_2 = F_3$ may be represented as a parallel-scale alignment chart, as shown in Fig. 11–76A. Note that all scales increase (functionally) in the same direction and that the function of the middle scale represents the *sum* of the other two. Reversing the direction of any scale changes the sign of its function in the formula, as for $F_1 - F_2 = F_3$ in Fig. 11–76B.

To illustrate this type of alignment chart, we shall use the formula $U + 2V = 3W$, where $0 \leq U \leq 14$ and $0 \leq V \leq 8$. First it is necessary to detemine and calibrate the two outer scales for U and V; we can make them any convenient length and position them any convenient distance apart, as shown in Fig. 11–77. These scales are used as the basis for the step-by-step construction shown in Fig. 11–78.

The limits of calibration for the middle scale are found by connecting the endpoints of the outer scales and substituting these values into the formula. Here, W is found to be 0 and 10 at the extreme ends (step 1). Two pairs of corresponding values of U and V are selected that will give the same value of W. For example, values of $U = 0$ and $V = 7.5$ give a value of 5 for W. We also find that $W = 5$ when $U = 14$ and $V = 0.5$. This should be verified by substitution before continuing with construction. We connect these corresponding pairs of values with isopleths to locate their intersection, which establishes the position of the W-scale.

Since the W-scale is linear ($3W$ is a linear function), it may be subdivided into uniform intervals by the methods commonly used to divide a line into equal parts (step 2). For a nonlinear scale, the scale modulus (and the scale equation) may be found in step 2 by substituting its length and its two end values into Eq. (1) of Article 11–36. The scales can be used to determine an infinite number of problem solutions when sets of two variables are known, as illustrated in step 3.

Parallel-Scale Graph with Logarithmic Scales. Problems involving formulas of the type $F_1 \times F_2 = F_3$ can be solved in a manner very similar to the example given in Fig. 11–78 when logarithmic scales are used. An example of this type of problem is the formula $R = S\sqrt{T}$, for $0.1 \leq S \leq 1.0$ and $1 \leq T \leq 100$. Assume the scales to be 6″ long. These scales need not be equal except for convenience. This formula may be converted into the required form by taking common logarithms of both sides, which gives

$$\log R = \log S + \tfrac{1}{2} \log T$$

FIGURE 11–78. PARALLEL-SCALE CHART (LINEAR)

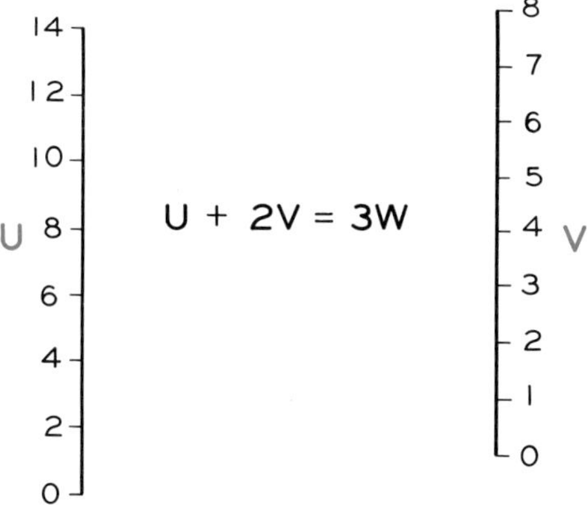

Given: The equation $U + 2V = 3W$ and the two scales constructed in Fig. 15–77.

Required: Construct a parallel-scale chart for determining various solutions of the given formula.

References: Articles 11–36 to 11–40.

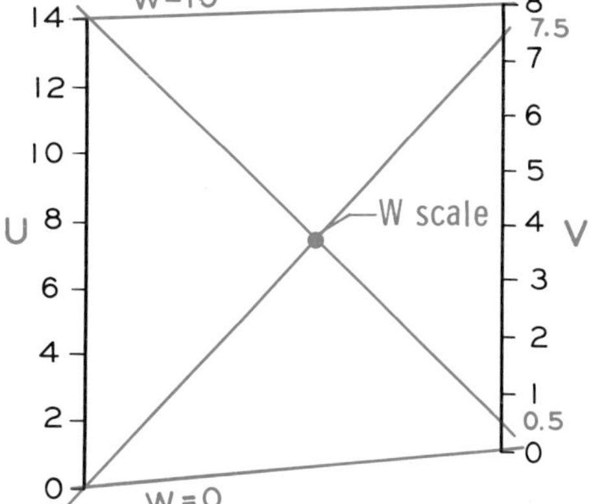

Step 1: Substitute the end values of the *U*- and *V*-scales into the formula to establish the extreme values of the *W*-scale. These values are found to be $W = 10$ and $W = 0$. Select two sets of corresponding values of *U* and *V* that will give the same value of *W*. *Example:* When $U = 0$ and $V = 7.5$, *W* will equal 5, and when $U = 14$ and $V = 0.5$, *W* will equal 5. Connect these sets of values; the intersection of their lines locates the position of the *W*-scale.

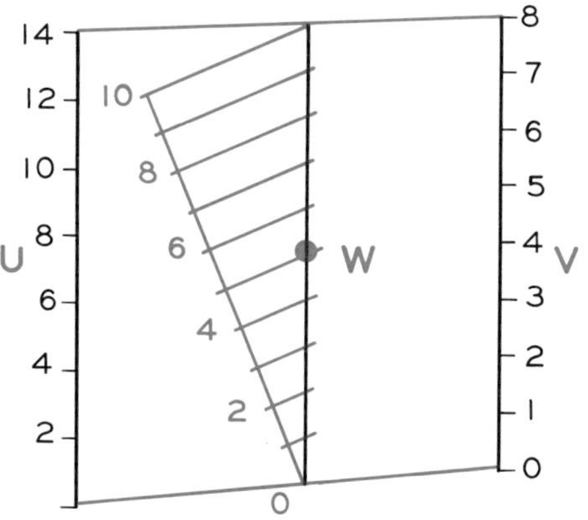

Step 2: Draw the *W*-scale parallel to the outer scales; its length is controlled by the previously established lines of $W = 10$ and $W = 0$. Since this scale is 10 linear divisions long, divide it graphically into ten units as shown. This will be a linear scale. See Article 3–8 to review the method of dividing the *W*-scale.

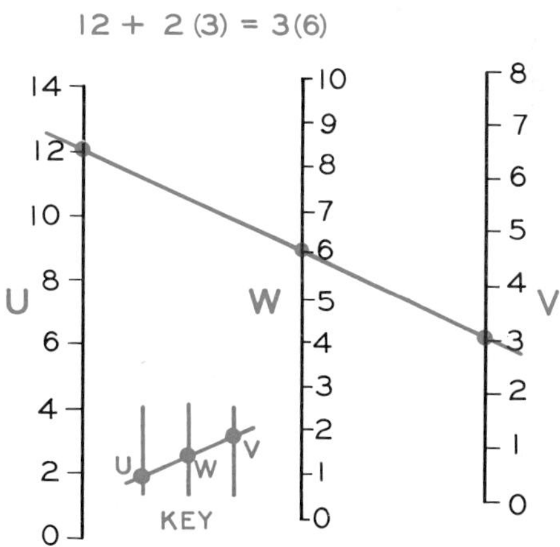

Step 3: The completed nomogram can be used as illustrated by selecting any two known variables and connecting them with an isopleth to determine the third unknown. A key is always included to illustrate how the nomogram is intended to be used. An example of $U = 12$ and $V = 3$ is shown to verify the accuracy of the graph.

Thus we have

$$F_1(S) + F_2(T) = F_3(R), \quad (1)$$

where $F_1(S) = \log S$, $F_2(T) = \frac{1}{2}\log T$, and $F_3(R) = \log R$. The scale modulus for $F_1(S)$ is, from Eq. (1),

$$m_S = \frac{6}{\log 1.0 - \log 0.1} = \frac{6}{0-(-1)} = 6 \quad (2)$$

Choosing the scale measuring point from $S = 0.1$, we find from Eq. (2) that the scale equation for $F_1(S)$ is

$$X_S = 6(\log S - \log 0.1) = 6(\log S + 1) \quad (3)$$

Similarly, the scale modulus for $F_2(T)$ is

$$m_T = \frac{6}{\frac{1}{2}\log 100 - \frac{1}{2}\log 1} = \frac{6}{\frac{1}{2}(2) - \frac{1}{2}(0)} = 6 \quad (4)$$

Thus, the scale equation, measuring from $T = 1$, is:

$$X_T = 6(\tfrac{1}{2}\log T - \tfrac{1}{2}\log 1) = 3\log T \quad (5)$$

The corresponding tables for the two scale equations may be computed as shown in Tables 11–4 and 11–5. We shall position the two scales 5″ apart, as shown in Fig. 11–79. The logarithmic scales are graduated using the values in Tables 11–4 and 11–5. The step-by-step procedure for constructing the remainder of the nomogram is given in Fig. 11–80 using the two outer scales determined here.

The end values of the middle (R) scale are found from the formula $R = S\sqrt{T}$ to be $R = 1.0\sqrt{100} = 10$ and $R = 0.1\sqrt{1} = 0.1$. Choosing a value of $R = 1.0$, we find that corresponding value pairs of S and T might be $S = 0.1$, $T = 100$ and $S = 1.0$, $T = 1.0$. We connect these pairs with isopleths in step 1 and position the middle scale at the intersection of the lines connecting the corresponding values. The R-scale is drawn parallel to the outer scales and is calibrated by deriving its scale modulus:

$$m_R = \frac{6}{\log 10 - \log 0.1} = \frac{6}{1-(-1)} = 3$$

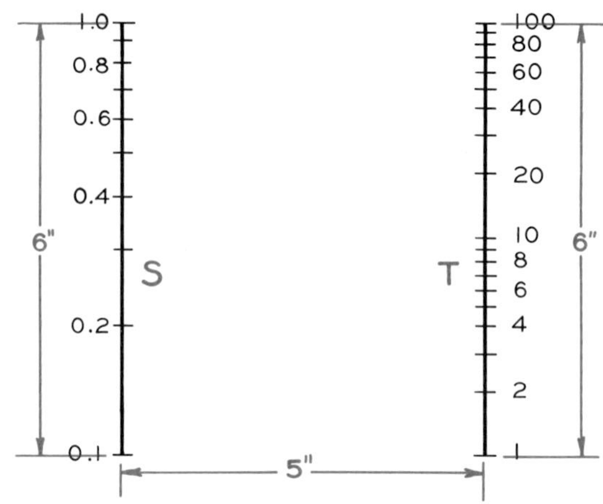

Fig. 11–79. Calibration of the outer scales for the formula $R = S\sqrt{T}$, where $0.1 \leq S \leq 1.0$ and $1 \leq T \leq 100$.

Table 11–4

S	0.1	0.2	0.3	0.4	0.5	0.6	0.7	0.8	0.9	1.0
X_S	0	1.80	2.88	3.61	4.19	4.67	5.07	5.42	5.72	6.00

Table 11–5

T	1	2	4	6	8	10	20	40	60	80	100
X_T	0	0.91	1.80	2.33	2.71	3.00	3.91	4.81	5.33	5.77	6.00

FIGURE 11–80. PARALLEL-SCALE CHART (LOGARITHMIC)

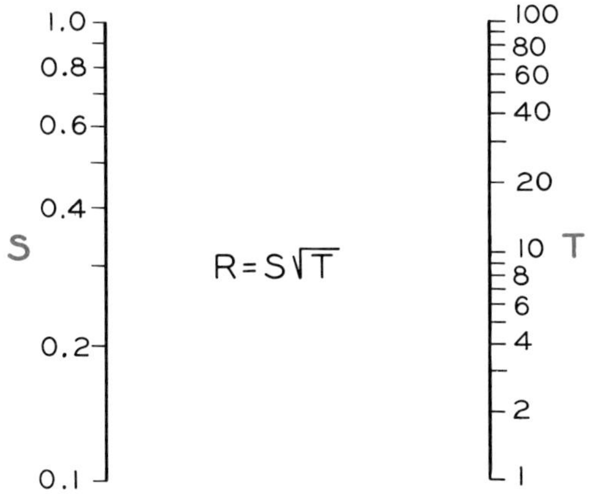

Given: The outer scales of a parallel-scale alignment graph as determined in Fig. 11–79.

Required: Construct a parallel alignment chart for solving the equation $R = S\sqrt{T}$.

Reference: Article 11–40.

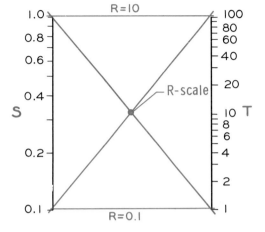

Step 1: Connect the end values of the outer scales to determine the extreme values of the *R*-scale, $R = 10$ and $R = 0.1$. Select corresponding values of *S* and *T* that will give the same value of *R*. Values of $S = 0.1$, $T = 100$ and $S = 1.0$, $T = 1.0$ give a value of $R = 1.0$. Connect the pairs to locate the position of the *R*-scale.

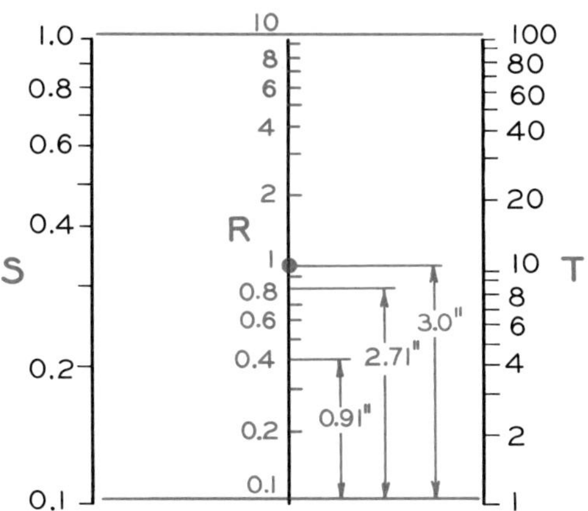
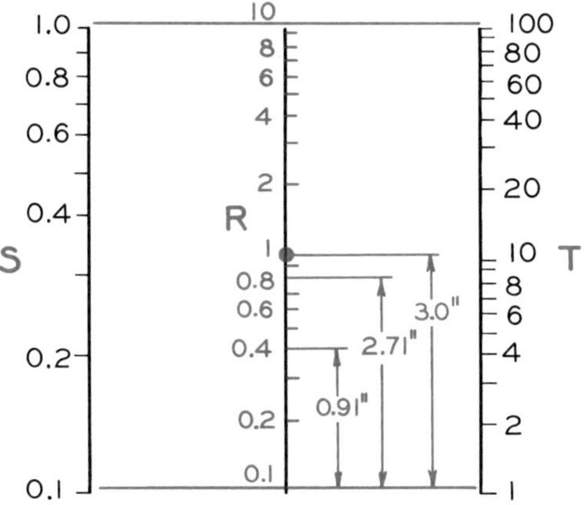

Step 2: Draw the *R*-scale to extend from 0.1 to 10. Calibrate it by substituting values determined from its scale equation. These values have been computed and tabulated in Table 11–6. The resulting tabulation is a logarithmic, two-cycle scale.

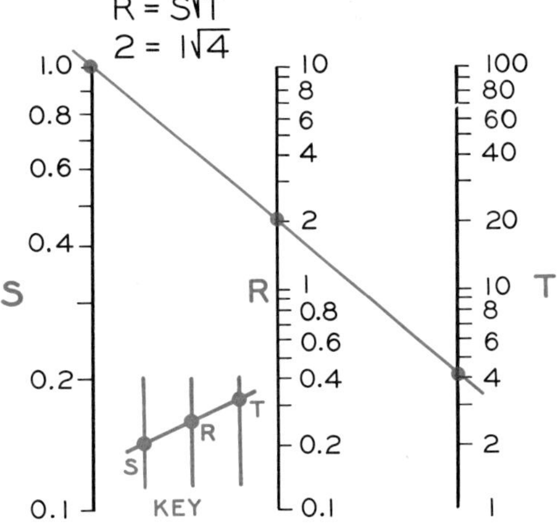

Step 3: Add labels to the finished nomogram and draw a key to indicate how it is to be used. An isopleth has been used to determine *R* when $S = 1.0$ and $T = 4$. The result of 2 is the same as that obtained mathematically, thus verifying the accuracy of the chart. Other combinations can be solved in this same manner.

Table 11–6

R	0.1	0.2	0.4	0.6	0.8	1.0	2.0	4.0	6.0	8.0	10.0
X_R	0	0.91	1.80	2.33	2.71	3.00	3.91	4.81	5.33	5.71	6.00

Thus its scale equation (measuring from $R = 0.1$) is

$$X_R = 3(\log R - \log 0.1) = 3(\log R + 1.0)$$

Table 11–6 is computed to give the values for the scale. These values are applied to the R-scale as shown in step 2. The finished nomogram can be used as illustrated in step 3 to compute the unknown variables when two variables are given.

Note that this example illustrates a general method of creating a parallel-scale graph for all formulas of the type $F_1 + F_2 = F_3$ through the use of a table of values computed from the scale equation. As an alternative method, this center scale may be found graphically once the end values of the scales have been determined. An example of the graphical method is illustrated in Fig. 11–81, in which the scale is calibrated by graphical enlargement or reduction of the desired portions of printed logarithmic scales as found in common logarithmic graph paper.

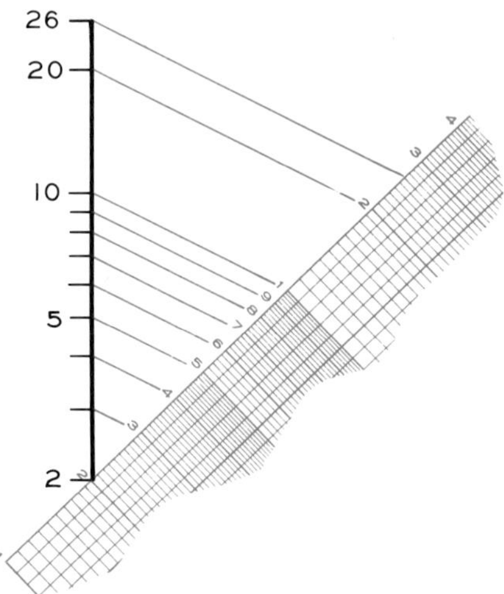

Fig. 11–81. Graphical calibration of a scale using logarithmic paper.

11–40 N- OR Z-CHARTS

Whenever F_2 and F_3 are linear functions, we can partially avoid using logarithmic scales for formulas of the type

$$F_1 = \frac{F_2}{F_3}$$

Instead, we use an N-chart, as shown in Fig. 11–82. The outer scales, or "legs" of the N are functional scales and will therefore be linear if F_2 and F_3 are linear, whereas if the same formula were drawn as a parallel-scale chart, all scales would have to be logarithmic.

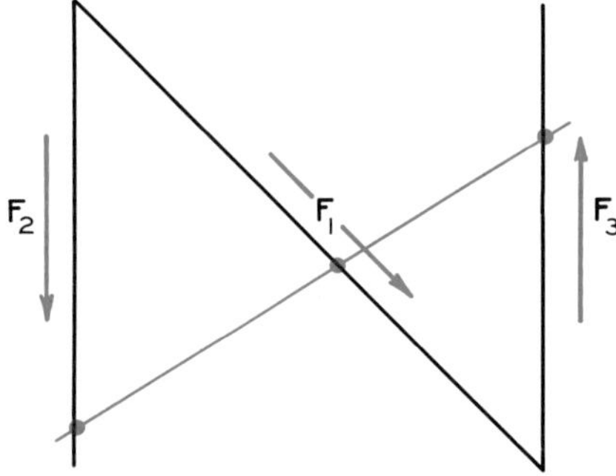

Fig. 11–82. An N-chart for solving an equation of the form $F_1 = F_2/F_3$.

Some main features of the N-chart are:

1. The outer scales are parallel functional scales of F_2 and F_3.
2. They increase (functionally) in *opposite* directions.
3. The diagonal scale connects the (functional) *zeros* of the outer scale
4. In general, the diagonal scale is not a functional scale for the function F_1 and is generally nonlinear.

Construction of an N-chart is simplified by the fact that locating the middle (diagonal) scale is usually less of a problem than it is for a parallel-scale chart. Calibration of the diagonal scale is most easily accomplished by graphical methods. To illustrate, an N-chart is constructed for the equation

$$A = \frac{B + 2}{C + 5}$$

where $0 \leq B \leq 8$ and $0 \leq C \leq 15$. This equation follows the form of

$$F_1 = \frac{F_2}{F_3}$$

where $F_1(A) = A, F_2(B) = B + 2,$ and $F_3(C) = C + 5$. Thus the outer scales will be for $B + 2$ and $C + 5$, and the diagonal scale will be for A.

The construction is begun in the same manner as for a parallel-scale chart by selecting the layout of the outer scales (Fig. 11–83). As before, the limits of the diagonal scale are determined by connecting the endpoints on the outer scales, giving $A = 0.1$ for $B = 0$, $C = 15$ and $A = 2.0$ for $B = 8$, $C = 0$, as shown in the given portion of Fig. 11–84. The remainder of the construction is given in step form in the figure.

The diagonal scale is located by finding the *function zeros* of the outer scales, i.e., the points where $B + 2 = 0$ or $B = -2$, and $C + 5 = 0$ or $C = -5$. The diagonal scale may then be drawn by connecting these points as shown in step 1. Calibration of the diagonal scale is most easily accomplished by substituting into the formula. Select the upper limit of an outer scale, for example, $B = 8$. This gives the formula

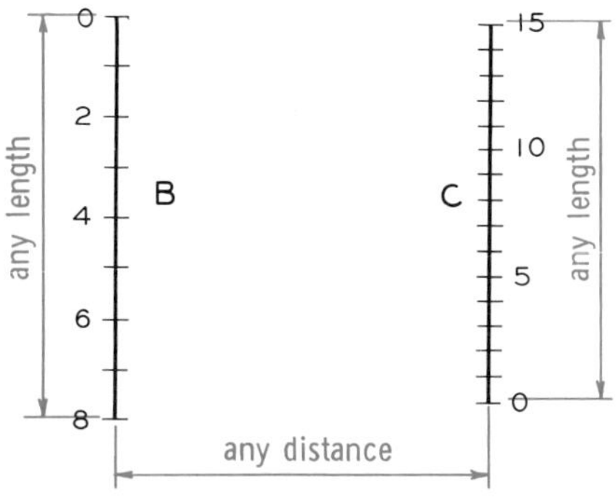

Fig. 11–83. Calibration of the outer scales of an N-chart for the equation $A = (B+2)/(C+5)$.'

$$A = \frac{10}{C + 5}$$

Solve this equation for the other outer scale variable,

$$C = \frac{10}{A} - 5$$

Using this as a "scale equation," make a table of values for the desired values of A and corresponding values of C (up to the limit of C in the chart), as shown in Table 11–7. Connect isopleths from $B = 8$ to the tabulated values of C. Their intersections with the diagonal scale give the required calibrations for approximately half the diagonal scale, as shown in step 2 of Fig. 11–84.

The remainder of the diagonal scale is calibrated by substituting the end value of the other outer scale ($C = 15$) into the formula, giving

$$A = \frac{B + 2}{20}$$

Solving this for B yields

$$B = 20A - 2$$

Table 11–7

A	2.0	1.5	1.0	0.9	0.8	0.7	0.6	0.5
C	0	1.67	5.0	6.11	7.50	9.28	11.7	15.0

A table for the desired values of A can be constructed as shown in Table 11–8. Isopleths connecting $C = 15$ with the tabulated values of B will locate the remaining calibrations on the A-scale, as shown in step 3.

Table 11–8

A	0.5	0.4	0.3	0.2	0.1
B	8.0	6.0	4.0	2.0	0

11–41 COMBINATION FORMS OF ALIGNMENT CHARTS

The types of charts discussed above may be used in combination to handle different types of formulas. For example, formulas of the type $F_1/F_2 = F_3/F_4$ (four variables) may be represented as *two* N-charts by the insertion of a "dummy" function. To do this, let

$$\frac{F_1}{F_2} = S \qquad \text{and then} \qquad S = \frac{F_3}{F_4}$$

Each of these may be represented as shown in part A of Fig. 11–85, where one N-chart is inverted and rotated 90°. In this way, the charts may be superimposed as shown in part B if the S-scales are of equal length. The S-scale, being a "dummy" scale, does not need to be calibrated; it is merely a "turning" scale for intermediate values of S which do not actually enter into the formula itself. The chart is read with *two* isopleths which connect the four variable values and cross on the S-scale as shown in part C. Charts of this form are commonly called *ratio charts.*

Formulas of the type $F_1 + F_2 = F_3F_4$ are handled similarly. As in the preceding example, a "dummy" function is used: $F_1 + F_2 = S$ and $S = F_3F_4$. In order to apply the superimposition principle, a more equitable arrangement is obtained by rewriting the equations as $F_2 = S - F_1$ and $F_3 = S/F_4$. These two equations then take the form of a parallel-scale chart and an N-chart, respectively, as shown in part A of Fig. 11–86. Again the S-scales must be identical but need not be calibrated. The charts are superimposed in part B. The S-scale is used as a "turning" scale for the two isopleths, as shown in part C. Many other combinations are possible, limited only by the ingenuity of the nomographer in adapting formulas and scale arrangements to his needs.

11–42 SUMMARY

Throughout the design process, graphics has been the primary agent of creativity. Sketches, graphs, and diagrams were used to identify the problem; freehand sketches were made to determine preliminary ideas; the ideas were refined graphically; and the designs were then analyzed for feasibility prior to selection of the best solution. The development process would be virtually impossible without the thinking process being recorded, and even stimulated, by the applications of graphics. Throughout this process, the designer is guided by his knowledge of physical properties, engineering fundamentals, and manufacturing limitations. fine his thinking or imagination during the creative development stage. The real confrontation However, he does not let these limitations conwith the physical, mathematical, and scientific principles is in the analysis stage of the design process. Many of the data derived from experiments with prototypes and laboratory experiments will be in a numerical form that must be analyzed to provide a total picture of the proposed design.

FIGURE 11–84. CONSTRUCTION OF AN N-CHART

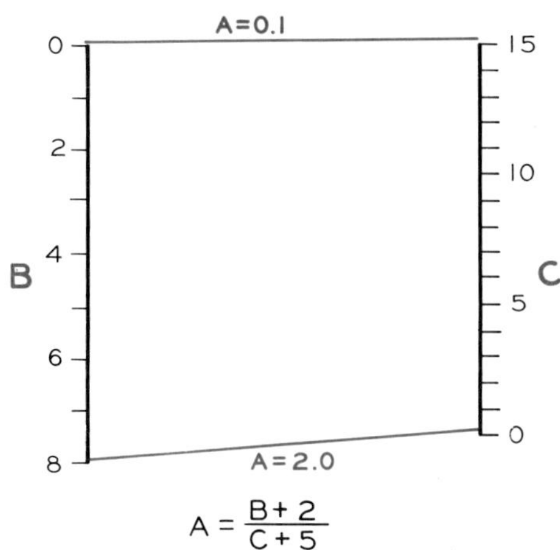

Given: The outer scales of an N-chart as determined in Fig. 15–83.
Required: Complete the N-chart for the formula $A = (B+2)/(C+5)$.
Reference: Article 11–41.

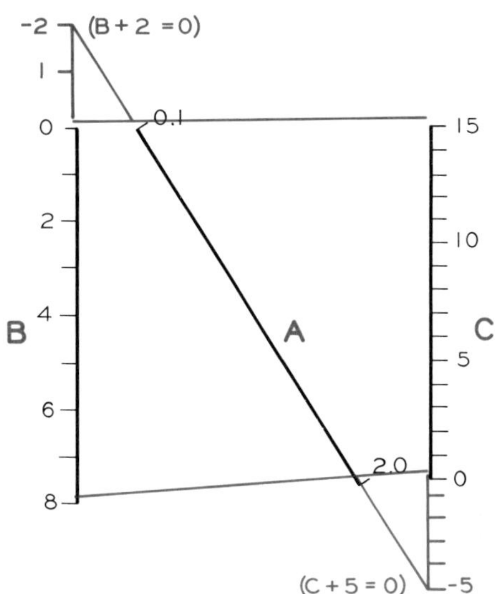

Step 1: Locate the diagonal scale by finding the functional zeros of the outer scales. This is done by setting $B + 2 = 0$ and $C + 5 = 0$, which gives a zero value for A when $B = -2$ and $C = -5$. Connect these points with diagonal scale A.

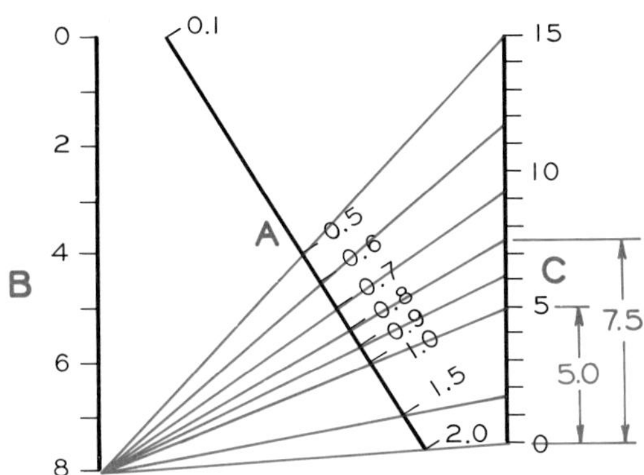

Step 2: Select the upper limit of one of the outer scales, $B = 8$ in this case, and substitute it into the given equation to find a series of values of C for the desired values of A, as shown in Table 11–7. Draw isopleths from $B = 8$ to the values of C to calibrate the A-scale.

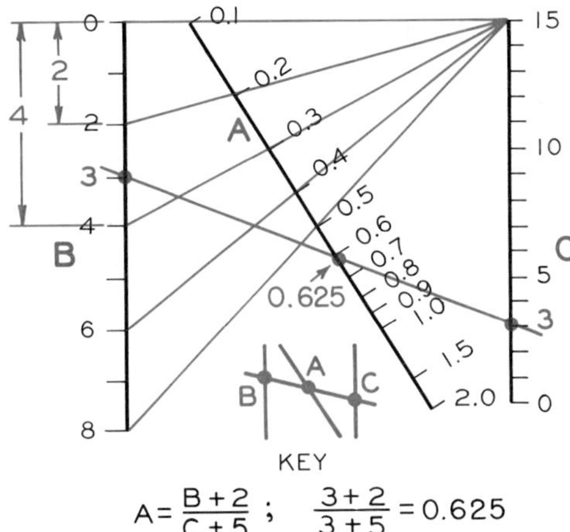

Step 3: Calibrate the remainder of the A-scale in the same manner by substituting the upper limit of the other outer scale ($C = 15$) into the equation to determine a series of values on the B-scale for desired values on the A-scale, as listed in Table 11–8. Draw isopleths from $C = 15$ to calibrate the A-scale as shown. Draw a key to indicate how the nomogram is to be used. Solve an example problem to verify its accuracy.

FIGURE 11–85. FOUR-VARIABLE CHART

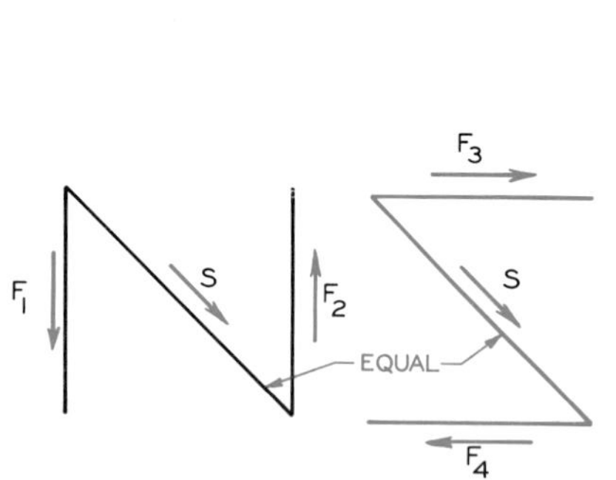

A. A combination chart can be developed to handle four variables in the form $F_1/F_2 = F_3/F_4$ by developing two N-charts in the forms $F_1/F_2 = S$ and $F_3/F4 = S$, where S is a dummy scale of equal length in both charts.

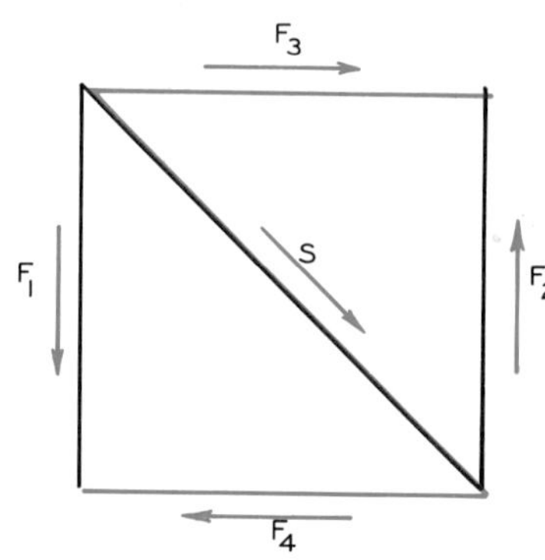

B. If equal-length scales are used in each of the N-charts and if the *S*-scales are equal, then the charts can be overlapped so that each is common to the *S*-scale.

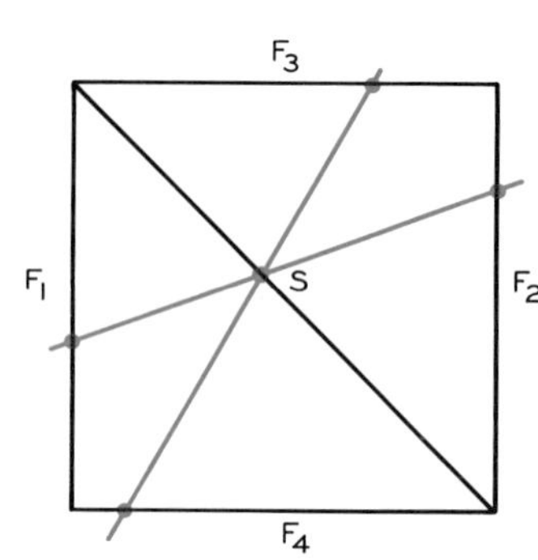

C. Two lines (isopleths) are drawn to cross at a common point on the *S*-scale. Numerous combinations of the four variables can be read on the surrounding scales. The *S*-scale need not be calibrated, since no values are read from it.

FIGURE 11–86. COMBINATION PARALLEL-SCALE CHART AND N-CHART

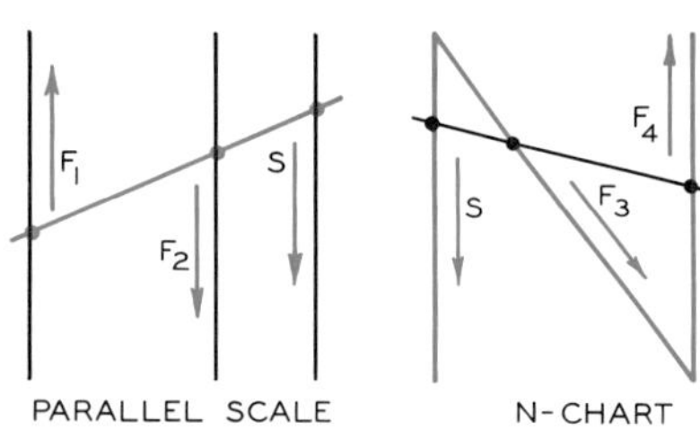

A. Formulas of the type $F_1 + F_2 = F_3F_4$ can be combined into one nomogram by constructing a parallel-scale chart and an N-chart with an equal *S*-scale (a dummy scale).

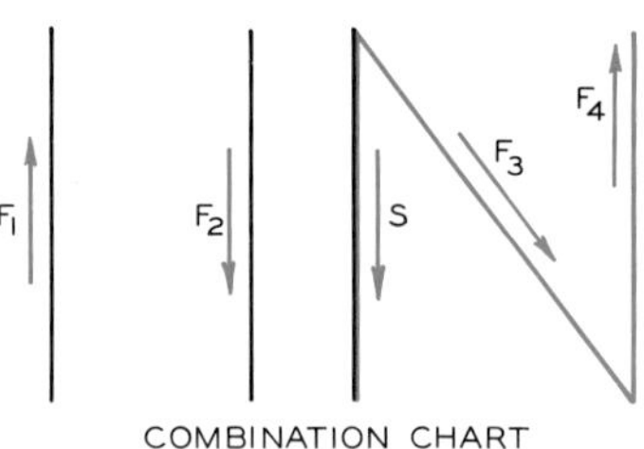

B. By superimposing the two equal *S*-scales, the two nomograms are combined into a combination chart. The *S*-scale need not be calibrated, since values are not read from it.

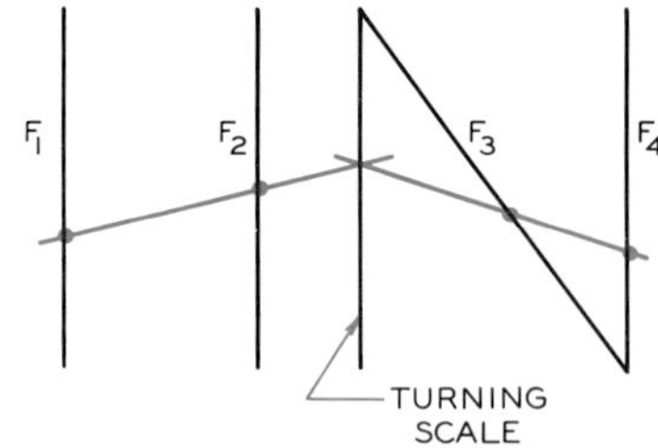

C. The addition of the variables can be handled at the left of the chart. The *S*-scale is the turning scale from which the N-chart can be used to find the unknown variables.

Various aspects of graphical analysis of data and components were covered in this chapter, including graphs, empirical data, linkages and mechanisms, graphical calculus, and nomography. Each of these areas can be evaluated using formal mathematical and analytical approaches; however, graphical methods are applicable to many problems. In some cases, the graphical technique will offer the best method of solution.

Although conventional engineering methods are probably used more in the analysis phase of the design process than any other, graphical methods can be applied to good advantage in this phase also. The engineer, technologist, or technician should be sufficiently aware of the applications of graphics to utilize this valuable method of problem solution.

PROBLEMS

General

The following problems are to be solved on $8\frac{1}{2}'' \times 11''$ or on $11'' \times 15''$ paper. The graph problems may be solved on commercially prepared paper or the grid may be constructed by the student. Problems involving geometric construction and mathematical calculations should show the construction and calculations as part of the problem for future reference. If the mathematical calculations are extensive, it may be desirable to include these on a separate sheet. Legible lettering practices and principles of good layout should be followed in these problems, and all notes and constructions should be provided to explain fully the method of solution.

Rectangular Graphs

1. Using the data given in Table 11–9, prepare a rectangular graph to compare the number of Master's and Doctor's degrees granted in the United States from 1920 to 1965. Use these data as a basis for predicting the trend for the period from 1965 to 1980.

Table 11–9

	Master's	Doctor's
1920	5,100	1,700
1930	14,700	3,100
1940	24,500	4,500
1950	64,000	7,800
1960	73,000	9,700
1965	102,000	15,200

2. Using the data given in Table 11–10, prepare a rectangular graph to compare the supply and demand of water in the United States from 1890 to 1980. Supply and demand are given in units of billions of gallons of water per day.

Table 11–10

	1890	1900	1910	1920	1930	1940	1950	1960	1970	1980
Supply	80	80	110	135	155	240	270	315	380	450
Demand	35	35	60	80	110	125	200	320	410	550

3. Analyze the data given in Table 11–11 in a rectangular graph to decide which lamps should be selected to provide economical lighting for an industrial plant. You are to decide whether to use single 1000-watt bulbs or twin 400-watt bulbs. The tables gives the candlepower directly under the lamps (0°) and at the various angles from the vertical when the lamps are mounted at a height of 25 feet, as shown in Fig. 11–87.

4. Construct a rectangular grid graph that shows the relationship in energy costs (mills per kilowatt-hour) and the percent capacity of two types of power plants. Plot energy costs along the Y-axis and the capacity factor along the X-axis. The plotted curve will compare the costs of a nuclear plant with a gas- or oil-fired plant. Data for a gas-fired plant: 17 mills, 10%; 12 mills, 20%; 8 mills 40%; 7 mills, 60%; 6

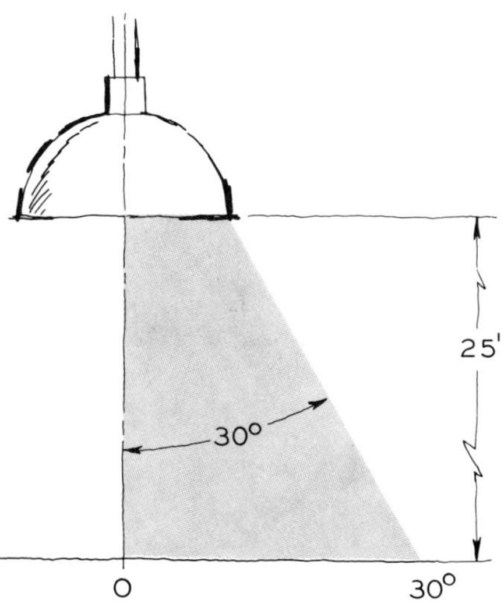

Fig. 11–87. The angular measurement of illumination under a lamp (Problem 3).

Table 11–11

Angle with vertical	0	10	20	30	40	50	60	70	80	90
Candlepower (thous.) 2–400W	37	34	25	12	5.5	2.5	2	0.5	0.5	0.5
Candlepower (thous.) 1–1000W	22	21	19	16	12.3	7	3	2	0.5	0.5

mills, 80%; 5.8 mills, 100%. Nuclear plant data: 24 mills, 10%; 14 mills, 20%; 7 mills, 40%; 5 mills, 60%; 4.2 mills, 80%; 3.7 mills, 100%.

5. Construct a rectangular grid graph to show the accident experience of Company A from 1953 to 1966. Plot the numbers of disabling accidents per million man-hours of work on the Y-axis. Years will be plotted on the X-axis. Data: 1953, 1.21; 1954, 0.97; 1955, 0.86; 1956, 0.63; 1957, 0.76; 1958, 0.99; 1959, 0.95; 1960, 0.55; 1961, 0.76; 1962, 0.68; 1963, 0.55; 1964, 0.73; 1965, 0.52; 1966, 0.46.

6. Construct a rectilinear graph that shows the relationship between the transverse resilience in inch-pounds (Y-axis) and the single-blow impact in foot-pounds (X-axis of gray iron. Data: 21 fp, 375 ip; 22 fp, 350 ip; 23 fp, 380 ip; 30 fp, 400 ip; 32 fp, 420 ip; 33 fp, 410 ip; 38 fp, 510 ip; 45 fp, 615 ip; 50 fp, 585 ip; 60 fp, 785 ip; 70 fp, 900 ip; 75 fp, 920 ip.

7. Using the data from a 1967 survey (Table 11–12), construct a rectangular graph to reflect the salaries of engineers during successive years after graduation with a bachelor's degree. These salaries are given in averages for deciles (10% intervals) and quartiles (25% intervals). Plot each of these as a separate curve.

Break-Even Graphs

8. Construct a break-even graph that shows the earnings for a new product that has a development cost of $12,000. The first 8000 will cost 50¢ each to manufacture, and you wish to break even at this quantity. What would be the profit at a volume of 20,000 and at 25,000?

9. Same as Problem 8 except that the development costs are $80,000, the manufacturing cost of the first 10,000 is $2.30 each, and the desired break-even point is at this quantity. What would be the profit at a volume of 20,000 and at 30,000? What sales price would be required to break even at 10,000 units?

10. A manufacturer has incorporated the manufacturing and development costs into a cost-per-unit estimate. He wishes to sell the product at $1.50 each. Construct a graph of the following data. On the Y-axis plot cost per unit in dollars; on the X-axis, number of units in thousands. Data: 1000, $2.55; 2000, $2.01; 3000, $1.55; 4000, $1.20; 5000, $0.98; 6000, $0.81; 7000, $0.80; 8000, $0.75; 9000, $0.73; 10,000, $0.70. How many must be sold to break even? What will be the total profit when 9000 are sold?

11. The cost per unit to produce a product by a manufacturing plant is given below. Construct a

Table 11–12

Years after graduation	0	5	10	15	20	25	30
Upper decile	9,800	13,000	17,500	20,500	22,500	23,800	25,000
Upper quartile	8,700	11,900	15,000	17,000	19,000	19,200	18,700
Median	8,000	10,500	13,000	14,500	15,800	15,200	15,000
Lower quartile	7,500	9,750	11,500	12,800	13,000	12,700	12,200
Lower decile	7,000	8,900	10,000	11,000	11,400	11,000	10,500

break-even graph with the cost per unit plotted on the Y-axis and the number of units on the X-axis. Data: 1000, \$5.90; 2000, \$4.50; 3000, \$3.80; 4000, \$3.20; 5000, \$2.85; 6000, \$2.55; 7000, \$2.30; 8000, \$2.17; 9000, \$2.00; 10,000, \$0.95.

Logarithmic Graphs

12. Using the data given in Table 11–13, construct a logarithmic graph where the vibration amplitude (A) is plotted as the ordinate and vibration frequency (F) as the abscissa. The data for curve 1 represent the maximum limits of machinery in good condition with no danger from vibration. The data for curve 2 are the lower limits of machinery that is being vibrated excessively to the danger point. The vertical scale should be three cycles and the horizontal scale two cycles.

13. Plot the data below on a two-cycle log graph to show the current in amperes (Y-axis) versus the voltage in volts (X-axis) of precision temperature-sensing resistors. Data: 1 volt, 1.9 amps; 2 volts, 4 amps; 4 volts, 8 amps; 8 volts, 17 amps; 10 volts, 20 amps; 20 volts, 30 amps; 40 volts, 36 amps; 80 volts, 31 amps; 100 volts, 30 amps.

Semilogarithmic Graphs

14. Construct a semilogarithmic graph of the data in Problem 7, using the median values to determine the ratios of increase during this tme period.

15. Construct a semilogarithmic graph to compare the relative ratios of the Master's and Doctor's degrees granted, as given in Problem 1.

16. Construct a semilog graph with the Y-axis a two-cycle log scale from 1 to 100 and the X-axis a linear scale from 1 to 7. Plot the data below to show the survivability of a shelter at varying distances from a one-megaton air burst. The data consist of overpressure in psi along the Y-axis, and distance from ground zero in miles along the X-axis. The data points represent an 80% chance of survival of the shelter. Data: 1 mi, 55 psi; 2 mi, 11 psi; 3 mi, 4.5 psi; 4 mi, 2.5 psi, 5 mi, 2.0 psi; 6 mi, 1.3 psi.

17. The growth of two divisions of a company, Division A and Division B, is given in the data below. Plot the data on a rectilinear graph and on a semilog graph. The semilog graph should have a one-cycle log scale on the Y-axis for sales in thousands of dollars, and a linear scale on the X-axis showing years for a six-year period. Data in dollars: 1 yr, A = \$11,700 and B = \$44,000; 2 yr, A = \$19,500 and B = \$50,000; 3 yr, A = \$25,000 and B = \$55,000; 4 yr, A = \$32,000 and B = \$64,000; 5 yr, A = \$42,000 and B = \$66,000; 6 yr, A = \$48,000 and B = \$75,000. Which division has the better growth rate?

Percentage Graphs

18. Plot the data given in Problem 7 on a semilog graph in order to determine the rates of increase of salaries. What is the rate of increase of the upper decile between 5 years and 20 years? What percent of the upper-decile salary at 20 years is the lower-decile salary at 20 years?

19. Plot the data given in Problem 2 on a semilog graph in order to determine the percentages and ratios of the data. What is the rate

Table 11–13

F	100	200	500	1000	2000	5000	10,000
$A(1)$	0.0028	0.002	0.0015	0.001	0.0006	0.0003	0.00013
$A(2)$	0.06	0.05	0.04	0.03	0.018	0.005	0.001

of increase in the demand for water from 1890 to 1920? from 1920 to 1970? What percent of the demand is the supply for the following years: 1900, 1930, and 1970?

20. Using the graph plotted in Problem 17, determine the rate of increase of Division A and Division B from year 1 to year 4. Also, what percent of the sales of Division A are the sales of Division B at the end of year 2 and at the end of year 6?

Bar Graphs

21. Construct a bar graph to depict the unemployment rate of high-school graduates and dropouts in various age categories. The age groups and the percent of unemployment of each group are given in Table 11–14.

Table 11–14

Ages	Percent of labor force	
	Graduates	Dropouts
16–17	18	22
18–19	12.5	17.5
20–21	8	13
22–24	5	9

22. Prepare a bar graph to compare the number of skilled workers employed in various occupations. Arrange the graph for ease of interpretation and comparison of occupations. Use the following data: carpenters, 82,000; all-round machinists, 310,000; plumbers, 350,000; bricklayers, 200,000; appliance service men, 185,000; automotive mechanics, 760,000; electricians, 380,000; painters, 400,000.

23. Construct a bar graph of the data shown in Table 11–10.

24. Using the data in Table 11–12, construct a bar graph showing the upper- and lower-decile earnings of engineers.

Pie Graphs

25. Prepare a pie chart to compare the areas of employment of male youth between the ages of 16 and 21 as tabulated in 1964: operatives, 25%; craftsmen, 9%; professions, technicals, and managers, 6%; clerical and sales, 17%; service, 11%; farm workers, 11%; laborers, 19%.

26. Make a pie graph to give the relationship between the following members of the scientific and technical team, as listed in 1965: engineers, 985,000; technicians, 932,000; scientists, 410,000.

27. Construct a circle graph of the following percentages of the employment status of the 1969 graduates of two-year technician programs one year after graduation: employed, 63%; continuing full-time study, 23%; considering job offers, 6%; military, 6%; other, 2%.

28. Construct a circle graph that shows the relationship between the types of degrees held by engineers in aeronautical engineering: bachelor's degree, 65%; master's degree, 29%; Ph.D. degrees, 6%.

Polar Graphs

29. Construct a polar graph of the data given in Problem 3.

30. Construct a polar graph of the following illumination, in lumens at various angles, emitted from a luminaire. The zero-degrees position is vertically under the overhead lamp. Data: 0°, 12,000; 10°, 15,000; 20°, 10,000; 30°, 8000; 40°, 4200; 50°, 2500; 60°, 1000; 70°, 0. The illumination is symmetrical about the vertical.

Empirical Data—General Types

31. The data shown in Table 11–15, A through F, have been tabulated from experimental laboratory tests. Plot these data on rectangular, logarithmic, and semilogarithmic graphs and determine the empirical equations of the data. Select the proper graph needed for each set of data.

Table 11–15

A	X	0	40	80	120	160	200	240	280			
	Y	4.0	7.0	9.8	12.5	15.3	17.2	21.0	24.0			
B	X	1	2	5	10	20	50	100	200	500	1000	
	Y	1.5	2.4	3.3	6.0	9.2	15.0	23.0	24.0	60.0	85.0	
C	X	1	5	10	50	100	500	1000				
	Y	3	10	19	70	110	400	700				
D	X	2	4	6	8	10	12	14				
	Y	6.5	14.0	32.0	75.0	115.0	320	710				
E	X	0	2	4	6	8	10	12	14			
	Y	20	34	53	96	115	270	430	730			
F	X	0	1	2	3	4	5	6	7	8	9	10
	Y	1.8	2.1	2.2	2.5	2.7	3.0	3.4	3.7	4.1	4.5	5.0

Empirical Data—Linear

32. Construct a linear graph to determine the equation for the yearly cost of a compressor in relationship to the compressor's size in horsepower. The yearly cost should be plotted on the Y-axis and the compressor's size in horsepower on the X-axis. Data: 0 hp, $0; 50 hp, $2100; 100 hp, $4500; 150 hp, $6700; 200 hp, $9000; 250 hp, $11,400. What is the equation of these data?

33. Construct a linear graph to determine the equation for the cost of soil investigation by boring to determine the proper foundation design for varying sizes of buildings. Plot the cost of borings in dollars along the Y-axis and the building area in sq ft along the X-axis. Data: 0 sq ft, $0; 25,000 sq ft, $35,000; 50,000 sq ft, $70,000; 750,000 sq ft, $100,000; 1,000,000 sq ft, $130,000.

34. Determine the equation of the empirical data plotted in Fig. 11–35.

Empirical Data—Logarithmic Graphs

35. The following empirical data compare in put voltage V with the input current I to a heat pump. Find the equation of the data given in Table 11–16.

36. Table 11–17 lists empirical data giving the relationship between the peak allowable current in amperes (I) versus the overload operating time in cycles at 60 cycles per second (C).

Table 11–16

Y-axis	V	0.8	1.3	1.75	1.85
X-axis	I	20	30	40	45

Table 11–17

Y-axis	I	2000	1840	1640	1480	1300	1120	1000
X-axis	C	1	2	5	10	20	50	100

Table 11–18

Y-axis	rms	7500	5200	4400	3400	2300	1700
X-axis	pdc	3	6	9	15	30	60

Place *I* on the *Y*-axis and *C* on the *X*-axis. Determine the equation for these data.

37. The empirical data given in Table 11–18 for a low-voltage circuit breaker used on a welding machine give the maximum loading during weld in amperes (rms) for the percent of duty cycle (pdc). Determine the equation for these data. Place rms along the *Y*-axis and pdc along the *X*-axis.

38. Construct a three-cycle × three-cycle logarithmic graph to find the equation of a machine's vibration during operation. Plot vibration displacement in mills along the *Y*-axis and vibration frequency in cycles per minute (cpm) along the *X*-axis. Data: 100 cpm, 0.80 mills; 400 cpm, 0.22 mills; 1000 cpm, 0.09 mills; 10,000 cpm, 0.009 mills; 5000 cpm, 0.0017 mills.

39. Determine the equation of the empirical data plotted in Fig. 11–35.

Empirical Data—Semilogarithmic Graphs

40. Construct a semilog graph of the following data to determine their equation. The *Y*-axis should be a two-cycle log scale and the *X*-axis a 10-unit linear scale. Plot the voltage (*E*) along the *Y*-axis and time (*T*) in sixteenths of a second along the *X*-axis to represent resistor voltage during capacitor charging. Data: 0, 10 volts; 2, 6 volts; 4, 3.6 volts; 6, 2.2 volts; 8, 1.4 volts; 10, 0.8 volts.

41. Find the equation of the data in Fig. 11–37.

42. Construct a semilog graph of the following data to determine their equation. The *Y*-axis should be a three-cycle log scale and the *X*-axis a linear scale from 0 to 250. These data give a comparison of the reduction factor, *R* (*Y*-axis), with the mass thickness per square foot (*X*-axis) of a nuclear protection barrier. Data: 0, 1.0*R*; 100, 0.9*R*; 150, 0.028*R*; 200, 0.009*R*; 300, 0.0011*R*.

Empirical Analysis

43. By determining stresses in the cable at intervals, construct an empirical curve that shows the stresses in relationship to the position of the loading boom (Fig. 11–88). This problem is very similar to the one illustrated in Fig. 11–58. Construct stress diagrams to arrive at the stresses in a number of positions. The boom must permit upward and downward movement of 45° with the horizontal.

Cams

The following cam problems are to be solved on Size B sheets (11″ × 17″) with the following standard dimensions: base circle, 4″; roller follower 0.75″ diameter; shaft, 0.75″ diameter; hub, 1.25″ diameter. The direction of rotation is clockwise. The follower is positioned vertically over the center of the base circle except in Problems 52 and 53. Lay out the problems and displacement diagrams as shown in Fig. 11–89.

44. Make a drawing of a plate cam with a knife-edge follower for uniform motion and a rise of $1\frac{1}{4}$″.

45. Make a displacement diagram and a drawing of the cam that will give a modified uniform motion to a knife-edge follower with a rise of 1.5″. Use an arc of one-quarter of the rise to modify the uniform motion in the displacement diagram.

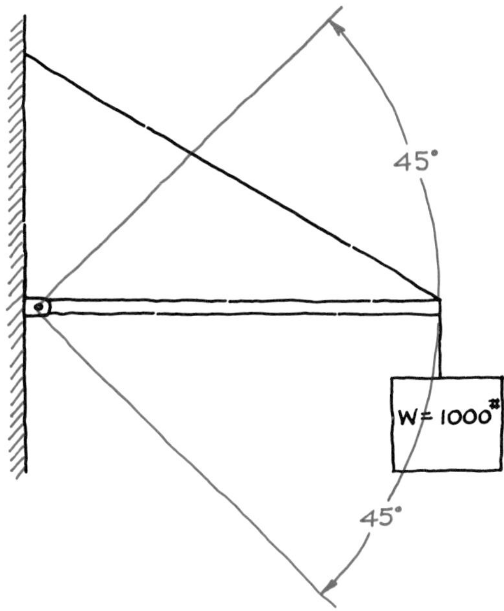

Fig. 11–88. Empirical analysis of stresses (Problem 43).

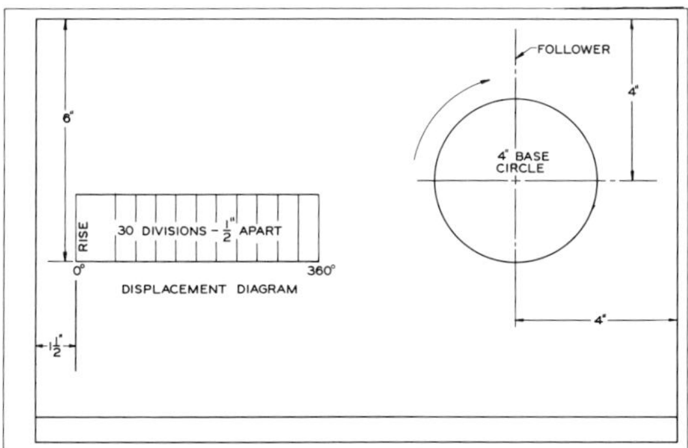

Fig. 11–89. Problem layout for the cam problems on Size B sheets.

46. Make a displacement diagram and a drawing of the cam that will give a harmonic motion to a roller follower with a rise of $1\frac{3}{8}''$.

47. Make a displacement diagram and a drawing of the cam that will give harmonic motion to a knife-edge follower with a rise of 0.75″.

48. Make a displacement diagram and a drawing of the cam that will give uniform acceleration to a knife-edge follower with a rise of 1.5″.

49. Make a displacement diagram and a drawing of the cam that will give uniform acceleration to a roller follower with a rise of 1.25″.

50. Make a displacement diagram and a drawing of the cam that will give the following motion to a knife-edge follower: dwell for 90°; rise 1″ with harmonic motion in 100°; fall 1″ with a modified uniform motion in 100°; and dwell for 70°.

51. Make a displacement diagram and a drawing of the cam that will give the following motion to a roller follower: rise 1.25″ with uniform acceleration in 120°; dwell for 120°; and fall 1.25″ with a harmonic motion in 120°.

52. Repeat Problem 44, but offset the follower 0.75″ to the right of the vertical center line.

53. Repeat Problem 46, but offset the follower 0.75″ to the left of the vertical center line.

Linkages

54. Design and analyze the linkage system for the internal safety valve shown in Fig. 11–90. Use the dimensions given in the table for the particular problem assigned. Determine the *other dimensions* that will permit the operation for the limits given. Show all graphical analysis and the design of the linkage system on 11″ × 15″ tracing paper.

55. Determine the location of the pivot points and the lengths of the linkage members in the clamping device shown in Fig. 11–91 that will raise the bar 90° when the handle is raised 60°. Show all graphical construction and analysis on 11″ × 15″ tracing paper. Disregard the details of the specific shapes of the members; be concerned with center-to-center dimensions.

56. Analyze the motion of the piston shown in Fig. 11–51 using the following dimensions: radius of driver, 3.5″; length of piston rod, 9.7″. Plot the distance of travel of the piston for each degree of rotation. The driver is turning at a rate of one revolution per second. Show all construction on a 11″ × 15″ tracing paper.

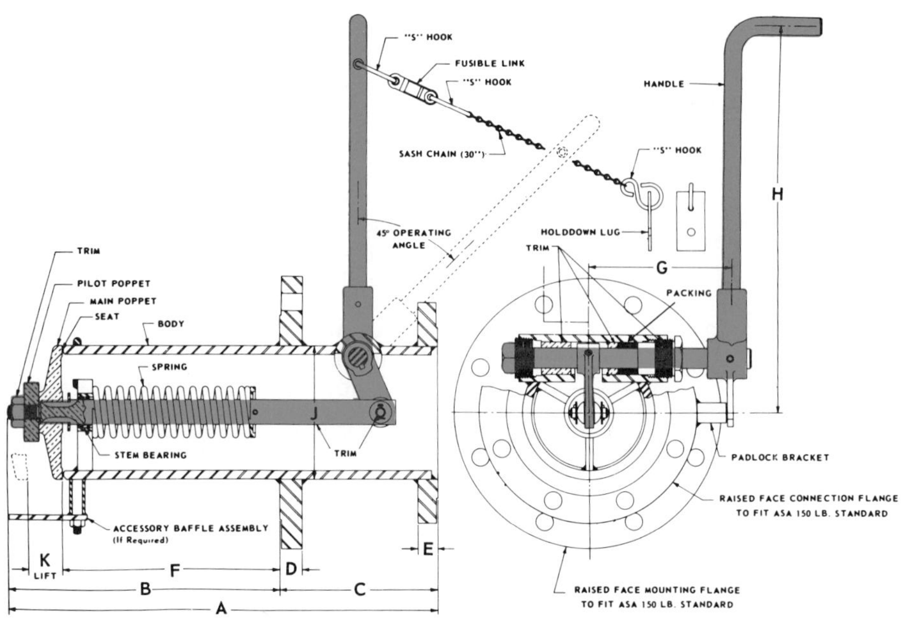

TABLE OF DIMENSIONS

VALVE SIZE	CONN. FLANGE SIZE	MOUNTING FLANGE SIZE		A	B	C	D	E	F	G	H	J	K	NET WT. LBS.	SHPG. WT. LBS.	BOX SIZE INCHES
		STANDARD	OVERSIZE													
4"	4"	6"	8"	17	11 5/8	5 3/8	1	15/16	10	6 1/8	12 1/4	4 1/2	1 7/16	78	82	16x17x19
6"	6"	8"	10"	21 3/4	13 3/4	8	1 1/8	1	11	7 1/4	19 1/4	6 5/8	1 3/4	130	136	18x16x23
8"	8"	10"	12"	23 3/4	14 3/4	9	1 3/16	1 1/8	12	8 5/8	31 3/4	8 5/8	2 1/8	173	208	20x17x32
10"	10"	12"	16"	24 3/4	14 3/4	10	1 1/4	1 3/16	12	10 1/8	32 3/4	10 3/4	3 1/2	255	297	24x20x32
12"	12"	14"	18"	24 1/4	14 7/8	9 3/8	1 3/8	1 1/4	12	11 1/8	37	12 3/4	4 3/8	350	402	26x22x32

Fig. 11–90. The linkage system of an internal safety valve (Problem 54). (Courtesy of General Precision Systems, Inc.)

57. A preliminary sketch and a pictorial of a linkage system to control the pitch of a helicopter blade are shown in Fig. 11–92. The variation in pitch is to be controlled by moving the vertical stick between the vertical and the left. The extreme positions of the stick are vertical and 85° to the left of vertical. The desired output of the linkage is at the high and low positions of point *A*. Modify this linkage as necessary to give this result; the stop screws are used to establish the high and low positions. Give the center-line dimensions of the linkage system after it has been modified to fulfill the desired needs. Solve the problem on 11″ × 15″ tracing paper at a convenient scale.

58. Design a steering linkage for the toy racer shown in Fig. 11–93 that will be steered by a child. The wheels are 8.5″ in diameter and are 20″ apart. The linkage should be as simple and

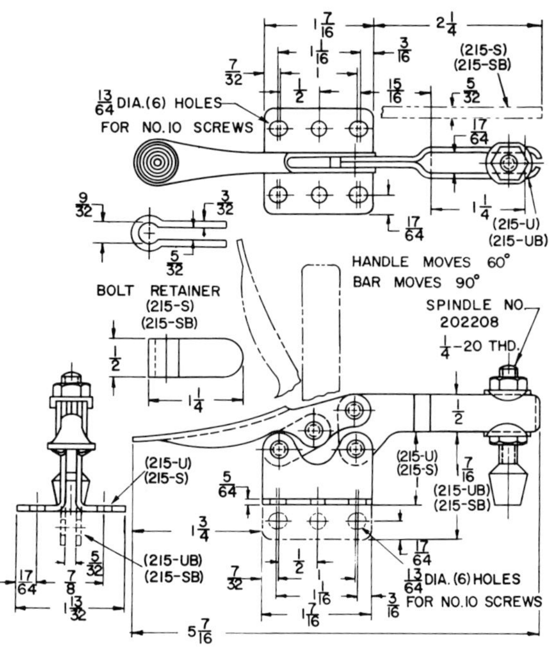

Fig. 11–91. A clamping device linkage (Problem 55). (Courtesy of De-Sta-Co Corporation.)

economical as possible to provide a system that could be mass-produced and assembled cheaply. Show your construction on a 11″ × 15″ sheet of tracing paper.

59. Design a linkage system for the child's racer shown in Fig. 11–94 that will enable the child to pedal the car. The front wheels are separated from the rear wheels by 26″. Estimate other dimensions to complete the linkage. Show your analysis and construction on an 11″ × 15″ sheet of tracing paper.

60. Construct a graph giving the travel of point *P* in Fig. 11–63 versus time in seconds. Use the following dimensions: $R_1 - 20''$, $R_2 - 40''$. Show all construction and analysis on a 11″ × 15″ sheet of tracing paper.

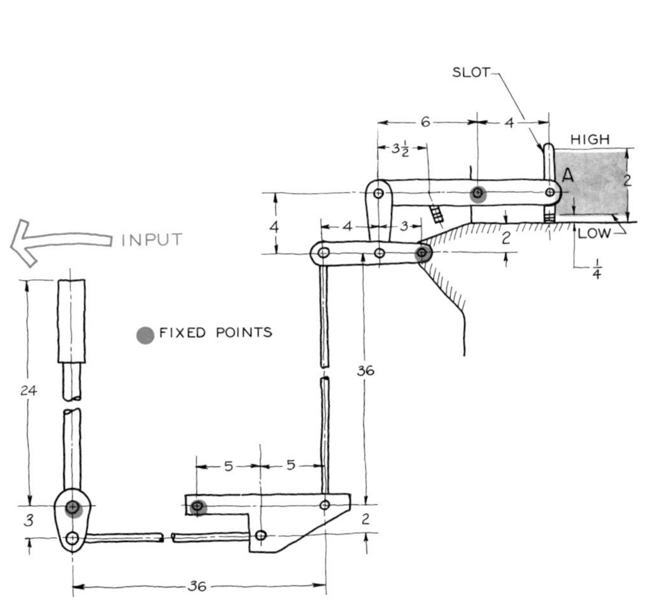

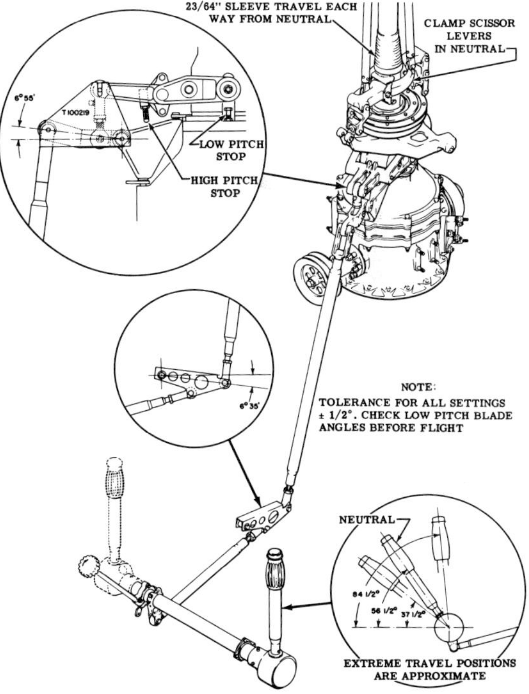

Fig. 11–92. Preliminary sketch and pictorial of the linkage system to control the pitch on a helicopter (Problem 57). (Courtesy of Bell Helicopter Corporation.)

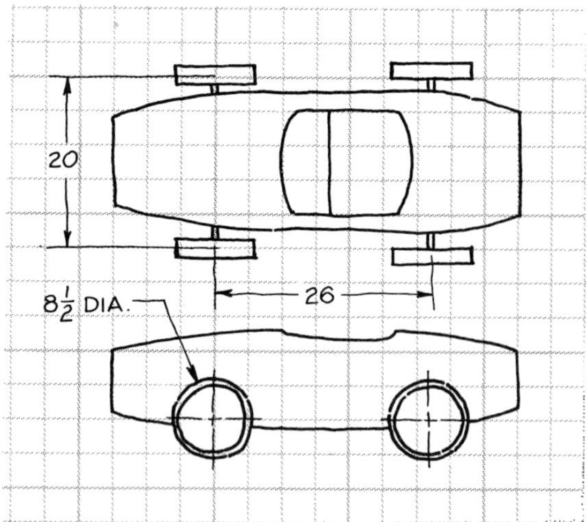

Fig. 11–93. Overall dimensions of a toy racer (Problem 58).

Fig. 11–94. Photograph of the toy car in Problems 58 and 59.

Differentiation

61. Plot the equation $Y = X^3/6$ as a rectangular graph. Graphically differentiate the curve to determine the first and second derivatives.

62. Analyze the motion of the piston discussed in Problem 57. Assume that the crackshaft is turning at a rate of one revolution per second. The piston rod is 9.7″ long, the radius of the input crank is 3.5″. Determine the velocity and acceleration at all intervals.

63. Analyze the motion of the shuttle plotted in Problem 60. Using the rate of revolution as one revolution per three seconds, determine the velocity and acceleration by graphical differentiation.

Integration

64. Plot the equation $Y = X + 2$ on a rectangular graph. Graphically integrate this curve to determine the first and second integrals. Find the area under the curve between $X = 1$ and $X = 7$.

65. Using graphical calculus, analyze a vertical strip 12″ wide on the inside face of the dam in Fig. 11–95. The force on this strip will be 52.0 lb/in. at the bottom of the dam. The first graph will be pounds per inch (ordinate) versus height in inches (abscissa). The second graph will be the integral of the first to give shear in pounds (ordinate) versus height in inches (abscissa). The third will be the integral of the second

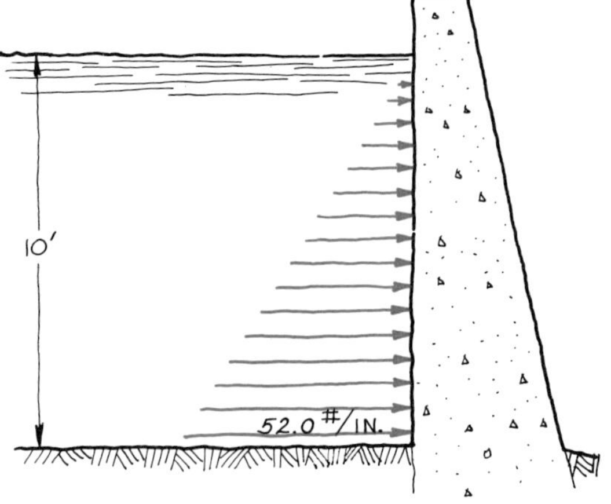

Fig. 11–95. Pressure on a 12″-wide section of a dam (Problem 65).

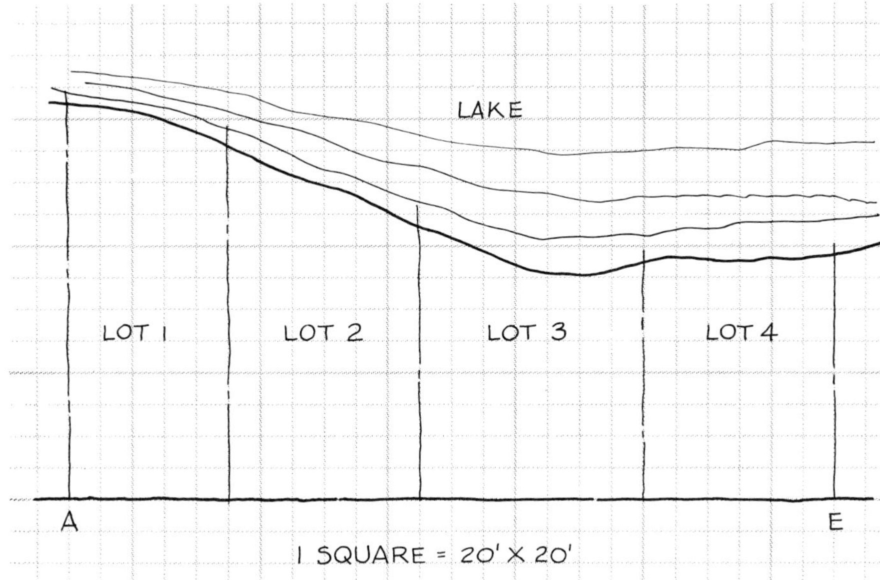

Fig. 11–96. Plot plan of a tract bounded by a lake front (Problem 66).

graph to give the moment in inch-pounds (ordinate) versus height in inches (abscissa). Convert these scales to give feet instead of inches.

66. A plot plan shows that a tract of land is bounded by a lake front (Fig. 11–96). By graphical integration, determine a graph that will represent the cumulative area of the land from point *A* to *E*. What is the total area? What is the area of each lot?

Nomography—General

The following problems are to be solved on an $8\frac{1}{2}'' \times 11''$ sheet with the scales selected to be most appropriate for the particular construction. Show all calculations and construction as part of the problems.

Nomography—Conversion Scales

67. Construct a chart that will convert inches to centimeters from 0 to 100 cm given that $1'' = 2.54$ cm.

68. Construct a chart for converting degrees in Fahrenheit to degrees in Celsius from 32°F to 212°F using the formula, $°C = \frac{4}{9}(°F - 32)$.

69. Construct a concurrent scale that will convert values of a radius (R) from 1 to 10 to an area (A) of a circle when $A = \pi R^2$.

70. Construct a concurrent scale that will convert numbers from 1 to 10 to their logarithms.

Parallel-Scale Nomograms

71. Construct a parallel-scale chart that will give solutions for the equation $A^2 + B^2 = C^2$ where values of A and B range from 0 to 20.

72. Construct a parallel-scale chart that will give the volume of cones varying in radius r from 1″ to 10″ and in height from 1″ to 30″. The equation for volume of a cone is $V = (\pi/3)r^2h$.

73. Construct a parallel-scale graph to determine your course grade based on the following equation:

$$\text{Grade }(G) = \underset{\text{(test average)}}{0.75} + \underset{\text{(final exam)}}{0.25}$$

(*Note*: Use factors that correspond to those used at your school if different ones are used.)

74. Construct a parallel-scale chart that can be used to find how many miles per gallon an automobile will travel. Use the following equation:

$$\text{mpg} = \frac{\text{miles}}{\text{gallon}}$$

where miles vary from 1 to 500 and gallons from 1 to 24.

75. Construct a parallel-scale chart that can be used to find the cost per mile (cpm) of an automobile. Use the following equation:

$$\text{cpm} = \frac{\text{cost}}{\text{miles}}$$

where miles vary from 1 to 500 and cost varies from 1 to \$10.00.

N-Charts

76. Using the equation stress = P/A, where P ranges from 0 to 1000 psi and A ranges from 0 to 15 in.^{2}, construct an N-chart to determine the stresses on a load-carrying member.

77. Construct an N-chart to determine the relationship of the equation $I = E/R$, where I = amperes of current, E = voltage, and R = resistance in ohms. Ranges: E = 0 to 1000 volts and R = 0 to 200 ohms.

78. Solve Problem 73 or 74 by using an N-chart.

79. Construct an N-chart to show the volume per 100-foot length of a ditch that is dug as the depth (D in foot) and width (W in foot) vary. The equation of these data is as follows:

$$\text{Volume} = 100D \times 100W$$

(*Hint*: Rewrite this equation in the standard form on an N-chart.)

Combination Charts

80. Construct a combination chart to express the law of sines as expressed in the equation $a/\sin A = b/\sin B$. Assume that a and b vary from 0 to 10, and that A and B vary from 0° to 90°.

81. Construct a combination chart to determine the velocity of sound in a solid, using the formula

$$C = \sqrt{\frac{E + 4\mu/3}{\rho}}$$

where E varies from 10^6 to 10^7 psi, μ varies from 1×10^6 to 2×10^6 psi, and C varies from 1000 to 1500 fps. (*Hint*: Rewrite the formula as $C^2\rho = E + \frac{4}{3}\mu$.)

BIBLIOGRAPHY AND SUGGESTED READING

1. Chambers, S. D., and V. M. Faires, *Analytic Mechanics.* New York: Macmillan, 1949.
2. Hammond, R. H., C. P. Buck, W. B. Rogers, G. W. Walsh, Jr., and H. P. Ackert, *Engineering Graphics for Design and Analysis.* New York: Ronald, 1964.
3. Luzadder, W. J., *Basic Graphics.* Englewood Cliffs, N. J.: Prentice-Hall, 1968.
4. Wellman, B. L., *Introduction to Graphical Analysis and Design.* New York: McGraw-Hill, 1966.
5. Woodward, Forrest, *Graphical Simulation.* Scranton, Pa.: International Textbook, 1967.

Appendixes

CONTENTS

APPENDIX 1. LOGARITHMS OF NUMBERS

N	0	1	2	3	4	5	6	7	8	9
1.0	.0000	.0043	.0086	.0128	.0170	.0212	.0253	.0294	.0334	.0374
1.1	.0414	.0453	.0492	.0531	.0569	.0607	.0645	.0682	.0719	.0755
1.2	.0792	.0828	.0864	.0899	.0934	.0969	.1004	.1038	.1072	.1106
1.3	.1139	.1173	.1206	.1239	.1271	.1303	.1335	.1367	.1399	.1430
1.4	.1461	.1492	.1523	.1553	.1584	.1614	.1644	.1673	.1703	.1732
1.5	.1761	.1790	.1818	.1847	.1875	.1903	.1931	.1959	.1987	.2014
1.6	.2041	.2068	.2095	.2122	.2148	.2175	.2201	.2227	.2253	.2279
1.7	.2304	.2330	.2355	.2380	.2405	.2430	.2455	.2480	.2504	.2529
1.8	.2553	.2577	.2601	.2625	.2648	.2672	.2695	.2718	.2742	.2765
1.9	.2788	.2810	.2833	.2856	.2878	.2900	.2923	.2945	.2967	.2989
2.0	.3010	.3032	.3054	.3075	.3096	.3118	.3139	.3160	.3181	.3201
2.1	.3222	.3243	.3263	.3284	.3304	.3324	.3345	.3365	.3385	.3404
2.2	.3424	.3444	.3464	.3483	.3502	.3522	.3541	.3560	.3579	.3598
2.3	.3617	.3636	.3655	.3674	.3692	.3711	.3729	.3747	.3766	.3784
2.4	.3802	.3820	.3838	.3856	.3874	.3892	.3909	.3927	.3945	.3962
2.5	.3979	.3997	.4014	.4031	.4048	.4065	.4082	.4099	.4116	.4133
2.6	.4150	.4166	.4183	.4200	.4216	.4232	.4249	.4265	.4281	.4298
2.7	.4314	.4330	.4346	.4362	.4378	.4393	.4409	.4425	.4440	.4456
2.8	.4472	.4487	.4502	.4518	.4533	.4548	.4564	.4579	.4594	.4609
2.9	.4624	.4639	.4654	.4669	.4683	.4698	.4713	.4728	.4742	.4757
3.0	.4771	.4786	.4800	.4814	.4829	.4843	.4857	.4871	.4886	.4900
3.1	.4914	.4928	.4942	.4955	.4969	.4983	.4997	.5011	.5024	.5038
3.2	.5051	.5065	.5079	.5092	.5105	.5119	.5132	.5145	.5159	.5172
3.3	.5185	.5198	.5211	.5224	.5237	.5250	.5263	.5276	.5289	.5302
3.4	.5315	.5328	.5340	.5353	.5366	.5378	.5391	.5403	.5416	.5428
3.5	.5441	.5453	.5465	.5478	.5490	.5502	.5514	.5527	.5539	.5551
3.6	.5563	.5575	.5587	.5599	.5611	.5623	.5635	.5647	.5658	.5670
3.7	.5682	.5694	.5705	.5717	.5729	.5740	.5752	.5763	.5775	.5786
3.8	.5798	.5809	.5821	.5832	.5843	.5855	.5866	.5877	.5888	.5899
3.9	.5911	.5922	.5933	.5944	.5955	.5966	.5977	.5988	.5999	.6010
4.0	.6021	.6031	.6042	.6053	.6064	.6075	.6085	.6096	.6107	.6117
4.1	.6128	.6138	.6149	.6160	.6170	.6180	.6191	.6201	.6212	.6222
4.2	.6232	.6243	.6253	.6263	.6274	.6284	.6294	.6304	.6314	.6325
4.3	.6335	.6345	.6355	.6365	.6375	.6385	.6395	.6405	.6415	.6425
4.4	.6435	.6444	.6454	.6464	.6474	.6484	.6493	.6503	.6513	.6522
4.5	.6532	.6542	.6551	.6561	.6571	.6580	.6590	.6599	.6609	.6618
4.6	.6628	.6637	.6646	.6656	.6665	.6675	.6684	.6693	.6702	.6712
4.7	.6721	.6730	.6739	.6749	.6758	.6767	.6776	.6785	.6794	.6803
4.8	.6812	.6821	.6830	.6839	.6848	.6857	.6866	.6875	.6884	.6893
4.9	.6902	.6911	.6920	.6928	.6937	.6946	.6955	.6964	.6972	.6981
5.0	.6990	.6998	.7007	.7016	.7024	.7033	.7042	.7050	.7059	.7067
5.1	.7076	.7084	.7093	.7101	.7110	.7118	.7126	.7135	.7143	.7152
5.2	.7160	.7168	.7177	.7185	.7193	.7202	.7210	.7218	.7226	.7235
5.3	.7243	.7251	.7259	.7267	.7275	.7284	.7292	.7300	.7308	.7316
5.4	.7324	.7332	.7340	.7348	.7356	.7364	.7372	.7380	.7388	.7396
N	0	1	2	3	4	5	6	7	8	9

APPENDIX 1. LOGARITHMS OF NUMBERS (Cont.)

N	0	1	2	3	4	5	6	7	8	9
5.5	.7404	.7412	.7419	.7427	.7435	.7443	.7451	.7459	.7466	.7474
5.6	.7482	.7490	.7497	.7505	.7513	.7520	.7528	.7536	.7543	.7551
5.7	.7559	.7566	.7574	.7582	.7589	.7597	.7604	.7612	.7619	.7627
5.8	.7634	.7642	.7649	.7657	.7664	.7672	.7679	.7686	.7694	.7701
5.9	.7709	.7716	.7723	.7731	.7738	.7745	.7752	.7760	.7767	.7774
6.0	.7782	.7789	.7796	.7803	.7810	.7818	.7825	.7832	.7839	.7846
6.1	.7853	.7860	.7868	.7875	.7882	.7889	.7896	.7903	.7910	.7917
6.2	.7924	.7931	.7938	.7945	.7952	.7959	.7966	.7973	.7980	.7987
6.3	.7993	.8000	.8007	.8014	.8021	.8028	.8035	.8041	.8048	.8055
6.4	.8062	.8069	.8075	.8082	.8089	.8096	.8102	.8109	.8116	.8122
6.5	.8129	.8136	.8142	.8149	.8156	.8162	.8169	.8176	.8182	.8189
6.6	.8195	.8202	.8209	.8215	.8222	.8228	.8235	.8241	.8248	.8254
6.7	.8261	.8267	.8274	.8280	.8287	.8293	.8299	.8306	.8312	.8319
6.8	.8325	.8331	.8338	.8344	.8351	.8357	.8363	.8370	.8376	.8382
6.9	.8388	.8395	.8401	.8407	.8414	.8420	.8426	.8432	.8439	.8445
7.0	.8451	.8457	.8463	.8470	.8476	.8482	.8488	.8494	.8500	.8506
7.1	.8513	.8519	.8525	.8531	.8537	.8543	.8549	.8555	.8561	.8567
7.2	.8573	.8579	.8585	.8591	.8597	.8603	.8609	.8615	.8621	.8627
7.3	.8633	.8639	.8645	.8651	.8657	.8663	.8669	.8675	.8681	.8686
7.4	.8692	.8698	.8704	.8710	.8716	.8722	.8727	.8733	.8739	.8745
7.5	.8751	.8756	.8762	.8768	.8774	.8779	.8785	.8791	.8797	.8802
7.6	.8808	.8814	.8820	.8825	.8831	.8837	.8842	.8848	.8854	.8859
7.7	.8865	.8871	.8876	.8882	.8887	.8893	.8899	.8904	.8910	.8915
7.8	.8921	.8927	.8932	.8938	.8943	.8949	.8954	.8960	.8965	.8971
7.9	.8976	.8982	.8987	.8993	.8998	.9004	.9009	.9015	.9020	.9025
8.0	.9031	.9036	.9042	.9047	.9053	.9058	.9063	.9069	.9074	.9079
8.1	.9085	.9090	.9096	.9101	.9106	.9112	.9117	.9122	.9128	.9133
8.2	.9138	.9143	.9149	.9154	.9159	.9165	.9170	.9175	.9180	.9186
8.3	.9191	.9196	.9201	.9206	.9212	.9217	.9222	.9227	.9232	.9238
8.4	.9243	.9248	.9253	.9258	.9263	.9269	.9274	.9279	.9284	.9289
8.5	.9294	.9299	.9304	.9309	.9315	.9320	.9325	.9330	.9335	.9340
8.6	.9345	.9350	.9355	.9360	.9365	.9370	.9375	.9380	.9385	.9390
8.7	.9395	.9400	.9405	.9410	.9415	.9420	.9425	.9430	.9435	.9440
8.8	.9445	.9450	.9455	.9460	.9465	.9469	.9474	.9479	.9484	.9489
8.9	.9494	.9499	.9504	.9509	.9513	.9518	.9523	.9528	.9533	.9538
9.0	.9542	.9547	.9552	.9557	.9562	.9566	.9571	.9576	.9581	.9586
9.1	.9590	.9595	.9600	.9605	.9609	.9614	.9619	.9624	.9628	.9633
9.2	.9638	.9643	.9647	.9652	.9657	.9661	.9666	.9671	.9675	.9680
9.3	.9685	.9689	.9694	.9699	.9703	.9708	.9713	.9717	.9722	.9727
9.4	.9731	.9736	.9741	.9745	.9750	.9754	.9759	.9763	.9768	.9773
9.5	.9777	.9782	.9786	.9791	.9795	.9800	.9805	.9809	.9814	.9818
9.6	.9823	.9827	.9832	.9836	.9841	.9845	.9850	.9854	.9859	.9863
9.7	.9868	.9872	.9877	.9881	.9886	.9890	.9894	.9899	.9903	.9908
9.8	.9912	.9917	.9921	.9926	.9930	.9934	.9939	.9943	.9948	.9952
9.9	.9956	.9961	.9965	.9969	.9974	.9978	.9983	.9987	.9991	.9996
N	0	1	2	3	4	5	6	7	8	9

APPENDIX 2. VALUES OF TRIGONOMETRIC FUNCTIONS

Degrees	Radians	Sine	Tangent	Cotangent	Cosine		
0° 00′	.0000	.0000	.0000		1.0000	1.5708	90° 00′
10′	.0029	.0029	.0029	343.77	1.0000	1.5679	50′
20′	.0058	.0058	.0058	171.89	1.0000	1.5650	40′
30′	.0087	.0087	.0087	114.59	1.0000	1.5621	30′
40′	.0116	.0116	.0116	85.940	.9999	1.5592	20′
50′	.0145	.0145	.0145	68.750	.9999	1.5563	10′
1° 00′	.0175	.0175	.0175	57.290	.9998	1.5533	89° 00′
10′	.0204	.0204	.0204	49.104	.9998	1.5504	50′
20′	.0233	.0233	.0233	42.964	.9997	1.5475	40′
30′	.0262	.0262	.0262	38.188	.9997	1.5446	30′
40′	.0291	.0291	.0291	34.368	.9996	1.5417	20′
50′	.0320	.0320	.0320	31.242	.9995	1.5388	10′
2° 00′	.0349	.0349	.0349	28.636	.9994	1.5359	88° 00′
10′	.0378	.0378	.0378	26.432	.9993	1.5330	50′
20′	.0407	.0407	.0407	24.542	.9992	1.5301	40′
30′	.0436	.0436	.0437	22.904	.9990	1.5272	30′
40′	.0465	.0465	.0466	21.470	.9989	1.5243	20′
50′	.0495	.0494	.0495	20.206	.9988	1.5213	10′
3° 00′	.0524	.0523	.0524	19.081	.9986	1.5184	87° 00′
10′	.0553	.0552	.0553	18.075	.9985	1.5155	50′
20′	.0582	.0581	.0582	17.169	.9983	1.5126	40′
30′	.0611	.0610	.0612	16.350	.9981	1.5097	30′
40′	.0640	.0640	.0641	15.605	.9980	1.5068	20′
50′	.0669	.0669	.0670	14.924	.9978	1.5039	10′
4° 00′	.0698	.0698	.0699	14.301	.9976	1.5010	86° 00′
10′	.0727	.0727	.0729	13.727	.9974	1.4981	50′
20′	.0756	.0756	.0758	13.197	.9971	1.4952	40′
30′	.0785	.0785	.0787	12.706	.9969	1.4923	30′
40′	.0814	.0814	.0816	12.251	.9967	1.4893	20′
50′	.0844	.0843	.0846	11.826	.9964	1.4864	10′
5° 00′	.0873	.0872	.0875	11.430	.9962	1.4835	85° 00′
10′	.0902	.0901	.0904	11.059	.9959	1.4806	50′
20′	.0931	.0929	.0934	10.712	.9957	1.4777	40′
30′	.0960	.0958	.0963	10.385	.9954	1.4748	30′
40′	.0989	.0987	.0992	10.078	.9951	1.4719	20′
50′	.1018	.1016	.1022	9.7882	.9948	1.4690	10′
6° 00′	.1047	.1045	.1051	9.5144	.9945	1.4661	84° 00′
10′	.1076	.1074	.1080	9.2553	.9942	1.4632	50′
20′	.1105	.1103	.1110	9.0098	.9939	1.4603	40′
30′	.1134	.1132	.1139	8.7769	.9936	1.4573	30′
40′	.1164	.1161	.1169	8.5555	.9932	1.4544	20′
50′	.1193	.1190	.1198	8.3450	.9929	1.4515	10′
7° 00′	.1222	.1219	.1228	8.1443	.9925	1.4486	83° 00′
10′	.1251	.1248	.1257	7.9530	.9922	1.4457	50′
20′	.1280	.1276	.1287	7.7704	.9918	1.4428	40′
30′	.1309	.1305	.1317	7.5958	.9914	1.4399	30′
40′	.1338	.1334	.1346	7.4287	.9911	1.4370	20′
50′	.1367	.1363	.1376	7.2687	.9907	1.4341	10′
8° 00′	.1396	.1392	.1405	7.1154	.9903	1.4312	82° 00′
10′	.1425	.1421	.1435	6.9682	.9899	1.4283	50′
20′	.1454	.1449	.1465	6.8269	.9894	1.4254	40′
30′	.1484	.1478	.1495	6.6912	.9890	1.4224	30′
40′	.1513	.1507	.1524	6.5606	.9886	1.4195	20′
50′	.1542	.1536	.1554	6.4348	.9881	1.4166	10′
9° 00′	.1571	.1564	.1584	6.3138	.9877	1.4137	81° 00′
		Cosine	Cotangent	Tangent	Sine	Radians	Degrees

APPENDIX 2. VALUES OF TRIGONOMETRIC FUNCTIONS (Cont.)

Degrees	Radians	Sine	Tangent	Cotangent	Cosine		
9° 00′	.1571	.1564	.1584	6.3138	.9877	1.4137	81° 00′
10′	.1600	.1593	.1614	6.1970	.9872	1.4108	50′
20′	.1629	.1622	.1644	6.0844	.9868	1.4079	40′
30′	.1658	.1650	.1673	5.9758	.9863	1.4050	30′
40′	.1687	.1679	.1703	5.8708	.9858	1.4021	20′
50′	.1716	.1708	.1733	5.7694	.9853	1.3992	10′
10° 00′	.1745	.1736	.1763	5.6713	.9848	1.3963	80° 00′
10′	.1774	.1765	.1793	5.5764	.9843	1.3934	50′
20′	.1804	.1794	.1823	5.4845	.9838	1.3904	40′
30′	.1833	.1822	.1853	5.3955	.9833	1.3875	30′
40′	.1862	.1851	.1883	5.3093	.9827	1.3846	20′
50′	.1891	.1880	.1914	5.2257	.9822	1.3817	10′
11° 00′	.1920	.1908	.1944	5.1446	.9816	1.3788	79° 00′
10′	.1949	.1937	.1974	5.0658	.9811	1.3759	50′
20′	.1978	.1965	.2004	4.9894	.9805	1.3730	40′
30′	.2007	.1994	.2035	4.9152	.9799	1.3701	30′
40′	.2036	.2022	.2065	4.8430	.9793	1.3672	20′
50′	.2065	.2051	.2095	4.7729	.9787	1.3643	10′
12° 00′	.2094	.2079	.2126	4.7046	.9781	1.3614	78° 00′
10′	.2123	.2108	.2156	4.6382	.9775	1.3584	50′
20′	.2153	.2136	.2186	4.5736	.9769	1.3555	40′
30′	.2182	.2164	.2217	4.5107	.9763	1.3526	30′
40′	.2211	.2193	.2247	4.4494	.9757	1.3497	20′
50′	.2240	.2221	.2278	4.3897	.9750	1.3468	10′
13° 00′	.2269	.2250	.2309	4.3315	.9744	1.3439	77° 00′
10′	.2298	.2278	.2339	4.2747	.9737	1.3410	50′
20′	.2327	.2306	.2370	4.2193	.9730	1.3381	40′
30′	.2356	.2334	.2401	4.1653	.9724	1.3352	30′
40′	.2385	.2363	.2432	4.1126	.9717	1.3323	20′
50′	.2414	.2391	.2462	4.0611	.9710	1.3294	10′
14° 00′	.2443	.2419	.2493	4.0108	.9703	1.3265	76° 00′
10′	.2473	.2447	.2524	3.9617	.9696	1.3235	50′
20′	.2502	.2476	.2555	3.9136	.9689	1.3206	40′
30′	.2531	.2504	.2586	3.8667	.9681	1.3177	30′
40′	.2560	.2532	.2617	3.8208	.9674	1.3148	20′
50′	.2589	.2560	.2648	3.7760	.9667	1.3119	10′
15° 00′	.2618	.2588	.2679	3.7321	.9659	1.3090	75° 00′
10′	.2647	.2616	.2711	3.6891	.9652	1.3061	50′
20′	.2676	.2644	.2742	3.6470	.9644	1.3032	40′
30′	.2705	.2672	.2773	3.6059	.9636	1.3003	30′
40′	.2734	.2700	.2805	3.5656	.9628	1.2974	20′
50′	.2763	.2728	.2836	3.5261	.9621	1.2945	10′
16° 00′	.2793	.2756	.2867	3.4874	.9613	1.2915	74° 00′
10′	.2822	.2784	.2899	3.4495	.9605	1.2886	50′
20′	.2851	.2812	.2931	3.4124	.9596	1.2857	40′
30′	.2880	.2840	.2962	3.3759	.9588	1.2828	30′
40′	.2909	.2868	.2994	3.3402	.9580	1.2799	20′
50′	.2938	.2896	.3026	3.3052	.9572	1.2770	10′
17° 00′	.2967	.2924	.3057	3.2709	.9563	1.2741	73° 00′
10′	.2996	.2952	.3089	3.2371	.9555	1.2712	50′
20′	.3025	.2979	.3121	3.2041	.9546	1.2683	40′
30′	.3054	.3007	.3153	3.1716	.9537	1.2654	30′
40′	.3083	.3035	.3185	3.1397	.9528	1.2625	20′
50′	.3113	.3062	.3217	3.1084	.9520	1.2595	10′
18° 00′	.3142	.3090	.3249	3.0777	.9511	1.2566	72° 00′
		Cosine	Cotangent	Tangent	Sine	Radians	Degrees

(Cont.)

APPENDIX 2. VALUES OF TRIGONOMETRIC FUNCTIONS (Cont.)

Degrees	Radians	Sine	Tangent	Cotangent	Cosine		
18° 00′	.3142	.3090	.3249	3.0777	.9511	1.2566	72° 00′
10′	.3171	.3118	.3281	3.0475	.9502	1.2537	50′
20′	.3200	.3145	.3314	3.0178	.9492	1.2508	40′
30′	.3229	.3173	.3346	2.9887	.9483	1.2479	30′
40′	.3258	.3201	.3378	2.9600	.9474	1.2450	20′
50′	.3287	.3228	.3411	2.9319	.9465	1.2421	10′
19° 00′	.3316	.3256	.3443	2.9042	.9455	1.2392	71° 00′
10′	.3345	.3283	.3476	2.8770	.9446	1.2363	50′
20′	.3374	.3311	.3508	2.8502	.9436	1.2334	40′
30′	.3403	.3338	.3541	2.8239	.9426	1.2305	30′
40′	.3432	.3365	.3574	2.7980	.9417	1.2275	20′
50′	.3462	.3393	.3607	2.7725	.9407	1.2246	10′
20° 00′	.3491	.3420	.3640	2.7475	.9397	1.2217	70° 00′
10′	.3520	.3448	.3673	2.7228	.9387	1.2188	50′
20′	.3549	.3475	.3706	2.6985	.9377	1.2159	40′
30′	.3578	.3502	.3739	2.6746	.9367	1.2130	30′
40′	.3607	.3529	.3772	2.6511	.9356	1.2101	20′
50′	.3636	.3557	.3805	2.6279	.9346	1.2072	10′
21° 00′	.3665	.3584	.3839	2.6051	.9336	1.2043	69° 00′
10′	.3694	.3611	.3872	2.5826	.9325	1.2014	50′
20′	.3723	.3638	.3906	2.5605	.9315	1.1985	40′
30′	.3752	.3665	.3939	2.5386	.9304	1.1956	30′
40′	.3782	.3692	.3973	2.5172	.9293	1.1926	20′
50′	.3811	.3719	.4006	2.4960	.9283	1.1897	10′
22° 00′	.3840	.3746	.4040	2.4751	.9272	1.1868	68° 00′
10′	.3869	.3773	.4074	2.4545	.9261	1.1839	50′
20′	.3898	.3800	.4108	2.4342	.9250	1.1810	40′
30′	.3927	.3827	.4142	2.4142	.9239	1.1781	30′
40′	.3956	.3854	.4176	2.3945	.9228	1.1752	20′
50′	.3985	.3881	.4210	2.3750	.9216	1.1723	10′
23° 00′	.4014	.3907	.4245	2.3559	.9205	1.1694	67° 00′
10′	.4043	.3934	.4279	2.3369	.9194	1.1665	50′
20′	.4072	.3961	.4314	2.3183	.9182	1.1636	40′
30′	.4102	.3987	.4348	2.2998	.9171	1.1606	30′
40′	.4131	.4014	.4383	2.2817	.9159	1.1577	20′
50′	.4160	.4041	.4417	2.2637	.9147	1.1548	10′
24° 00′	.4189	.4067	.4452	2.2460	.9135	1.1519	66° 00′
10′	.4218	.4094	.4487	2.2286	.9124	1.1490	50′
20′	.4247	.4120	.4522	2.2113	.9112	1.1461	40′
30′	.4276	.4147	.4557	2.1943	.9100	1.1432	30′
40′	.4305	.4173	.4592	2.1775	.9088	1.1403	20′
50′	.4334	.4200	.4628	2.1609	.9075	1.1374	10′
25° 00′	.4363	.4226	.4663	2.1445	.9063	1.1345	65° 00′
10′	.4392	.4253	.4699	2.1283	.9051	1.1316	50′
20′	.4422	.4279	.4734	2.1123	.9038	1.1286	40′
30′	.4451	.4305	.4770	2.0965	.9026	1.1257	30′
40′	.4480	.4331	.4806	2.0809	.9013	1.1228	20′
50′	.4509	.4358	.4841	2.0655	.9001	1.1199	10′
26° 00′	.4538	.4384	.4877	2.0503	.8988	1.1170	64° 00′
10′	.4567	.4410	.4913	2.0353	.8975	1.1141	50′
20′	.4596	.4436	.4950	2.0204	.8962	1.1112	40′
30′	4625	.4462	.4986	2.0057	.8949	1.1083	30′
40′	.4654	.4488	.5022	1.9912	.8936	1.1054	20′
50′	.4683	.4514	.5059	1.9768	.8923	1.1025	10′
27° 00′	.4712	.4540	.5095	1.9626	.8910	1.0996	63° 00′
		Cosine	Cotangent	Tangent	Sine	Radians	Degrees

APPENDIX 2. VALUES OF TRIGONOMETRIC FUNCTIONS (Cont.)

Degrees	Radians	Sine	Tangent	Cotangent	Cosine		
27° 00′	.4712	.4540	.5095	1.9626	.8910	1.0996	63° 00′
10′	.4741	.4566	.5132	1.9486	.8897	1.0966	50′
20′	.4771	.4592	.5169	1.9347	.8884	1.0937	40′
30′	.4800	.4617	.5206	1.9210	.8870	1.0908	30′
40′	.4829	.4643	.5243	1.9074	.8857	1.0879	20′
50′	.4858	.4669	.5280	1.8940	.8843	1.0850	10′
28° 00′	.4887	.4695	.5317	1.8807	.8829	1.0821	62° 00′
10′	.4916	.4720	.5354	1.8676	.8816	1.0792	50′
20′	.4945	.4746	.5392	1.8546	.8802	1.0763	40′
30′	.4974	.4772	.5430	1.8418	.8788	1.0734	30′
40′	.5003	.4797	.5467	1.8291	.8774	1.0705	20′
50′	.5032	.4823	.5505	1.8165	.8760	1.0676	10′
29° 00′	.5061	.4848	.5543	1.8040	.8746	1.0647	61° 00′
10′	.5091	.4874	.5581	1.7917	.8732	1.0617	50′
20′	.5120	.4899	.5619	1.7796	.8718	1.0588	40′
30′	.5149	.4924	.5658	1.7675	.8704	1.0559	30′
40′	.5178	.4950	.5696	1.7556	.8689	1.0530	20′
50′	.5207	.4975	.5735	1.7437	.8675	1.0501	10′
30° 00′	.5236	.5000	.5774	1.7321	.8660	1.0472	60° 00′
10′	.5265	.5025	.5812	1.7205	.8646	1.0443	50′
20′	.5294	.5050	.5851	1.7090	.8631	1.0414	40′
30′	.5323	.5075	.5890	1.6977	.8616	1.0385	30′
40′	.5352	.5100	.5930	1.6864	.8601	1.0356	20′
50′	.5381	.5125	.5969	1.6753	.8587	1.0327	10′
31° 00′	.5411	.5150	.6009	1.6643	.8572	1.0297	59° 00′
10′	.5440	.5175	.6048	1.6534	.8557	1.0268	50′
20′	.5469	.5200	.6088	1.6426	.8542	1.0239	40′
30′	.5498	.5225	.6128	1.6319	.8526	1.0210	30′
40′	.5527	.5250	.6168	1.6212	.8511	1.0181	20′
50′	.5556	.5275	.6208	1.6107	.8496	1.0152	10′
32° 00′	.5585	.5299	.6249	1.6003	.8480	1.0123	58° 00′
10′	.5614	.5324	.6289	1.5900	.8465	1.0094	50′
20′	.5643	.5348	.6330	1.5798	.8450	1.0065	40′
30′	.5672	.5373	.6371	1.5697	.8434	1.0036	30′
40′	.5701	.5398	.6412	1.5597	.8418	1.0007	20′
50′	.5730	.5422	.6453	1.5497	.8403	.9977	10′
33° 00′	.5760	.5446	.6494	1.5399	.8387	.9948	57° 00′
10′	.5789	.5471	.6536	1.5301	.8371	.9919	50′
20′	.5818	.5495	.6577	1.5204	.8355	.9890	40′
30′	.5847	.5519	.6619	1.5108	.8339	.9861	30′
40′	.5876	.5544	.6661	1.5013	.8323	.9832	20′
50′	.5905	.5568	.6703	1.4919	.8307	.9803	10′
34° 00′	.5934	.5592	.6745	1.4826	.8290	.9774	56° 00′
10′	.5963	.5616	.6787	1.4733	.8274	.9745	50′
20′	.5992	.5640	.6830	1.4641	.8258	.9716	40′
30′	.6021	.5664	.6873	1.4550	.8241	.9687	30′
40′	.6050	.5688	.6916	1.4460	.8225	.9657	20′
50′	.6080	.5712	.6959	1.4370	.8208	.9628	10′
35° 00′	.6109	.5736	.7002	1.4281	.8192	.9599	55° 00′
10′	.6138	.5760	.7046	1.4193	.8175	.9570	50′
20′	.6167	.5783	.7089	1.4106	.8158	.9541	40′
30′	.6196	.5807	.7133	1.4019	.8141	.9512	30′
40′	.6225	.5831	.7177	1.3934	.8124	.9483	20′
50′	.6254	.5854	.7221	1.3848	.8107	.9454	10′
36° 00′	.6283	.5878	.7265	1.3764	.8090	.9425	54° 00′
		Cosine	Cotangent	Tangent	Sine	Radians	Degrees

(Cont.)

APPENDIX 2. VALUES OF TRIGONOMETRIC FUNCTIONS (Cont.)

Degrees	Radians	Sine	Tangent	Cotangent	Cosine		
36° 00′	.6283	.5878	.7265	1.3764	.8090	.9425	**54° 00′**
10′	.6312	.5901	.7310	1.3680	.8073	.9396	50′
20′	.6341	.5925	.7355	1.3597	.8056	.9367	40′
30′	.6370	.5948	.7400	1.3514	.8039	.9338	30′
40′	.6400	.5972	.7445	1.3432	.8021	.9308	20′
50′	.6429	.5995	.7490	1.3351	.8004	.9279	10′
37° 00′	.6458	.6018	.7536	1.3270	.7986	.9250	**53° 00′**
10′	.6487	.6041	.7581	1.3190	.7969	.9221	50′
20′	.6516	.6065	.7627	1.3111	.7951	.9192	40′
30′	.6545	.6088	.7673	1.3032	.7934	.9163	30′
40′	.6574	.6111	.7720	1.2954	.7916	.9134	20′
50′	.6603	.6134	.7766	1.2876	.7898	.9105	10′
38° 00′	.6632	.6157	.7813	1.2799	.7880	.9076	**52° 00′**
10′	.6661	.6180	.7860	1.2723	.7862	.9047	50′
20′	.6690	.6202	.7907	1.2647	.7844	.9018	40′
30′	.6720	.6225	.7954	1.2572	.7826	.8988	30′
40′	.6749	.6248	.8002	1.2497	.7808	.8959	20′
50′	.6778	.6271	.8050	1.2423	.7790	.8930	10′
39° 00′	.6807	.6293	.8098	1.2349	.7771	.8901	**51° 00′**
10′	.6836	.6316	.8146	1.2276	.7753	.8872	50′
20′	.6865	.6338	.8195	1.2203	.7735	.8843	40′
30′	.6894	.6361	.8243	1.2131	.7716	.8814	30′
40′	.6923	.6383	.8292	1.2059	.7698	.8785	20′
50′	.6952	.6406	.8342	1.1988	.7679	.8756	10′
40° 00′	.6981	.6428	.8391	1.1918	.7660	.8727	**50° 00′**
10′	.7010	.6450	.8441	1.1847	.7642	.8698	50′
20′	.7039	.6472	.8491	1.1778	.7623	.8668	40′
30′	.7069	.6494	.8541	1.1708	.7604	.8639	30′
40′	.7098	.6517	.8591	1.1640	.7585	.8610	20′
50′	.7127	.6539	.8642	1.1571	.7566	.8581	10′
41° 00′	.7156	.6561	.8693	1.1504	.7547	.8552	**49° 00′**
10′	.7185	.6583	.8744	1.1436	.7528	.8523	50′
20′	.7214	.6604	.8796	1.1369	.7509	.8494	40′
30′	.7243	.6626	.8847	1.1303	.7490	.8465	30′
40′	.7272	.6648	.8899	1.1237	.7470	.8436	20′
50′	.7301	.6670	.8952	1.1171	.7451	.8407	10′
42° 00′	.7330	.6691	.9004	1.1106	.7431	.8378	**48° 00′**
10′	.7359	.6713	.9057	1.1041	.7412	.8348	50′
20′	.7389	.6734	.9110	1.0977	.7392	.8319	40′
30′	.7418	.6756	.9163	1.0913	.7373	.8290	30′
40′	.7447	.6777	.9217	1.0850	.7353	.8261	20′
50′	.7476	.6799	.9271	1.0786	.7333	.8232	10′
43° 00′	.7505	.6820	.9325	1.0724	.7314	.8203	**47° 00′**
10′	.7534	.6841	.9380	1.0661	.7294	.8174	50′
20′	.7563	.6862	.9435	1.0599	.7274	.8145	40′
30′	.7592	.6884	.9490	1.0538	.7254	.8116	30′
40′	.7621	.6905	.9545	1.0477	.7234	.8087	20′
50′	.7650	.6926	.9601	1.0416	.7214	.8058	10′
44° 00′	.7679	.6947	.9657	1.0355	.7193	.8029	**46° 00′**
10′	.7709	.6967	.9713	1.0295	.7173	.7999	50′
20′	.7738	.6988	.9770	1.0235	.7153	.7970	40′
30′	.7767	.7009	.9827	1.0176	.7133	.7941	30′
40′	.7796	.7030	.9884	1.0117	.7112	.7912	20′
50′	.7825	.7050	.9942	1.0058	.7092	.7883	10′
45° 00′	.7854	.7071	1.0000	1.0000	.7071	.7854	**45° 00′**
		Cosine	Cotangent	Tangent	Sine	Radians	Degrees

APPENDIX 3. WEIGHTS AND MEASURES

UNITED STATES SYSTEM

LINEAR MEASURE

Inches	Feet	Yards	Rods	Furlongs	Miles
1.0 =	.08333 =	.02778 =	.0050505 =	.00012626 =	.00001578
12.0 =	1.0 =	.33333 =	.0606061 =	.00151515 =	.00018939
36.0 =	3.0 =	1.0 =	.1818182 =	.00454545 =	.00056818
198.0 =	16.5 =	5.5 =	1.0 =	.025 =	.003125
7920.0 =	660.0 =	220.0 =	40.0 =	1.0 =	.125
63360.0 =	5280.0 =	1760.0 =	320.0 =	8.0 =	1.0

SQUARE AND LAND MEASURE

Sq. Inches	Square Feet	Square Yards	Sq. Rods	Acres	Sq. Miles
1.0 =	.006944 =	.000772			
144.0 =	1.0 =	.111111			
1296.0 =	9.0 =	1.0 =	.03306 =	.000207	
39204.0 =	272.25 =	30.25 =	1.0 =	.00625 =	.0000098
	43560.0 =	4840.0 =	160.0 =	1.0 =	.0015625
		3097600.0 =	102400.0 =	640.0 =	1.0

AVOIRDUPOIS WEIGHTS

Grains	Drams	Ounces	Pounds	Tons
1.0 =	.03657 =	.002286 =	.000143 =	.0000000714
27.34375 =	1.0 =	.0625 =	.003906 =	.00000195
437.5 =	16.0 =	1.0 =	.0625 =	.00003125
7000.0 =	256.0 =	16.0 =	1.0 =	.0005
14000000.0 =	512000.0 =	32000.0 =	2000.0 =	1.0

DRY MEASURE

Pints	Quarts	Pecks	Cubic Feet	Bushels
1.0 =	.5 =	.0625 =	.01945 =	.01563
2.0 =	1.0 =	.125 =	.03891 =	.03125
16.0 =	8.0 =	1.0 =	.31112 =	.25
51.42627 =	25.71314 =	3.21414 =	1.0 =	.80354
64.0 =	32.0 =	4.0 =	1.2445 =	1.0

LIQUID MEASURE

Gills	Pints	Quarts	U. S. Gallons	Cubic Feet
1.0 =	.25 =	.125 =	.03125 =	.00418
4.0 =	1.0 =	.5 =	.125 =	.01671
8.0 =	2.0 =	1.0 =	.250 =	.03342
32.0 =	8.0 =	4.0 =	1.0 =	.1337
			7.48052 =	1.0

METRIC SYSTEM

UNITS

Length—Meter : Mass—Gram : Capacity—Liter

for pure water at 4°C. (39.2°F.)

1 cubic decimeter or 1 liter = 1 kilogram

$$1000\ \text{Milli}\begin{Bmatrix}\textit{meters}\ (\text{mm})\\ \textit{grams}\ (\text{mg})\\ \textit{liters}\ (\text{ml})\end{Bmatrix} = 100\ \text{Centi}\begin{Bmatrix}\textit{meters}\ (\text{cm})\\ \textit{grams}\ (\text{cg})\\ \textit{liters}\ (\text{cl})\end{Bmatrix} = 10\ \text{Deci}\begin{Bmatrix}\textit{meters}\ (\text{dm})\\ \textit{grams}\ (\text{dg})\\ \textit{liters}\ (\text{dl})\end{Bmatrix} = 1\begin{Bmatrix}\text{meter}\\ \text{gram}\\ \text{liter}\end{Bmatrix}$$

$$1000\begin{Bmatrix}\text{meters}\\ \text{grams}\\ \text{liters}\end{Bmatrix} = 100\ \text{Deka}\begin{Bmatrix}\textit{meters}\ (\text{dkm})\\ \textit{grams}\ (\text{dkg})\\ \textit{liters}\ (\text{dkl})\end{Bmatrix} = 10\ \text{Hecto}\begin{Bmatrix}\textit{meters}\ (\text{hm})\\ \textit{grams}\ (\text{hg})\\ \textit{liters}\ (\text{hl})\end{Bmatrix} = 1\ \text{Kilo}\begin{Bmatrix}\textit{meter}\ (\text{km})\\ \textit{gram}\ (\text{kg})\\ \textit{liter}\ (\text{kl})\end{Bmatrix}$$

1 Metric Ton	=	1000 Kilograms
100 Square Meters	=	1 Are
100 Ares	=	1 Hectare
100 Hectares	=	1 Square Kilometer

(Courtesy of the American Institute of Steel Construction.)

APPENDIX 4. DECIMAL EQUIVALENTS AND TEMPERATURE CONVERSION

DECIMAL EQUIVALENTS—INCH-MILLIMETER CONVERSION TABLE

½	¼	⅛	1/16	1/32	1/64	Decimals	Millimeters	½	¼	⅛	1/16	1/32	1/64	Decimals	Millimeters
					1	.015625	.396875						33	.515625	13.096875
				1		.031250	.793750					17		.531250	13.493750
					3	.046875	1.190625						35	.546875	13.890625
			1			.062500	1.587500				9			.562500	14.287500
					5	.078125	1.984375						37	.578125	14.684375
				3		.093750	2.381250					19		.593750	15.081250
					7	.109375	2.778125						39	.609375	15.478125
		1				.125000	3.175000			5				.625000	15.875000
					9	.140625	3.571875						41	.640625	16.271875
				5		.156250	3.968750					21		.656250	16.668750
					11	.171875	4.365625						43	.671875	17.065625
			3			.187500	4.762500				11			.687500	17.462500
					13	.203125	5.159375						45	.703125	17.859375
				7		.218750	5.556250					23		.718750	18.256250
					15	.234375	5.953125						47	.734375	18.653125
	1					.250000	6.350000		3					.750000	19.050000
					17	.265625	6.746875						49	.765625	19.446875
				9		.281250	7.143750					25		.781250	19.843750
					19	.296875	7.540625						51	.796875	20.240625
			5			.312500	7.937500				13			.812500	20.637500
					21	.328125	8.334375						53	.828125	21.034375
				11		.343750	8.731250					27		.843750	21.431250
					23	.359375	9.128125						55	.859375	21.828125
		3				.375000	9.525000			7				.875000	22.225000
					25	.390625	9.921875						57	.890625	22.621875
				13		.406250	10.318750					29		.906250	23.018750
					27	.421875	10.715625						59	.921875	23.415625
			7			.437500	11.112500				15			.937500	23.812500
					29	.453125	11.509375						61	.953125	24.209375
				15		.468750	11.906250					31		.968750	24.606250
					31	.484375	12.303125						63	.984375	25.003125
1						.500000	12.700000	2	4	8	16	32	64	1.000000	25.400000

APPENDIX 4. DECIMAL EQUIVALENTS AND TEMPERATURE CONVERSION (Cont.)

TEMPERATURE CONVERSION

-210 to 0			1 to 25			26 to 50			51 to 75			76 to 100			101 to 340			341 to 490			491 to 750		
C.	C. or F.	F.	C.	C. or F.	F.	C.	C. or F.	F.	C.	C. or F.	F.	C.	C. or F.	F.	C.	C. or F.	F.	C.	C. or F.	F.	C.	C. or F.	F.
-134	**-210**	-346	-17.2	**1**	33.8	-3.33	**26**	78.8	10.6	**51**	123.8	24.4	**76**	168.8	43	**110**	230	177	**350**	662	260	**500**	932
-129	**-200**	-328	-16.7	**2**	35.6	-2.78	**27**	80.6	11.1	**52**	125.6	25.0	**77**	170.6	49	**120**	248	182	**360**	680	266	**510**	950
-123	**-190**	-310	-16.1	**3**	37.4	-2.22	**28**	82.4	11.7	**53**	127.4	25.6	**78**	172.4	54	**130**	266	188	**370**	698	271	**520**	968
-118	**-180**	-292	-15.6	**4**	39.2	-1.67	**29**	84.2	12.2	**54**	129.2	26.1	**79**	174.2	60	**140**	284	193	**380**	716	277	**530**	986
-112	**-170**	-274	-15.0	**5**	41.0	-1.11	**30**	86.0	12.8	**55**	131.0	26.7	**80**	176.0	66	**150**	302	199	**390**	734	282	**540**	1004
-107	**-160**	-256	-14.4	**6**	42.8	-0.56	**31**	87.8	13.3	**56**	132.8	27.2	**81**	177.8	71	**160**	320	204	**400**	752	288	**550**	1022
-101	**-150**	-238	-13.9	**7**	44.6	0	**32**	89.6	13.9	**57**	134.6	27.8	**82**	179.6	77	**170**	338	210	**410**	770	293	**560**	1040
-95.6	**-140**	-220	-13.3	**8**	46.4	0.56	**33**	91.4	14.4	**58**	136.4	28.3	**83**	181.4	82	**180**	356	216	**420**	788	299	**570**	1058
-90.0	**-130**	-202	-12.8	**9**	48.2	1.11	**34**	93.2	15.0	**59**	138.2	28.9	**84**	183.2	88	**190**	374	221	**430**	806	304	**580**	1076
-84.4	**-120**	-184	-12.2	**10**	50.0	1.67	**35**	95.0	15.6	**60**	140.0	29.4	**85**	185.0	93	**200**	392	227	**440**	824	310	**590**	1094
-78.9	**-110**	-166	-11.7	**11**	51.8	2.22	**36**	96.8	16.1	**61**	141.8	30.0	**86**	186.8	99	**210**	410	232	**450**	842	316	**600**	1112
-73.3	**-100**	-148	-11.1	**12**	53.6	2.78	**37**	98.6	16.7	**62**	143.6	30.6	**87**	188.6	100	**212**	413	238	**460**	860	321	**610**	1130
-67.8	**-90**	-130	-10.6	**13**	55.4	3.33	**38**	100.4	17.2	**63**	145.4	31.1	**88**	190.4	104	**220**	428	243	**470**	878	327	**620**	1148
-62.2	**-80**	-112	-10.0	**14**	57.2	3.89	**39**	102.2	17.8	**64**	147.2	31.7	**89**	192.2	110	**230**	446	249	**480**	896	332	**630**	1166
-56.7	**-70**	-94	-9.44	**15**	59.0	4.44	**40**	104.0	18.3	**65**	149.0	32.2	**90**	194.0	116	**240**	464	254	**490**	914	338	**640**	1184
-51.1	**-60**	-76	-8.89	**16**	60.8	5.00	**41**	105.8	18.9	**66**	150.8	32.8	**91**	195.8	121	**250**	482				343	**650**	1202
-45.6	**-50**	-58	-8.33	**17**	62.6	5.56	**42**	107.6	19.4	**67**	152.6	33.3	**92**	197.6	127	**260**	500				349	**660**	1220
-40.0	**-40**	-40	-7.78	**18**	64.4	6.11	**43**	109.4	20.0	**68**	154.4	33.9	**93**	199.4	132	**270**	518				354	**670**	1238
-34.4	**-30**	-22	-7.22	**19**	66.2	6.67	**44**	111.2	20.6	**69**	156.2	34.4	**94**	201.2	138	**280**	536				360	**680**	1256
-28.9	**-20**	-4	-6.67	**20**	68.0	7.22	**45**	113.0	21.1	**70**	158.0	35.0	**95**	203.0	143	**290**	554				366	**690**	1274
-23.3	**-10**	14	-6.11	**21**	69.8	7.78	**46**	114.8	21.7	**71**	159.8	35.6	**96**	204.8	149	**300**	572				371	**700**	1292
-17.8	**0**	32	-5.56	**22**	71.6	8.33	**47**	116.6	22.2	**72**	161.6	36.1	**97**	206.6	154	**310**	590				377	**710**	1310
			-5.00	**23**	73.4	8.89	**48**	118.4	22.8	**73**	163.4	36.7	**98**	208.4	160	**320**	608				382	**720**	1328
			-4.44	**24**	75.2	9.44	**49**	120.2	23.3	**74**	165.2	37.2	**99**	210.2	166	**330**	626				388	**730**	1346
			-3.89	**25**	77.0	10.0	**50**	122.0	23.9	**75**	167.0	37.8	**100**	212.0	171	**340**	644				393	**740**	1364
																					399	**750**	1382

NOTE:—The numbers in bold face type refer to the temperature either in degrees Centigrade or Fahrenheit which it is desired to convert into the other scale. If converting from Fahrenheit degrees to Centigrade degrees the equivalent temperature will be found in the left column, while if converting from degrees Centigrade to degrees Fahrenheit, the answer will be found in the column on the right.

$$°F = \frac{9}{5}(°C) + 32$$

$$°C = \frac{5}{9}(°F - 32)$$

INTERPOLATION FACTORS

C.		F.	C.		F.
0.56	**1**	1.8	3.33	**6**	10.8
1.11	**2**	3.6	3.89	**7**	12.6
1.67	**3**	5.4	4.44	**8**	14.4
2.22	**4**	7.2	5.00	**9**	16.2
2.78	**5**	9.0	5.56	**10**	18.0

(Courtesy of Stephens–Adamson Manufacturing Company.)

APPENDIX 5. WEIGHTS AND SPECIFIC GRAVITIES

Substance	Weight Lb. per Cu. Ft.	Specific Gravity
METALS, ALLOYS, ORES		
Aluminum, cast, hammered	165	2.55-2.75
Brass, cast, rolled	534	8.4-8.7
Bronze, 7.9 to 14% Sn	509	7.4-8.9
Bronze, aluminum	481	7.7
Copper, cast, rolled	556	8.8-9.0
Copper ore, pyrites	262	4.1-4.3
Gold, cast, hammered	1205	19.25-19.3
Iron, cast, pig	450	7.2
Iron, wrought	485	7.6-7.9
Iron, spiegel-eisen	468	7.5
Iron, ferro-silicon	437	6.7-7.3
Iron ore, hematite	325	5.2
Iron ore, hematite in bank	160-180	....
Iron ore, hematite loose	130-160	....
Iron ore, limonite	237	3.6-4.0
Iron ore, magnetite	315	4.9-5.2
Iron slag	172	2.5-3.0
Lead	710	11.37
Lead ore, galena	465	7.3-7.6
Magnesium, alloys	112	1.74-1.83
Manganese	475	7.2-8.0
Manganese ore, pyrolusite	259	3.7-4.6
Mercury	849	13.6
Monel Metal	556	8.8-9.0
Nickel	565	8.9-9.2
Platinum, cast, hammered	1330	21.1-21.5
Silver, cast, hammered	656	10.4-10.6
Steel, rolled	490	7.85
Tin, cast, hammered	459	7.2-7.5
Tin ore, cassiterite	418	6.4-7.0
Zinc, cast, rolled	440	6.9-7.2
Zinc ore, blende	253	3.9-4.2
VARIOUS SOLIDS		
Cereals, oats ... bulk	32	....
Cereals, barley ... bulk	39	....
Cereals, corn, rye ... bulk	48	....
Cereals, wheat ... bulk	48	....
Hay and Straw ... bales	20	....
Cotton, Flax, Hemp	93	1.47-1.50
Fats	58	0.90-0.97
Flour, loose	28	0.40-0.50
Flour, pressed	47	0.70-0.80
Glass, common	156	2.40-2.60
Glass, plate or crown	161	2.45-2.72
Glass, crystal	184	2.90-3.00
Leather	59	0.86-1.02
Paper	58	0.70-1.15
Potatoes, piled	42	....
Rubber, caoutchouc	59	0.92-0.96
Rubber goods	94	1.0-2.0
Salt, granulated, piled	48	....
Saltpeter	67	....
Starch	96	1.53
Sulphur	125	1.93-2.07
Wool	82	1.32
TIMBER, U. S. SEASONED		
Moisture Content by Weight:		
Seasoned timber 15 to 20%		
Green timber up to 50%		
Ash, white, red	40	0.62-0.65
Cedar, white, red	22	0.32-0.38
Chestnut	41	0.66
Cypress	30	0.48
Fir, Douglas spruce	32	0.51
Fir, eastern	25	0.40
Elm, white	45	0.72
Hemlock	29	0.42-0.52
Hickory	49	0.74-0.84
Locust	46	0.73
Maple, hard	43	0.68
Maple, white	33	0.53
Oak, chestnut	54	0.86
Oak, live	59	0.95
Oak, red, black	41	0.65
Oak, white	46	0.74
Pine, Oregon	32	0.51
Pine, red	30	0.48
Pine, white	26	0.41
Pine, yellow, long-leaf	44	0.70
Pine, yellow, short-leaf	38	0.61
Poplar	30	0.48
Redwood, California	26	0.42
Spruce, white, black	27	0.40-0.46
Walnut, black	38	0.61
VARIOUS LIQUIDS		
Alcohol, 100%	49	0.79
Acids, muriatic 40%	75	1.20
Acids, nitric 91%	94	1.50
Acids, sulphuric 87%	112	1.80
Lye, soda 66%	106	1.70
Oils, vegetable	58	0.91-0.94
Oils, mineral, lubricants	57	0.90-0.93
Water, 4°C. max. density	62.428	1.0
Water, 100°C.	59.830	0.9584
Water, ice	56	0.88-0.92
Water, snow, fresh fallen	8	.125
Water, sea water	64	1.02-1.03
GASES		
Air, 0°C. 760 mm.	.08071	1.0
Ammonia	.0478	0.5920
Carbon dioxide	.1234	1.5291
Carbon monoxide	.0781	0.9673
Gas, illuminating	.028-.036	0.35-0.45
Gas, natural	.038-.039	0.47-0.48
Hydrogen	.00559	0.0693
Nitrogen	.0784	0.9714
Oxygen	.0892	1.1056

The specific gravities of solids and liquids refer to water at 4°C., those of gases to air at 0°C. and 760 mm. pressure. The weights per cubic foot are derived from average specific gravities, except where stated that weights are for bulk, heaped or loose material, etc.

(Courtesy of the American Institute of Steel Construction.)

APPENDIX 5. WEIGHTS AND SPECIFIC GRAVITIES (Cont.)

Substance	Weight Lb. per Cu. Ft.	Specific Gravity
ASHLAR MASONRY		
Granite, syenite, gneiss	165	2.3-3.0
Limestone, marble	160	2.3-2.8
Sandstone, bluestone	140	2.1-2.4
MORTAR RUBBLE MASONRY		
Granite, syenite, gneiss	155	2.2-2.8
Limestone, marble	150	2.2-2.6
Sandstone, bluestone	130	2.0-2.2
DRY RUBBLE MASONRY		
Granite, syenite, gneiss	130	1.9-2.3
Limestone, marble	125	1.9-2.1
Sandstone, bluestone	110	1.8-1.9
BRICK MASONRY		
Pressed brick	140	2.2-2.3
Common brick	120	1.8-2.0
Soft brick	100	1.5-1.7
CONCRETE MASONRY		
Cement, stone, sand	144	2.2-2.4
Cement, slag, etc.	130	1.9-2.3
Cement, cinder, etc.	100	1.5-1.7
VARIOUS BUILDING MATERIALS		
Ashes, cinders	40-45	....
Cement, portland, loose	90	....
Cement, portland, set	183	2.7-3.2
Lime, gypsum, loose	53-64	....
Mortar, set	103	1.4-1.9
Slags, bank slag	67-72	....
Slags, bank screenings	98-117	....
Slags, machine slag	96	....
Slags, slag sand	49-55	....
EARTH, ETC., EXCAVATED		
Clay, dry	63	....
Clay, damp, plastic	110	....
Clay and gravel, dry	100	....
Earth, dry, loose	76	....
Earth, dry, packed	95	....
Earth, moist, loose	78	....
Earth, moist, packed	96	....
Earth, mud, flowing	108	....
Earth, mud, packed	115	....
Riprap, limestone	80-85	....
Riprap, sandstone	90	....
Riprap, shale	105	....
Sand, gravel, dry, loose	90-105	....
Sand, gravel, dry, packed	100-120	....
Sand, gravel, dry, wet	118-120	....
EXCAVATIONS IN WATER		
Sand or gravel	60	....
Sand or gravel and clay	65	....
Clay	80	....
River mud	90	....
Soil	70	....
Stone riprap	65	....
MINERALS		
Asbestos	153	2.1-2.8
Barytes	281	4.50
Basalt	184	2.7-3.2
Bauxite	159	2.55
Borax	109	1.7-1.8
Chalk	137	1.8-2.6
Clay, marl	137	1.8-2.6
Dolomite	181	2.9
Feldspar, orthoclase	159	2.5-2.6
Gneiss, serpentine	159	2.4-2.7
Granite, syenite	175	2.5-3.1
Greenstone, trap	187	2.8-3.2
Gypsum, alabaster	159	2.3-2.8
Hornblende	187	3.0
Limestone, marble	165	2.5-2.8
Magnesite	187	3.0
Phosphate rock, apatite	200	3.2
Porphyry	172	2.6-2.9
Pumice, natural	40	0.37-0.90
Quartz, flint	165	2.5-2.8
Sandstone, bluestone	147	2.2-2.5
Shale, slate	175	2.7-2.9
Soapstone, talc	169	2.6-2.8
STONE, QUARRIED, PILED		
Basalt, granite, gneiss	96	....
Limestone, marble, quartz	95	....
Sandstone	82	....
Shale	92	....
Greenstone, hornblende	107	....
BITUMINOUS SUBSTANCES		
Asphaltum	81	1.1-1.5
Coal, anthracite	97	1.4-1.7
Coal, bituminous	84	1.2-1.5
Coal, lignite	78	1.1-1.4
Coal, peat, turf, dry	47	0.65-0.85
Coal, charcoal, pine	23	0.28-0.44
Coal, charcoal, oak	33	0.47-0.57
Coal, coke	75	1.0-1.4
Graphite	131	1.9-2.3
Paraffine	56	0.87-0.91
Petroleum	54	0.87
Petroleum, refined	50	0.79-0.82
Petroleum, benzine	46	0.73-0.75
Petroleum, gasoline	42	0.66-0.69
Pitch	69	1.07-1.15
Tar, bituminous	75	1.20
COAL AND COKE, PILED		
Coal, anthracite	47-58	....
Coal, bituminous, lignite	40-54	....
Coal, peat, turf	20-26	....
Coal, charcoal	10-14	....
Coal, coke	23-32	....

The specific gravities of solids and liquids refer to water at 4°C., those of gases to air at 0°C. and 760 mm. pressure. The weights per cubic foot are derived from average specific gravities, except where stated that weights are for bulk, heaped or loose material, etc.

(Courtesy of the American Institute of Steel Construction.)

APPENDIX 6. WIRE AND SHEET METAL GAGES

WIRE AND SHEET METAL GAGES IN DECIMALS OF AN INCH

Name of Gage	United States Standard Gage*		The United States Steel Wire Gage	American or Brown & Sharpe Wire Gage	New Birmingham Standard Sheet & Hoop Gage	British Imperial or English Legal Standard Wire Gage	Birmingham or Stubs Iron Wire Gage	Name of Gage
Principal Use	Uncoated Steel Sheets and Light Plates		Steel Wire except Music Wire	Non-Ferrous Sheets and Wire	Iron and Steel Sheets and Hoops	Wire	Strips, Bands, Hoops and Wire	Principal Use
Gage No.	Weight Oz. per Sq. Ft.	Approx. Thickness Inches	Thickness, Inches					Gage No.
7/0's			.4900		.6666	.500		7/0's
6/0's			.4615	.5800	.625	.464		6/0's
5/0's			.4305	.5165	.5883	.432	.500	5/0's
4/0's			.3938	.4600	.5416	.400	.454	4/0's
3/0's			.3625	.4096	.500	.372	.425	3/0's
2/0's			.3310	.3648	.4452	.348	.380	2/0's
0			.3065	.3249	.3964	.324	.340	0
1			.2830	.2893	.3532	.300	.300	1
2			.2625	.2576	.3147	.276	.284	2
3	160	.2391	.2437	.2294	.2804	.252	.259	3
4	150	.2242	.2253	.2043	.250	.232	.238	4
5	140	.2092	.2070	.1819	.2225	.212	.220	5
6	130	.1943	.1920	.1620	.1981	.192	.203	6
7	120	.1793	.1770	.1443	.1764	.176	.180	7
8	110	.1644	.1620	.1285	.1570	.160	.165	8
9	100	.1495	.1483	.1144	.1398	.144	.148	9
10	90	.1345	.1350	.1019	.1250	.128	.134	10
11	80	.1196	.1205	.0907	.1113	.116	.120	11
12	70	.1046	.1055	.0808	.0991	.104	.109	12
13	60	.0897	.0915	.0720	.0882	.092	.095	13
14	50	.0747	.0800	.0641	.0785	.080	.083	14
15	45	.0673	.0720	.0571	.0699	.072	.072	15
16	40	.0598	.0625	.0508	.0625	.064	.065	16
17	36	.0538	.0540	.0453	.0556	.056	.058	17
18	32	.0478	.0475	.0403	.0495	.048	.049	18
19	28	.0418	.0410	.0359	.0440	.040	.042	19
20	24	.0359	.0348	.0320	.0392	.036	.035	20
21	22	.0329	.0318	.0285	.0349	.032	.032	21
22	20	.0299	.0286	.0253	.0313	.028	.028	22
23	18	.0269	.0258	.0226	.0278	.024	.025	23
24	16	.0239	.0230	.0201	.0248	.022	.022	24
25	14	.0209	.0204	.0179	.0220	.020	.020	25
26	12	.0179	.0181	.0159	.0196	.018	.018	26
27	11	.0164	.0173	.0142	.0175	.0164	.016	27
28	10	.0149	.0162	.0126	.0156	.0148	.014	28
29	9	.0135	.0150	.0113	.0139	.0136	.013	29
30	8	.0120	.0140	.0100	.0123	.0124	.012	30
31	7	.0105	.0132	.0089	.0110	.0116	.010	31
32	6.5	.0097	.0128	.0080	.0098	.0108	.009	32
33	6	.0090	.0118	.0071	.0087	.0100	.008	33
34	5.5	.0082	.0104	.0063	.0077	.0092	.007	34
35	5	.0075	.0095	.0056	.0069	.0084	.005	35
36	4.5	.0067	.0090	.0050	.0061	.0076	.004	36
37	4.25	.0064	.0085	.0045	.0054	.0068		37
38	4	.0060	.0080	.0040	.0048	.0060		38
39			.0075	.0035	.0043	.0052		39
40			.0070	.0031	.0039	.0048		40

* U. S. Standard Gage is officially a weight gage, in oz. per sq. ft. as tabulated. The Approx. Thickness shown is the "Manufacturers' Standard" of the American Iron and Steel Institute, based on steel as weighing 501.81 lbs. per cu. ft. (489.6 true weight plus 2.5 percent for average over-run in area and thickness). The A.I.S.I. standard nomenclature for flat rolled carbon steel is as follows:

Widths, Inches	Thicknesses, Inch							
	0.2500 and thicker	0.2499 to 0.2031	0.2030 to 0.1875	0.1874 to 0.0568	0.0567 to 0.0344	0.0343 to 0.0255	0.0254 to 0.0142	0.0141 and thinner
To 3½ incl.	Bar	Bar	Strip	Strip	Strip	Strip	Sheet	Sheet
Over 3½ to 6 incl.	Bar	Bar	Strip	Strip	Strip	Sheet	Sheet	Sheet
" 6 to 12 "	Plate	Strip	Strip	Strip	Sheet	Sheet	Sheet	Sheet
" 12 to 32 "	Plate	Sheet	Sheet	Sheet	Sheet	Sheet	Sheet	Black Plate
" 32 to 48 "	Plate	Sheet	Sheet	Sheet	Sheet	Sheet	Sheet	Sheet
" 48	Plate	Plate	Plate	Sheet	Sheet	Sheet	Sheet	——

(Courtesy of the American Institute of Steel Construction.)

APPENDIX 7. AMERICAN WELDING SOCIETY STANDARD WELDING SYMBOLS

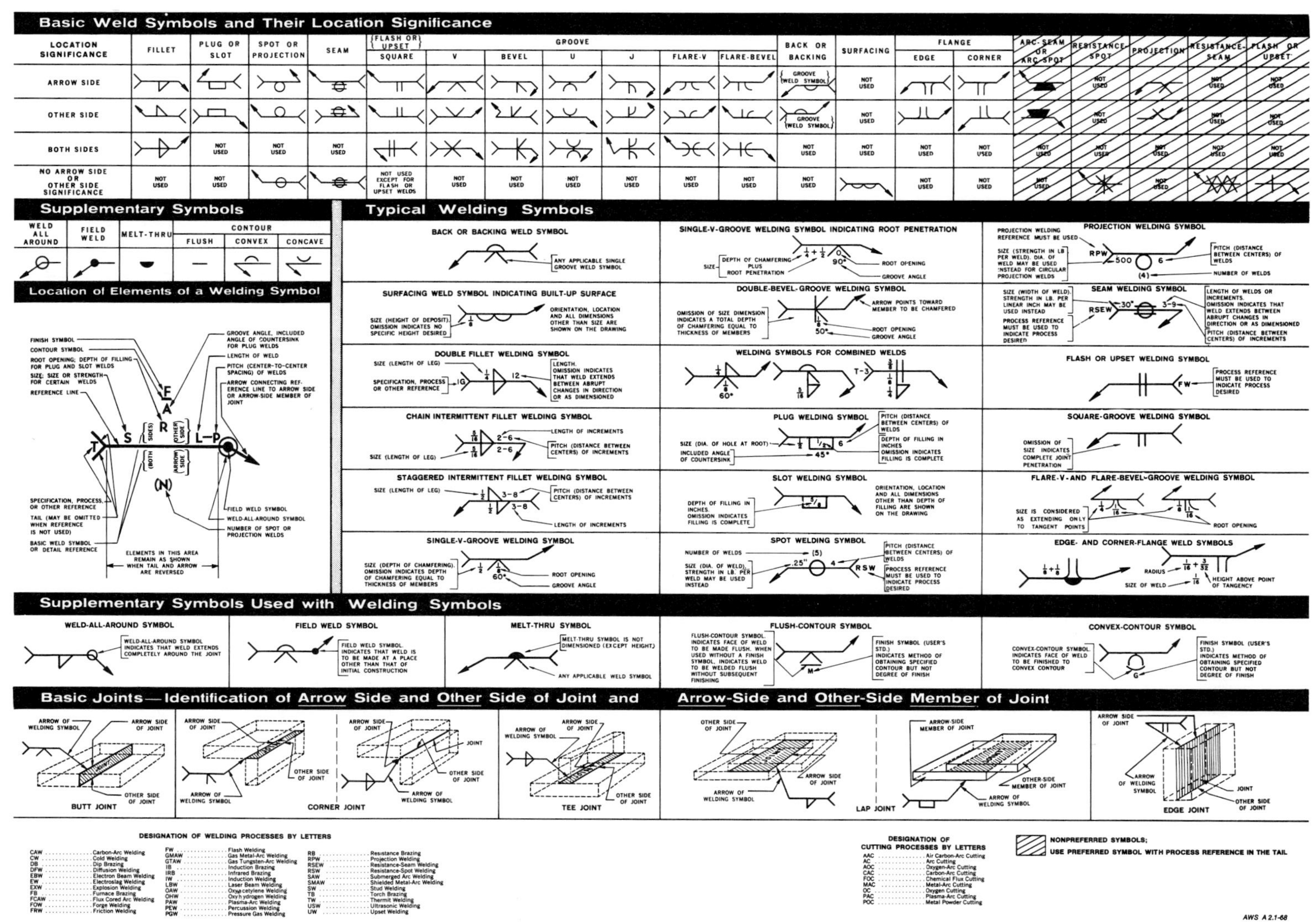

APPENDIX 8. BASIC WELDING OPERATIONS AND SYMBOLS

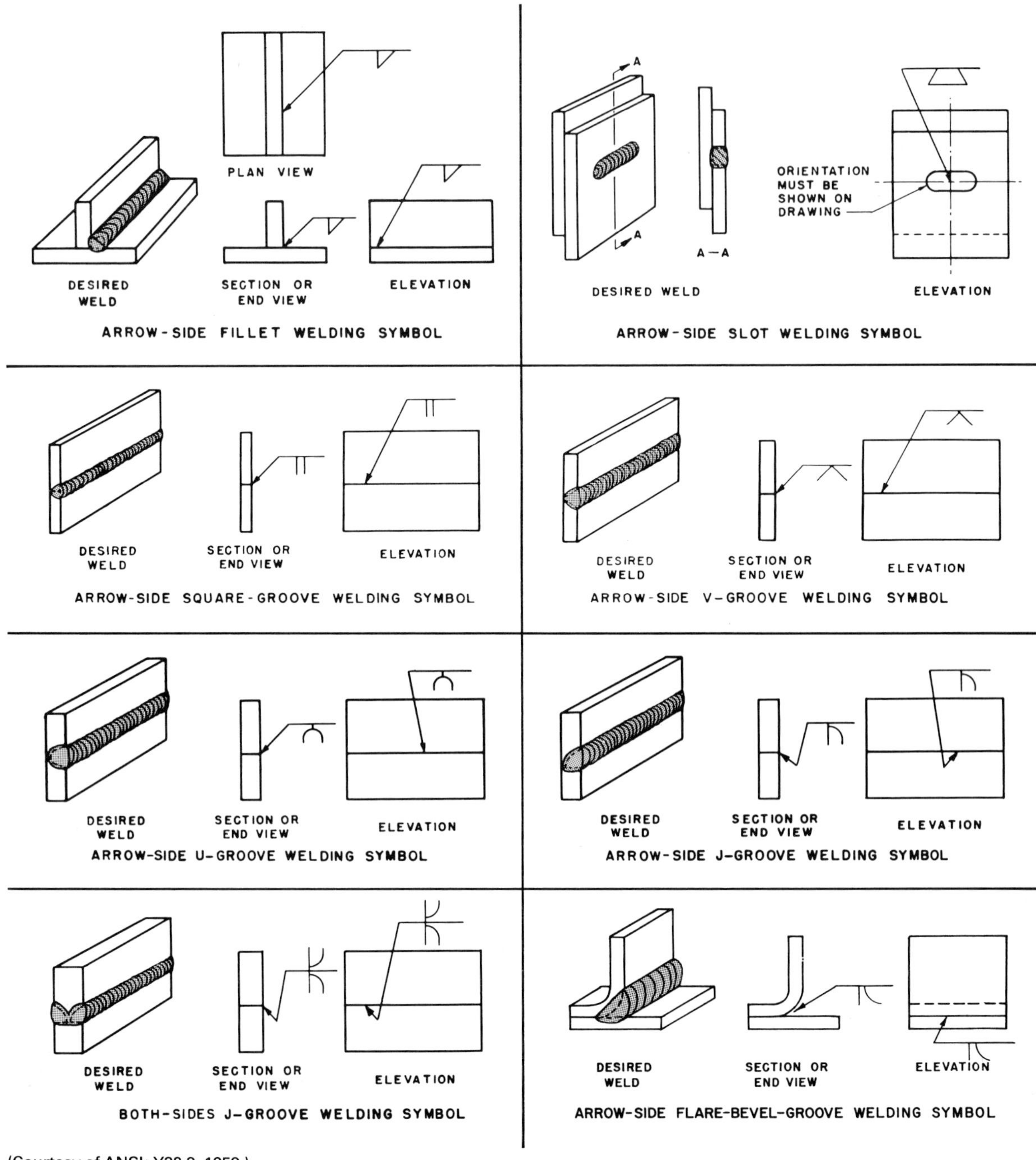

(Courtesy of ANSI; Y32.3–1959.)

APPENDIX 9. AMERICAN STANDARD GRAPHICAL SYMBOLS FOR ELECTRONIC DIAGRAMS

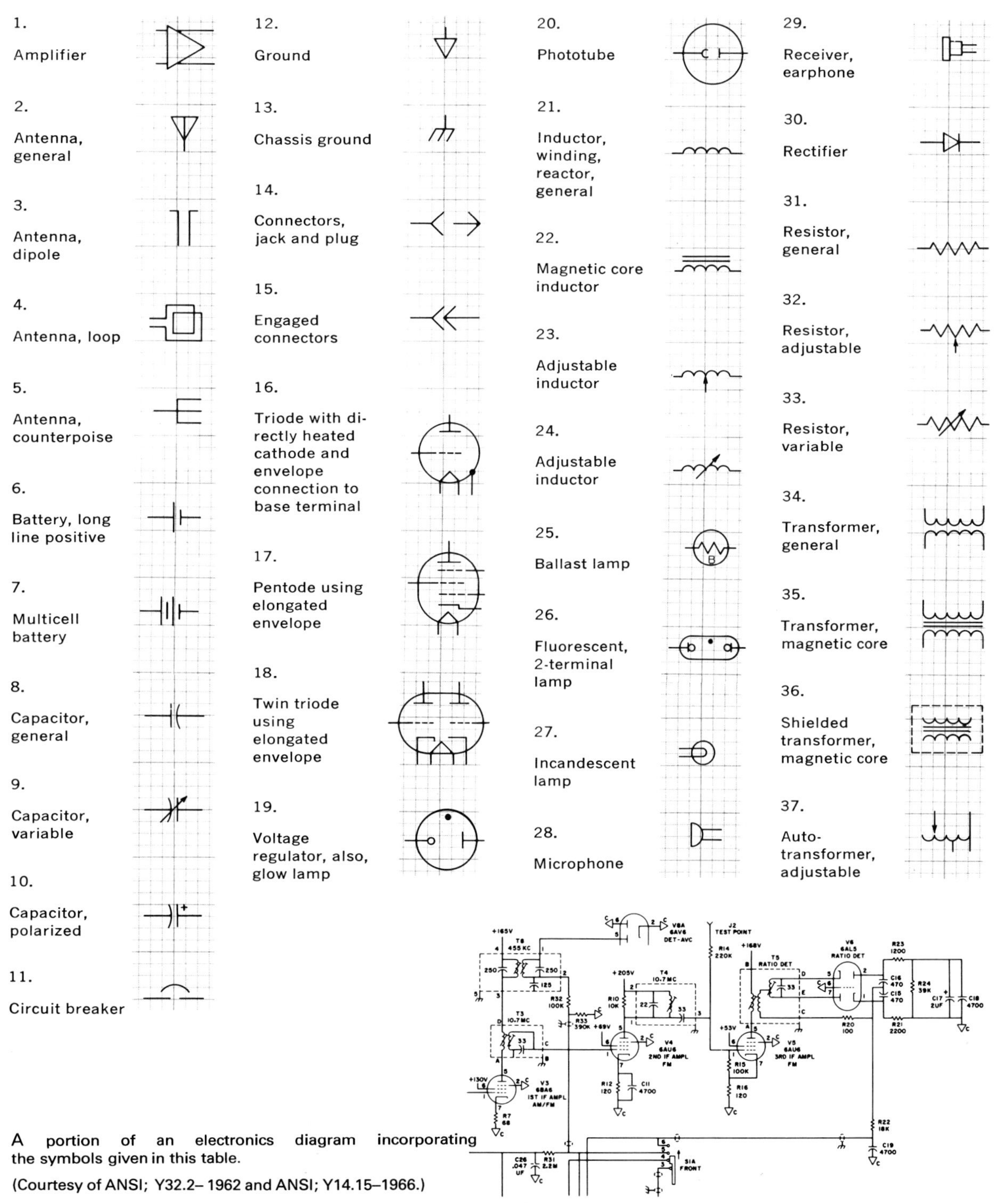

A portion of an electronics diagram incorporating the symbols given in this table.

(Courtesy of ANSI; Y32.2–1962 and ANSI; Y14.15–1966.)

APPENDIX 10. PIPING SYMBOLS

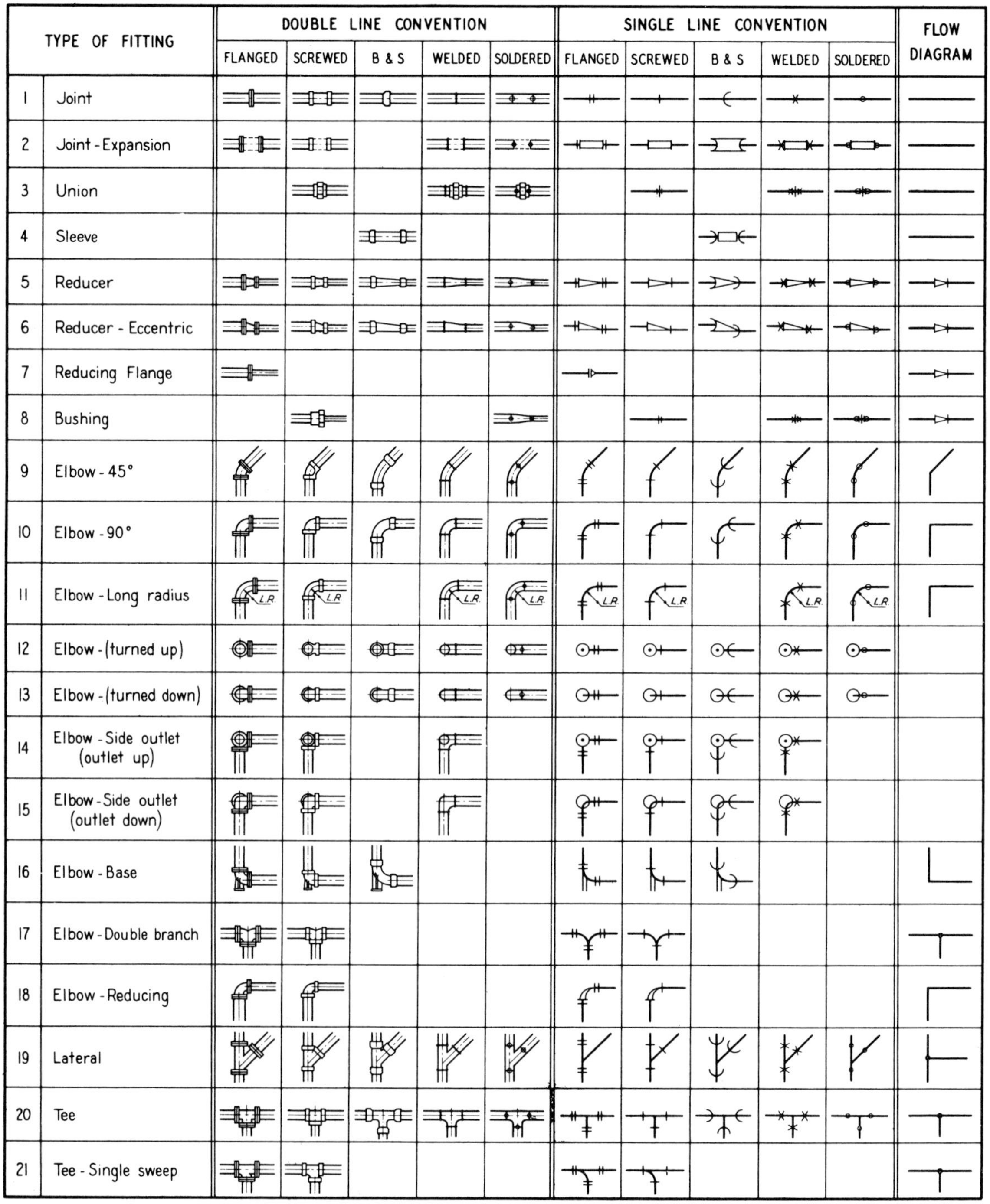

TYPE OF FITTING		DOUBLE LINE CONVENTION					SINGLE LINE CONVENTION					FLOW DIAGRAM
		FLANGED	SCREWED	B & S	WELDED	SOLDERED	FLANGED	SCREWED	B & S	WELDED	SOLDERED	
1	Joint											
2	Joint - Expansion											
3	Union											
4	Sleeve											
5	Reducer											
6	Reducer - Eccentric											
7	Reducing Flange											
8	Bushing											
9	Elbow - 45°											
10	Elbow - 90°											
11	Elbow - Long radius											
12	Elbow - (turned up)											
13	Elbow - (turned down)											
14	Elbow - Side outlet (outlet up)											
15	Elbow - Side outlet (outlet down)											
16	Elbow - Base											
17	Elbow - Double branch											
18	Elbow - Reducing											
19	Lateral											
20	Tee											
21	Tee - Single sweep											

APPENDIX 10. PIPING SYMBOLS (Cont.)

TYPE OF FITTING		DOUBLE LINE CONVENTION					SINGLE LINE CONVENTION					FLOW DIAGRAM
		FLANGED	SCREWED	B & S	WELDED	SOLDERED	FLANGED	SCREWED	B & S	WELDED	SOLDERED	
22	Tee-Double sweep	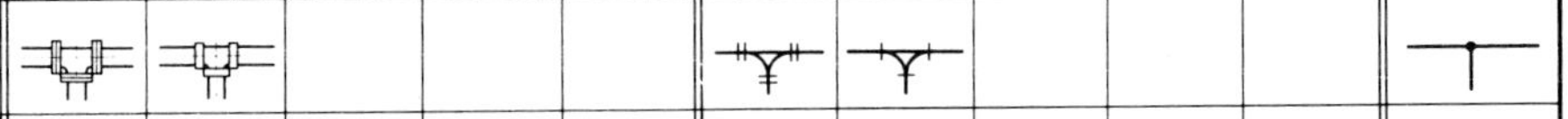										
23	Tee-(outlet up)	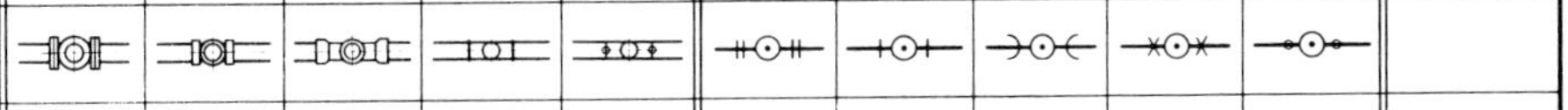										
24	Tee-(outlet down)	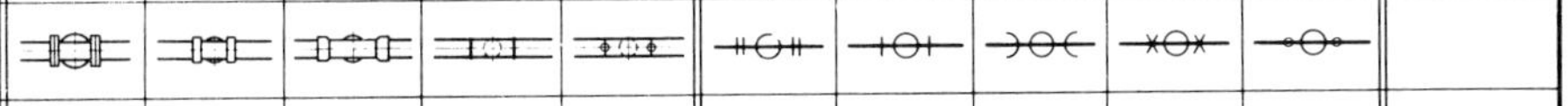										
25	Tee-Side outlet (outlet up)	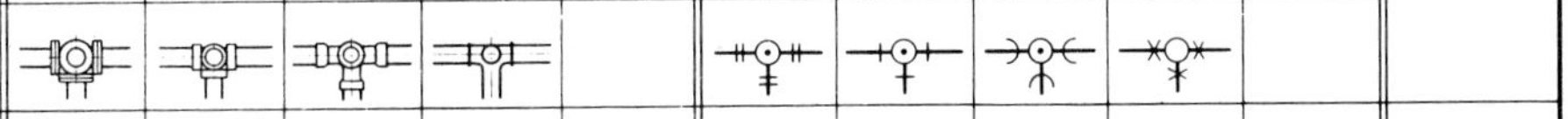										
26	Tee-Side outlet (outlet down)	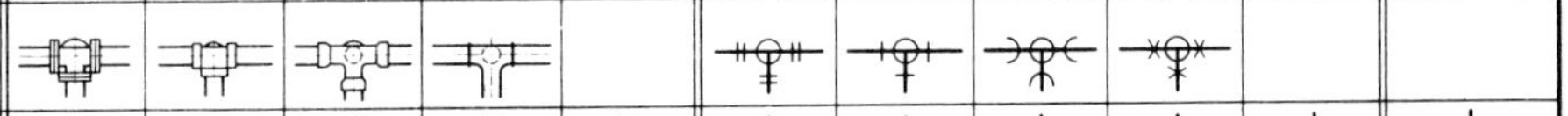										
27	Cross	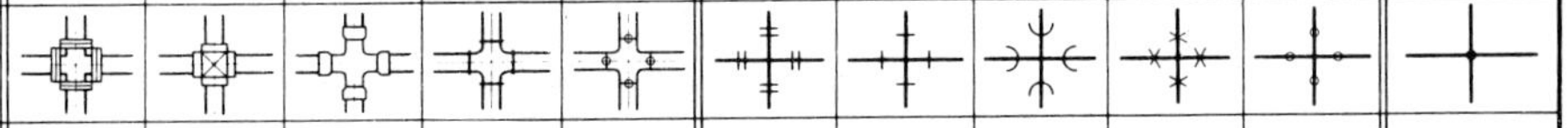										
28	Valve-Globe	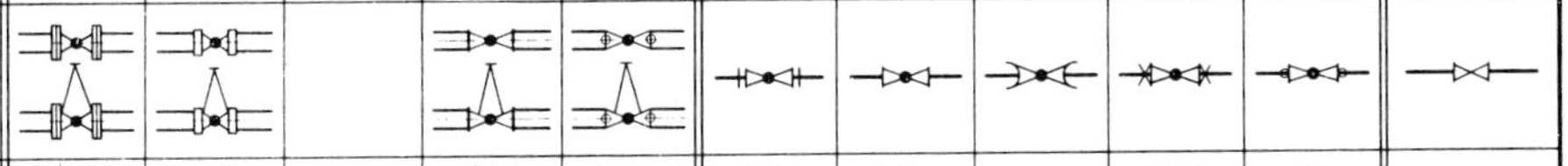										
29	Valve-Angle	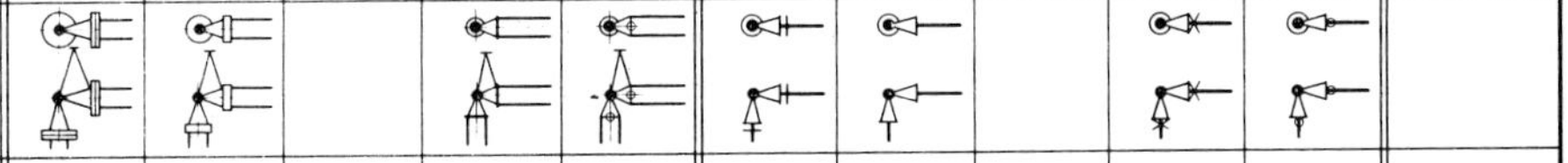										
30	Valve-Motor operated globe	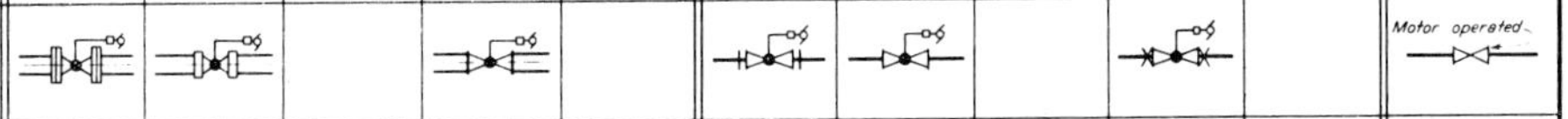										Motor operated
31	Valve-Gate	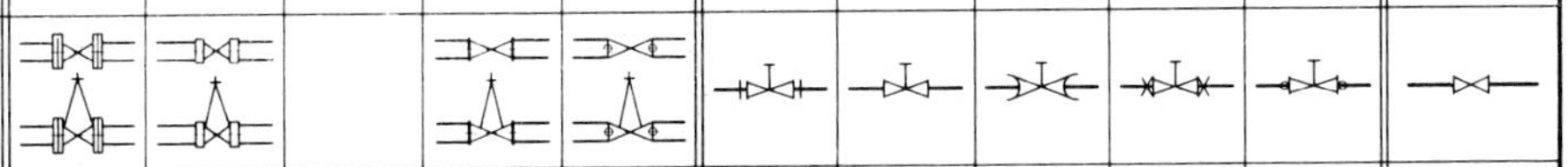										
32	Valve-Angle gate	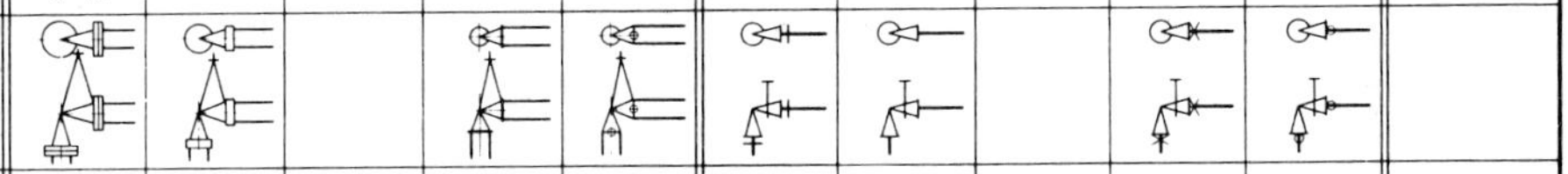										
33	Valve-Motor operated gate	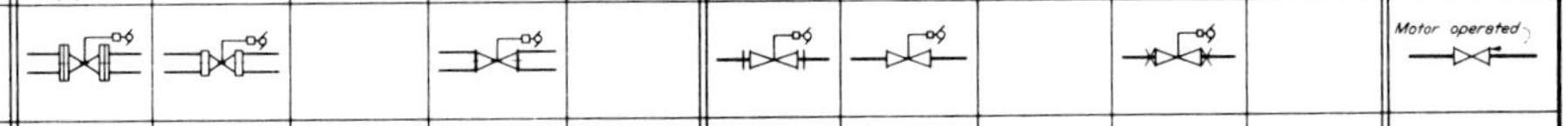										Motor operated
34	Valve-Check	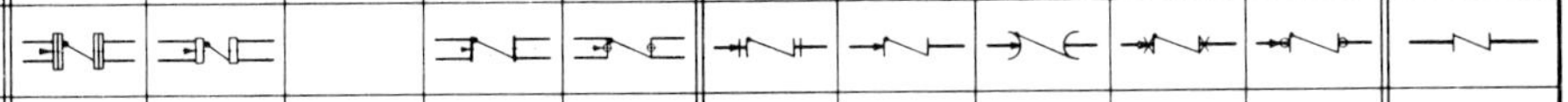										
35	Valve-Angle check	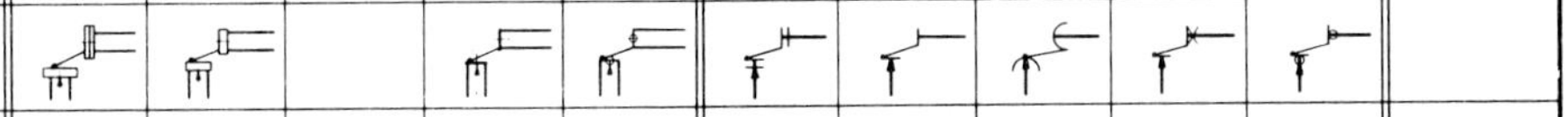										
36	Valve-Safety	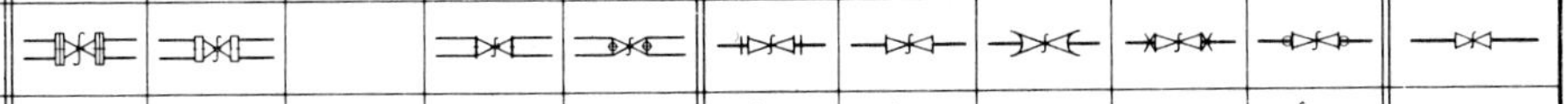										
37	Valve-Angle safety	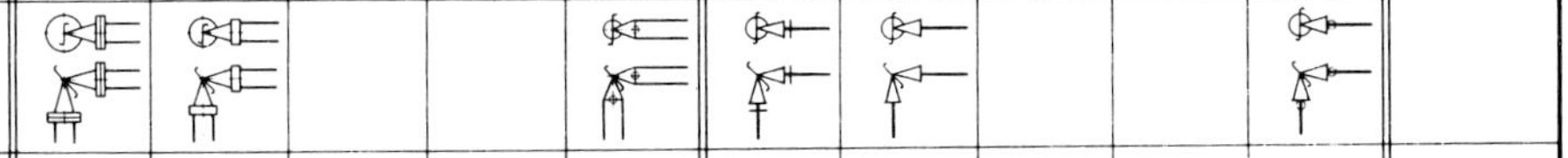										
38	Valve-Quick opening	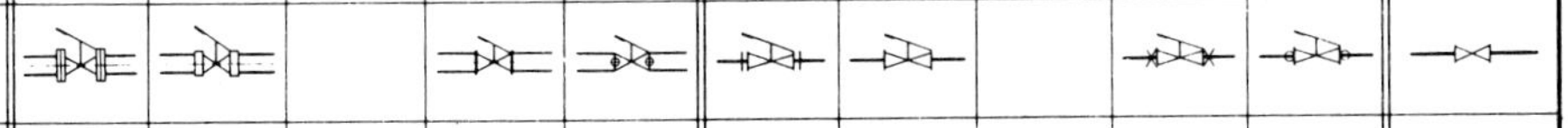										
39	Valve-Float operating	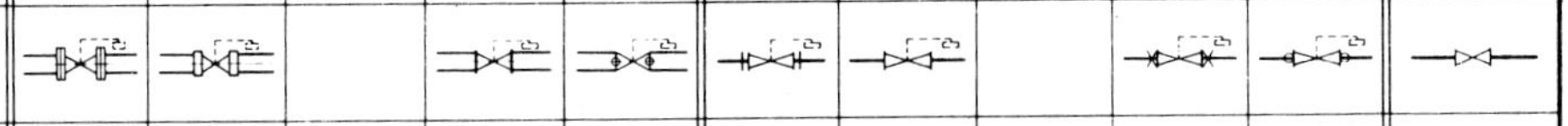										
40	Stop Cock	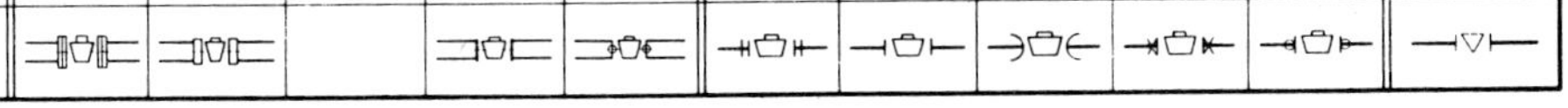										

APPENDIX 11. AMERICAN STANDARD TAPER PIPE THREADS, NPT[1]

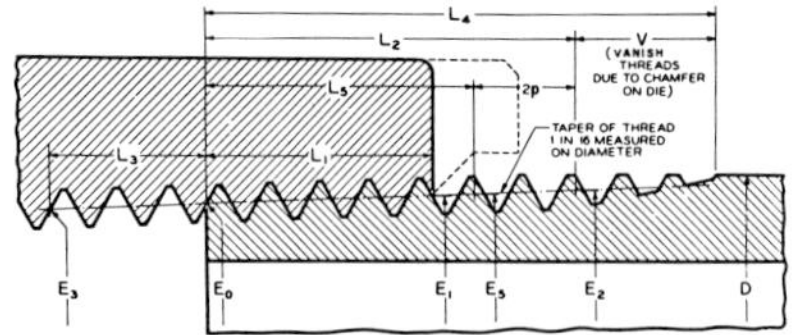

1	2	3	4	5	6	7	8	9	10	11
				Pitch Diameter at Beginning of External Thread	Hand-Tight Engagement			Effective Thread, External		
					Length[2] L_1			Length L_2		Dia E_2
Nominal Pipe Size	Outside Diameter of Pipe D	Threads per Inch n	Pitch of Thread p	E_0	In.	Thds	Dia E_1	In.	Thds	In.
1/16	0.3125	27	0.03704	0.27118	0.160	4.32	0.28118	0.2611	7.05	0.28750
1/8	0.405	27	0.03704	0.36351	0.180	4.86	0.37476	0.2639	7.12	0.38000
1/4	0.540	18	0.05556	0.47739	0.200	3.60	0.48989	0.4018	7.23	0.50250
3/8	0.675	18	0.05556	0.61201	0.240	4.32	0.62701	0.4078	7.34	0.63750
1/2	0.840	14	0.07143	0.75843	0.320	4.48	0.77843	0.5337	7.47	0.79179
3/4	1.050	14	0.07143	0.96768	0.339	4.75	0.98887	0.5457	7.64	1.00179
1	1.315	11 1/2	0.08696	1.21363	0.400	4.60	1.23863	0.6828	7.85	1.25630
1 1/4	1.660	11 1/2	0.08696	1.55713	0.420	4.83	1.58338	0.7068	8.13	1.60130
1 1/2	1.900	11 1/2	0.08696	1.79609	0.420	4.83	1.82234	0.7235	8.32	1.84130
2	2.375	11 1/2	0.08696	2.26902	0.436	5.01	2.29627	0.7565	8.70	2.31630
2 1/2	2.875	8	0.12500	2.71953	0.682	5.46	2.76216	1.1375	9.10	2.79062
3	3.500	8	0.12500	3.34062	0.766	6.13	3.38850	1.2000	9.60	3.41562
3 1/2	4.000	8	0.12500	3.83750	0.821	6.57	3.88881	1.2500	10.00	3.91562
4	4.500	8	0.12500	4.33438	0.844	6.75	4.38712	1.3000	10.40	4.41562
5	5.563	8	0.12500	5.39073	0.937	7.50	5.44929	1.4063	11.25	5.47862
6	6.625	8	0.12500	6.44609	0.958	7.66	6.50597	1.5125	12.10	6.54062
8	8.625	8	0.12500	8.43359	1.063	8.50	8.50003	1.7125	13.70	8.54062
10	10.750	8	0.12500	10.54531	1.210	9.68	10.62094	1.9250	15.40	10.66562
12	12.750	8	0.12500	12.53281	1.360	10.88	12.61781	2.1250	17.00	12.66562
14 OD	14.000	8	0.12500	13.77500	1.562	12.50	13.87262	2.2500	18.90	13.91562
16 OD	16.000	8	0.12500	15.76250	1.812	14.50	15.87575	2.4500	19.60	15.91562
18 OD	18.000	8	0.12500	17.75000	2.000	16.00	17.87500	2.6500	21.20	17.91562
20 OD	20.000	8	0.12500	19.73750	2.125	17.00	19.87031	2.8500	22.80	19.91562
24 OD	24.000	8	0.12500	23.71250	2.375	19.00	23.86094	3.2500	26.00	23.91562

All dimensions are given in inches.

[1] The basic dimensions of the American Standard Taper Pipe Thread are given in inches to four or five decimal places. While this implies a greater degree of precision than is ordinarily attained, these dimensions are the basis of gage dimensions and are so expressed for the purpose of eliminating errors in computations.

[2] Also length of thin ring gage and length from gaging notch to small end of plug gage.

(Courtesy of ANSI; B2.1–1960.)

APPENDIX 12. AMERICAN STANDARD 250-LB CAST IRON FLANGED FITTINGS

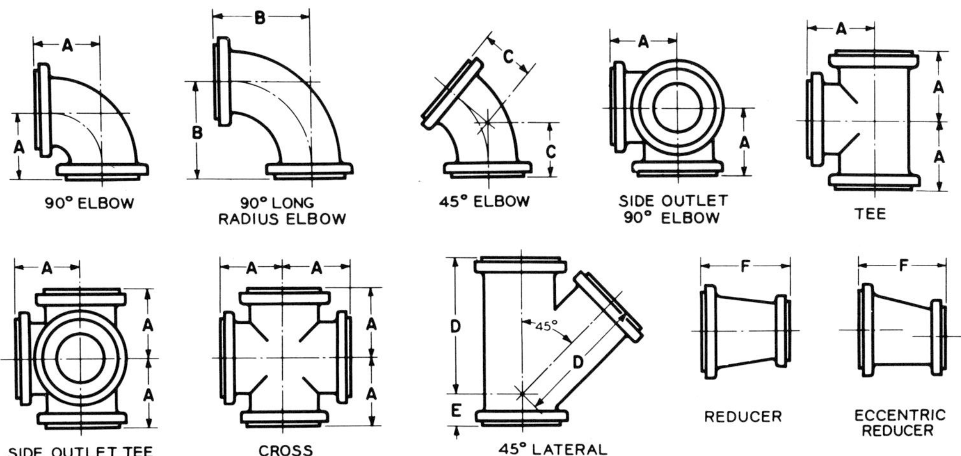

Dimensions of 250-lb Cast Iron Flanged Fittings

Nominal Pipe Size	Flanges: Dia of Flange	Flanges: Thickness of Flange (Min)	Flanges: Dia of Raised Face	Fittings: Inside Dia of Fittings (Min)	Fittings: Wall Thickness	Straight: Center to Face 90 Deg Elbow Tees, Crosses and True "Y" A	Straight: Center to Face 90 Deg Long Radius Elbow B	Straight: Center to Face 45 Deg Elbow C	Straight: Center to Face Lateral D	Straight: Short Center to Face True "Y" and Lateral E	Straight: Face to Face Reducer F
1	4 7/8	11/16	2 11/16	1	7/16	4	5	2	6 1/2	2	....
1 1/4	5 1/4	3/4	3 1/16	1 1/4	7/16	4 1/4	5 1/2	2 1/2	7 1/4	2 1/4	....
1 1/2	6 1/8	13/16	3 9/16	1 1/2	7/16	4 1/2	6	2 3/4	8 1/2	2 1/2	....
2	6 1/2	7/8	4 3/16	2	7/16	5	6 1/2	3	9	2 1/2	5
2 1/2	7 1/2	1	4 15/16	2 1/2	1/2	5 1/2	7	3 1/2	10 1/2	2 1/2	5 1/2
3	8 1/4	1 1/8	5 11/16	3	9/16	6	7 3/4	3 1/2	11	3	6
3 1/2	9	1 3/16	6 5/16	3 1/2	9/16	6 1/2	8 1/2	4	12 1/2	3	6 1/2
4	10	1 1/4	6 15/16	4	5/8	7	9	4 1/2	13 1/2	3	7
5	11	1 3/8	8 5/16	5	11/16	8	10 1/4	5	15	3 1/2	8
6	12 1/2	1 7/16	9 11/16	6	3/4	8 1/2	11 1/2	5 1/2	17 1/2	4	9
8	15	1 5/8	11 15/16	8	13/16	10	14	6	20 1/2	5	11
10	17 1/2	1 7/8	14 1/16	10	15/16	11 1/2	16 1/2	7	24	5 1/2	12
12	20 1/2	2	16 7/16	12	1	13	19	8	27 1/2	6	14
14	23	2 1/8	18 15/16	13 1/4	1 1/8	15	21 1/2	8 1/2	31	6 1/2	16
16	25 1/2	2 1/4	21 1/16	15 1/4	1 1/4	16 1/2	24	9 1/2	34 1/2	7 1/2	18
18	28	2 3/8	23 5/16	17	1 3/8	18	26 1/2	10	37 1/2	8	19
20	30 1/2	2 1/2	25 9/16	19	1 1/2	19 1/2	29	10 1/2	40 1/2	8 1/2	20
24	36	2 3/4	30 5/16	23	1 5/8	22 1/2	34	12	47 1/2	10	24
30	43	3	37 3/16	29	2	27 1/2	41 1/2	15	....	....	30

All dimensions are given in inches.
(Courtesy of ANSI; B16.1–1967.)

APPENDIX 13. AMERICAN STANDARD 125-LB CAST IRON FLANGED FITTINGS

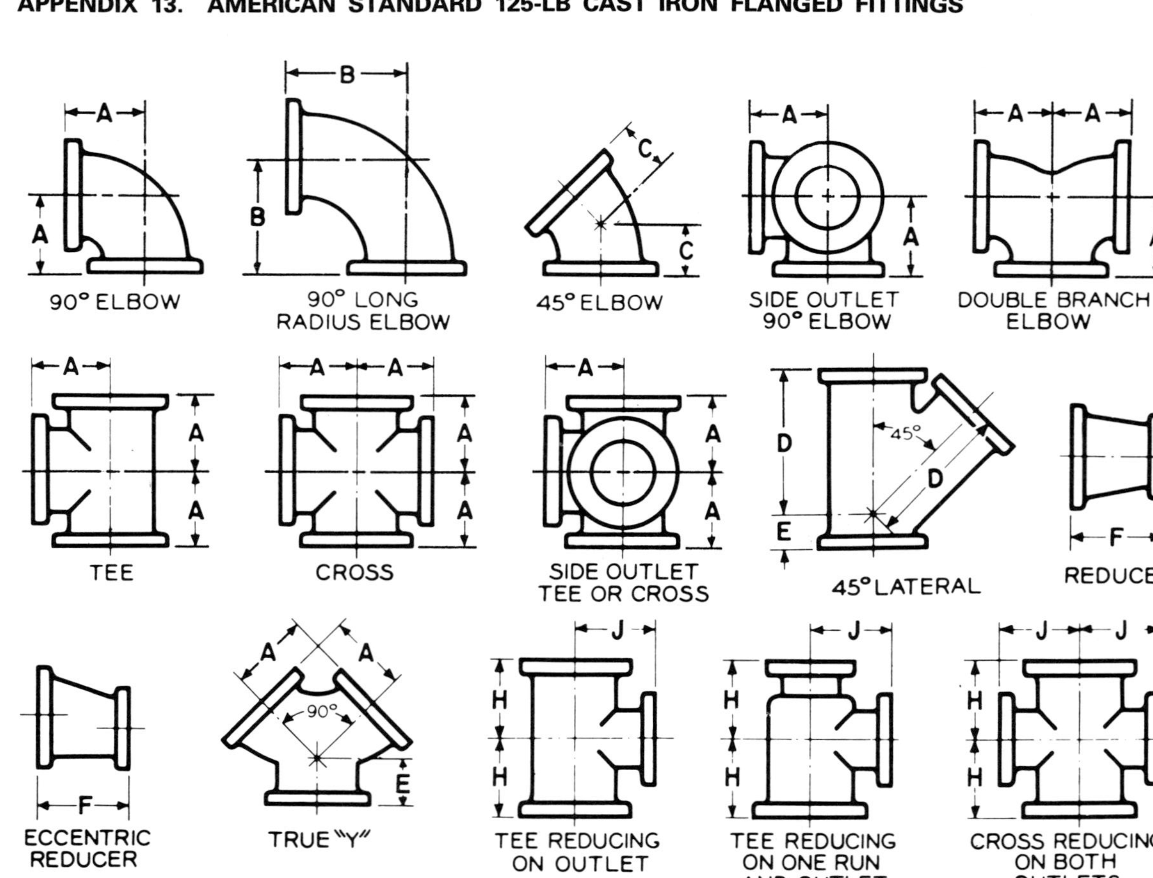

APPENDIX 13. AMERICAN STANDARD 125-LB CAST IRON FLANGED FITTINGS (Cont.)

Nominal Pipe Size	Flanges		General		Straight Fittings						Reducing Fittings (Short Body Patterns)		
											Tees and Crosses		
	Dia of Flange	Thickness of Flange (Min)	Inside Dia of Flange Fittings	Wall Thickness	Center to Face 90 deg Elbow Tees, Crosses True "Y" and Double Branch Elbow A	Center to Face 90 deg Long Radius El-bow B	Center to Face 45 deg Elbow C	Center to Face Lateral D	Short Center to Face True "Y" and Lateral E	Face to Face Reducer F	Size of Outlet and Smaller	Center to Face Run H	Center to Face Outlet or Side Outlet J
1	4¼	7/16	1	5/16	3½	5	1¾	5¾	1¾	. . .			
1¼	4⅝	½	1¼	5/16	3¾	5½	2	6¼	1¾	. . .			
1½	5	9/16	1½	5/16	4	6	2¼	7	2				
2	6	5/8	2	5/16	4½	6½	2½	8	2½	5			
2½	7	11/16	2½	5/16	5	7	3	9½	2½	5½			
3	7½	¾	3	3/8	5½	7¾	3	10	3	6			
3½	8½	13/16	3½	7/16	6	8½	3½	11½	3	6½			
4	9	15/16	4	½	6½	9	4	12	3	7	All reducing tees and crosses, sizes 16 in. and smaller, shall have same center to face dimensions as straight size fittings, corresponding to the size of the largest opening.		
5	10	15/16	5	½	7½	10¼	4½	13½	3½	8			
6	11	1	6	9/16	8	11½	5	14½	3½	9			
8	13½	1⅛	8	5/8	9	14	5½	17½	4½	11			
10	16	1 3/16	10	¾	11	16½	6½	20½	5	12			
12	19	1¼	12	13/16	12	19	7½	24½	5½	14			
14	21	1⅜	14	7/8	14	21½	7½	27	6	16			
16	23½	1 7/16	16	1	15	24	8	30	6½	18			
18	25	1 9/16	18	1 1/16	16½	26½	8½	32	7	19	12	13	15½
20	27½	1 11/16	20	1⅛	18	29	9½	35	8	20	14	14	17
24	32	1⅞	24	1¼	22	34	11	40½	9	24	16	15	19
30	38¾	2⅛	30	1 7/16	25	41½	15	49	10	30	20	18	23
36	46	2⅜	36	1⅝	28*	49	18	. . .	. . .	36	24	20	26
42	53	2⅝	42	1 13/16	31*	56½	21	. . .	. . .	42	24	23	30
48	59½	2¾	48	2	34*	64	24	. . .	. . .	48	30	26	34

All dimensions are given in inches.
(Courtesy of ANSI; B16.1–1967.)

APPENDIX 14. AMERICAN NATIONAL STANDARD UNIFIED INCH SCREW THREADS (UN AND UNR THREAD FORM)

Sizes			Threads per Inch											
			Series with Graded Pitches			Series with Constant Pitches								
Primary	Secondary	Basic Major Diameter	Coarse UNC	Fine UNF	Extra Fine UNEF	4UN	6UN	8UN	12UN	16UN	20UN	28UN	32UN	Sizes
0		0.0600	—	80	—	—	—	—	—	—	—	—	—	0
	1	0.0730	64	72	—	—	—	—	—	—	—	—	—	1
2		0.0860	56	64	—	—	—	—	—	—	—	—	—	2
	3	0.0990	48	56	—	—	—	—	—	—	—	—	—	3
4		0.1120	40	48	—	—	—	—	—	—	—	—	—	4
5		0.1250	40	44	—	—	—	—	—	—	—	—	—	5
6		0.1380	32	40	—	—	—	—	—	—	—	—	UNC	6
8		0.1640	32	36	—	—	—	—	—	—	—	—	UNC	8
10		0.1900	24	32	—	—	—	—	—	—	—	—	UNF	10
	12	0.2160	24	28	32	—	—	—	—	—	—	UNF	UNEF	12
$\frac{1}{4}$		0.2500	20	28	32	—	—	—	—	—	UNC	UNF	UNEF	$\frac{1}{4}$
$\frac{5}{16}$		0.3125	18	24	32	—	—	—	—	—	20	28	UNEF	$\frac{5}{16}$
$\frac{3}{8}$		0.3750	16	24	32	—	—	—	—	UNC	20	28	UNEF	$\frac{3}{8}$
$\frac{7}{16}$		0.4375	14	20	28	—	—	—	—	16	UNF	UNEF	32	$\frac{7}{16}$
$\frac{1}{2}$		0.5000	13	20	28	—	—	—	—	16	UNF	UNEF	32	$\frac{1}{2}$
$\frac{9}{16}$		0.5625	12	18	24	—	—	—	UNC	16	20	28	32	$\frac{9}{16}$
$\frac{5}{8}$		0.6250	11	18	24	—	—	—	12	16	20	28	32	$\frac{5}{8}$
	$\frac{11}{16}$	0.6875	—	—	24	—	—	—	12	16	20	28	32	$\frac{11}{16}$
$\frac{3}{4}$		0.7500	10	16	20	—	—	—	12	UNF	UNEF	28	32	$\frac{3}{4}$
	$\frac{13}{16}$	0.8125	—	—	20	—	—	—	12	16	UNEF	28	32	$\frac{13}{16}$
$\frac{7}{8}$		0.8750	9	14	20	—	—	—	12	16	UNEF	28	32	$\frac{7}{8}$
	$\frac{15}{16}$	0.9375	—	—	20	—	—	—	12	16	UNEF	28	32	$\frac{15}{16}$
1		1.0000	8	12	20	—	—	UNC	UNF	16	UNEF	28	32	1
	$1\frac{1}{16}$	1.0625	—	—	18	—	—	8	12	16	20	28	—	$1\frac{1}{16}$
$1\frac{1}{8}$		1.1250	7	12	18	—	—	8	UNF	16	20	28	—	$1\frac{1}{8}$
	$1\frac{3}{16}$	1.1875	—	—	18	—	—	8	12	16	20	28	—	$1\frac{3}{16}$
$1\frac{1}{4}$		1.2500	7	12	18	—	—	8	UNF	16	20	28	—	$1\frac{1}{4}$
	$1\frac{5}{16}$	1.3125	—	—	18	—	—	8	12	16	20	28	—	$1\frac{5}{16}$
$1\frac{3}{8}$		1.3750	6	12	18	—	UNC	8	UNF	16	20	28	—	$1\frac{3}{8}$
	$1\frac{7}{16}$	1.4375	—	—	18	—	6	8	12	16	20	28	—	$1\frac{7}{16}$
$1\frac{1}{2}$		1.5000	6	12	18	—	UNC	8	UNF	16	20	28	—	$1\frac{1}{2}$
	$1\frac{9}{16}$	1.5625	—	—	18	—	6	8	12	16	20	—	—	$1\frac{9}{16}$
$1\frac{5}{8}$		1.6250	—	—	18	—	6	8	12	16	20	—	—	$1\frac{5}{8}$
	$1\frac{11}{16}$	1.6875	—	—	18	—	6	8	12	16	20	—	—	$1\frac{11}{16}$
$1\frac{3}{4}$		1.7500	5	—	—	—	6	8	12	16	20	—	—	$1\frac{3}{4}$
	$1\frac{13}{16}$	1.8125	—	—	—	—	6	8	12	16	20	—	—	$1\frac{13}{16}$
$1\frac{7}{8}$		1.8750	—	—	—	—	6	8	12	16	20	—	—	$1\frac{7}{8}$
	$1\frac{15}{16}$	1.9375	—	—	—	—	6	8	12	16	20	—	—	$1\frac{15}{16}$
2		2.0000	$4\frac{1}{2}$	—	—	—	6	8	12	16	20	—	—	2
	$2\frac{1}{8}$	2.1250	—	—	—	—	6	8	12	16	20	—	—	$2\frac{1}{8}$
$2\frac{1}{4}$		2.2500	$4\frac{1}{2}$	—	—	—	6	8	12	16	20	—	—	$2\frac{1}{4}$
	$2\frac{3}{8}$	2.3750	—	—	—	—	6	8	12	16	20	—	—	$2\frac{3}{8}$
$2\frac{1}{2}$		2.5000	4	—	—	UNC	6	8	12	16	20	—	—	$2\frac{1}{2}$
	$2\frac{5}{8}$	2.6250	—	—	—	4	6	8	12	16	20	—	—	$2\frac{5}{8}$
$2\frac{3}{4}$		2.7500	4	—	—	UNC	6	8	12	16	20	—	—	$2\frac{3}{4}$
	$2\frac{7}{8}$	2.8750	—	—	—	4	6	8	12	16	20	—	—	$2\frac{7}{8}$

* Series designation shown indicates the UN thread form; however, the UNR thread form may be specified by substituting UNR in place of UN in all designations for external use only.

APPENDIX 14. AMERICAN NATIONAL STANDARD UNIFIED INCH SCREW THREADS (UN AND UNR THREAD FORM)

Sizes			Threads per Inch											
			Series with Graded Pitches			Series with Constant Pitches								
Primary	Secondary	Basic Major Diameter	Coarse UNC	Fine UNF	Extra Fine UNEF	4UN	6UN	8UN	12UN	16UN	20UN	28UN	32UN	Sizes
3		3.0000	4	—	—	UNC	6	8	12	16	20	—	—	3
	$3\frac{1}{8}$	3.1250	—	—	—	4	6	8	12	16	—	—	—	$3\frac{1}{8}$
$3\frac{1}{4}$		3.2500	4	—	—	UNC	6	8	12	16	—	—	—	$3\frac{1}{4}$
	$3\frac{3}{8}$	3.3750	—	—	—	4	6	8	12	16	—	—	—	$3\frac{3}{8}$
$3\frac{1}{2}$		3.5000	4	—	—	UNC	6	8	12	16	—	—	—	$3\frac{1}{2}$
	$3\frac{5}{8}$	3.6250	—	—	—	4	6	8	12	16	—	—	—	$3\frac{5}{8}$
$3\frac{3}{4}$		3.7500	4	—	—	UNC	6	8	12	16	—	—	—	$3\frac{3}{4}$
	$3\frac{7}{8}$	3.8750	—	—	—	4	6	8	12	16	—	—	—	$3\frac{7}{8}$
4		4.0000	4	—	—	UNC	6	8	12	16	—	—	—	4
	$4\frac{1}{8}$	4.1250	—	—	—	4	6	8	12	16	—	—	—	$4\frac{1}{8}$
$4\frac{1}{4}$		4.2500	—	—	—	4	6	8	12	16	—	—	—	$4\frac{1}{4}$
	$4\frac{3}{8}$	4.3750	—	—	—	4	6	8	12	16	—	—	—	$4\frac{3}{8}$
$4\frac{1}{2}$		4.5000	—	—	—	4	6	8	12	16	—	—	—	$4\frac{1}{2}$
	$4\frac{5}{8}$	4.6250	—	—	—	4	6	8	12	16	—	—	—	$4\frac{5}{8}$
$4\frac{3}{4}$		4.7500	—	—	—	4	6	8	12	16	—	—	—	$4\frac{3}{4}$
	$4\frac{7}{8}$	4.8750	—	—	—	4	6	8	12	16	—	—	—	$4\frac{7}{8}$
5		5.0000	—	—	—	4	6	8	12	16	—	—	—	5
	$5\frac{1}{8}$	5.1250	—	—	—	4	6	8	12	16	—	—	—	$5\frac{1}{8}$
$5\frac{1}{4}$		5.2500	—	—	—	4	6	8	12	16	—	—	—	$5\frac{1}{4}$
	$5\frac{3}{8}$	5.3750	—	—	—	4	6	8	12	16	—	—	—	$5\frac{3}{8}$
$5\frac{1}{2}$		5.5000	—	—	—	4	6	8	12	16	—	—	—	$5\frac{1}{2}$
	$5\frac{5}{8}$	5.6250	—	—	—	4	6	8	12	16	—	—	—	$5\frac{5}{8}$
$5\frac{3}{4}$		5.7500	—	—	—	4	6	8	12	16	—	—	—	$5\frac{3}{4}$
	$5\frac{7}{8}$	5.8750	—	—	—	4	6	8	12	16	—	—	—	$5\frac{7}{8}$
6		6.0000	—	—	—	4	6	8	12	16	—	—	—	6

(Courtesy of ANSI; B1.1–1974.)

APPENDIX 15. TAP DRILL SIZES FOR AMERICAN NATIONAL AND UNIFIED COARSE AND FINE THREADS

$$p = \text{pitch} = \frac{1}{\text{No. thrd. per in.}}$$

$$d = \text{depth} = p \times .649519$$

$$f = \text{flat} = \frac{p}{8}$$

$$\text{pitch diameter} = d - \frac{.6495}{N}$$

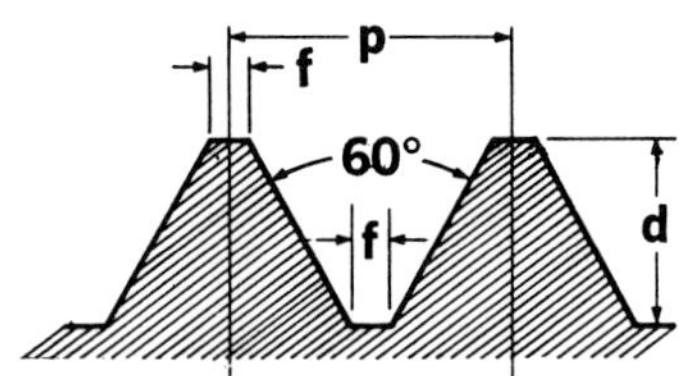

For Nos. 575 and 585 Screw Thread Micrometers

Size	Threads per inch NC UNC	Threads per inch NF UNF	Outside Diameter Inches	Pitch Diameter Inches	Root Diameter Inches	Tap Drill Approx. 75% Full Thread	Decimal Equiv. of Tap Drill
0	..	80	.0600	.0519	.0438	3/64	.0469
1	64	..	.0730	.0629	.0527	53	.0595
1	..	72	.0730	.0640	.0550	53	.0595
2	56	..	.0860	.0744	.0628	50	.0700
2	..	64	.0860	.0759	.0657	50	.0700
3	48	..	.0990	.0855	.0719	47	.0785
3	..	56	.0990	.0874	.0758	46	.0810
4	40	..	.1120	.0958	.0795	43	.0890
4	..	48	.1120	.0985	.0849	42	.0935
5	40	..	.1250	.1088	.0925	38	.1015
5	..	44	.1250	.1102	.0955	37	.1040
6	32	..	.1380	.1177	.0974	36	.1065
6	..	40	.1380	.1218	.1055	33	.1130
8	32	..	.1640	.1437	.1234	29	.1360
8	..	36	.1640	.1460	.1279	29	.1360
10	24	..	.1900	.1629	.1359	26	.1470
10	..	32	.1900	.1697	.1494	21	.1590
12	24	..	.2160	.1889	.1619	16	.1770
12	..	28	.2160	.1928	.1696	15	.1800
1/4	20	..	.2500	.2175	.1850	7	.2010
1/4	..	28	.2500	.2268	.2036	3	.2130
5/16	18	..	.3125	.2764	.2403	F	.2570
5/16	..	24	.3125	.2854	.2584	I	.2720
3/8	16	..	.3750	.3344	.2938	5/16	.3125
3/8	..	24	.3750	.3479	.3209	Q	.3320
7/16	14	..	.4375	.3911	.3447	U	.3680
7/16	..	20	.4375	.4050	.3726	25/64	.3906
1/2	13	..	.5000	.4500	.4001	27/64	.4219
1/2	..	20	.5000	.4675	.4351	29/64	.4531
9/16	12	..	.5625	.5084	.4542	31/64	.4844
9/16	..	18	.5625	.5264	.4903	33/64	.5156
5/8	11	..	.6250	.5660	.5069	17/32	.5312
5/8	..	18	.6250	.5889	.5528	37/64	.5781
3/4	10	..	.7500	.6850	.6201	21/32	.6562
3/4	..	16	.7500	.7094	.6688	11/16	.6875
7/8	9	..	.8750	.8028	.7307	49/64	.7656
7/8	..	14	.8750	.8286	.7822	13/16	.8125

APPENDIX 15. TAP DRILL SIZES FOR AMERICAN NATIONAL AND UNIFIED COARSE AND FINE THREADS (Cont.)

Size	Threads per inch NC UNC	Threads per inch NF UNF	Outside Diameter Inches	Pitch Diameter Inches	Root Diameter Inches	Tap Drill Approx. 75% Full Thread	Decimal Equiv. of Tap Drill
1	8	..	1.0000	.9188	.8376	7/8	.8750
1	..	12	1.0000	.9459	.8917	59/64	.9219
1 1/8	7	..	1.1250	1.0322	.9394	63/64	.9844
1 1/8	..	12	1.1250	1.0709	1.0168	1 3/64	1.0469
1 1/4	7	..	1.2500	1.1572	1.0644	1 7/64	1.1094
1 1/4	..	12	1.2500	1.1959	1.1418	1 11/64	1.1719
1 3/8	6	..	1.3750	1.2667	1.1585	1 7/32	1.2187
1 3/8	..	12	1.3750	1.3209	1.2668	1 19/64	1.2969
1 1/2	6	..	1.5000	1.3917	1.2835	1 11/32	1.3437
1 1/2	..	12	1.5000	1.4459	1.3918	1 27/64	1.4219
1 3/4	5	..	1.7500	1.6201	1.4902	1 9/16	1.5625
2	4 1/2	..	2.0000	1.8557	1.7113	1 25/32	1.7812
2 1/4	4 1/2	..	2.2500	2.1057	1.9613	2 1/32	2.0313
2 1/2	4	..	2.5000	2.3376	2.1752	2 1/4	2.2500
2 3/4	4	..	2.7500	2.5876	2.4252	2 1/2	2.5000
3	4	..	3.0000	3.8376	2.6752	2 3/4	2.7500
3 1/4	4	..	3.2500	3.0876	2.9252	3	3.0000
3 1/2	4	..	3.5000	3.3376	3.1752	3 1/4	3.2500
3 3/4	4	..	3.7500	3.5876	3.4252	3 1/2	3.5000
4	4	..	4.0000	3.3786	3.6752	3 3/4	3.7500

(Courtesy of the L. S. Starrett Company.)

APPENDIX 16. ISO METRIC SCREW THREAD STANDARD SERIES

Nominal Size Dia. (mm)			Pitches (mm)														Nom-inal Size Dia. (mm)
Column[a]			Series with Graded Pitches		Series with Constant Pitches												
1	2	3	Coarse	Fine	6	4	3	2	1.5	1.25	1	0.75	0.5	0.35	0.25	0.2	
0.25			0.075	—	—	—	—	—	—	—	—	—	—	—	—	—	0.25
0.3			0.08	—	—	—	—	—	—	—	—	—	—	—	—	—	0.3
	0.35		0.09	—	—	—	—	—	—	—	—	—	—	—	—	—	0.35
0.4			0.1	—	—	—	—	—	—	—	—	—	—	—	—	—	0.4
	0.45		0.1	—	—	—	—	—	—	—	—	—	—	—	—	—	0.45
0.5			0.125	—	—	—	—	—	—	—	—	—	—	—	—	—	0.5
	0.55		0.125	—	—	—	—	—	—	—	—	—	—	—	—	—	0.55
0.6			0.15	—	—	—	—	—	—	—	—	—	—	—	—	—	0.6
	0.7		0.175	—	—	—	—	—	—	—	—	—	—	—	—	—	0.7
0.8			0.2	—	—	—	—	—	—	—	—	—	—	—	—	—	0.8
	0.9		0.225	—	—	—	—	—	—	—	—	—	—	—	—	—	0.9
			0.25	—	—	—	—	—	—	—	—	—	—	—	—	0.2	1
	1.1		0.25	—	—	—	—	—	—	—	—	—	—	—	—	0.2	1.1
1.2			0.25	—	—	—	—	—	—	—	—	—	—	—	—	0.2	1.2
	1.4		0.3	—	—	—	—	—	—	—	—	—	—	—	—	0.2	1.4
1.6			0.35	—	—	—	—	—	—	—	—	—	—	—	—	0.2	1.6
	1.8		0.35	—	—	—	—	—	—	—	—	—	—	—	—	0.2	1.8
2			0.4	—	—	—	—	—	—	—	—	—	—	—	0.25	—	2
	2.2		0.45	—	—	—	—	—	—	—	—	—	—	—	0.25	—	2.2
2.5			0.45	—	—	—	—	—	—	—	—	—	—	0.35	—	—	2.5
3			0.5	—	—	—	—	—	—	—	—	—	—	0.35	—	—	3
	3.5		0.6	—	—	—	—	—	—	—	—	—	—	0.35	—	—	3.5
4			0.7	—	—	—	—	—	—	—	—	—	0.5	—	—	—	4
	4.5		0.75	—	—	—	—	—	—	—	—	—	0.5	—	—	—	4.5
5			0.8	—	—	—	—	—	—	—	—	—	0.5	—	—	—	5
		5.5	—	—	—	—	—	—	—	—	—	—	0.5	—	—	—	5.5
6			1	—	—	—	—	—	—	—	—	0.75	—	—	—	—	6
		7	1	—	—	—	—	—	—	—	—	0.75	—	—	—	—	7
8			1.25	1	—	—	—	—	—	—	1	0.75	—	—	—	—	8
		9	1.25	—	—	—	—	—	—	—	1	0.75	—	—	—	—	9
10			1.5	1.25	—	—	—	—	—	1.25	1	0.75	—	—	—	—	10
		11	1.5	—	—	—	—	—	—	—	1	0.75	—	—	—	—	11
12			1.75	1.25	—	—	—	—	1.5	1.25	1	—	—	—	—	—	12
	14		2	1.5	—	—	—	—	1.5	1.25[b]	1	—	—	—	—	—	14
		15	—	—	—	—	—	—	1.5	—	1	—	—	—	—	—	15
16			2	1.5	—	—	—	—	1.5	—	1	—	—	—	—	—	16
		17	—	—	—	—	—	—	1.5	—	1	—	—	—	—	—	17
	18		2.5	1.5	—	—	—	2	1.5	—	1	—	—	—	—	—	18
20			2.5	1.5	—	—	—	2	1.5	—	1	—	—	—	—	—	20
	22		2.5	1.5	—	—	—	2	1.5	—	1	—	—	—	—	—	22

[a] Thread diameter should be selected from columns 1, 2 or 3, with preference being in that order.
[b] Pitch 1.25 mm in combination with diameter 14 mm has been included for sparkplug applications.
[c] Diameter 35 mm has been included for bearing locknut applications.

The use of pitches shown in parentheses should be avoided wherever possible.

The pitches enclosed in the bold frame, together with the corresponding nominal diameters in columns 1 and 2, are those combinations which have been established by ISO Recommendations as a selected "coarse" and "fine" series for commercial fasteners.

APPENDIX 16. ISO METRIC SCREW THREAD STANDARD SERIES (Cont.)

Nominal Size Dia. (mm)			Pitches (mm)														Nominal Size Dia. (mm)
Column[a]			Series with Graded Pitches		Series with Constant Pitches												
1	2	3	Coarse	Fine	6	4	3	2	1.5	1.25	1	0.75	0.5	0.35	0.25	0.2	
24			3	2	—	—	—	2	1.5	—	1	—	—	—	—	—	24
		25	—	—	—	—	—	2	1.5	—	1	—	—	—	—	—	25
		26	—	—	—	—	—	—	1.5	—	1	—	—	—	—	—	26
	27		3	2	—	—	—	2	1.5	—	1	—	—	—	—	—	27
		28	—	—	—	—	—	2	1.5	—	1	—	—	—	—	—	28
30			3.5	2	—	—	(3)	2	1.5	—	1	—	—	—	—	—	30
		32	—	—	—	—	—	2	1.5	—	—	—	—	—	—	—	32
	33		3.5	2	—	—	(3)	2	1.5	—	—	—	—	—	—	—	33
		35[c]	—	—	—	—	—	—	1.5	—	—	—	—	—	—	—	35[c]
36			4	3	—	—	—	2	1.5	—	—	—	—	—	—	—	36
		38	—	—	—	—	—	—	1.5	—	—	—	—	—	—	—	38
	39		4	3	—	—	—	2	1.5	—	—	—	—	—	—	—	39
		40	—	—	—	—	3	2	1.5	—	—	—	—	—	—	—	40
42			4.5	3	—	4	3	2	1.5	—	—	—	—	—	—	—	42
	45		4.5	3	—	4	3	2	1.5	—	—	—	—	—	—	—	45
48			5	3	—	4	3	2	1.5	—	—	—	—	—	—	—	48
		50	—	—	—	—	3	2	1.5	—	—	—	—	—	—	—	50
	52		5	3	—	4	3	2	1.5	—	—	—	—	—	—	—	52
		55	—	—	—	4	3	2	1.5	—	—	—	—	—	—	—	55
56			5.5	4	—	4	3	2	1.5	—	—	—	—	—	—	—	56
		58	—	—	—	4	3	2	1.5	—	—	—	—	—	—	—	58
	60		5.5	4	—	4	3	2	1.5	—	—	—	—	—	—	—	60
		62	—	—	—	4	3	2	1.5	—	—	—	—	—	—	—	62
64			6	4	—	4	3	2	1.5	—	—	—	—	—	—	—	64
		65	—	—	—	4	3	2	1.5	—	—	—	—	—	—	—	65
	68		6	4	—	4	3	2	1.5	—	—	—	—	—	—	—	68
		70	—	—	6	4	3	2	1.5	—	—	—	—	—	—	—	70
72			—	—	6	4	3	2	1.5	—	—	—	—	—	—	—	72
		75	—	—	—	4	3	2	1.5	—	—	—	—	—	—	—	75
	76		—	—	6	4	3	2	1.5	—	—	—	—	—	—	—	76
		78	—	—	—	—	—	2	—	—	—	—	—	—	—	—	78
80			—	—	6	4	3	2	1.5	—	—	—	—	—	—	—	80
		82	—	—	—	—	—	2	—	—	—	—	—	—	—	—	82
	85		—	—	6	4	3	2	—	—	—	—	—	—	—	—	85
90			—	—	6	4	3	2	—	—	—	—	—	—	—	—	90

APPENDIX 16. ISO METRIC SCREW THREAD STANDARD SERIES (Cont.)

Nominal Size Dia. (mm)			Pitches (mm)														Nom- inal Size Dia. (mm)
Column[a]			Series with Graded Pitches		Series with Constant Pitches												
1	2	3	Coarse	Fine	6	·	3	2	1.5	1.25	1	0.75	0.5	0.35	0.25	0.2	
	95		—	—	6	4	3	2	—	—	—	—	—	—	—	—	95
100			—	—	6	4	3	2	—	—	—	—	—	—	—	—	100
	105		—	—	6	4	3	2	—	—	—	—	—	—	—	—	105
110			—	—	6	4	3	2	—	—	—	—	—	—	—	—	110
	115		—	—	6	4	3	2	—	—	—	—	—	—	—	—	115
	120		—	—	6	4	3	2	—	—	—	—	—	—	—	—	120
125			—	—	6	4	3	2	—	—	—	—	—	—	—	—	125
	130		—	—	6	4	3	2	—	—	—	—	—	—	—	—	130
		135	—	—	6	4	3	2	—	—	—	—	—	—	—	—	135
140			—	—	6	4	3	2	—	—	—	—	—	—	—	—	140
		145	—	—	6	4	3	2	—	—	—	—	—	—	—	—	145
	150		—	—	6	4	3	2	—	—	—	—	—	—	—	—	150
		155	—	—	6	4	3	—	—	—	—	—	—	—	—	—	155
160			—	—	6	4	3	—	—	—	—	—	—	—	—	—	160
		165	—	—	6	4	3	—	—	—	—	—	—	—	—	—	165
	170		—	—	6	4	3	—	—	—	—	—	—	—	—	—	170
		175	—	—	6	4	3	—	—	—	—	—	—	—	—	—	175
180			—	—	6	4	3	—	—	—	—	—	—	—	—	—	180
		185	—	—	6	4	3	—	—	—	—	—	—	—	—	—	185
	190		—	—	6	4	3	—	—	—	—	—	—	—	—	—	190
		195	—	—	6	4	3	—	—	—	—	—	—	—	—	—	195
200			—	—	6	4	3	—	—	—	—	—	—	—	—	—	200
		205	—	—	6	4	3	—	—	—	—	—	—	—	—	—	205
	210		—	—	6	4	3	—	—	—	—	—	—	—	—	—	210
220			—	—	6	4	3	—	—	—	—	—	—	—	—	—	220
		225	—	—	6	4	3	—	—	—	—	—	—	—	—	—	225
		230	—	—	6	4	3	—	—	—	—	—	—	—	—	—	230
		235	—	—	6	4	3	—	—	—	—	—	—	—	—	—	235
	240		—	—	6	4	3	—	—	—	—	—	—	—	—	—	240
		245	—	—	6	4	3	—	—	—	—	—	—	—	—	—	245
250			—	—	6	4	3	—	—	—	—	—	—	—	—	—	250
		255	—	—	6	4	—	—	—	—	—	—	—	—	—	—	255
	260		—	—	6	4	—	—	—	—	—	—	—	—	—	—	260
		265	—	—	6	4	—	—	—	—	—	—	—	—	—	—	265
		270	—	—	6	4	—	—	—	—	—	—	—	—	—	—	270
		275	—	—	6	4	—	—	—	—	—	—	—	—	—	—	275
280			—	—	6	4	—	—	—	—	—	—	—	—	—	—	280
		285	—	—	6	4	—	—	—	—	—	—	—	—	—	—	285
		290	—	—	6	4	—	—	—	—	—	—	—	—	—	—	290
		295	—	—	6	4	—	—	—	—	—	—	—	—	—	—	295
	300		—	—	6	4	—	—	—	—	—	—	—	—	—	—	300

[1] Thread diameter should be selected from columns 1, 2, or 3; with preference being in that order.

APPENDIX 17. SQUARE AND ACME THREADS

Size	Threads per Inch	Size	Threads per Inch
$\frac{3}{8}$	12	2	$2\frac{1}{2}$
$\frac{7}{16}$	10	$2\frac{1}{4}$	2
$\frac{1}{2}$	10	$2\frac{1}{2}$	2
$\frac{9}{16}$	8	$2\frac{3}{4}$	2
$\frac{5}{8}$	8	3	$1\frac{1}{2}$
$\frac{3}{4}$	6	$3\frac{1}{4}$	$1\frac{1}{2}$
$\frac{7}{8}$	5	$3\frac{1}{2}$	$1\frac{1}{3}$
1	5	$3\frac{3}{4}$	$1\frac{1}{3}$
$1\frac{1}{8}$	4	4	$1\frac{1}{3}$
$1\frac{1}{4}$	4	$4\frac{1}{4}$	$1\frac{1}{3}$
$1\frac{1}{2}$	3	$4\frac{1}{2}$	1
$1\frac{3}{4}$	$2\frac{1}{2}$	over $4\frac{1}{2}$	1

APPENDIX 18. AMERICAN STANDARD SQUARE BOLTS AND NUTS

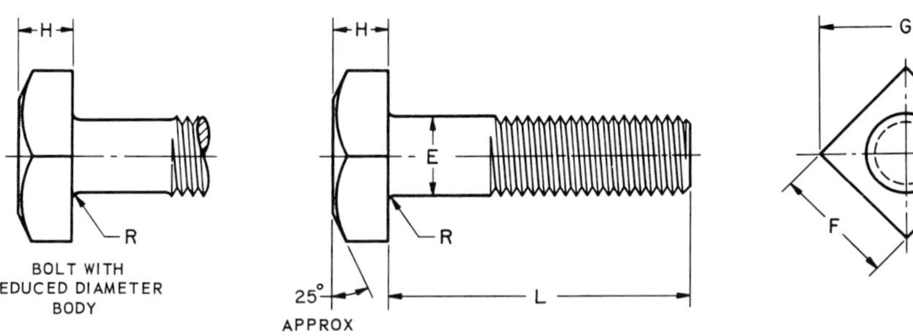

Dimensions of Square Bolts

Nominal Size or Basic Product Dia		Body Dia E	Width Across Flats F			Width Across Corners G		Height H			Radius of Fillet R
		Max	Basic	Max	Min	Max	Min	Basic	Max	Min	Max
1/4	0.2500	0.260	3/8	0.3750	0.362	0.530	0.498	11/64	0.188	0.156	0.031
5/16	0.3125	0.324	1/2	0.5000	0.484	0.707	0.665	13/64	0.220	0.186	0.031
3/8	0.3750	0.388	9/16	0.5625	0.544	0.795	0.747	1/4	0.268	0.232	0.031
7/16	0.4375	0.452	5/8	0.6250	0.603	0.884	0.828	19/64	0.316	0.278	0.031
1/2	0.5000	0.515	3/4	0.7500	0.725	1.061	0.995	21/64	0.348	0.308	0.031
5/8	0.6250	0.642	15/16	0.9375	0.906	1.326	1.244	27/64	0.444	0.400	0.062
3/4	0.7500	0.768	1 1/8	1.1250	1.088	1.591	1.494	1/2	0.524	0.476	0.062
7/8	0.8750	0.895	1 5/16	1.3125	1.269	1.856	1.742	19/32	0.620	0.568	0.062
1	1.0000	1.022	1 1/2	1.5000	1.450	2.121	1.991	21/32	0.684	0.628	0.093
1 1/8	1.1250	1.149	1 11/16	1.6875	1.631	2.386	2.239	3/4	0.780	0.720	0.093
1 1/4	1.2500	1.277	1 7/8	1.8750	1.812	2.652	2.489	27/32	0.876	0.812	0.093
1 3/8	1.3750	1.404	2 1/16	2.0625	1.994	2.917	2.738	29/32	0.940	0.872	0.093
1 1/2	1.5000	1.531	2 1/4	2.2500	2.175	3.182	2.986	1	1.036	0.964	0.093

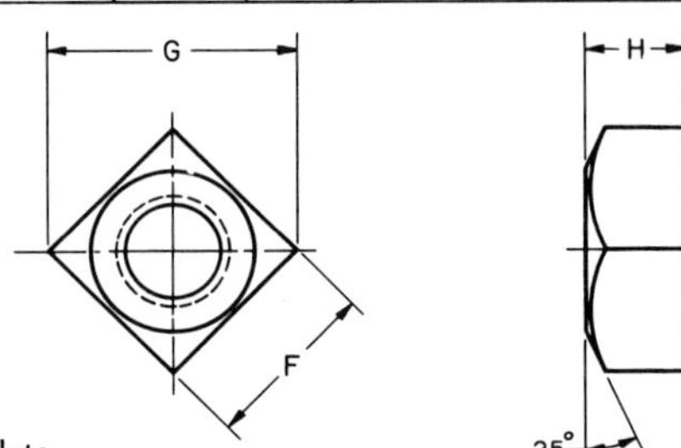

Dimensions of Square Nuts

Nominal Size or Basic Major Dia of Thread		Width Across Flats F			Width Across Corners G		Thickness H		
		Basic	Max	Min	Max	Min	Basic	Max	Min
1/4	0.2500	7/16	0.4375	0.425	0.619	0.584	7/32	0.235	0.203
5/16	0.3125	9/16	0.5625	0.547	0.795	0.751	17/64	0.283	0.249
3/8	0.3750	5/8	0.6250	0.606	0.884	0.832	21/64	0.346	0.310
7/16	0.4375	3/4	0.7500	0.728	1.061	1.000	3/8	0.394	0.356
1/2	0.5000	13/16	0.8125	0.788	1.149	1.082	7/16	0.458	0.418
5/8	0.6250	1	1.0000	0.969	1.414	1.330	35/64	0.569	0.525
3/4	0.7500	1 1/8	1.1250	1.088	1.591	1.494	21/32	0.680	0.632
7/8	0.8750	1 5/16	1.3125	1.269	1.856	1.742	49/64	0.792	0.740
1	1.0000	1 1/2	1.5000	1.450	2.121	1.991	7/8	0.903	0.847
1 1/8	1.1250	1 11/16	1.6875	1.631	2.386	2.239	1	1.030	0.970
1 1/4	1.2500	1 7/8	1.8750	1.812	2.652	2.489	1 3/32	1.126	1.062
1 3/8	1.3750	2 1/16	2.0625	1.994	2.917	2.738	1 13/64	1.237	1.169
1 1/2	1.5000	2 1/4	2.2500	2.175	3.182	2.986	1 5/16	1.348	1.276

(Courtesy of ANSI; B18.2.1–1965 and ANSI; B18.2.2–1965.)

APPENDIX 19. AMERICAN STANDARD HEXAGON HEAD BOLTS AND NUTS

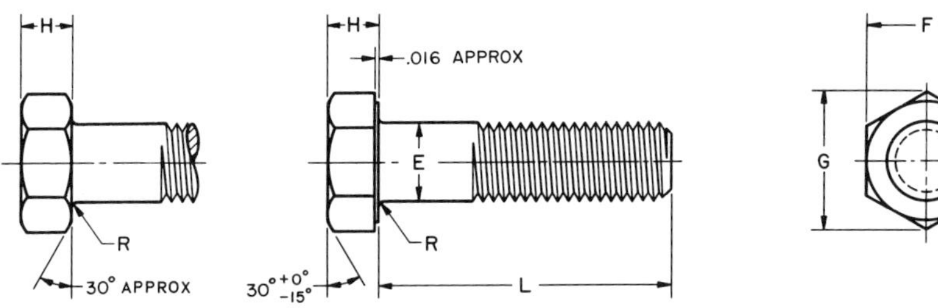

Dimensions of Hex Cap Screws (Finished Hex Bolts)

Nominal Size or Basic Product Dia	Body Dia E		Width Across Flats F			Width Across Corners G		Height H			Radius of Fillet R	
	Max	Min	Basic	Max	Min	Max	Min	Basic	Max	Min	Max	Min
1/4 0.2500	0.2500	0.2450	7/16	0.4375	0.428	0.505	0.488	5/32	0.163	0.150	0.025	0.015
5/16 0.3125	0.3125	0.3065	1/2	0.5000	0.489	0.577	0.557	13/64	0.211	0.195	0.025	0.015
3/8 0.3750	0.3750	0.3690	9/16	0.5625	0.551	0.650	0.628	15/64	0.243	0.226	0.025	0.015
7/16 0.4375	0.4375	0.4305	5/8	0.6250	0.612	0.722	0.698	9/32	0.291	0.272	0.025	0.015
1/2 0.5000	0.5000	0.4930	3/4	0.7500	0.736	0.866	0.840	5/16	0.323	0.302	0.025	0.015
9/16 0.5625	0.5625	0.5545	13/16	0.8125	0.798	0.938	0.910	23/64	0.371	0.348	0.045	0.020
5/8 0.6250	0.6250	0.6170	15/16	0.9375	0.922	1.083	1.051	25/64	0.403	0.378	0.045	0.020
3/4 0.7500	0.7500	0.7410	1 1/8	1.1250	1.100	1.299	1.254	15/32	0.483	0.455	0.045	0.020
7/8 0.8750	0.8750	0.8660	1 5/16	1.3125	1.285	1.516	1.465	35/64	0.563	0.531	0.065	0.040
1 1.0000	1.0000	0.9900	1 1/2	1.5000	1.469	1.732	1.675	39/64	0.627	0.591	0.095	0.060
1 1/8 1.1250	1.1250	1.1140	1 11/16	1.6875	1.631	1.949	1.859	11/16	0.718	0.658	0.095	0.060
1 1/4 1.2500	1.2500	1.2390	1 7/8	1.8750	1.812	2.165	2.066	25/32	0.813	0.749	0.095	0.060
1 3/8 1.3750	1.3750	1.3630	2 1/16	2.0625	1.994	2.382	2.273	27/32	0.878	0.810	0.095	0.060
1 1/2 1.5000	1.5000	1.4880	2 1/4	2.2500	2.175	2.598	2.480	15/16	0.974	0.902	0.095	0.060
1 3/4 1.7500	1.7500	1.7380	2 5/8	2.6250	2.538	3.031	2.893	1 3/32	1.134	1.054	0.095	0.060
2 2.0000	2.0000	1.9880	3	3.0000	2.900	3.464	3.306	1 7/32	1.263	1.175	0.095	0.060
2 1/4 2.2500	2.2500	2.2380	3 3/8	3.3750	3.262	3.897	3.719	1 3/8	1.423	1.327	0.095	0.060
2 1/2 2.5000	2.5000	2.4880	3 3/4	3.7500	3.625	4.330	4.133	1 17/32	1.583	1.479	0.095	0.060
2 3/4 2.7500	2.7500	2.7380	4 1/8	4.1250	3.988	4.763	4.546	1 11/16	1.744	1.632	0.095	0.060
3 3.0000	3.0000	2.9880	4 1/2	4.5000	4.350	5.196	4.959	1 7/8	1.935	1.815	0.095	0.060

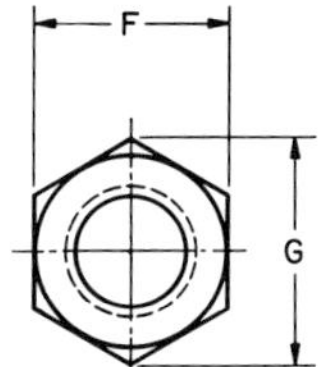

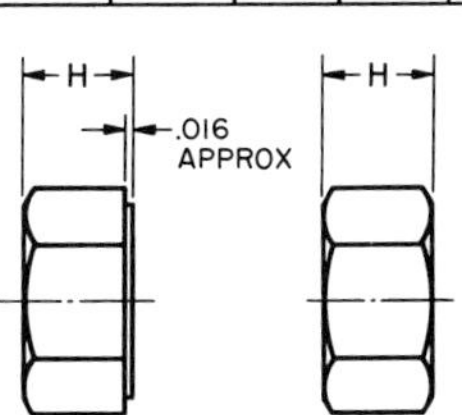

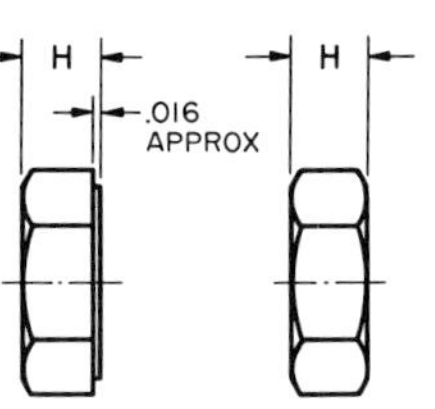

Dimensions of Hex Nuts and Hex Jam Nuts

Nominal Size or Basic Major Dia of Thread		Width Across Flats F			Width Across Corners G		Thickness Hex Nuts H			Thickness Hex Jam Nuts H		
		Basic	Max	Min	Max	Min	Basic	Max	Min	Basic	Max	Min
1/4	0.2500	7/16	0.4375	0.428	0.505	0.488	7/32	0.226	0.212	5/32	0.163	0.150
5/16	0.3125	1/2	0.5000	0.489	0.577	0.557	17/64	0.273	0.258	3/16	0.195	0.180
3/8	0.3750	9/16	0.5625	0.551	0.650	0.628	21/64	0.337	0.320	7/32	0.227	0.210
7/16	0.4375	11/16	0.6875	0.675	0.794	0.768	3/8	0.385	0.365	1/4	0.260	0.240
1/2	0.5000	3/4	0.7500	0.736	0.866	0.840	7/16	0.448	0.427	5/16	0.323	0.302
9/16	0.5625	7/8	0.8750	0.861	1.010	0.982	31/64	0.496	0.473	5/16	0.324	0.301
5/8	0.6250	15/16	0.9375	0.922	1.083	1.051	35/64	0.559	0.535	3/8	0.387	0.363
3/4	0.7500	1 1/8	1.1250	1.088	1.299	1.240	41/64	0.665	0.617	27/64	0.446	0.398
7/8	0.8750	1 5/16	1.3125	1.269	1.516	1.447	3/4	0.776	0.724	31/64	0.510	0.458
1	1.0000	1 1/2	1.5000	1.450	1.732	1.653	55/64	0.887	0.831	35/64	0.575	0.519
1 1/8	1.1250	1 11/16	1.6875	1.631	1.949	1.859	31/32	0.999	0.939	39/64	0.639	0.579
1 1/4	1.2500	1 7/8	1.8750	1.812	2.165	2.066	1 1/16	1.094	1.030	23/32	0.751	0.687
1 3/8	1.3750	2 1/16	2.0625	1.994	2.382	2.273	1 11/64	1.206	1.138	25/32	0.815	0.747
1 1/2	1.5000	2 1/4	2.2500	2.175	2.598	2.480	1 9/32	1.317	1.245	27/32	0.880	0.808

(Courtesy of ANSI; B18.2.1–1965 and ANSI; B18.2.2–1965.)

APPENDIX 20. FILLISTER HEAD AND ROUND HEAD CAP SCREWS

Fillister Head Cap Screws

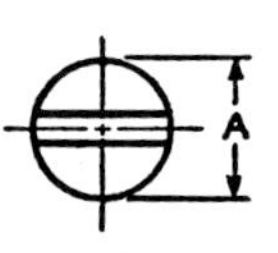

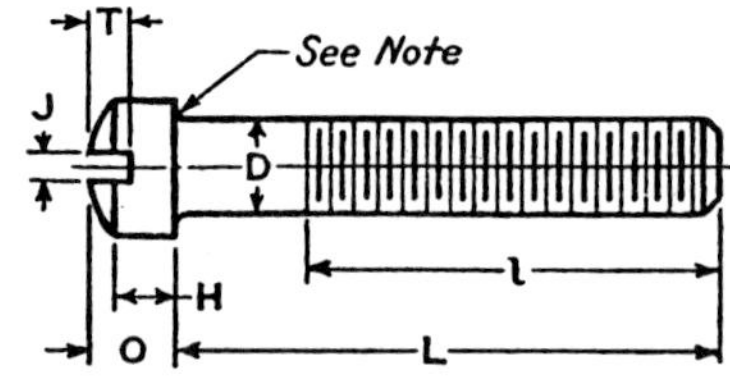

Nominal Size	D		A		H		O		J		T	
	Body Diameter		Head Diameter		Height of Head		Total Height of Head		Width of Slot		Depth of Slot	
	Max	Min	Max	Min	Max	Min	Max	Min	Max	Min	Max	Min
1/4	0.250	0.245	0.375	0.363	0.172	0.157	0.216	0.194	0.075	0.064	0.097	0.077
5/16	0.3125	0.307	0.437	0.424	0.203	0.186	0.253	0.230	0.084	0.072	0.115	0.090
3/8	0.375	0.369	0.562	0.547	0.250	0.229	0.314	0.284	0.094	0.081	0.142	0.112
7/16	0.4375	0.431	0.625	0.608	0.297	0.274	0.368	0.336	0.094	0.081	0.168	0.133
1/2	0.500	0.493	0.750	0.731	0.328	0.301	0.413	0.376	0.106	0.091	0.193	0.153
9/16	0.5625	0.555	0.812	0.792	0.375	0.346	0.467	0.427	0.118	0.102	0.213	0.168
5/8	0.625	0.617	0.875	0.853	0.422	0.391	0.521	0.478	0.133	0.116	0.239	0.189
3/4	0.750	0.742	1.000	0.976	0.500	0.466	0.612	0.566	0.149	0.131	0.283	0.223
7/8	0.875	0.866	1.125	1.098	0.594	0.556	0.720	0.668	0.167	0.147	0.334	0.264
1	1.000	0.990	1.312	1.282	0.656	0.612	0.803	0.743	0.188	0.166	0.371	0.291

All dimensions are given in inches.

The radius of the fillet at the base of the head:
For sizes 1/4 to 3/8 in. incl. is 0.016 min and 0.031 max,
7/16 to 9/16 in. incl. is 0.016 min and 0.047 max,
5/8 to 1 in. incl. is 0.031 min and 0.062 max.

Round Head Cap Screws

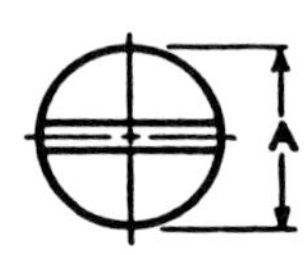

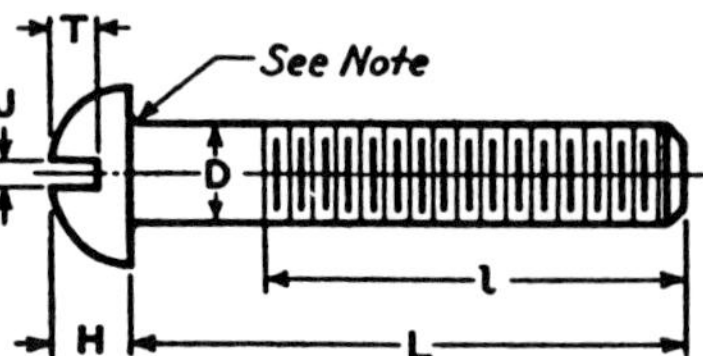

Nominal Size	D		A		H		J		T	
	Body Diameter		Head Diameter		Height of Head		Width of Slot		Depth of Slot	
	Max	Min	Max	Min	Max	Min	Max	Min	Max	Min
1/4	0.250	0.245	0.437	0.418	0.191	0.175	0.075	0.064	0.117	0.097
5/16	0.3125	0.307	0.562	0.540	0.245	0.226	0.084	0.072	0.151	0.126
3/8	0.375	0.369	0.625	0.603	0.273	0.252	0.094	0.081	0.168	0.138
7/16	0.4375	0.431	0.750	0.725	0.328	0.302	0.094	0.081	0.202	0.167
1/2	0.500	0.493	0.812	0.786	0.354	0.327	0.106	0.091	0.218	0.178
9/16	0.5625	0.555	0.937	0.909	0.409	0.378	0.118	0.102	0.252	0.207
5/8	0.625	0.617	1.000	0.970	0.437	0.405	0.133	0.116	0.270	0.220
3/4	0.750	0.742	1.250	1.215	0.546	0.507	0.149	0.131	0.338	0.278

All dimensions are given in inches.

Radius of the fillet at the base of the head:
For sizes 1/4 to 3/8 in. incl. is 0.016 min and 0.031 max,
7/16 to 9/16 in. incl..is 0.016 min and 0.047 max,
5/8 to 1 in..incl. is 0.031 min and 0.062 max.

(Courtesy of ANSI; B18.6.2–1956.)

Index

Harnessing the world's surging rivers for irrigation and power—a challenge met by engineers and construction men with many notable achievements. The first great masonry dam—the first Aswan—was constructed on the Nile, and completed in 1902. The Hoover Dam (726.4 feet high, 1200 feet long) was started in 1928, dedicated in 1935 and became one of the largest hydroelectric suppliers in the world.

Probing and observing, engineers begin to chart the course of a fabulous voyage—the penetration of outer space. To accomplish it, engineering on a massive scale, to unprecedented degrees of accuracy and reliability, produced the first manned satellites (1961), Telstar (1962), the first photographs of the moon and Mars. Giant radio telescopes now "see" to the outer edges of the universe.